HOW TECHNOLOGY WORKS

HOW TECHNOLOGY WORKS

DK | Penguin Random House

Editorial consultants
Alison Ahearn, Roger Bridgman, Giles Chapman, Caramel Quin, Josephine Roberts, Kristina Routh

Project Art Editors
Dave Ball, Mik Gates, Mark Lloyd, Francis Wong , Steve Woosnam-Savage

Designer
Gregory McCarthy

Design Assistant
Bianca Zambrea

Illustrators
Edwood Burn, Mark Clifton, Phil Gamble, Manjari Hooda, Rohit Rojal, Lakshmi Rao, Nain Rawat, Gus Scott, Alok Singh

Managing Art Editor
Michael Duffy

Jacket Editor
Emma Dawson

Jacket Designers
Surabhi Wadhwa-Gandhi, Tanya Mehrotra

Jackets Editorial Coordinator
Priyanka Sharma

Producer, Preproduction
Jacqueline Street-Elkayam

Art Director
Karen Self

Contributors
Jack Challoner, Clive Gifford, Ian Graham, Wendy Horobin, Andrew Humphreys, Hilary Lamb, Katie John

Senior Editors
Peter Frances, Rob Houston

Project Editors
Nathan Joyce, Ruth O'Rourke-Jones, Martyn Page, David Summers, Miezan van Zyl

Editors
Claire Gell, Kate Taylor

US Editor
Jennette ElNaggar

Managing Editor
Angeles Gavira Guerrero

Jackets Design Development Manager
Sophia MTT

Managing Jackets Editor
Saloni Singh

Senior Producer
Meskerem Berhane

Associate Publishing Director
Liz Wheeler

Publishing Director
Jonathan Metcalf

First American Edition, 2019
Published in the United States by DK Publishing
1450 Broadway, 8th Floor, New York, New York 10018

19 20 21 22 23 10 9 8 7 6 5 4 3 2 1
001–311448–Apr/2019

Published in Great Britain by Dorling Kindersley Limited

A catalog record for this book is available from the Library of Congress.
ISBN: 978-1-4654-7964-8

Printed and bound in China

A WORLD OF IDEAS:
SEE ALL THERE IS TO KNOW

www.dk.com

CONTENTS

POWER TECHNOLOGY

TRANSPORTATION TECHNOLOGY

MATERIALS AND CONSTRUCTION TECHNOLOGY

TECHNOLOGY IN THE HOME

SOUND AND VISION TECHNOLOGY

COMPUTER TECHNOLOGY

COMMUNICATIONS TECHNOLOGY

FARMING AND FOOD TECHNOLOGY

MEDICAL TECHNOLOGY

POWER
TECHNOLOGY

Power and energy

Energy makes things happen—from the smallest pulse of electricity to a blast of explosives. Energy is measured in joules. Power is the rate at which energy is converted from one form to another.

Measuring power

Power can be calculated by taking the amount of energy converted and dividing it by the time taken. The more energy that is converted in a set period or the quicker a specific amount of energy is converted, the greater the power. So an 1,800-watt electric heater can convert three times as much heat energy per second as a 600-watt model.

WHAT IS TORQUE?

Torque is a measure of the amount of twisting or turning force generated. It is most commonly used to describe an engine's "pulling power."

UNITS OF POWER

Power is measured in a number of ways and exhibited by many different things, including engines, appliances, and people.

Watt (W)
A watt is equal to 1 joule of work done or 1 joule of energy converted per second. The rate at which a light bulb converts electrical energy into light is measured in watts. The higher the number of watts, the higher the power.

Kilowatt (kW) and megawatt (MW)
A kilowatt is 1,000 watts and is a useful measure of electricity used by large engines and appliances. A megawatt (MW) is equal to 1,000,000 watts. Only huge machines can generate power on this scale, including power plants, aircraft carriers, and supercolliders used for particle physics experiments.

Kilowatt-hour (kWh)
One kWh is equal to 1,000 watts used for 60 minutes, or 3.6 million joules. The amount of electrical energy used in a house is usually monitored and measured in kilowatt-hours (kWh).

Horsepower (hp)
The power of vehicle engines is often measured in horsepower (hp). One horsepower is equal to 746 watts. Brake horsepower (bhp) simply means that an engine's energy loss due to friction has been accounted for.

Power production and usage

How we think about and measure power depends on the object or task carried out. For some objects, "power" refers to how much power is produced, while for others, it indicates the amount of power used.

Nuclear power station: 1,000 MW
Like a wind turbine, a nuclear facility's power is often considered in terms of how much electricity it can generate when running at optimal capacity.

Microwave oven: 1,000 W
Microwave ovens are measured in terms of how much power they consume (for example, 1,000 W) and how much energy they consume in a year (typically 62 kWh).

Gas-engine supercar: 1,479 hp
A car engine's peak horsepower refers to its maximum power output. Some supercars, such as the Bugatti Chiron, can reach up to 1,479 hp.

Wind turbine: 3.5 MW
A typical offshore wind turbine can produce up to 3.5 MW of electricity each year—enough to supply power to about 1,000 households.

LED TV: 60 W
Although an LED TV has a far lower power rating (typically 60 W) than a microwave, it is in use far more, so its annual energy consumption (around 54 kWh) is similar.

Electric car: 147 hp
Most electric cars produce far less power than gas engines, but their electric motors generate more torque at a standstill and at low speeds.

Energy conversion

The law of conservation of energy states that energy cannot be destroyed or lost. It can, however, be converted from one form to many other different forms. Electricity is a particularly valued energy source because it can be converted into sound energy, heat (thermal energy), light (radiant energy), and, in the case of a motor, movement (kinetic energy).

Chemical energy
Chemical energy is the energy stored in the bonds of chemical compounds, from foods and batteries to fossil fuels. It can be released through chemical reactions, which break the bonds between atoms. For example, burning a fossil fuel such as coal converts the chemical energy stored in the coal to light and heat.

Kinetic energy
Kinetic energy is the energy an object possesses because it is moving, such as a person sprinting or a skier traveling downhill. There are various types of kinetic energy, including rotational and vibrational energy. The amount of kinetic energy an object has depends on its speed and mass.

Mechanical energy
Mechanical energy is the kinetic energy of an object combined with its potential energy—energy that is not doing any work but can be converted—arising due to an object's position. An example is a compressed spring, which releases its potential energy when it bounces back to its original position.

Thermal energy
Thermal energy is technically a type of kinetic energy derived from the vibrating movement of atoms in a substance. Heat describes the flow of thermal energy from one place to another, such as the heat transferred from a flame to a cooking pot on a stove.

WASTED ENERGY

A machine always wastes a proportion of the energy it uses. Light bulbs convert only some of the electricity they receive to light, while some is wasted as heat. A poorly tuned or damaged machine, such as a fridge with a door seal that leaks cool air, can also waste further energy.

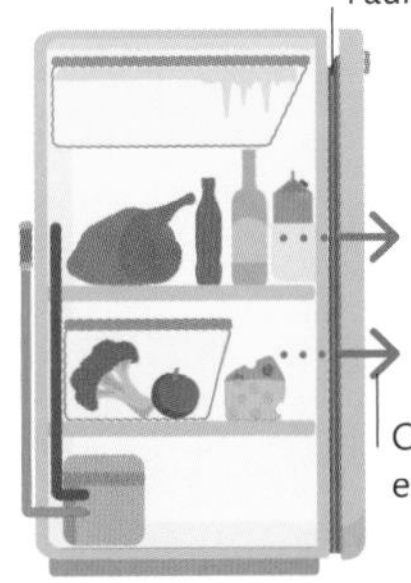

Energy conversion in a solar panel
A solar panel contains a series of photovoltaic cells (see p.30). These convert the radiant energy in sunlight into electrical energy in the form of a flow of electrons.

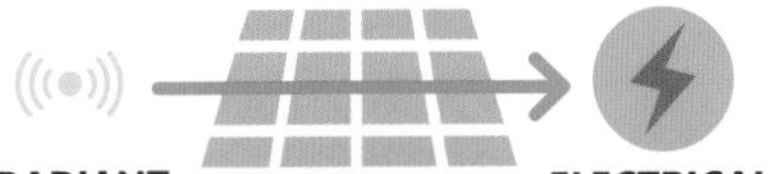

Fossil fuels

Around two-thirds of the world's electricity and more than a billion motor vehicles and other machines are powered by fuel derived from the fossilized remains of once living things. These fossil fuels (oil, coal, and natural gas) are nonrenewable resources with limited reserves. When burned, their chemical energy transforms mostly to heat energy but with significant emissions of greenhouse gases.

CHINA AND THE **US** GENERATE A TOTAL OF **40 PERCENT** OF THE WORLD'S **GREENHOUSE GAS EMISSIONS**

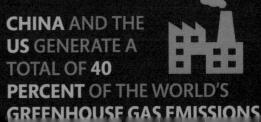

Water supply

A fresh, clean, plentiful water supply is taken for granted in many countries. Before it reaches your tap, water undergoes several types of treatment to make it fit for human consumption.

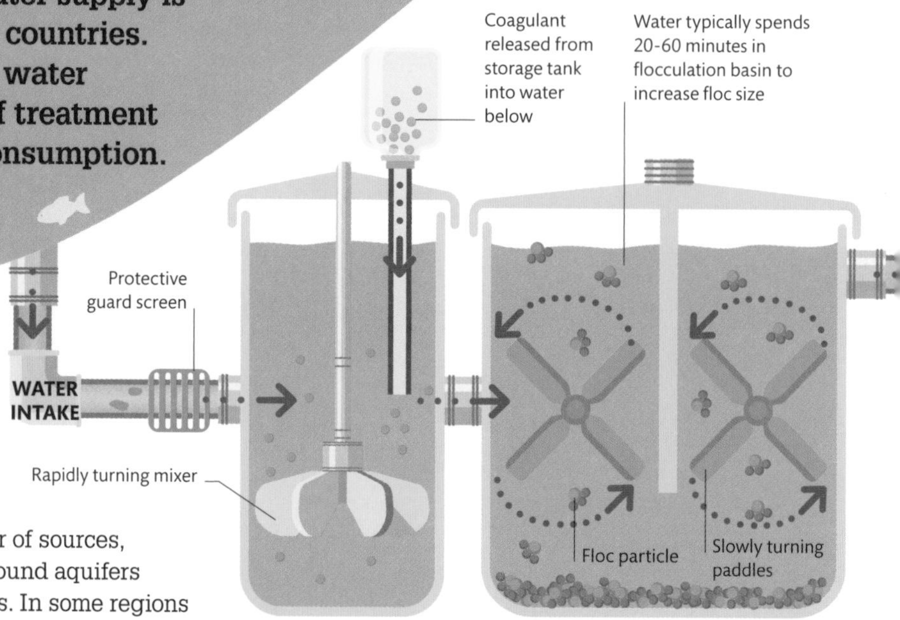

1 Water intake
Water flows through a series of screens to filter out fish and other waterborne creatures and debris, such as grit, litter, and leaves, preventing them from entering the water-treatment system.

2 Coagulation
The water is mixed rapidly with a coagulant such as ammonium sulphate to help particles suspended in the water collide with each other and clump together.

3 Flocculation
Slowly turning paddles encourage clumped-together particles, called floc, to bind together in larger deposits. These deposits, along with sediment and some bacteria, settle at the bottom of the flocculation basin, while the cleaner water moves to the next stage for further processing.

How water is processed

Fresh water is drawn from a number of sources, including lakes, rivers, and underground aquifers (water-bearing rocks), into reservoirs. In some regions lacking plentiful fresh water, desalination plants remove salt from seawater. Whatever the source, the water is purified to remove microorganisms, some of which can cause diseases. Purification also removes harmful chemicals and unwanted odors or tastes before the water is fit for consumption. The water is tested at each stage to monitor its quality.

FLUORIDATION

Fluoride is added to some public water supplies to help add minerals to tooth enamel lost during the process of tooth decay. However, critics claim that overexposure to fluoride in young children can cause "pitting" (small depressions or faults appearing in tooth enamel) and tooth discoloration.

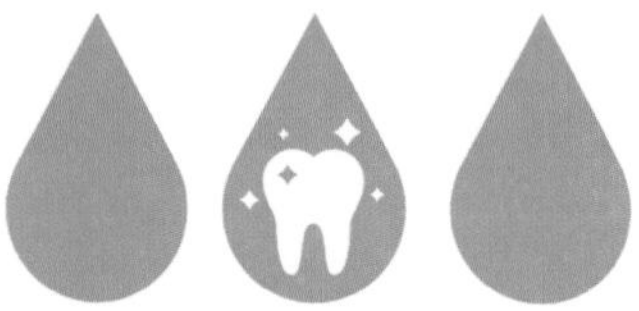

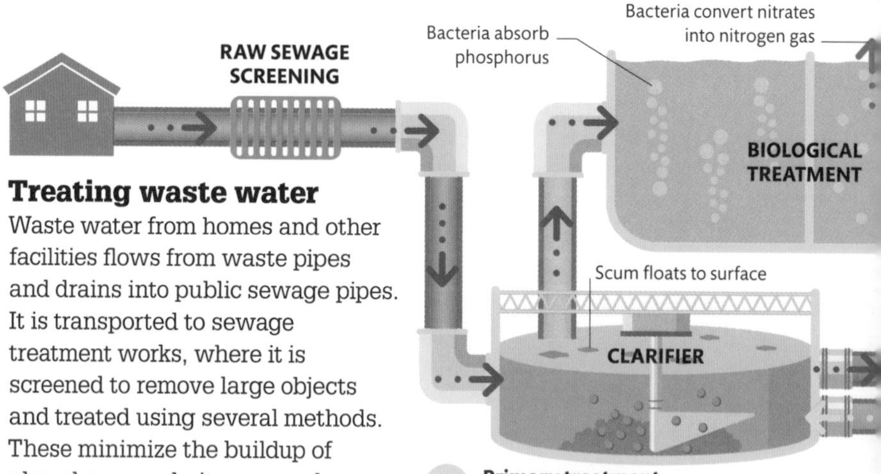

Treating waste water

Waste water from homes and other facilities flows from waste pipes and drains into public sewage pipes. It is transported to sewage treatment works, where it is screened to remove large objects and treated using several methods. These minimize the buildup of phosphorus and nitrogen and remove fats, solid waste particles, and harmful microorganisms.

1 Primary treatment
Solids, such as human waste, settle at the bottom of clarifier tanks and are pumped away; a skimmer removes oil and scum from the surface.

844 MILLION

THE NUMBER OF **PEOPLE** WHO LACK ACCESS TO CLEAN **DRINKING WATER**

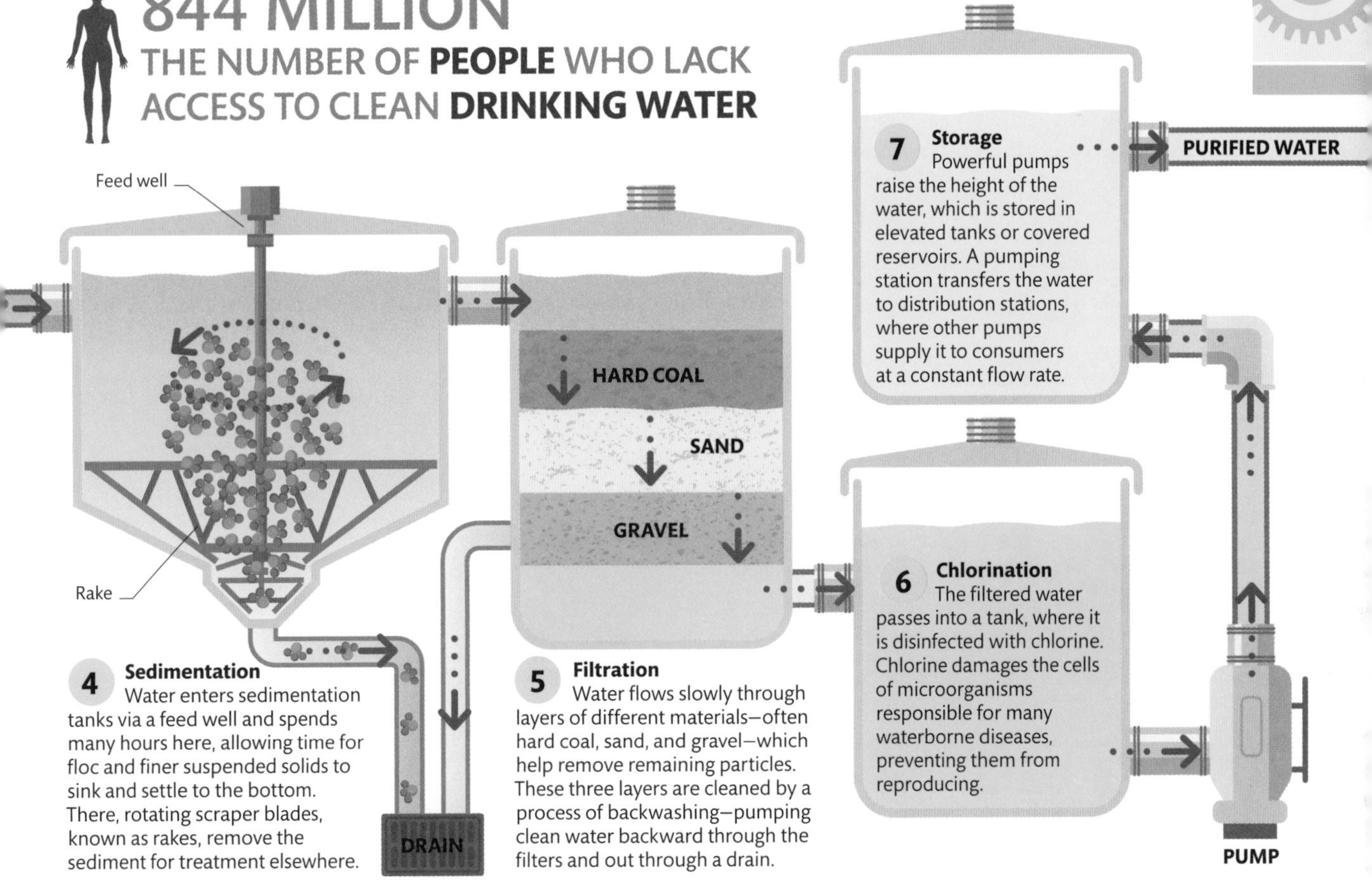

4 Sedimentation
Water enters sedimentation tanks via a feed well and spends many hours here, allowing time for floc and finer suspended solids to sink and settle to the bottom. There, rotating scraper blades, known as rakes, remove the sediment for treatment elsewhere.

5 Filtration
Water flows slowly through layers of different materials—often hard coal, sand, and gravel—which help remove remaining particles. These three layers are cleaned by a process of backwashing—pumping clean water backward through the filters and out through a drain.

6 Chlorination
The filtered water passes into a tank, where it is disinfected with chlorine. Chlorine damages the cells of microorganisms responsible for many waterborne diseases, preventing them from reproducing.

7 Storage
Powerful pumps raise the height of the water, which is stored in elevated tanks or covered reservoirs. A pumping station transfers the water to distribution stations, where other pumps supply it to consumers at a constant flow rate.

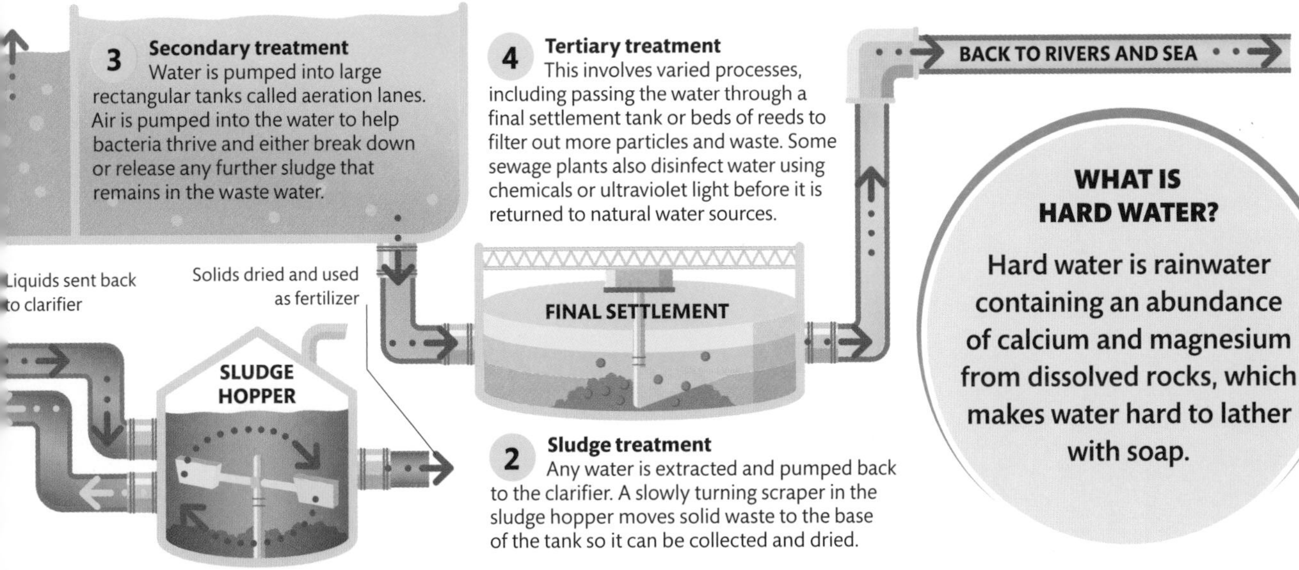

3 Secondary treatment
Water is pumped into large rectangular tanks called aeration lanes. Air is pumped into the water to help bacteria thrive and either break down or release any further sludge that remains in the waste water.

4 Tertiary treatment
This involves varied processes, including passing the water through a final settlement tank or beds of reeds to filter out more particles and waste. Some sewage plants also disinfect water using chemicals or ultraviolet light before it is returned to natural water sources.

2 Sludge treatment
Any water is extracted and pumped back to the clarifier. A slowly turning scraper in the sludge hopper moves solid waste to the base of the tank so it can be collected and dried.

WHAT IS HARD WATER?

Hard water is rainwater containing an abundance of calcium and magnesium from dissolved rocks, which makes water hard to lather with soap.

Oil refineries

Crude oil is extracted from oil deposits in Earth's crust and shipped or piped to refineries. It consists of a combination of many types of hydrocarbons. These can be separated out into various products that are used in different ways.

Fractional distillation

The different hydrocarbons in crude oil have varying boiling points. This means that they can be separated out by evaporating the oil then condensing the gases into different products at different temperatures. This takes place in a distillation tower. Substances with lower boiling points condense higher up the tower. Trays at calibrated heights collect the substances, known as fractions.

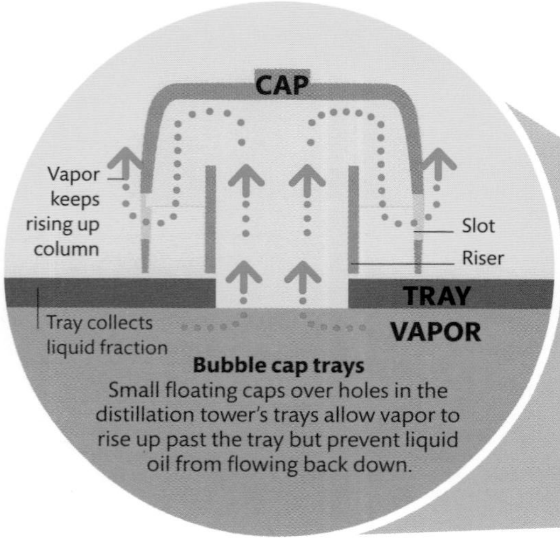

Bubble cap trays
Small floating caps over holes in the distillation tower's trays allow vapor to rise up past the tray but prevent liquid oil from flowing back down.

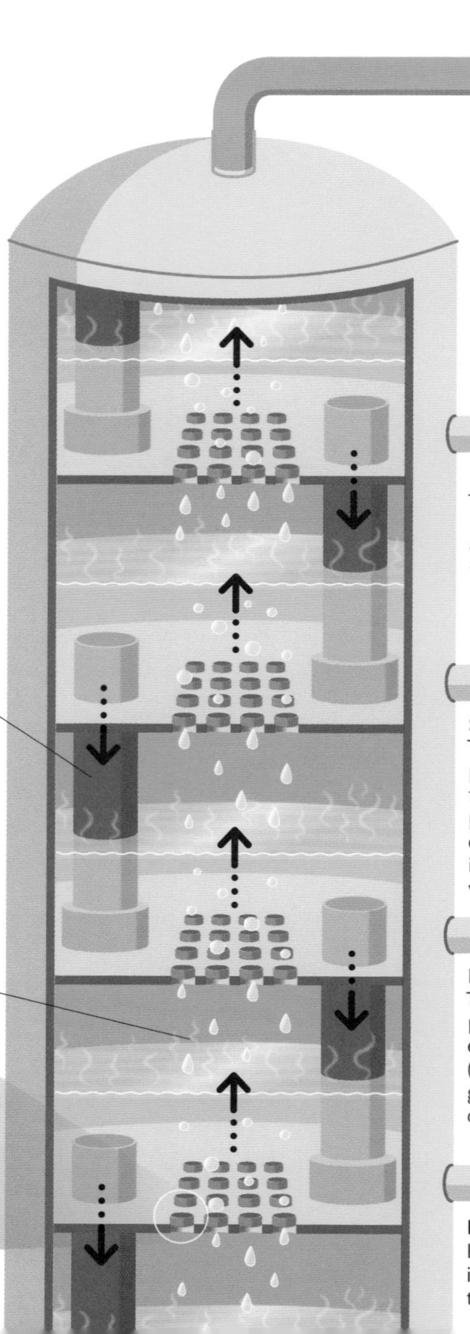

5 Tray collection
At each level in the column, as a fraction of the oil vapor cools and condenses to a liquid, it is collected on a tray and piped away for processing and storage.

Pipe called a downcomer channels liquid down from one tray to another

4 Vapor rises
Fractions of hydrocarbons with lower boiling points continue to travel up the column further than heavier fractions, passing through holes in trays as they rise.

Vapor rises, passing through holes in trays

3 Distillation
At a certain height and temperature within the tower, a fraction condenses into liquid, separating out from the rest of the oil vapor, which rises up the column.

Liquid petroleum gas
Lighter hydrocarbons, such as propane and butane, remain as vapor. These are processed into bottled gases used in heating and cooking.

Light naphtha
This fraction is often used to produce ethylene, which is used to make many plastics, including polyethylene.

Straight-run gasoline
This is gasoline produced without further chemical processing. Almost half of crude oil is refined into gasoline used as vehicle fuel.

Heavy naphtha
This fraction is often processed further, for example, by cracking (see below), to produce gasoline and other crude-oil products.

Kerosene
Kerosene is used as a fuel in heaters or refined further to produce potent jet fuel.

JAMNAGAR REFINERY, IN GUJARAT, INDIA, IS THE WORLD'S LARGEST, PRODUCING UP TO 1.24 MILLION BARRELS OF OIL A DAY

1 Desalted crude oil enters furnace
After salts and some other impurities have been removed, crude oil is sent to the distillation furnace. There, it is heated with superheated steam to around 750°F (400°C).

2 Oil enters column
The heated crude oil enters the distillation tower. Most of it rises up the column as gas, but some heavier fractions remain liquid.

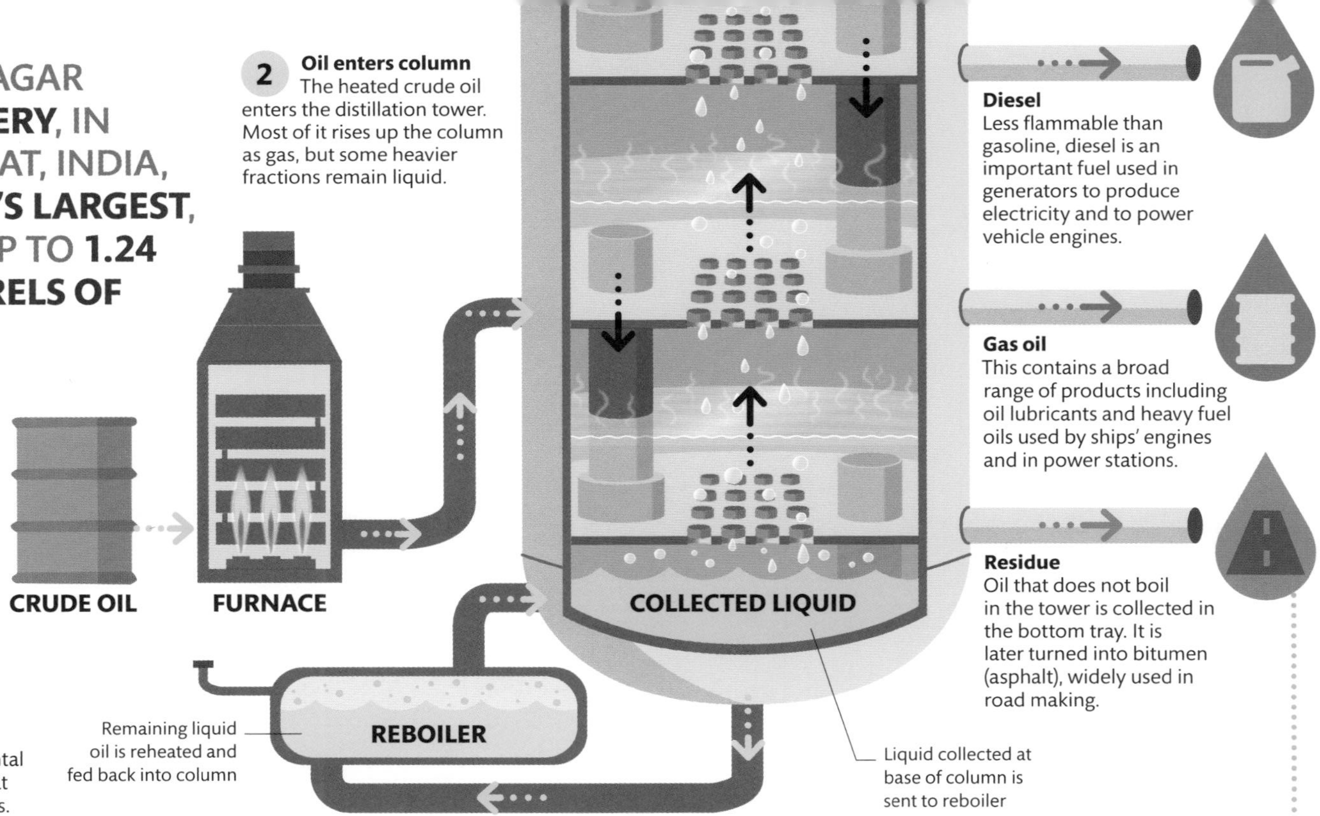
CRUDE OIL
FURNACE
COLLECTED LIQUID
REBOILER
Remaining liquid oil is reheated and fed back into column
Liquid collected at base of column is sent to reboiler
Diesel
Less flammable than gasoline, diesel is an important fuel used in generators to produce electricity and to power vehicle engines.
Gas oil
This contains a broad range of products including oil lubricants and heavy fuel oils used by ships' engines and in power stations.
Residue
Oil that does not boil in the tower is collected in the bottom tray. It is later turned into bitumen (asphalt), widely used in road making.

Distillation tower
An oil refinery's distillation tower contains a vertical column divided into horizontal sections containing trays that collect the different fractions.

TREATING OIL SPILLS

Oil tanker accidents and pipeline leaks can release crude oil into the environment, causing catastrophic damage to ecosystems. The clean-up process at sea can involve using long booms to skim the surface oil from the water as well as chemical treatments.

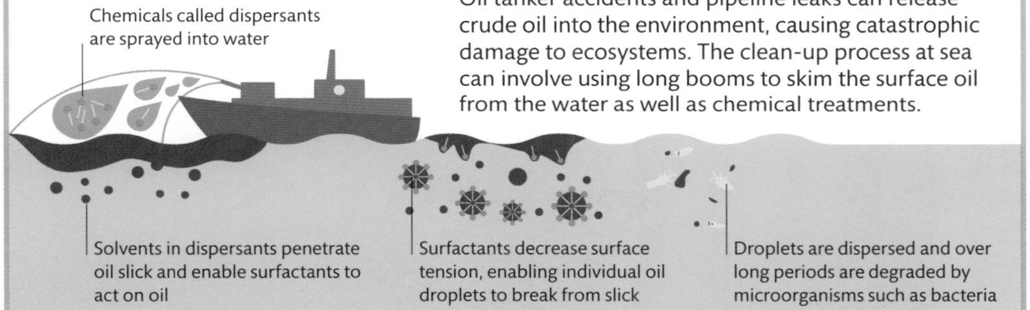
Chemicals called dispersants are sprayed into water
Solvents in dispersants penetrate oil slick and enable surfactants to act on oil
Surfactants decrease surface tension, enabling individual oil droplets to break from slick
Droplets are dispersed and over long periods are degraded by microorganisms such as bacteria

Processing and treatment

Crude oil fractions that have lower boiling points are more flammable and burn with a cleaner flame. As a result, they tend to be in higher demand than heavier fractions. To meet demand, some of the heavier fractions, which are composed of long chains of molecules, are converted into more useful and valuable products through a process called cracking. This often involves breaking down the molecules using heat or a catalyst, such as silicon dioxide or aluminum oxide.

Generators

Electrical generators work using the principle of electromagnetic induction. As a wire coil spins between the two poles of a magnet, an electric current is induced to flow through the wire and around a circuit.

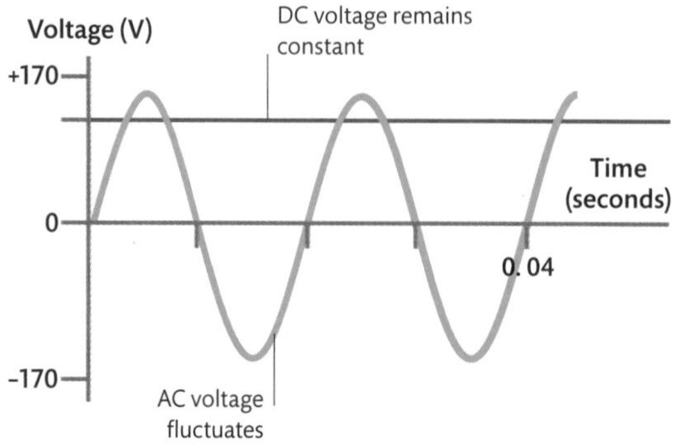

Contrasting current
DC produces a steady voltage, while AC's voltage alters as it reverses direction. AC's voltage must rise higher than DC's to transfer the same energy during the same time period.

Direct current and alternating current

Generators produce either alternating or direct electric current. Direct current (DC) flows in one direction only around an electrical circuit and is generated by batteries and cells. Alternating current (AC) reverses its direction many times per second. Its voltage can also be greatly increased or decreased by devices called transformers, so it travels more efficiently over distances, which is why AC is used for our main power grid.

AC generators

An AC generator is also known as an alternator. Its rotating wire coil is connected to an electric output circuit via slip rings and brushes. The brushes make continual contact with the output circuit, conducting current between the rotating slip ring and the fixed wires attached to the brushes. The current induced in an AC generator changes direction twice during each 360° revolution completed by the coil.

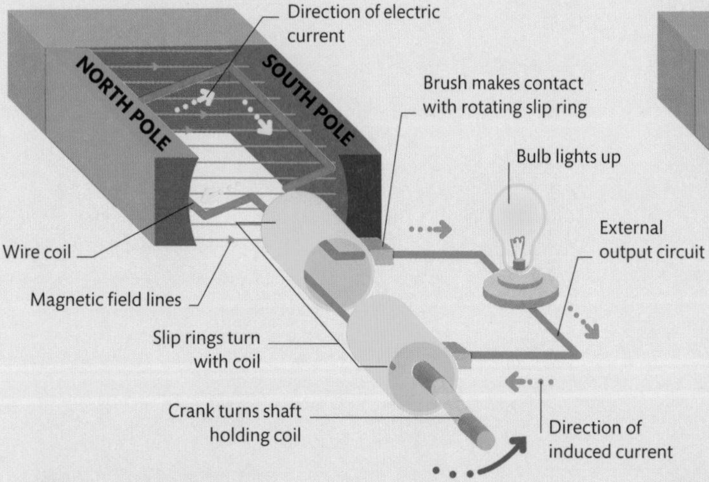

1 Coil begins to turn
The shaft of this experimental AC generator is rotated by mechanical force delivered by turning a handle. The shaft turns a wire coil through a magnetic field generated by a permanent magnet's north and south poles. As the coil cuts through the magnetic field, a current flowing in one direction is induced, reaching a peak when the coil lies horizontally through the field.

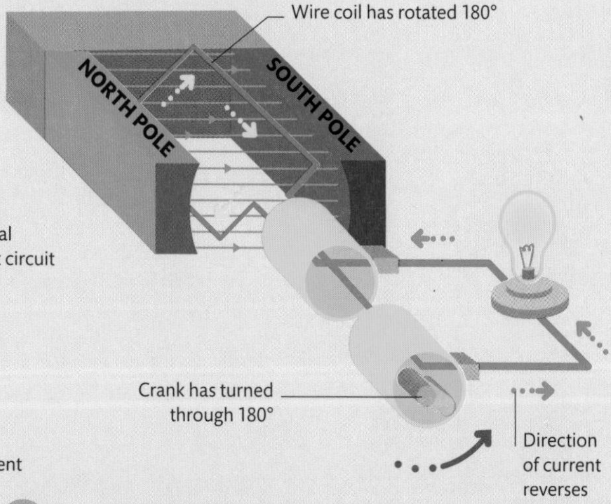

2 Current direction changed
As the coil completes another 180° turn through the magnetic field, the points that were initially face up are now face down; the coil's position relative to the north and south magnetic poles alters, its magnetic polarity changes, and the direction of the induced electric current reverses. The current reverses every half-turn, flowing to the slip rings and brushes and into the external output circuit.

BICYCLE DYNAMOS

A bicycle dynamo powers an electric light through the action of a spinning knurled wheel, which turns because it is connected to the moving tire wall. The spinning wheel rotates a driveshaft attached to a permanent magnet. As this magnet turns, its magnetic field changes, inducing an electric current in the wire coil of the dynamo's electromagnet.

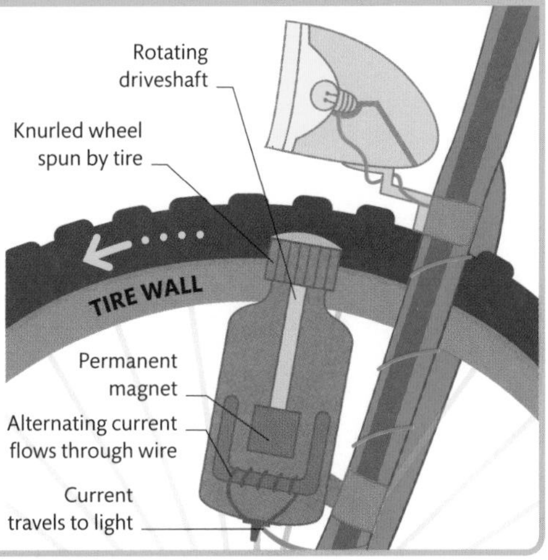

WHAT IS THE FREQUENCY OF AN ALTERNATING CURRENT?

It is how often AC changes direction, measured in hertz (Hz). One Hz is one change per second. Electricity is generated at 60 Hz in the US and often 50 Hz in Europe.

DC generators

DC generators use a device called a commutator to convert AC to DC. It is split into two segments insulated from one another so that no electricity flows between them. The commutator keeps the current flowing in one direction to the output circuit by switching polarity at the same moment that the AC signal reverses direction.

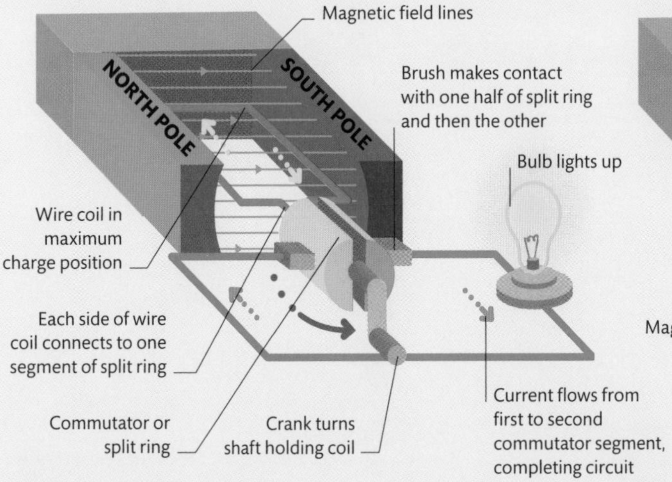

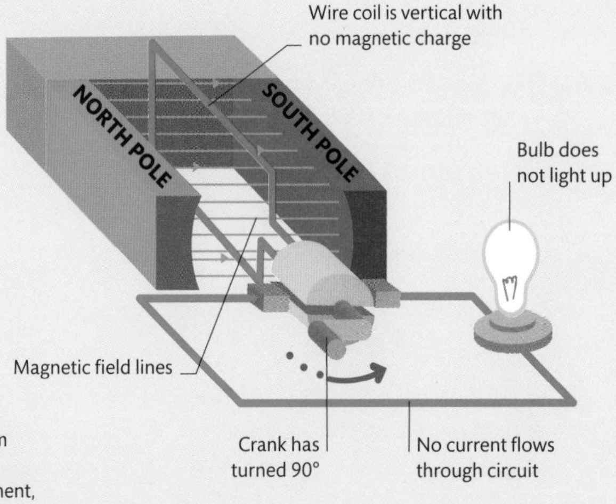

1 Reversing the connection
At peak position, the current flows to the first split ring segment, through the circuit to the second split ring segment, and into the coil, completing the circuit. When the coil turns another 180°, the brush breaks contact with the first segment, making contact with the second segment, on the opposite side of the circuit. The current flows the same way for both the coil's first and second 180° turns.

2 Inconsistent current
When the coil is vertical and not cutting through the horizontal magnetic field lines, no current is produced. This means DC electricity is produced in pulses rather than a steady flow. Most practical DC generators solve this problem by containing multiple coils (so one is always in a horizontal position when the other coils are in less optimal positions) and additional commutators.

Universal motors

In a universal electric motor, the permanent magnet is replaced by an electromagnet made of a number of windings of wire through which current can flow. This produces a magnetic field, inside which a coil, called an armature, rotates. Both the armature and the wire windings surrounding it receive the same electric current because they are wired in series. This means a universal motor can run on both DC and AC electricity.

Inside a power drill
Many power drills feature a universal motor that offers high levels of turning force (torque) and allows the user to select the best speed for a particular task.

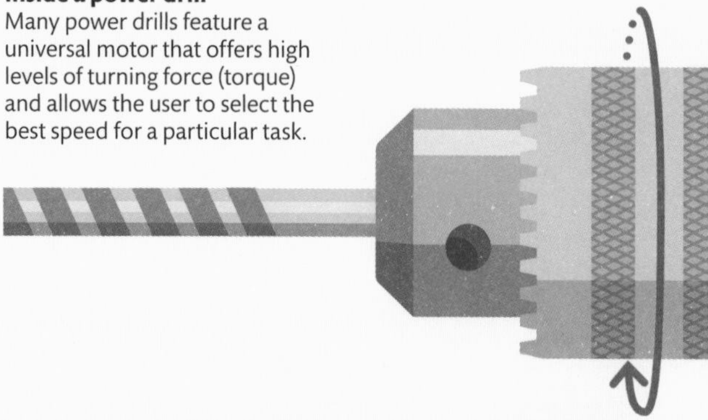

Motors

Electric motors use the forces of attraction and repulsion between an electric current and a magnetic field to create turning movement. Motors vary in size, from microscopic actuators inside electronic gadgets to giant power plants that propel large ships.

AROUND 45 PERCENT OF ALL **ELECTRICITY CONSUMED** POWERS **ELECTRIC MOTORS**

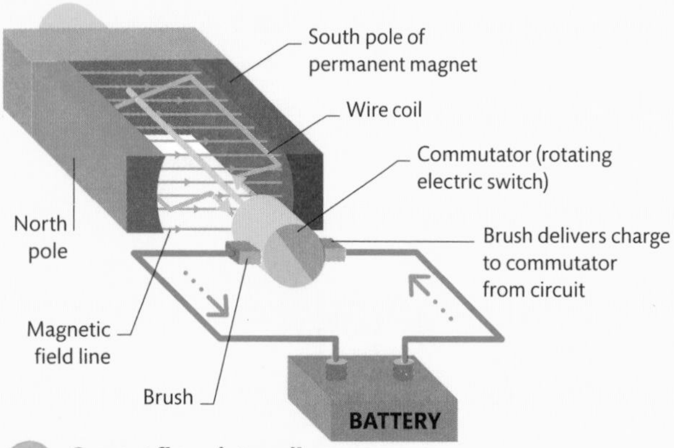

1 Current flows into coil
An electric current flows into a wire coil positioned between the poles of a permanent magnet. The coil becomes an electromagnet.

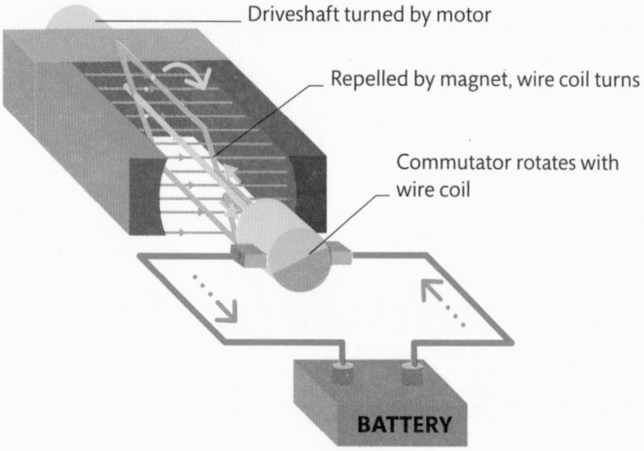

2 Wire coil turns
Repelled by the magnet's like poles, the coil turns away. After a quarter turn, the unlike poles attract, forcing the coil into a half-turn.

How an electric motor works

In many motors, a coil of wire moves through the magnetic field produced by a stationary magnet. When current flows through the coil, it becomes an electromagnet with north and south poles. The coil swings around to align its poles with those of the permanent magnet. A commutator reverses the coil's current every half-turn to switch the coil's poles and keep it spinning in the same direction. The coil is connected to a driveshaft, which transmits the motor's turning force to components, such as wheels.

HOW FAST DO DC MOTORS SPIN?

An average DC motor spins at 25,000 rpm, but some motors, such as those in vacuum cleaners, can reach up to 125,000 rpm.

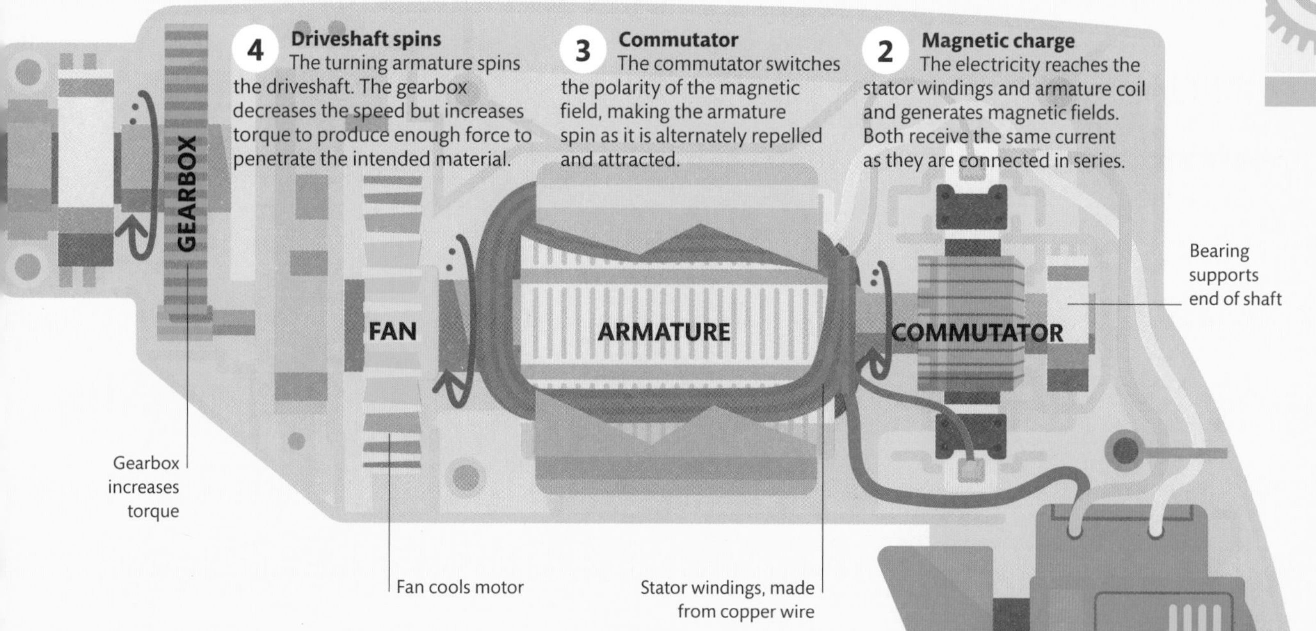

4 Driveshaft spins
The turning armature spins the driveshaft. The gearbox decreases the speed but increases torque to produce enough force to penetrate the intended material.

3 Commutator
The commutator switches the polarity of the magnetic field, making the armature spin as it is alternately repelled and attracted.

2 Magnetic charge
The electricity reaches the stator windings and armature coil and generates magnetic fields. Both receive the same current as they are connected in series.

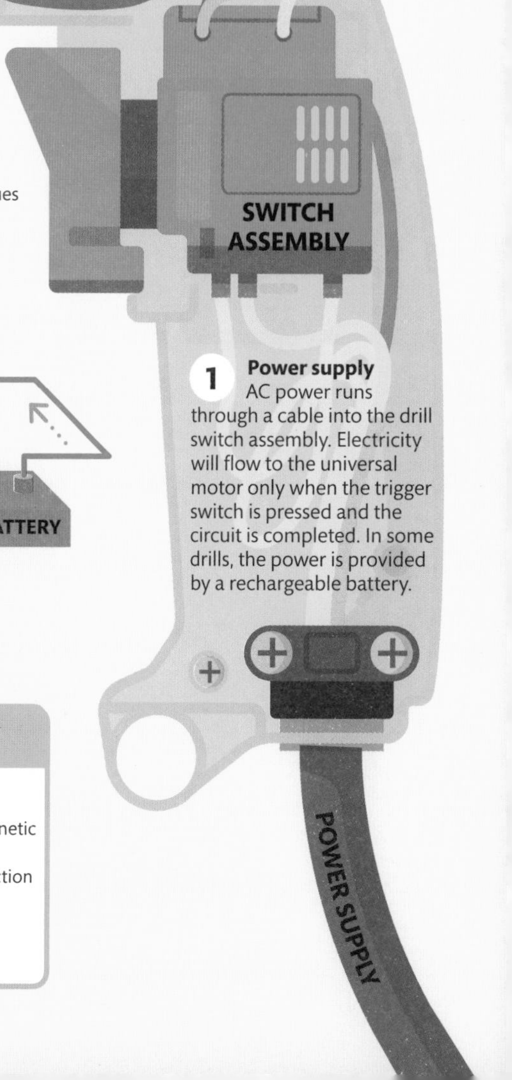

1 Power supply
AC power runs through a cable into the drill switch assembly. Electricity will flow to the universal motor only when the trigger switch is pressed and the circuit is completed. In some drills, the power is provided by a rechargeable battery.

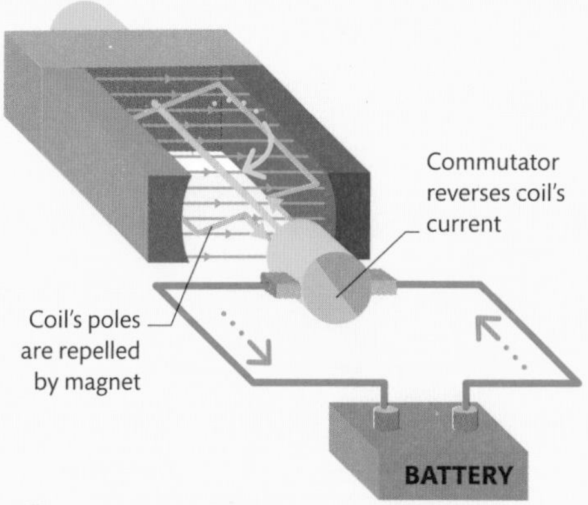

3 Reversing current
The commutator reverses the current's direction. This switches the polarity of the coil's magnetic field so its poles are again repelled.

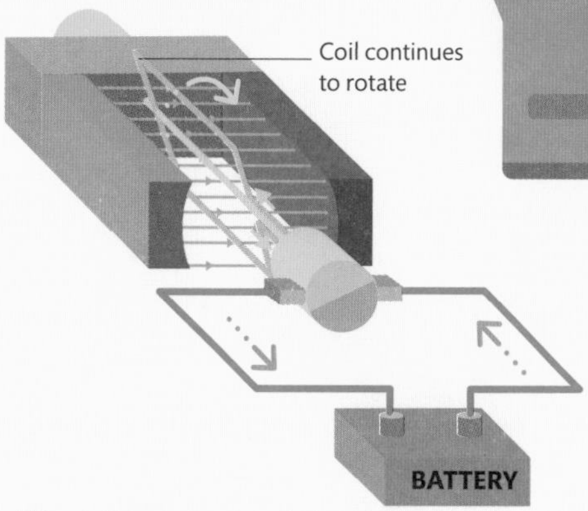

4 Poles repel
The coil continues to rotate, repelled and then attracted by the permanent magnet as the current is continually switched.

FLEMING'S LEFT-HAND RULE

This is a simple way of working out which way a motor's coil will turn. Stick out your left thumb, forefinger, and middle finger, at right angles. With the forefinger aligned with the magnetic field's direction, the middle finger shows the current direction, and the thumb shows the direction of the coil's turn.

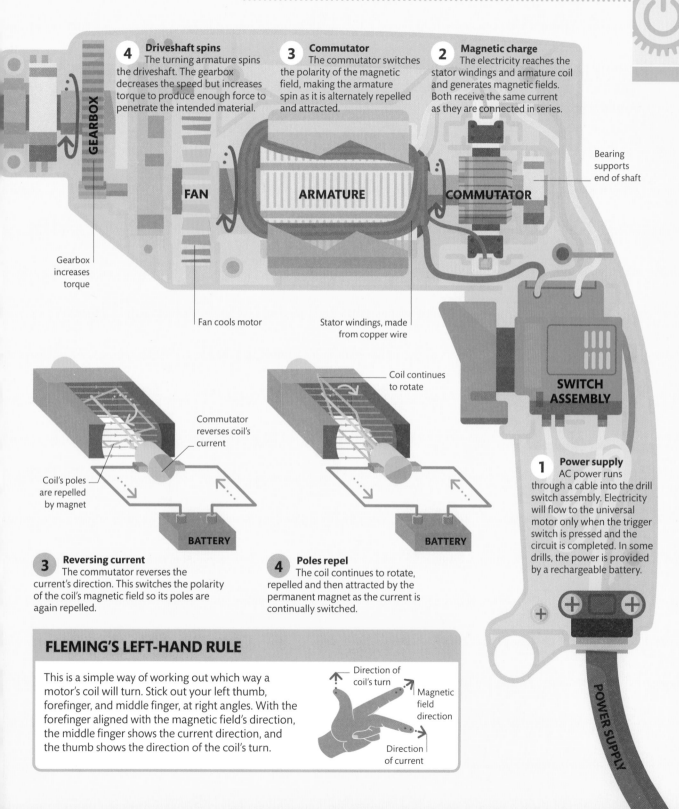

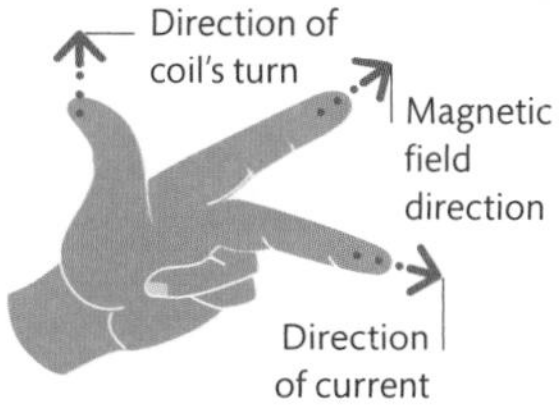

Power stations

Electricity is an extremely versatile source of energy, able to be transported long distances and used in countless applications. Vast amounts of electricity are generated by power stations, most burning fossil fuels such as coal.

66 PERCENT OF THE **WORLD'S ELECTRICITY SUPPLY** COMES FROM **FOSSIL FUELS**

How a power station works

In a conventional coal-fired power station, a furnace heats water to create superheated steam. This drives a turbine that, in turn, powers electricity generators. A large power station can generate 2,000 megawatts of electricity—enough to power more than a million households. The used steam is cooled, condensed back to water, and reused; waste gases are treated and cleaned; and the furnace's ash is often processed into cinder blocks.

HAS OUR RELIANCE ON COAL DECREASED?

On the contrary, coal use has been rocketing in recent decades. Since the 1970s, our annual consumption has increased by more than 200 percent.

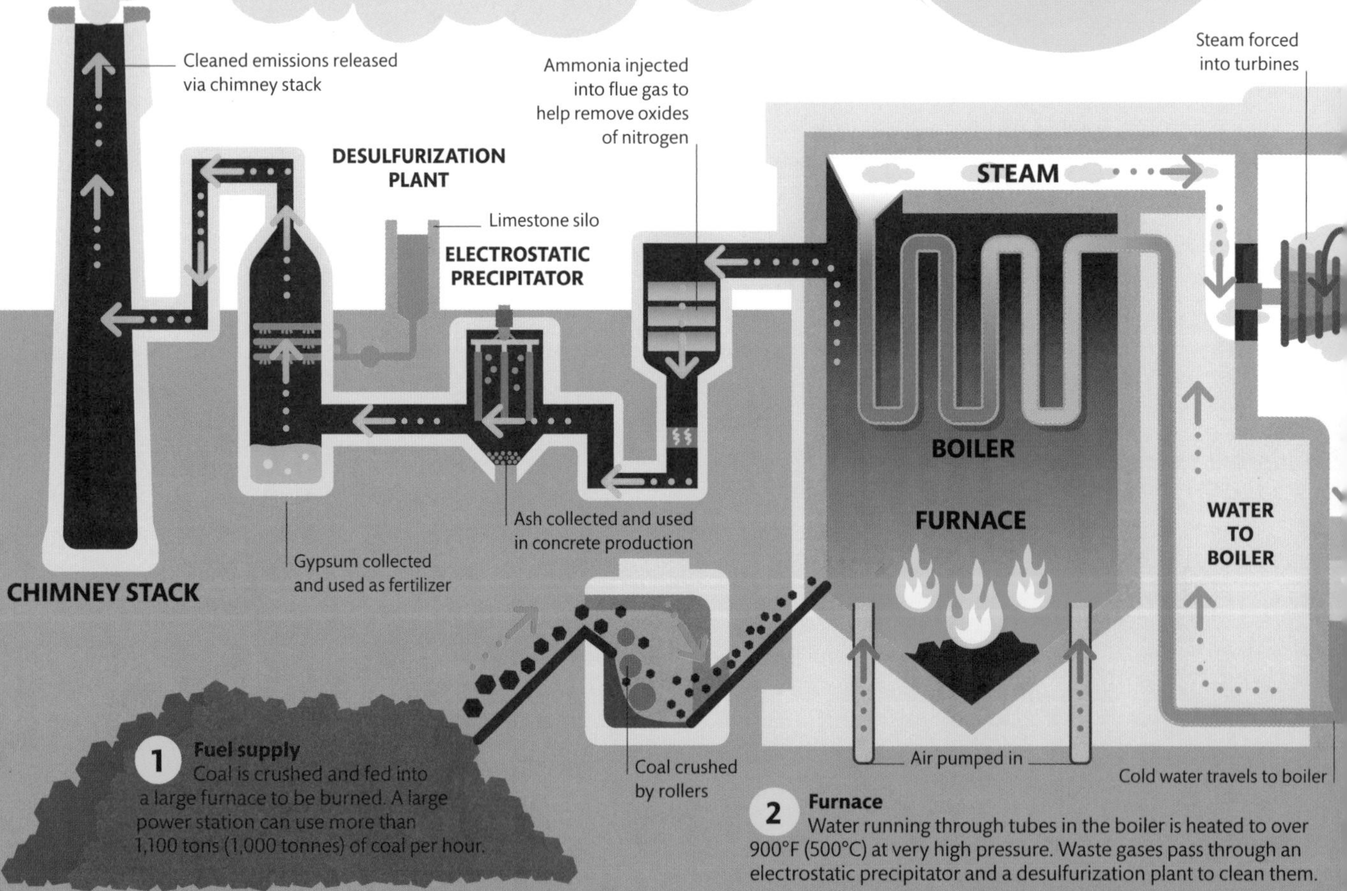

1 Fuel supply
Coal is crushed and fed into a large furnace to be burned. A large power station can use more than 1,100 tons (1,000 tonnes) of coal per hour.

2 Furnace
Water running through tubes in the boiler is heated to over 900°F (500°C) at very high pressure. Waste gases pass through an electrostatic precipitator and a desulfurization plant to clean them.

Cleaning emissions

Furnace gases are cleaned of harmful pollutants before release. Precipitators use electric charges to remove particulates (tiny particles) while more than 95 percent of sulfur is removed by flue gas desulfurization systems (see opposite page). However, harmful emissions still occur. Each year, US coal-fired plants emit about 1.1 million tons (1 million tonnes) of sulfur dioxide.

PRECIPITATOR

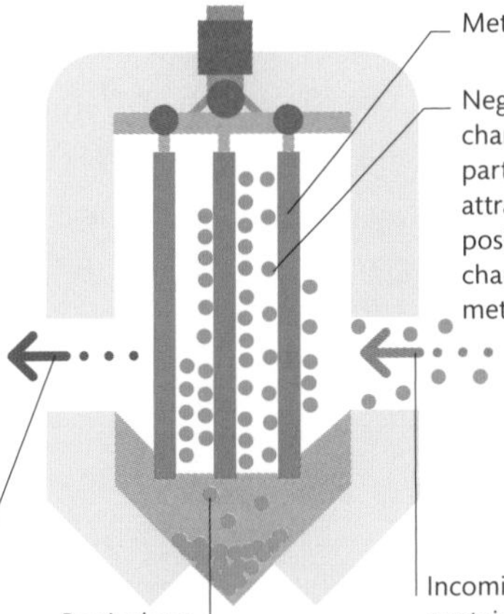

ENERGY EFFICIENCY

Only around one-third of all the energy in fuel reaches the user. More than 60 percent is lost at the power station.

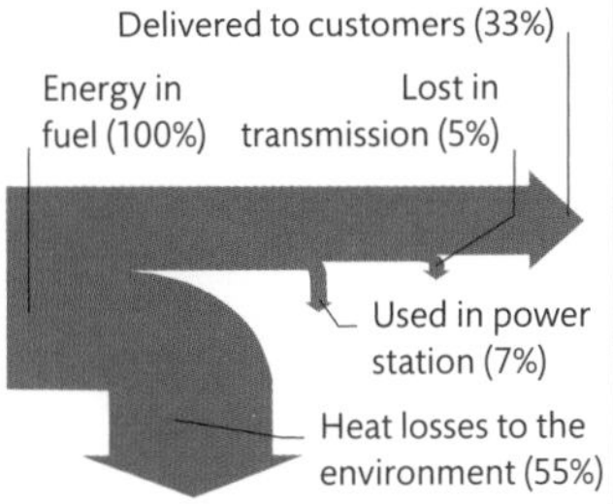

3 Turbine
The high-pressure steam turns the fan blades of steam turbines with great force and speed. This rotational motion is transmitted to the generator by a driveshaft.

5 Electricity supply
The voltage of the electricity is greatly increased by a step-up transformer. This improves efficiency as the electricity is transmitted away via power lines.

6 Cooling tower
Steam is cooled in the condenser then sprayed into cooling towers, where most of the water cools and is piped back for reuse. Some steam escapes, and much heat is lost.

4 Generator
The generator's rotor spins thousands of times each minute, producing an alternating current (AC), typically with a voltage of 25,000 volts.

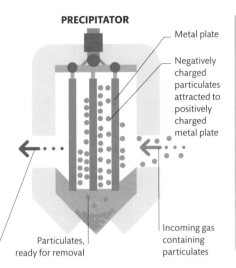

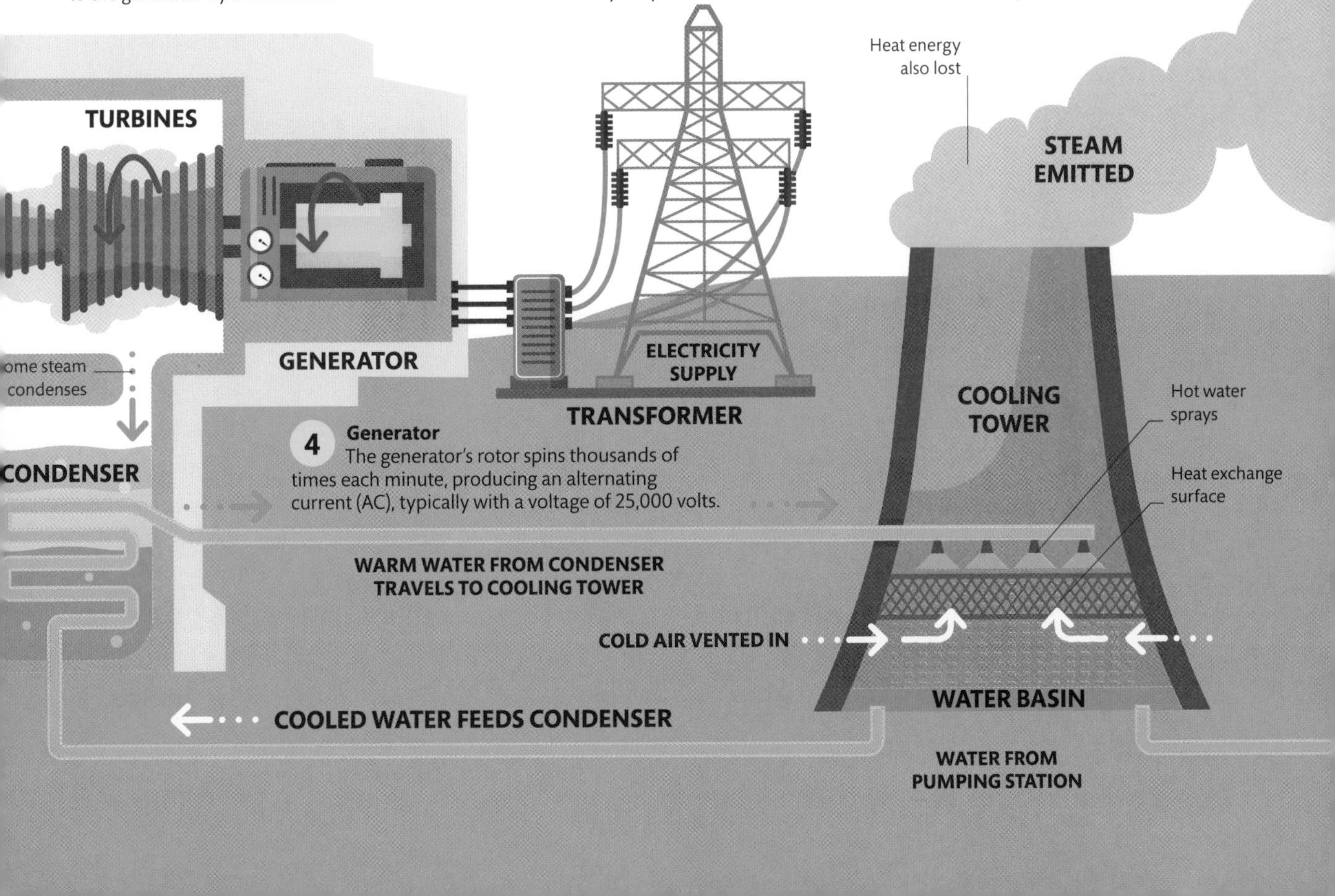

Electricity supply

Most electricity is generated at large power stations (see pp.20–21) and then distributed to consumers, such as factories and homes, sometimes over substantial distances. This involves a large and complex network of cables and facilities together known as a power grid.

Transmission towers
Transmission towers are usually tall steel and aluminum towers featuring a lattice or tubular frame. They carry power lines at safe heights above ground level and feature insulators to separate the high-voltage cable from the earthed tower.

Power transmission

The vast amount of electricity required by industry, businesses, and homes has to be distributed to precisely where it is needed. Power lines both above and below ground transmit the electricity, while transformers, some of which are located in substations, adjust the voltage. A network of sensors ensures that these vital pieces of equipment are working optimally.

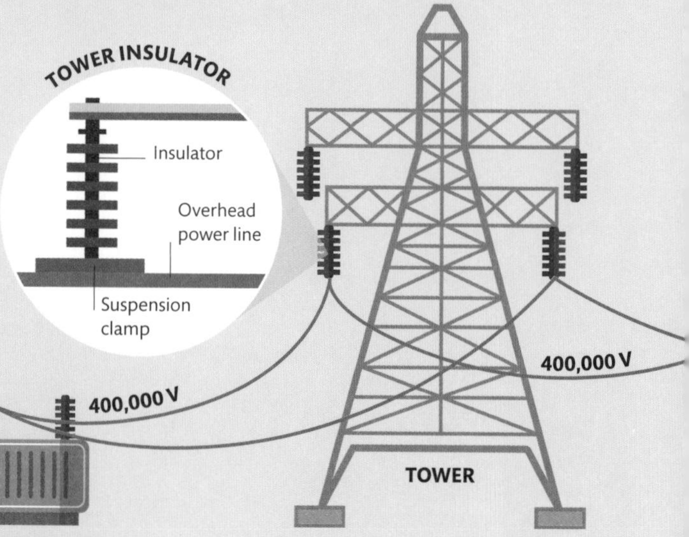

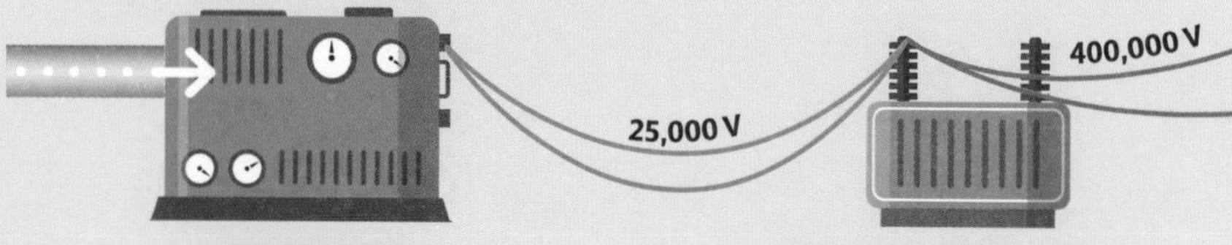

1 Power station
A generator located in a power station converts kinetic energy into electrical energy. This provides a supply of alternating current (AC) electricity (see p.16) typically at 25,000 volts.

2 Grid substation
A grid substation uses transformers to increase the voltage, typically to 400,000 volts. The higher the voltage, the less energy is lost as heat from resistance when the electricity travels along power lines.

3 High-voltage tower lines
Transmission towers are often made from steel-reinforced aluminum. They are fitted with glass or ceramic insulators to prevent electricity from the power lines traveling down the steel tower to the ground.

TRANSFORMERS

A transformer alters the voltage of electricity through the process of electromagnetic induction. First, alternating current (AC) flows through a transformer's primary coil, wound around an iron core. It produces a changing magnetic field, which induces a voltage in a secondary coil. If the secondary coil contains more wire than the primary, the voltage is increased, or stepped up. Less wire results in a voltage decrease, or step down.

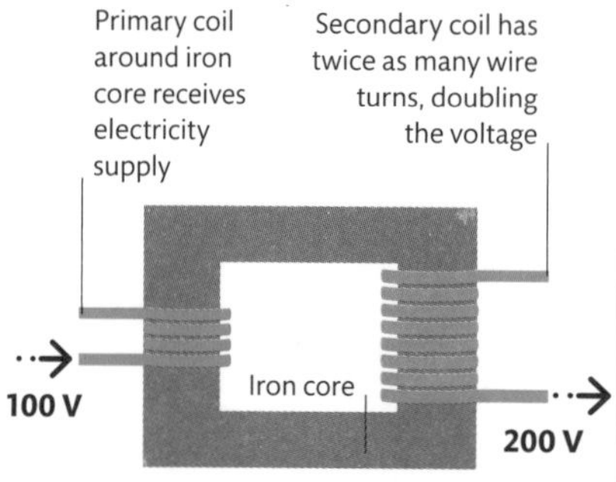

STEP-UP TRANSFORMER

HOW CAN BIRDS PERCH ON POWER LINES?

Electricity always flows along the path of least resistance. Birds do not conduct electricity well, so the electricity bypasses them, instead continuing to travel along the power line.

THE WORLD'S TALLEST ELECTRICITY PYLON LINE, IN CHINA, IS 1,200 FT (370 M) HIGH

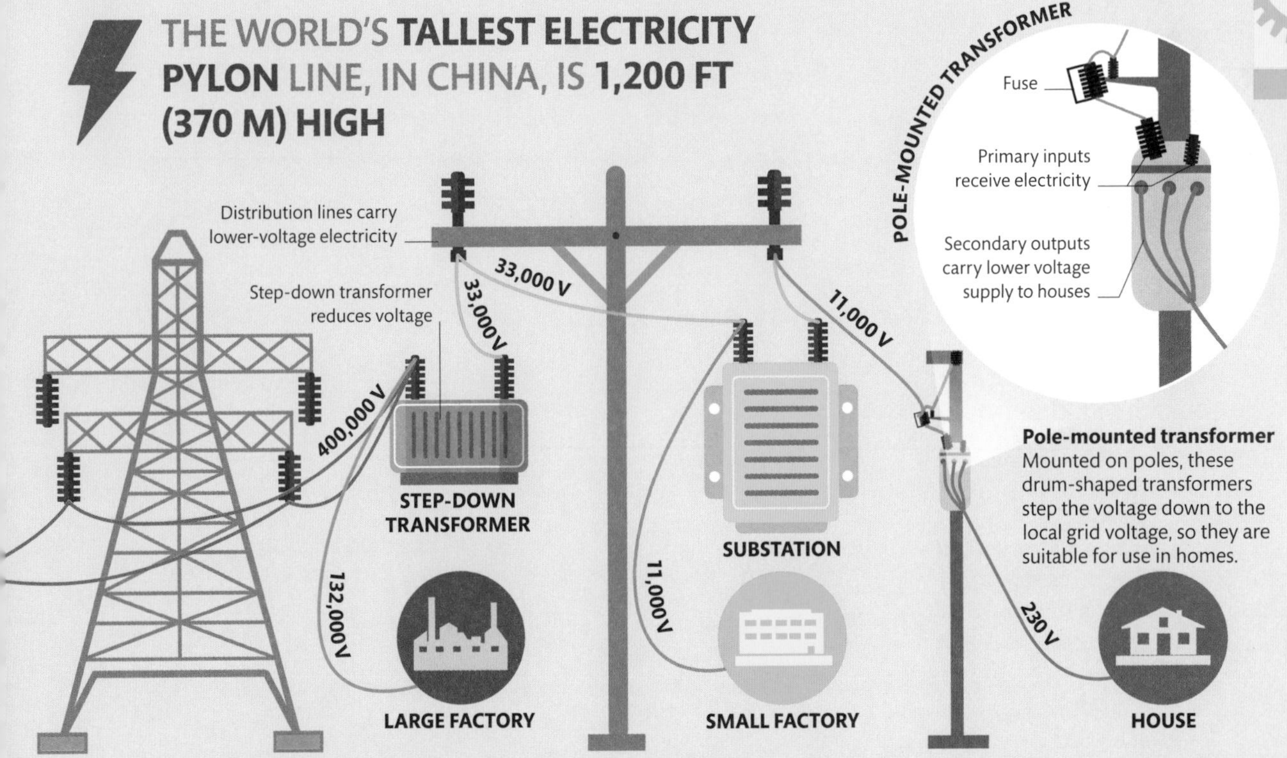

4 Direct supply to industry
Some factories with high electricity requirements may take power directly from the high-voltage lines. Other factories require a transformer to step the voltage down to approximately 132,000 volts.

5 Distribution substation
The high-voltage electricity is reduced to a much lower voltage at a substation, which usually includes several transformers. From here, it is supplied to smaller industrial and commercial customers.

6 Domestic supply
A network of distribution lines supplies electricity to houses. The final voltage reduction takes place at pole-mounted transformers, before the electricity supply passes into a home's fuse box.

Underground cables

To reduce the visual impact and land use of rows of towers and the power lines they support, many electricity supply cables are buried underground. These cables require multiple layers of protection and insulation. The cables are placed in trenches. Individual cables can be up to 0.6 miles (1 km) long, and extra reinforcement is added to the trenches at the points where the cables are joined. The cables are guarded by concrete covers that prevent accidental cable cutting.

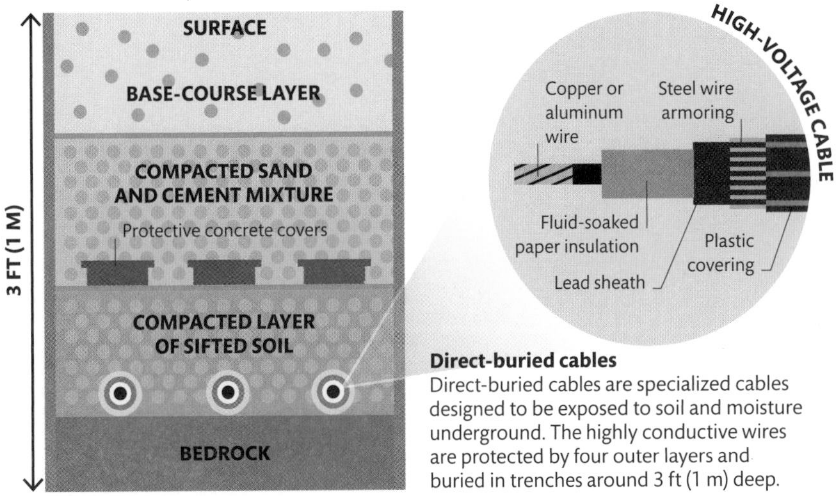

Direct-buried cables
Direct-buried cables are specialized cables designed to be exposed to soil and moisture underground. The highly conductive wires are protected by four outer layers and buried in trenches around 3 ft (1 m) deep.

Nuclear power

Nuclear energy is released when the nuclei of atoms are either split apart (nuclear fission) or fused together (nuclear fusion). A nuclear power plant harnesses the energy released from fission to generate electricity.

Nuclear fission

Nuclear power plants are fueled by radioactive elements such as uranium. When atoms of the fuel are split, a huge amount of energy is released as heat. This heat turns steam-driven turbines, which power electricity generators. Nuclear fission uses small amounts of fuel and produces far less greenhouse gas emissions than fossil fuels.

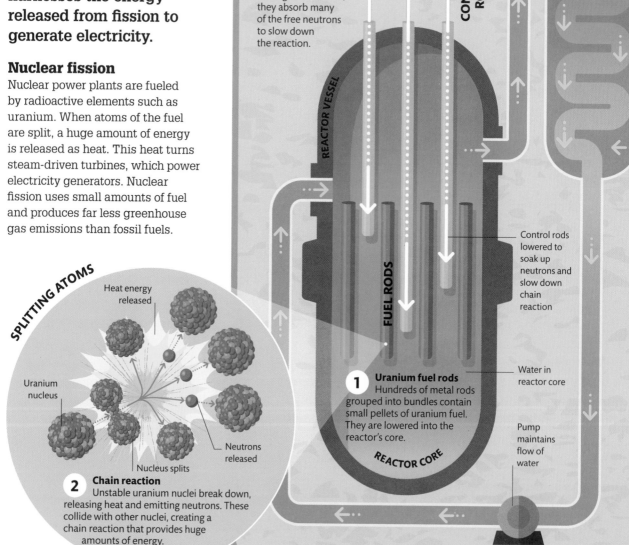

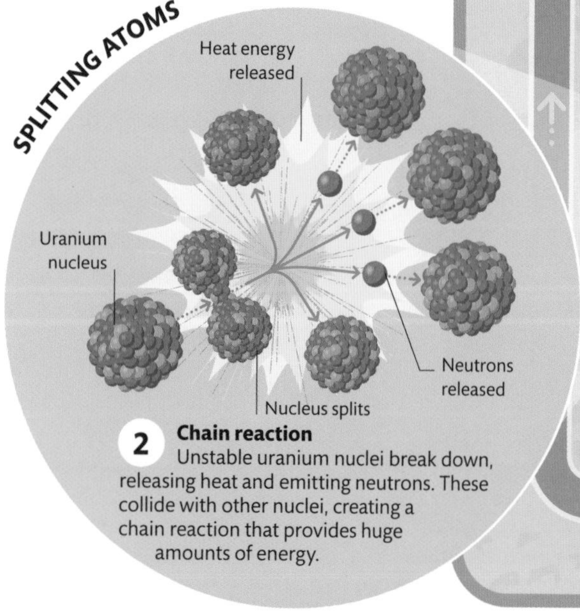

2 Chain reaction
Unstable uranium nuclei break down, releasing heat and emitting neutrons. These collide with other nuclei, creating a chain reaction that provides huge amounts of energy.

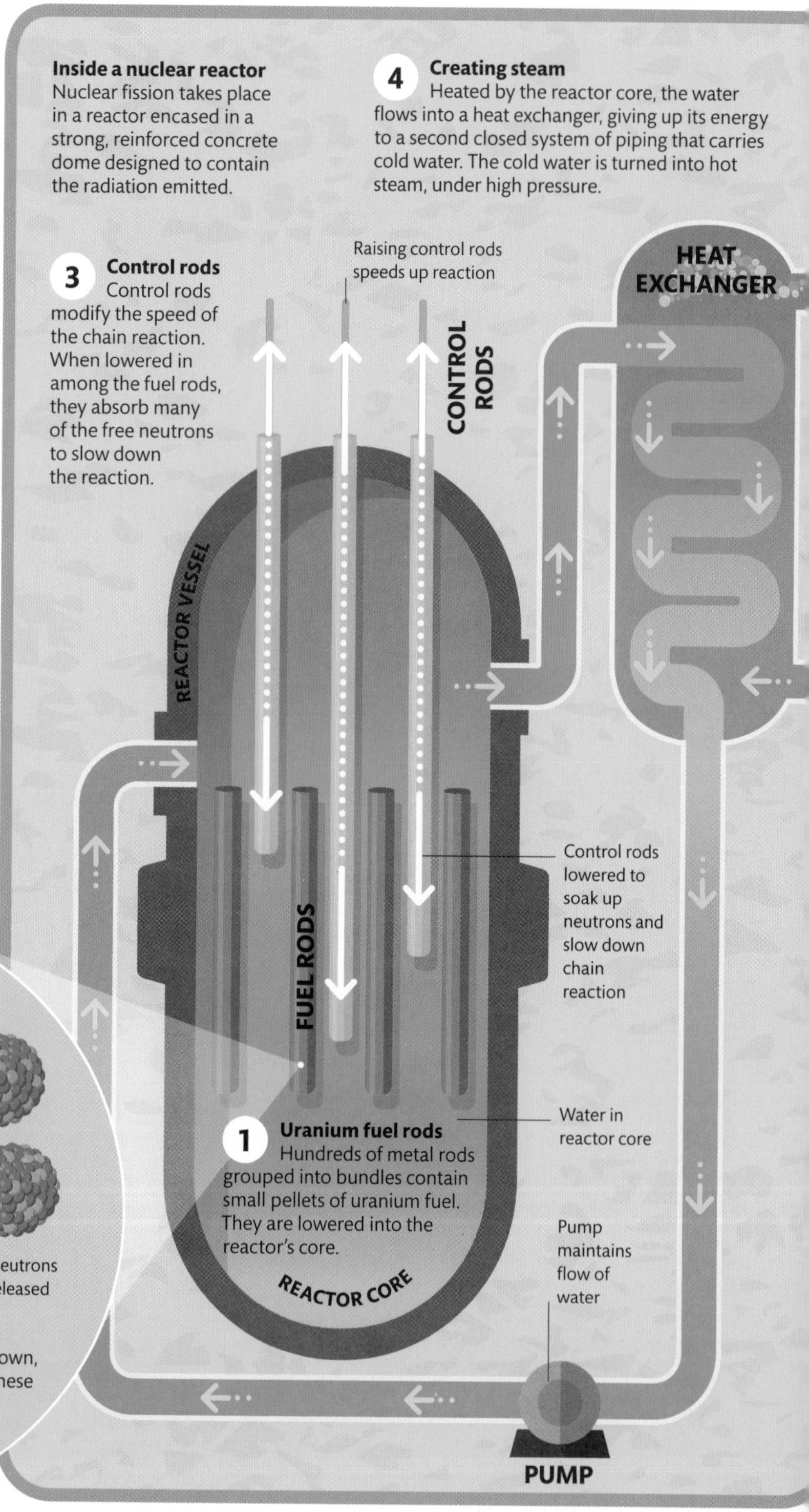

Inside a nuclear reactor
Nuclear fission takes place in a reactor encased in a strong, reinforced concrete dome designed to contain the radiation emitted.

1 Uranium fuel rods
Hundreds of metal rods grouped into bundles contain small pellets of uranium fuel. They are lowered into the reactor's core.

3 Control rods
Control rods modify the speed of the chain reaction. When lowered in among the fuel rods, they absorb many of the free neutrons to slow down the reaction.

4 Creating steam
Heated by the reactor core, the water flows into a heat exchanger, giving up its energy to a second closed system of piping that carries cold water. The cold water is turned into hot steam, under high pressure.

5 Turning the turbine Housed in a turbine hall, the turbine's fan blades are turned by the high-pressure steam from the heat exchanger. The turbines typically spin at speeds of 1,800–3,600 rpm.

6 Power supply The electricity generator is driven by the turbine's driveshaft. A transformer increases the electricity's voltage so that it can be carried by a local or regional power grid efficiently.

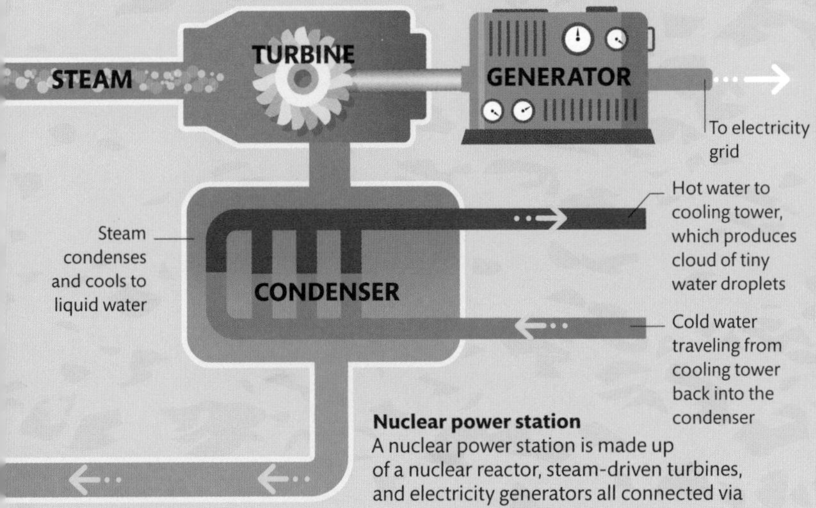

Nuclear power station
A nuclear power station is made up of a nuclear reactor, steam-driven turbines, and electricity generators all connected via numerous control and safety systems.

NUCLEAR MELTDOWN

The failure of a nuclear reactor's coolant system can lead to excessive amounts of heat building up in the fuel rods. In extreme cases, the rods can melt and burn through their containment structure. This can release huge amounts of radioactivity that may contaminate the environment.
In 2011, following an earthquake and tsunami, three of the reactors in Japan's Fukushima Daiichi plant suffered partial meltdowns.

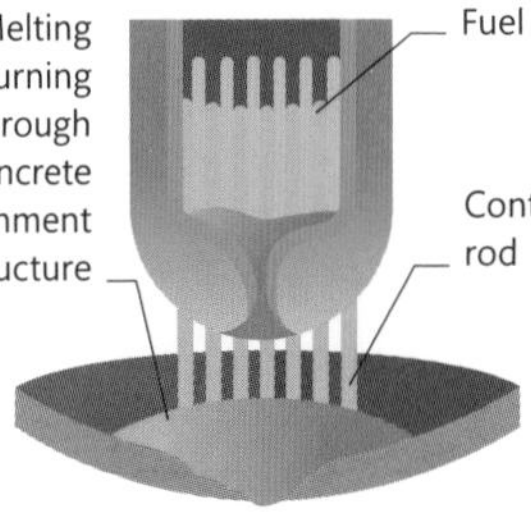

1 Fuel bundle Spent fuel rods that emit high levels of heat and radioactivity are left to cool for several years.

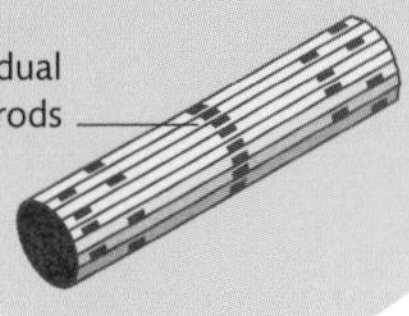

2 Disposed canister Radioactive waste is vitrified by mixing it with inert molten glass. The mixture solidifies within canisters or capsules.

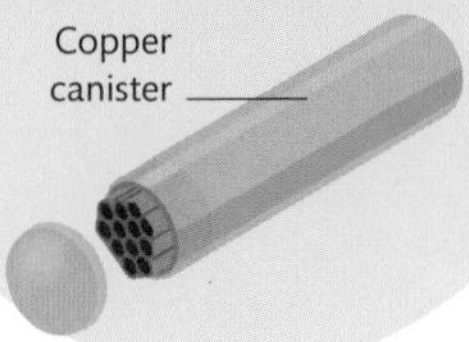

3 Sealed with clay The capsules are buried surrounded by a thick layer of impermeable clay acting as an additional barrier.

4 Burial site The containment burial site, 1,600–3,330 ft (500–1,000 m) below Earth's surface, is monitored and maintained.

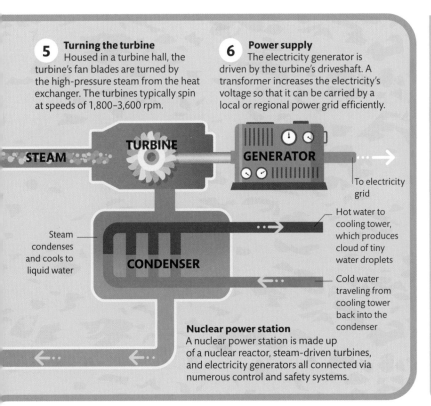

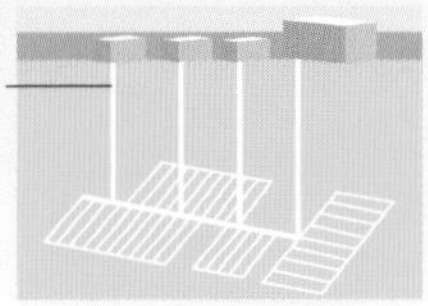

Radioactive waste disposal

Used (or spent) fuel rods are removed from a reactor every 2–5 years but still generate heat for decades and emit harmful levels of radioactivity for even longer. Most are initially stored in deep, cold-water storage pools for a number of years before they are either reprocessed or placed in concrete-encased casks. Several countries have proposed plans to store waste deep underground, but no sites are yet operational.

Geological repository plans
One proposed solution for radioactive waste disposal involves the already established technique of vitrification followed by burial in temperature-regulated deep bore holes.

A 1,000 MW NUCLEAR PLANT PRODUCES **ABOUT 30 TONS (27 TONNES) OF SPENT NUCLEAR FUEL** A YEAR

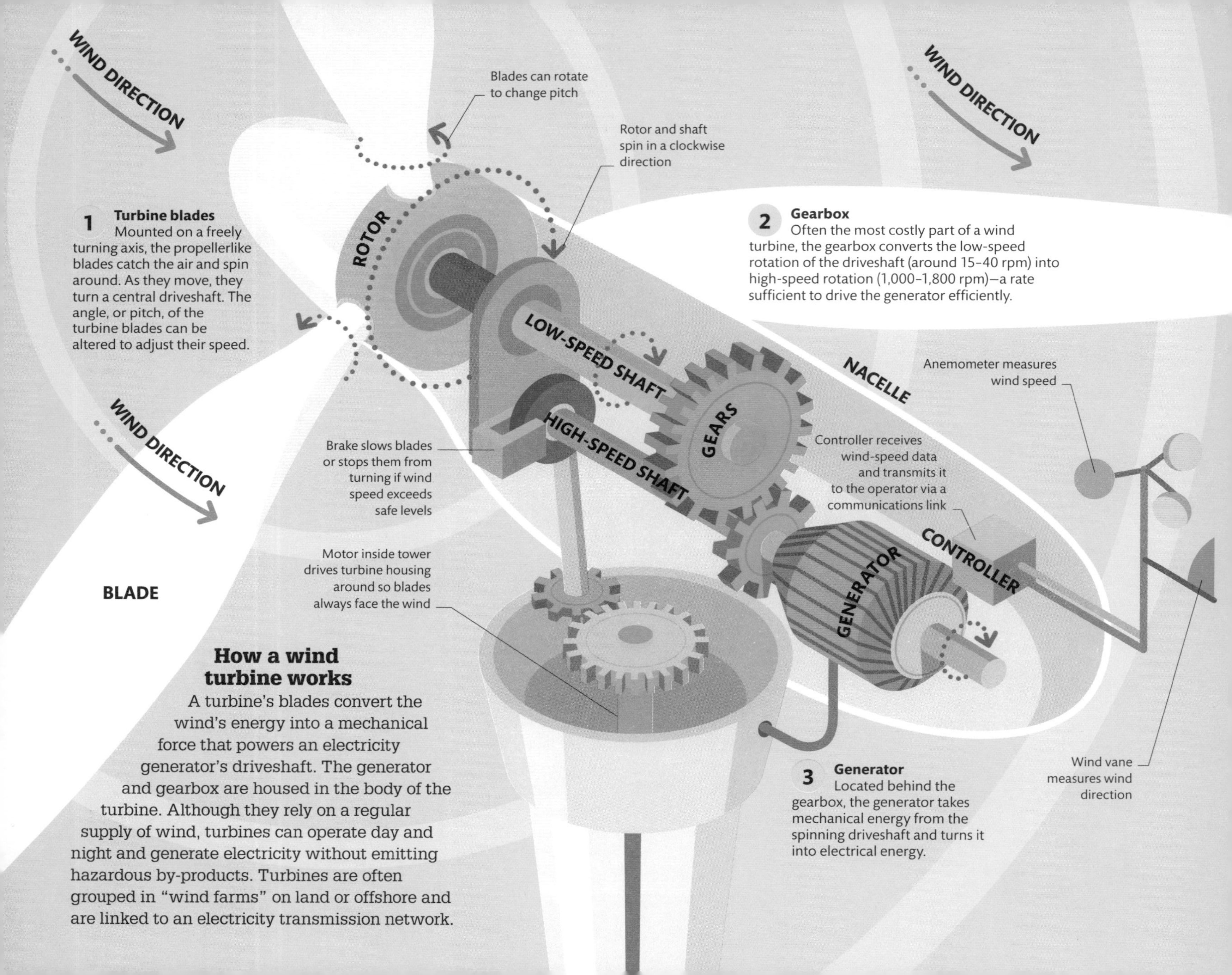
WIND DIRECTION
Blades can rotate to change pitch
Rotor and shaft spin in a clockwise direction
WIND DIRECTION
ROTOR
LOW-SPEED SHAFT
HIGH-SPEED SHAFT
GEARS
NACELLE
WIND DIRECTION
Anemometer measures wind speed
Brake slows blades or stops them from turning if wind speed exceeds safe levels
Controller receives wind-speed data and transmits it to the operator via a communications link
CONTROLLER
GENERATOR
Motor inside tower drives turbine housing around so blades always face the wind
BLADE
Wind vane measures wind direction

1 Turbine blades
Mounted on a freely turning axis, the propellerlike blades catch the air and spin around. As they move, they turn a central driveshaft. The angle, or pitch, of the turbine blades can be altered to adjust their speed.

2 Gearbox
Often the most costly part of a wind turbine, the gearbox converts the low-speed rotation of the driveshaft (around 15–40 rpm) into high-speed rotation (1,000–1,800 rpm)—a rate sufficient to drive the generator efficiently.

3 Generator
Located behind the gearbox, the generator takes mechanical energy from the spinning driveshaft and turns it into electrical energy.

How a wind turbine works

A turbine's blades convert the wind's energy into a mechanical force that powers an electricity generator's driveshaft. The generator and gearbox are housed in the body of the turbine. Although they rely on a regular supply of wind, turbines can operate day and night and generate electricity without emitting hazardous by-products. Turbines are often grouped in "wind farms" on land or offshore and are linked to an electricity transmission network.

Wind power

For centuries, the power of the wind has been harnessed to propel sailing ships and to drive windmills. Modern wind turbines provide a renewable source of energy by turning the kinetic energy of wind into electrical energy without consuming fossil fuels or emitting greenhouse gases.

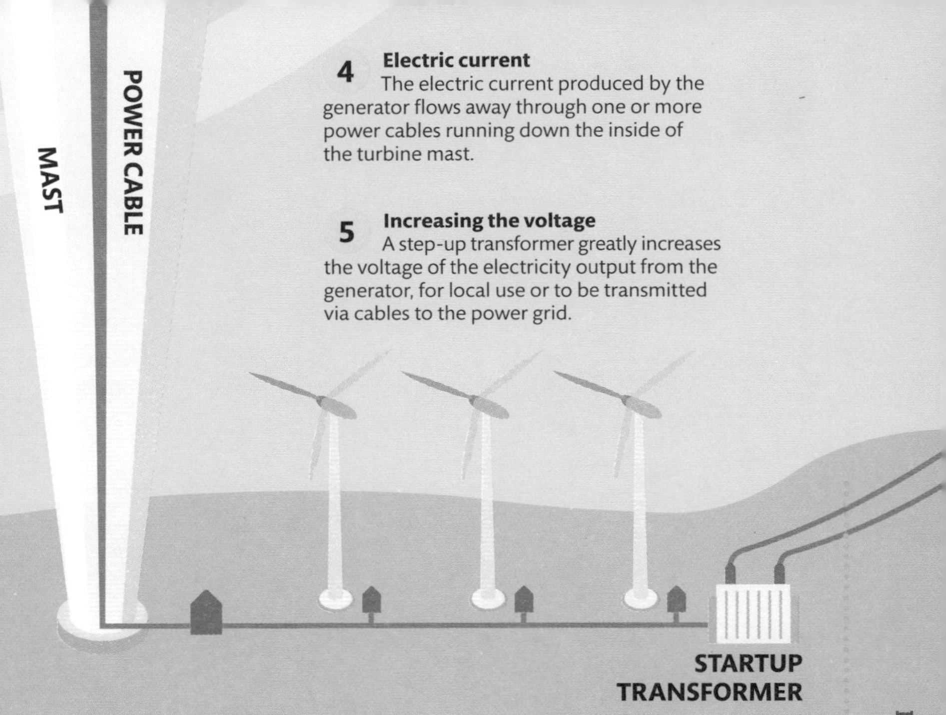
MAST
POWER CABLE
STARTUP TRANSFORMER

4 Electric current
The electric current produced by the generator flows away through one or more power cables running down the inside of the turbine mast.

5 Increasing the voltage
A step-up transformer greatly increases the voltage of the electricity output from the generator, for local use or to be transmitted via cables to the power grid.

AN AVERAGE **WIND TURBINE** CAN GENERATE ENOUGH **ELECTRICITY** TO POWER **1,000 HOMES**

Microgeneration

Small-scale renewable energy systems use freestanding or roof-mounted wind turbines to generate electricity, often in conjunction with other sustainable energy sources such as solar thermal collectors to heat water or photovoltaic cells. Together, they reduce the reliance on large, centralized power plants that often burn fossil fuels and emit harmful by-products.

Self-sufficiency
A wind turbine can fulfill a household's electricity needs. Excess electricity is supplied to the power grid, with a smart meter keeping track of flow both ways.

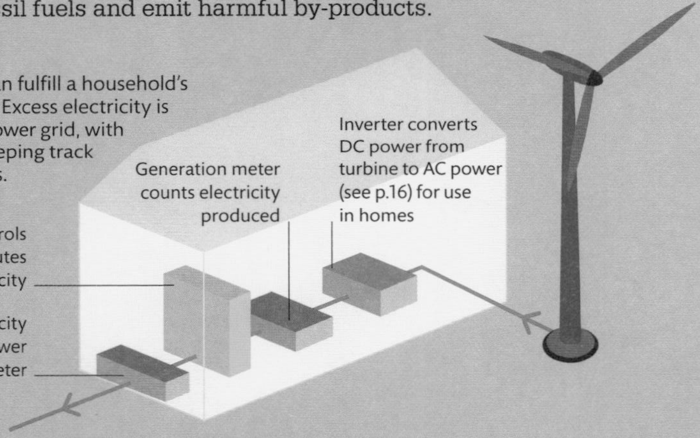
Generation meter counts electricity produced
Inverter converts DC power from turbine to AC power (see p.16) for use in homes
Fuse box controls and distributes electricity
Excess electricity supplied to power grid via smart meter

TURBINES AND WILDLIFE

Wind turbine construction can disturb ecosystems at sea and on land, but the most direct threat they pose is to birds and bats. Siting wind farms away from nesting sites and routes used by migrating birds is one solution. Another potential option might be "acoustic lighthouses"—devices placed close to wind turbines that emit loud sounds to warn birds.

Water and geothermal power

The energy in moving water and the heat in Earth's crust can be harnessed to produce electricity. Both offer clean, sustainable sources of energy but tend to involve significant infrastructural investment.

Tidal power

Tidal power captures the kinetic energy of the natural ebb and flow of ocean tides to turn turbines that power electricity generators. Some systems use freestanding turbines, similar to wind turbines, while tidal barrages involve multiple turbines in a large dam, usually constructed across a bay or estuary.

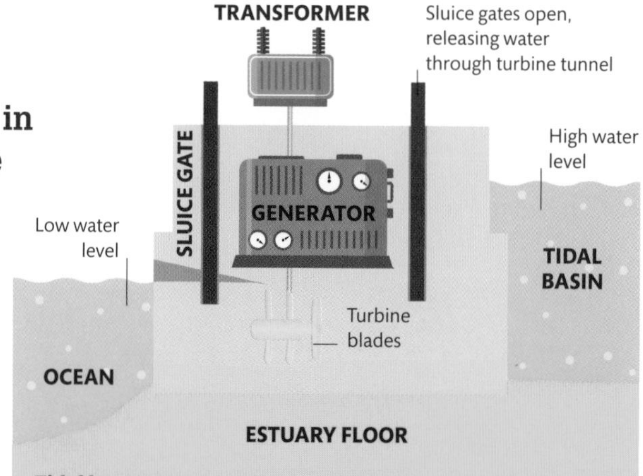

TRANSFORMER
Sluice gates open, releasing water through turbine tunnel
High water level
SLUICE GATE
GENERATOR
Low water level
TIDAL BASIN
Turbine blades
OCEAN
ESTUARY FLOOR

Tidal barrage
Water flowing into an estuary at high tide is held back behind a dam called a barrage. Sluice gates then control its flow through the turbine.

Hydroelectric power

Hydroelectric power (HEP) harnesses the power in falling or fast-flowing water to turn turbines that drive an electricity generator. Most commonly, water is collected at higher elevations behind a dam and then channeled past turbines.

2 Power generated
The water passes through a turbine at high speed, turning its blades with considerable force. The turbine powers a generator that produces an electric current.

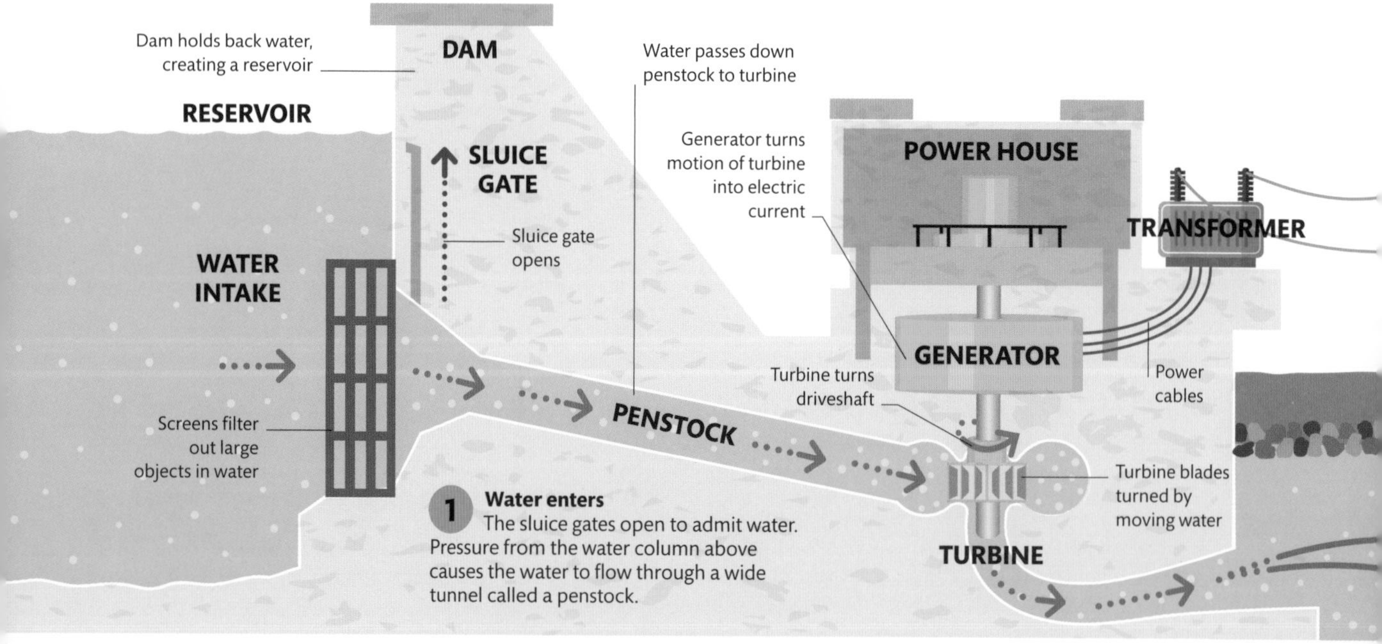

Dam holds back water, creating a reservoir
DAM
Water passes down penstock to turbine
RESERVOIR
SLUICE GATE
Generator turns motion of turbine into electric current
POWER HOUSE
TRANSFORMER
Sluice gate opens
WATER INTAKE
GENERATOR
Turbine turns driveshaft
Power cables
PENSTOCK
Screens filter out large objects in water
Turbine blades turned by moving water
TURBINE

1 Water enters
The sluice gates open to admit water. Pressure from the water column above causes the water to flow through a wide tunnel called a penstock.

THE DANGERS OF DRILLING

Enhanced geothermal systems (EGSs) inject fluid under high pressure to create cracks and fractures in rock so that the fluid can travel through a larger area and obtain more heat. There is some evidence that such fracturing could create uncontrollable seismic activity. In 2006, a geothermal plant in Basel, Switzerland, was held responsible for inducing a magnitude 3.4 earthquake. Eleven years later, a magnitude 5.4 earthquake in Pohang, South Korea, injured 82 people. Initial studies suggest a local geothermal plant may have been the cause.

THE **ITAIPU DAM** ON THE PARAGUAY–BRAZIL BORDER PROVIDES **76 PERCENT** OF **PARAGUAY'S ENERGY**

Channeling water
A continuous strong flow of water is needed for an HEP scheme to generate electricity constantly. Some schemes, known as pumped storage HEP systems, pump the outlet water back up to the reservoir at times of lower electricity demand, using surplus electricity.

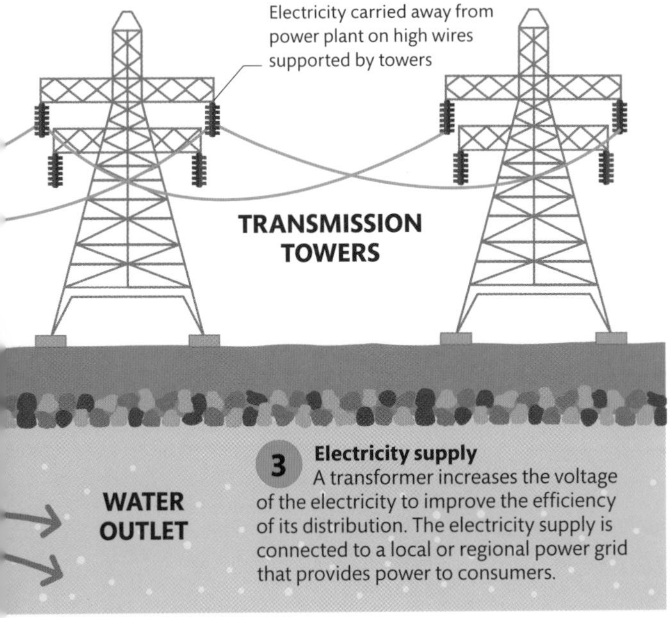
Electricity carried away from power plant on high wires supported by towers
TRANSMISSION TOWERS
WATER OUTLET
3 Electricity supply
A transformer increases the voltage of the electricity to improve the efficiency of its distribution. The electricity supply is connected to a local or regional power grid that provides power to consumers.

Geothermal power

Heat from hot underground rocks can be harnessed in different ways. Underground water can be tapped directly, or water can be pumped through a geothermal region to gain heat used to generate electricity. Geothermal power produces a small fraction of the harmful emissions of a coal-fired power station.

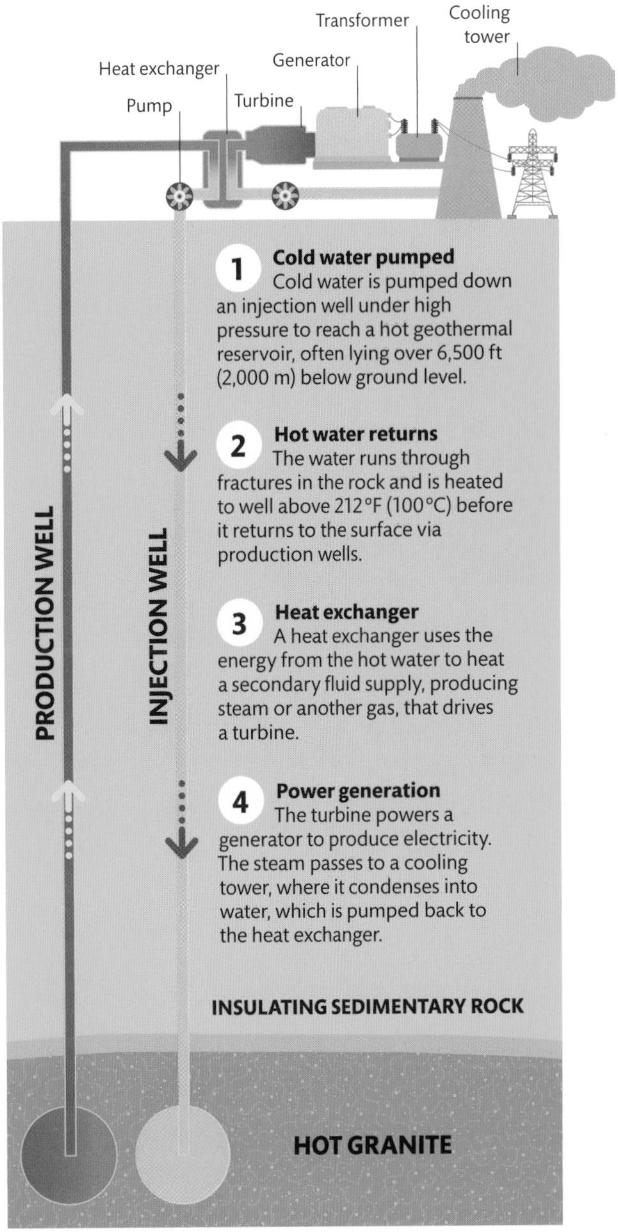
Transformer
Cooling tower
Heat exchanger
Generator
Pump
Turbine
1 Cold water pumped
Cold water is pumped down an injection well under high pressure to reach a hot geothermal reservoir, often lying over 6,500 ft (2,000 m) below ground level.
2 Hot water returns
The water runs through fractures in the rock and is heated to well above 212°F (100°C) before it returns to the surface via production wells.
3 Heat exchanger
A heat exchanger uses the energy from the hot water to heat a secondary fluid supply, producing steam or another gas, that drives a turbine.
4 Power generation
The turbine powers a generator to produce electricity. The steam passes to a cooling tower, where it condenses into water, which is pumped back to the heat exchanger.
PRODUCTION WELL
INJECTION WELL
INSULATING SEDIMENTARY ROCK
HOT GRANITE

Solar power and bioenergy

The power of the sun can be used, on various scales, to heat water directly or to generate large amounts of electricity using photovoltaic cells. Biomass—organic material derived from plants or animals—can also be a valuable source of energy.

Solar power

Energy from sunlight can be harnessed in several ways. Solar collectors use the sun to heat water contained in pipes. Photovoltaic (PV) panels—made of silicon combined, or "doped," with other elements to conduct electricity—convert the sun's energy into electricity via the photoelectric effect (see left). This renewable energy produces no harmful emissions, and it has been estimated that a domestic photovoltaic system can prevent up to 1.9 tons (1.7 metric tons) of carbon emissions per year.

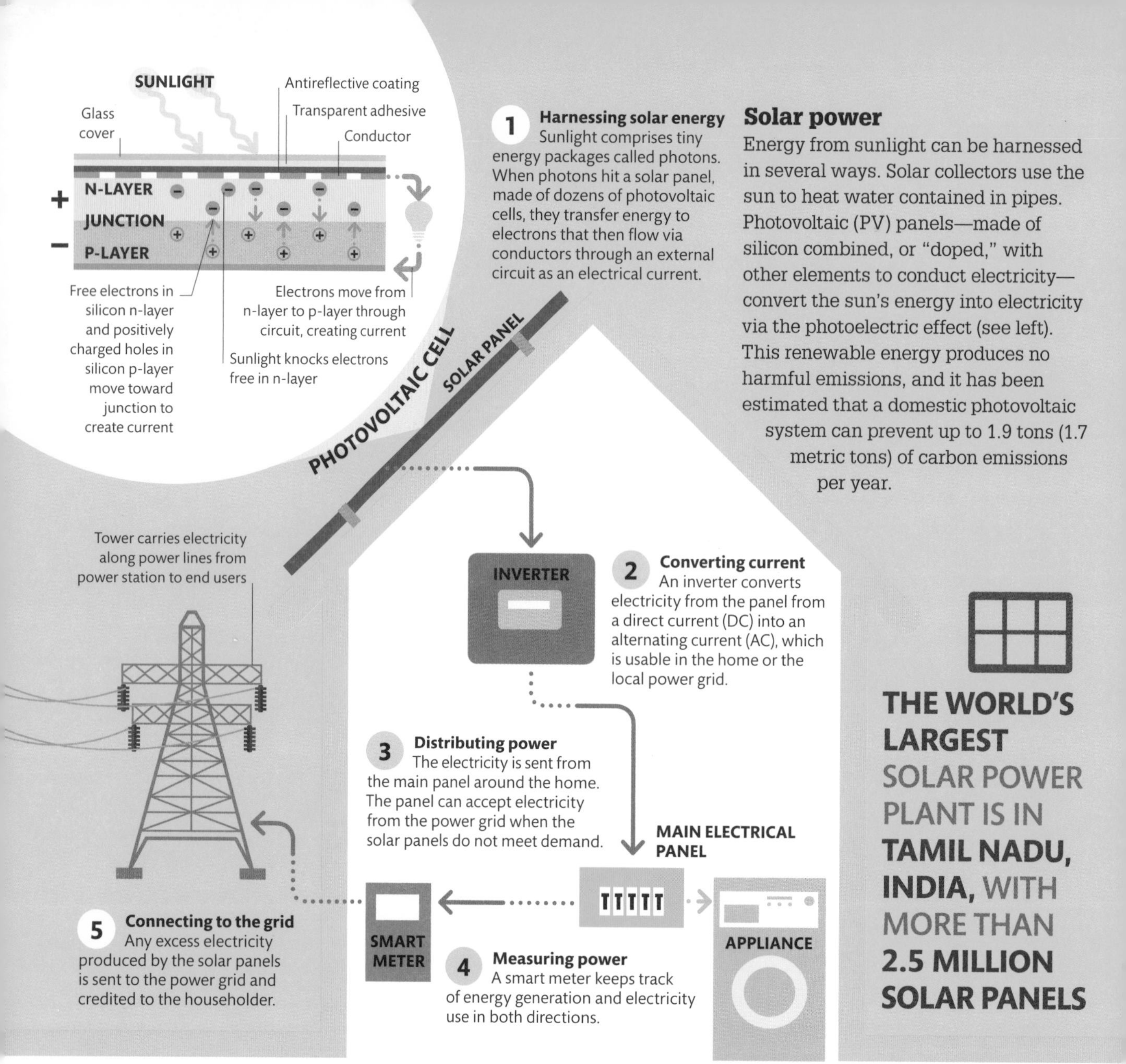

SUNLIGHT
Glass cover
Antireflective coating
Transparent adhesive
Conductor
+
N-LAYER
JUNCTION
–
P-LAYER
Free electrons in silicon n-layer and positively charged holes in silicon p-layer move toward junction to create current
Sunlight knocks electrons free in n-layer
Electrons move from n-layer to p-layer through circuit, creating current
PHOTOVOLTAIC CELL
SOLAR PANEL
INVERTER
Tower carries electricity along power lines from power station to end users
MAIN ELECTRICAL PANEL
SMART METER
APPLIANCE

1 Harnessing solar energy Sunlight comprises tiny energy packages called photons. When photons hit a solar panel, made of dozens of photovoltaic cells, they transfer energy to electrons that then flow via conductors through an external circuit as an electrical current.

2 Converting current An inverter converts electricity from the panel from a direct current (DC) into an alternating current (AC), which is usable in the home or the local power grid.

3 Distributing power The electricity is sent from the main panel around the home. The panel can accept electricity from the power grid when the solar panels do not meet demand.

4 Measuring power A smart meter keeps track of energy generation and electricity use in both directions.

5 Connecting to the grid Any excess electricity produced by the solar panels is sent to the power grid and credited to the householder.

THE WORLD'S LARGEST SOLAR POWER PLANT IS IN **TAMIL NADU, INDIA,** WITH MORE THAN **2.5 MILLION SOLAR PANELS**

Bioenergy

Bioenergy is generated by burning biomass—organic material including plant waste and animal matter—in power stations or by converting by-products into biofuels. Biomass is considered a renewable energy source because harvested crops and trees can be replaced. However, expanding the scale of bioenergy is problematic because it would require the conversion of arable land used for food production.

Sewage
Sludge from sewage treatment is broken down by anaerobic bacteria in digester tanks to produce methane and other gases, which can be purified and burned as fuel.

Industrial residue
Certain wastes left over from industrial processes—particularly black liquor from wood pulp and paper production—are rich in organic matter and can be burned as fuel to power electricity generators.

Agricultural
Crops, including rapeseed, sugar cane, and beet, are grown for processing into biofuels. Nonfood energy crops are sometimes grown on land that would otherwise have little agricultural value.

Forestry
Trees are the most ancient of fuels, burned for heat and light for thousands of years. Logs, chips, wood pellets, and sawdust account for over a third of all biomass energy used.

Animal waste
While animal remains can be burned as biomass, the manure produced by farmed animals, including cows, can also be treated to produce a methane-rich "biogas" that can be burned.

Municipal solid waste
Some of the vast amounts of solid waste produced is burned to generate heat or electricity. This has the added benefit of reducing the amount of space needed for landfill sites.

ETHANOL BIOFUEL

Ethanol is an alcohol that is produced from the sugars found in biomass crops, including sugar cane, corn, and sorghum. In Brazil, the world's leading producer of biofuel ethanol, more than 80 percent of all new cars and almost half of all motorcycles are able to run on ethanol or a gasoline-ethanol blend.

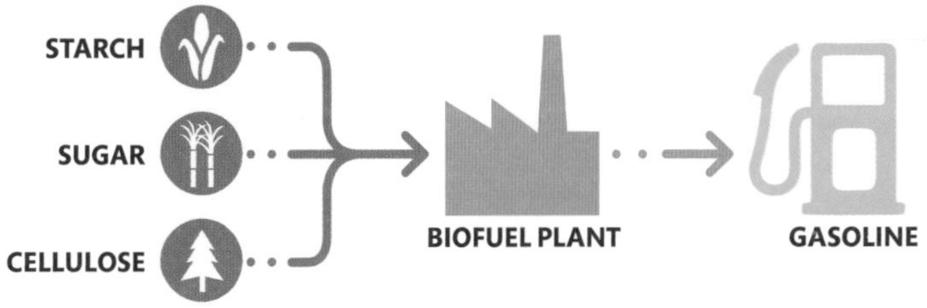

Batteries

A battery is a portable store of chemical energy that can be converted into electrical energy. Batteries are classified into two broad groups: primary (single use) and secondary (rechargeable).

RECYCLED BATTERIES CONTAIN **ZINC** AND **MANGANESE,** WHICH CAN BE USED **AS MICRONUTRIENTS** TO HELP **CORN GROW**

How a battery works

In a battery, chemical reactions take place that free electrons from metal atoms. The electrons flow to the anode through the electrolyte. When an electrical circuit connects the terminals, the electrons return to the cathode, flowing as an electric current. This conversion of chemical energy into electrical energy is known as discharging.

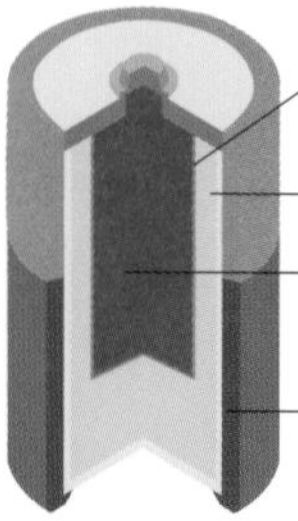

Separator
Electrolyte
Cathode (carbon rod)
Anode (usually zinc inner case)

Inside a battery
A battery comprises a positive electrode (cathode) and a negative electrode (anode), separated by a substance that conducts electricity called an electrolyte.

1 Chemical reactions
When a battery is connected to an electric circuit, a chemical reaction makes metal atoms lose electrons. These are attracted to and gained by a chemical paste called the electrolyte.

KEY Electron — Wire — Positive charge — Direction of current

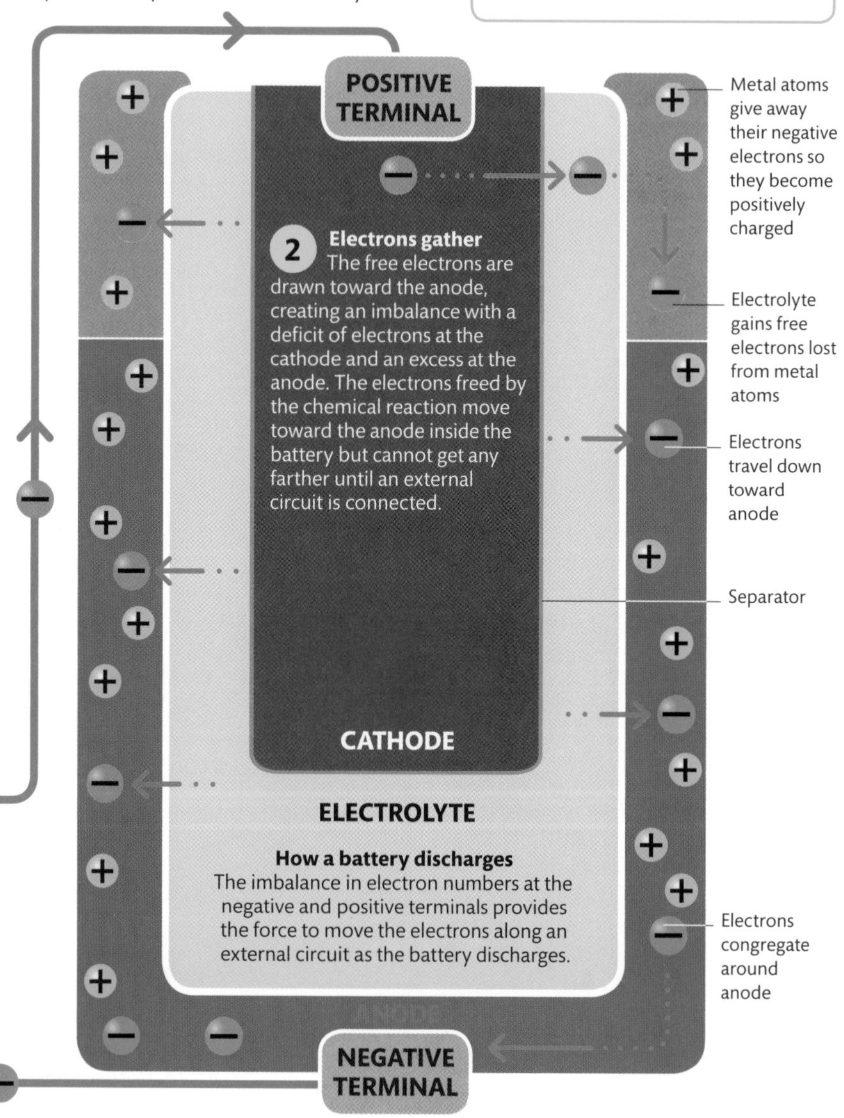

POSITIVE TERMINAL
Metal atoms give away their negative electrons so they become positively charged
Electrolyte gains free electrons lost from metal atoms
Electrons travel down toward anode
Separator
CATHODE
ELECTROLYTE
Electrons congregate around anode
ANODE
NEGATIVE TERMINAL

2 Electrons gather
The free electrons are drawn toward the anode, creating an imbalance with a deficit of electrons at the cathode and an excess at the anode. The electrons freed by the chemical reaction move toward the anode inside the battery but cannot get any farther until an external circuit is connected.

How a battery discharges
The imbalance in electron numbers at the negative and positive terminals provides the force to move the electrons along an external circuit as the battery discharges.

4 Incoming electrons
The electrons reenter the battery via the cathode. The flow continues until the store of chemicals is exhausted.

3 Migrating electrons
The external circuit connecting the anode and cathode provides a path along which the electrons can flow, producing an electric current. Along the way, this current can be used to power an electrical device.

Light bulb lit by flow of electric current

How a battery recharges
When a battery is plugged into a charger, a current passes through it in the opposite direction to that produced when the battery was discharging. This moves electrons back to where they started, recharging the battery.

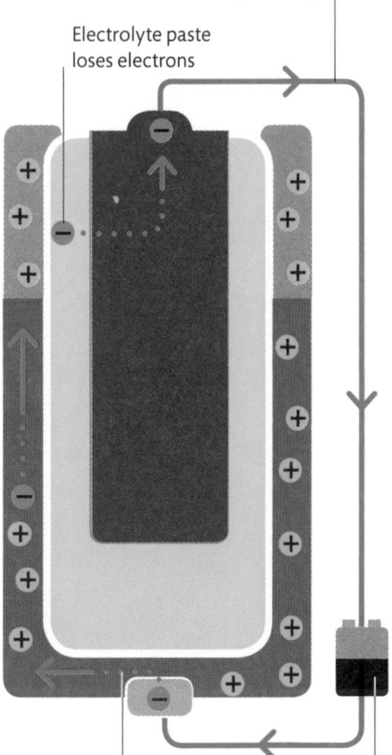

Lithium-ion batteries
Found in smartphones and many other devices and machines, including electric cars, lithium-ion (Li-ion) batteries use the large amounts of energy present in the highly reactive metal lithium. Lithium's low weight but high energy density produces a good power-to-weight ratio for batteries that can withstand hundreds of discharge and recharge cycles.

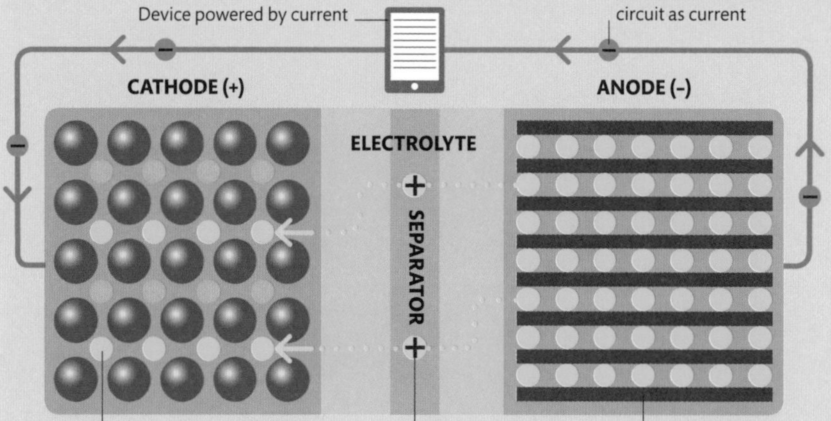

How a lithium-ion battery works
Lithium ions flow through the electrolyte to the cathode during discharge while the free electrons flow through the external circuit, providing power. Recharging returns the lithium ions and electrons.

KEY Metal · Oxygen · Lithium

WHAT IS THE WORLD'S LARGEST BATTERY?
Tesla's gigantic lithium-ion battery in South Australia covers 2.5 acres (1 hectare) and provides 129 MWh (see p.10) of electricity.

Batteries of the future
A great deal of research is focused on battery development. One innovation that may lead to faster-charging, longer-lasting batteries uses a solid state alkali metal rather than the liquid or gel electrolytes in Li-ion cells. Flexible batteries using devices called supercapacitors that can be recharged in seconds might revolutionize wearable and portable technology.

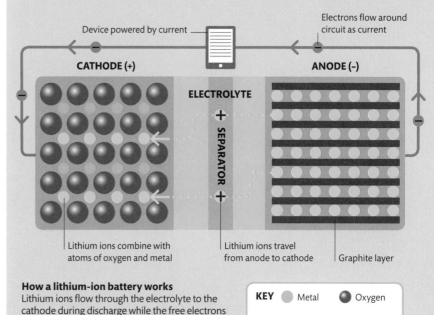

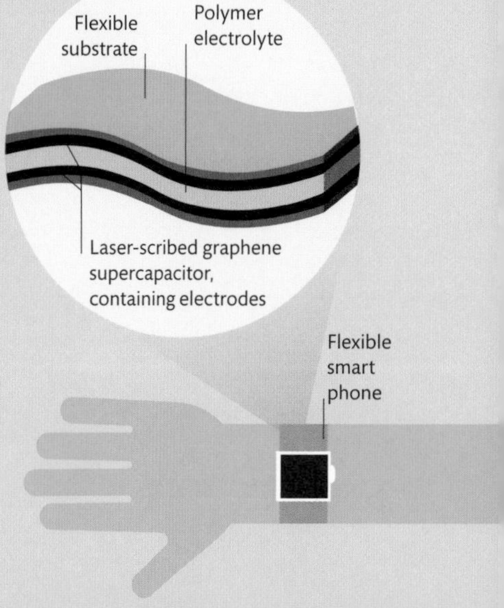

Supercapacitor
Electric charge is stored as a coating of ions on a supercapacitor's electrode layers, which are separated by an electrolyte made of a flexible polymer.

Fuel cells

Fuel cells generate electricity through a chemical reaction caused by mixing fuel with oxygen. There are a number of types, but cells using hydrogen are increasingly used in vehicles and electronic devices.

How a fuel cell works

A fuel cell is an electrochemical cell that produces an electric current, which is used to power motors or other electrical devices. Hydrogen fuel cells produce electricity without combustion and emit only water as a by-product. Obtaining oxygen from the air and hydrogen from its internal tank, a car's electric motor can typically run for about 300 miles (480 km).

Hydrogen-powered car
Fuel cells are usually deployed in stacks. These provide an electric current, which is increased by a boost converter before supplying the motor.

Inside a fuel cell
A fuel cell is similar in structure to a battery (see pp.32–33). The cell produces a flow of electrons from the anode out of the cell and back to the cathode. This external flow of current powers the car.

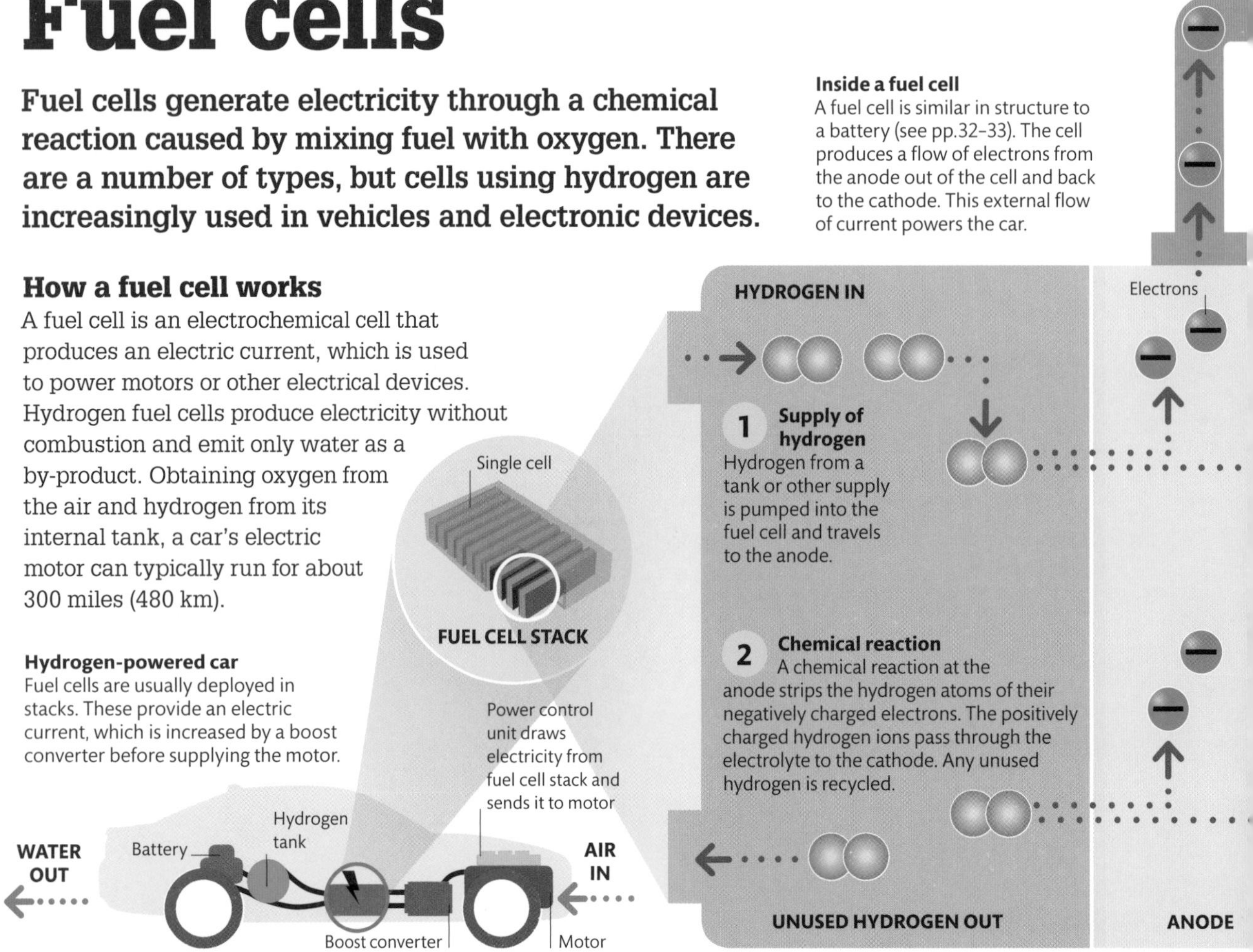

Single cell
FUEL CELL STACK
Power control unit draws electricity from fuel cell stack and sends it to motor
WATER OUT
Battery
Hydrogen tank
AIR IN
Boost converter
Motor
HYDROGEN IN
1 Supply of hydrogen
Hydrogen from a tank or other supply is pumped into the fuel cell and travels to the anode.
2 Chemical reaction
A chemical reaction at the anode strips the hydrogen atoms of their negatively charged electrons. The positively charged hydrogen ions pass through the electrolyte to the cathode. Any unused hydrogen is recycled.
Electrons
UNUSED HYDROGEN OUT
ANODE

Sources of hydrogen

Most hydrogen is produced from fossil fuels, particularly natural gas. The most common method used is a process called steam-methane reformation, which produces some carbon dioxide emissions. Other processes, such as electrolysis, harness hydrogen without harmful emissions but use lots of energy.

Steam-methane reformation
Methane and steam react to produce a mixture of gases, which are sent to a shift reactor where more hydrogen and carbon dioxide are created. A purification stage produces pure hydrogen.

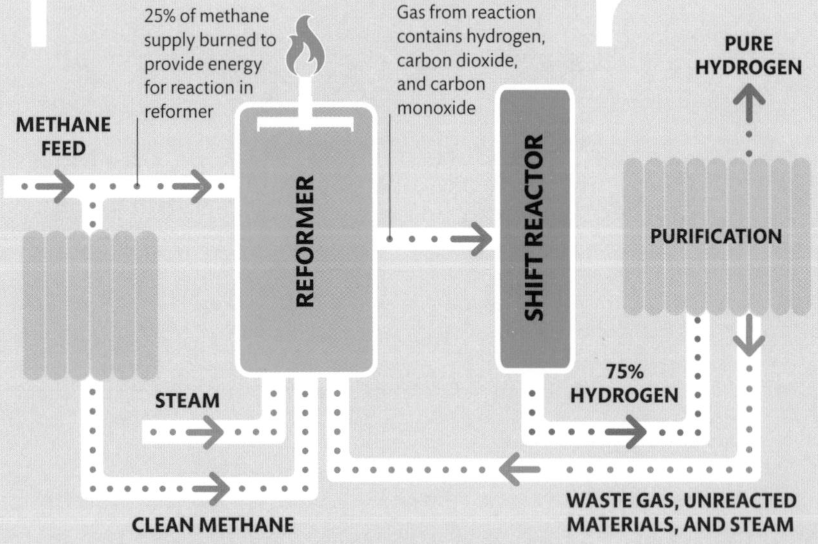

25% of methane supply burned to provide energy for reaction in reformer
Gas from reaction contains hydrogen, carbon dioxide, and carbon monoxide
PURE HYDROGEN
METHANE FEED
REFORMER
SHIFT REACTOR
PURIFICATION
STEAM
75% HYDROGEN
CLEAN METHANE
WASTE GAS, UNREACTED MATERIALS, AND STEAM

HYDROGEN FUEL CELLS USE 50 PERCENT LESS FUEL THAN GAS ENGINES

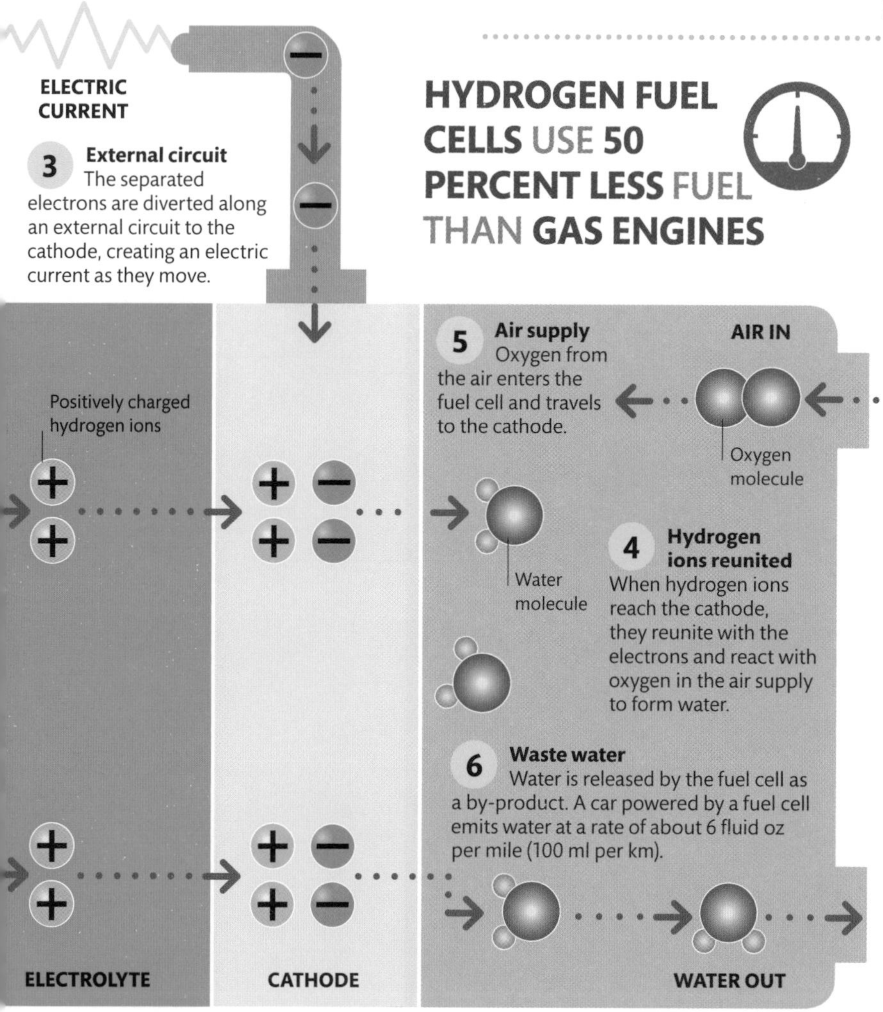
ELECTRIC CURRENT
3 External circuit
The separated electrons are diverted along an external circuit to the cathode, creating an electric current as they move.
5 Air supply
Oxygen from the air enters the fuel cell and travels to the cathode.
AIR IN
Positively charged hydrogen ions
Oxygen molecule
Water molecule
4 Hydrogen ions reunited
When hydrogen ions reach the cathode, they reunite with the electrons and react with oxygen in the air supply to form water.
6 Waste water
Water is released by the fuel cell as a by-product. A car powered by a fuel cell emits water at a rate of about 6 fluid oz per mile (100 ml per km).
ELECTROLYTE
CATHODE
WATER OUT

USES OF FUEL CELLS

Fuel cells remain an emerging technology but one with a vast range of potential applications that make use of its compact, convenient, and exhaust-free source of electricity.

Vehicles
Increasing numbers of fork-lift trucks, zero-emission buses, city trams, and some cars are powered by fuel cells.

Military
Small cells can power soldiers' electronic devices, while larger cells can keep drones in the air for long periods.

Portable electronics
Micro fuel cells are being developed to recharge smartphones, tablets, and other mobile devices.

Space
Fuel cells are a common source of power in spacecraft. Manned craft also use the fresh water they produce.

Aircraft
Experimental fuel-cell aircraft exist, but airliners are more likely to use them as back-up power supplies.

FUEL CELLS IN SPACE

Fuel cells first traveled into space on NASA's Gemini missions in 1965–1966. A stack of hydrogen cells housed in the service module also provided electrical power for the Apollo manned missions to the Moon (1969–1972). Each of the fuel cells contained 31 separate cells connected in series. The fuel cells used by Apollo proved highly successful, producing up to 2,300 watts of power while being less bulky than batteries and more efficient than solar panels.

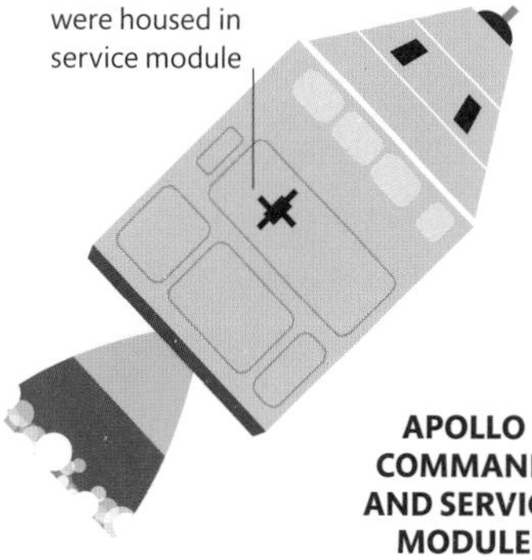
Fuel cell stacks were housed in service module

APOLLO COMMAND AND SERVICE MODULE

ARE HYDROGEN FUEL CELLS SAFE?

Fears persist since hydrogen is extremely flammable, but fuel cells are manufactured with rigorous safeguards, and hydrogen tanks in vehicles are very tough and crushproof.

TRANSPORTATION TECHNOLOGY

Moving machines

Business, industry, leisure, and tourism rely on fast, long-distance transportation for the movement of goods and people. Transportation technology depends on the use of energy and the application of many different forces to produce motion.

THE WHEEL

The wheel is one of the world's most important inventions. A wheel and axle work like a rotating lever, transmitting force in a circular direction. Turning the wheel around the axle moves the rim of the wheel a greater distance with less force. Turning the rim of the wheel turns the axle with greater force.

Rim moves further and faster than axle
AXLE
Wheel rotates around axle

Combining forces

An object such as a vehicle moves when it is acted on by one or more forces. As a force is applied, a transfer of energy takes place, either setting the vehicle in motion or changing its speed and direction. Several forces usually act on a vehicle at the same time. Some of the forces may work together, while others work against each other. The combined effect is a single force called the resultant force.

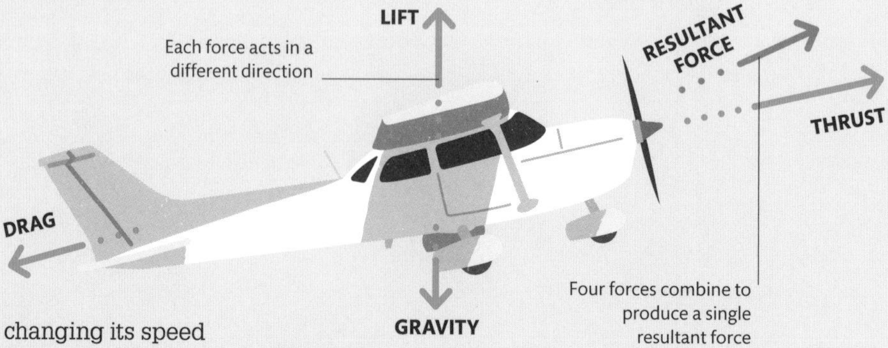

LIFT
Each force acts in a different direction
RESULTANT FORCE
THRUST
DRAG
GRAVITY
Four forces combine to produce a single resultant force

Forces of flight
Four forces are at work when an airplane is in flight. It is pulled down by gravity, pushed upward by lift from its wings, forced forward by thrust from its spinning propeller, and pulled backward by drag. When an airplane is accelerating and climbing, there is an upward resultant force.

Friction

Friction is a force that resists the motion of surfaces sliding across each other. Some friction is necessary; for example, rubber tires rely on friction for grip. However, friction also causes wear and generates heat. Both of these effects are damaging to machines with moving parts. Friction levels depend on the roughness of the surfaces in contact and the amount of force pressing them together. Adding a lubricant reduces friction, because it forms a thin film between the surfaces, keeping them apart.

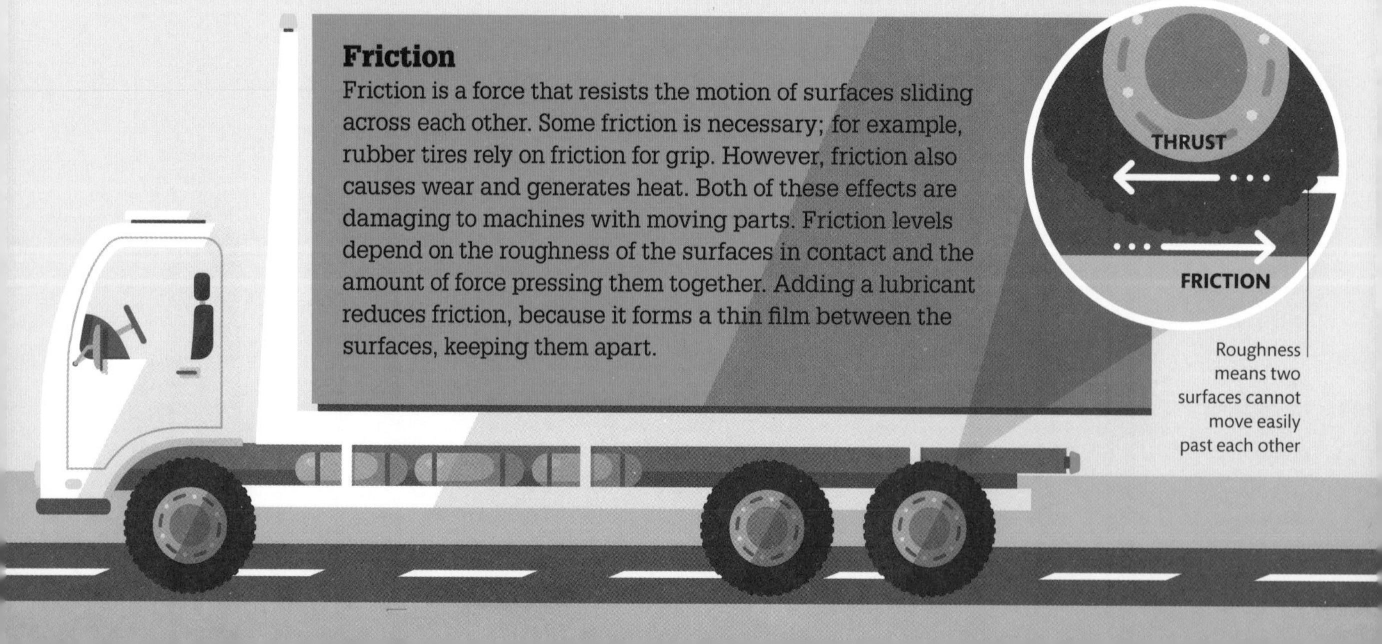

THRUST
FRICTION
Roughness means two surfaces cannot move easily past each other

THERE ARE MORE THAN **1 BILLION BICYCLES** IN THE WORLD, WITH **UP TO 100 MILLION ADDED** EVERY YEAR

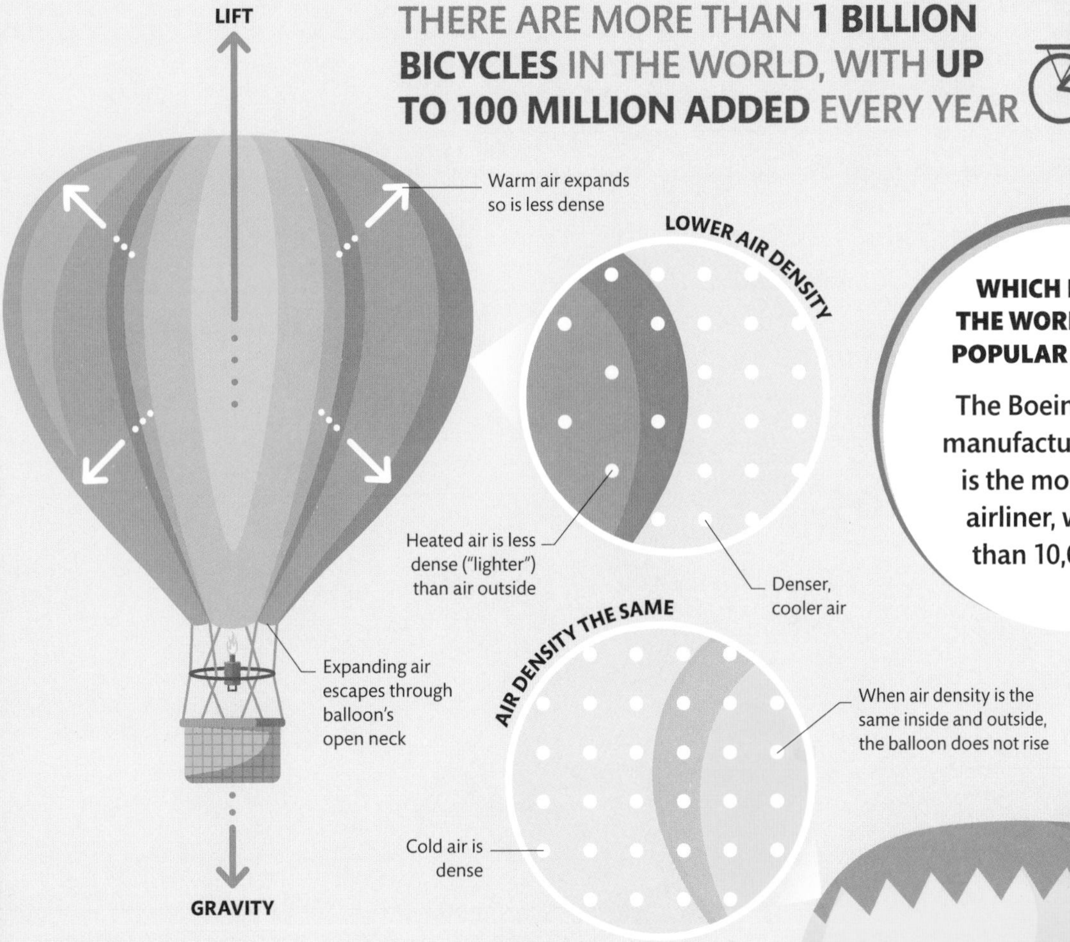
LIFT
Warm air expands so is less dense
LOWER AIR DENSITY
Heated air is less dense ("lighter") than air outside
Denser, cooler air
Expanding air escapes through balloon's open neck
AIR DENSITY THE SAME
When air density is the same inside and outside, the balloon does not rise
Cold air is dense
GRAVITY

WHICH PLANE IS THE WORLD'S MOST POPULAR AIRLINER?

The Boeing 737, first manufactured in 1967, is the most popular airliner, with more than 10,000 built.

Gas power

Most transportation technology relies on one simple scientific principle—gas expands when it is heated. Gasoline, diesel, turbine, and rocket engines are all activated by expanding gas. When gas expands inside an engine, it does so with great force that can be used to turn wheels or propellers or produce a powerful jet. Most often the gas involved is air. Burning fuel usually provides the heat to make the air expand, but other energy sources are sometimes used. Some warships, submarines, and ice breakers are nuclear-powered. They use the heat produced by radioactive elements such as uranium to generate the expanding gas that powers their propellers.

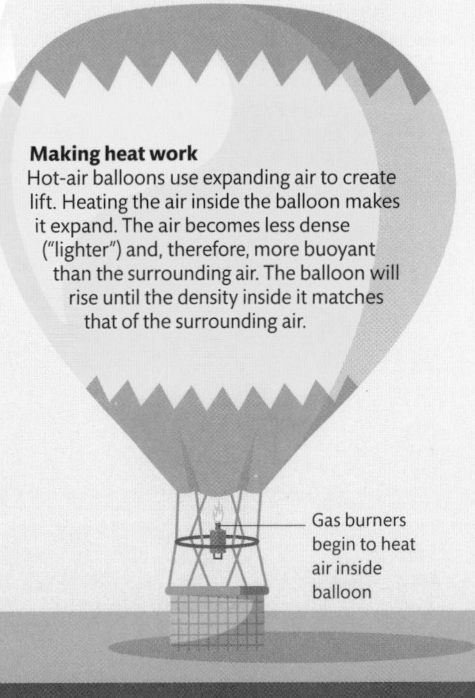
Gas burners begin to heat air inside balloon

Making heat work
Hot-air balloons use expanding air to create lift. Heating the air inside the balloon makes it expand. The air becomes less dense ("lighter") and, therefore, more buoyant than the surrounding air. The balloon will rise until the density inside it matches that of the surrounding air.

Bicycles

The invention of the bicycle was the biggest advance in personal transportation since the domestication of the horse. Bicycles are still one of the most energy-efficient forms of transportation.

Transmitting power

A bicycle rider's muscle power is transmitted to the back wheel by a chain connected, via levers called cranks, to the pedals. The rider can pedal efficiently only within a narrow speed range. Gears enable the rider to stay within this range by turning the back wheel faster or slower for the same pedaling rate.

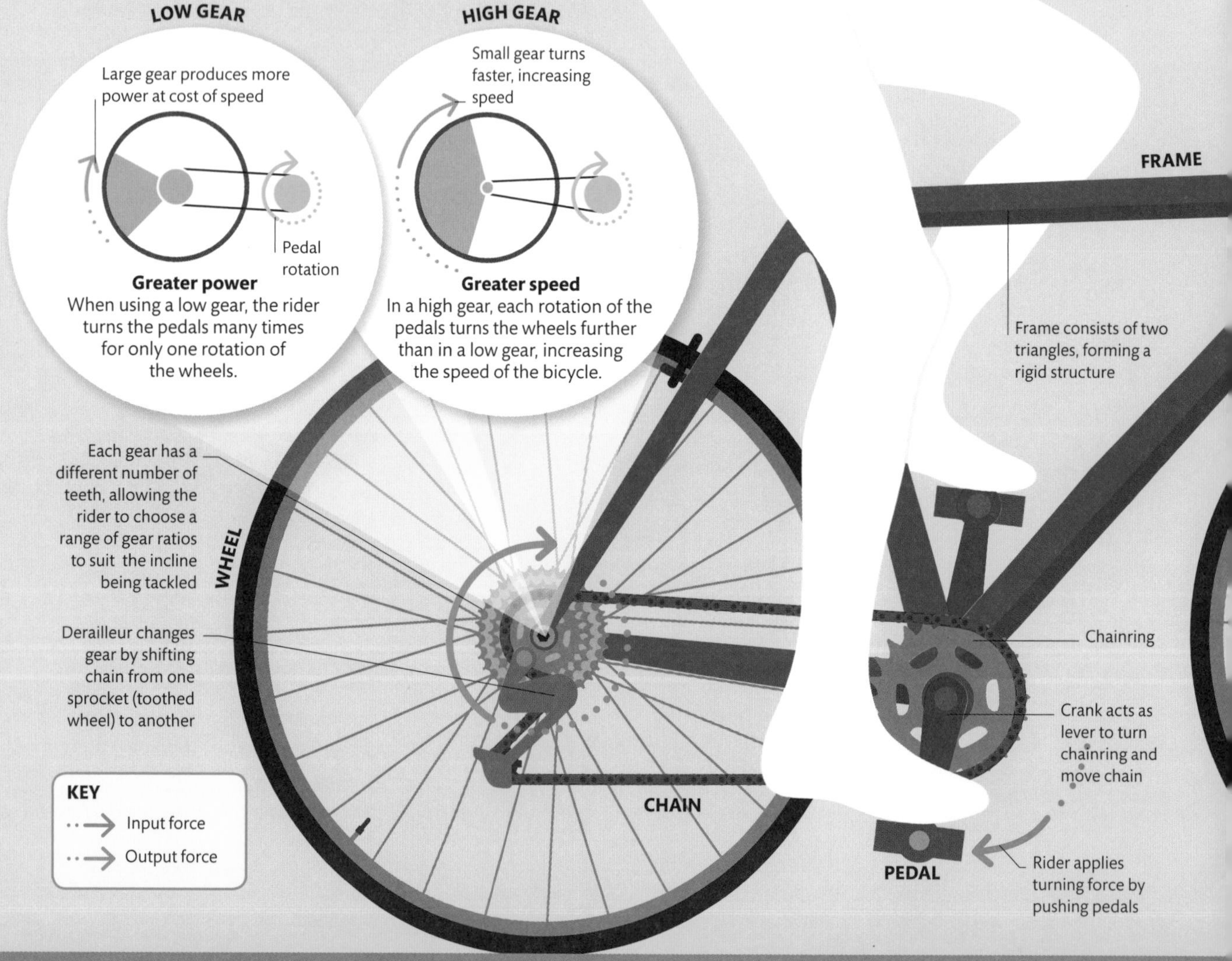

LOW GEAR
Large gear produces more power at cost of speed
Pedal rotation
Greater power
When using a low gear, the rider turns the pedals many times for only one rotation of the wheels.
HIGH GEAR
Small gear turns faster, increasing speed
Greater speed
In a high gear, each rotation of the pedals turns the wheels further than in a low gear, increasing the speed of the bicycle.
FRAME
Frame consists of two triangles, forming a rigid structure
Each gear has a different number of teeth, allowing the rider to choose a range of gear ratios to suit the incline being tackled
WHEEL
Derailleur changes gear by shifting chain from one sprocket (toothed wheel) to another
Chainring
Crank acts as lever to turn chainring and move chain
CHAIN
PEDAL
Rider applies turning force by pushing pedals
KEY
Input force
Output force

HEADSET

MAINTAINING BALANCE

To balance on a bicycle, a rider must control his or her center of gravity. When cycling in a straight line, the rider steers toward a fall to make sure that the center of gravity is always over the wheels, forming the base of support.

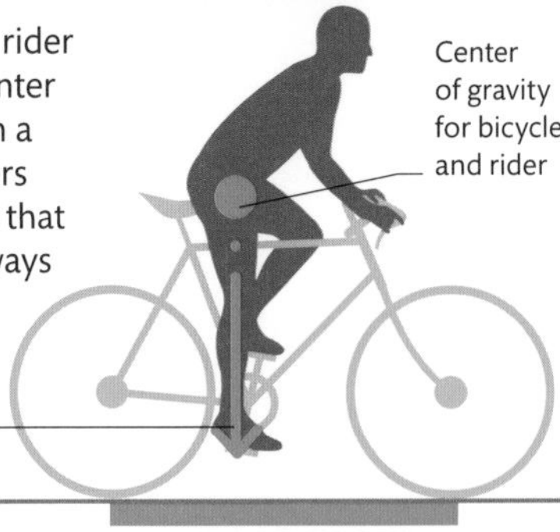

HANDLEBARS

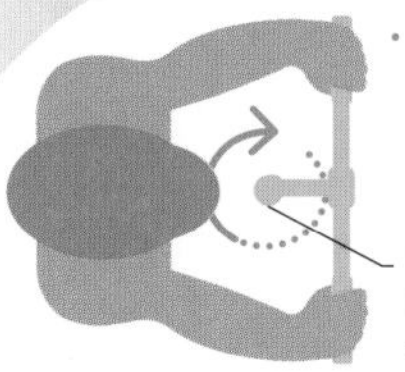

Handlebars are levers that magnify the input force to turn the front wheel. Some bicycles have drop handlebars. These make the rider bend lower, into a more aerodynamic position.

BRAKES

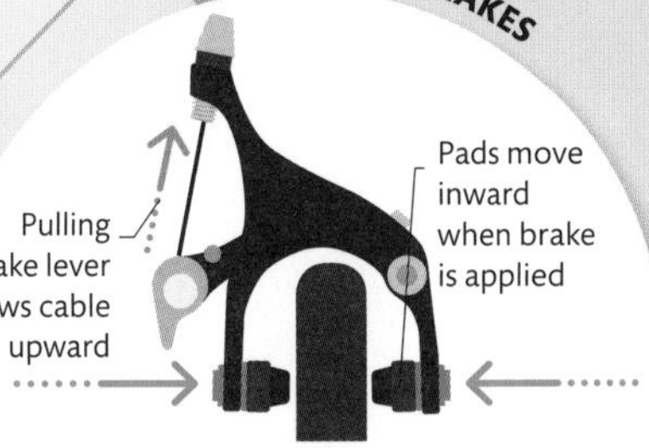

Caliper brakes consist of pads on either side of each wheel. Squeezing the brake lever pulls a cable that makes the pads grip the wheel, increasing friction and slowing the wheel.

Freewheeling

Two mechanical principles usually help to explain why a bicycle can stay upright: the gyroscopic effect and the caster effect. Recent research suggests another important influence is the fact that the front of the bicycle has a center of gravity that is lower than the rear and forward of the steering axis. During a fall, the front of the bicycle falls faster than the rear, turning the front wheel toward the fall and keeping the bicycle upright.

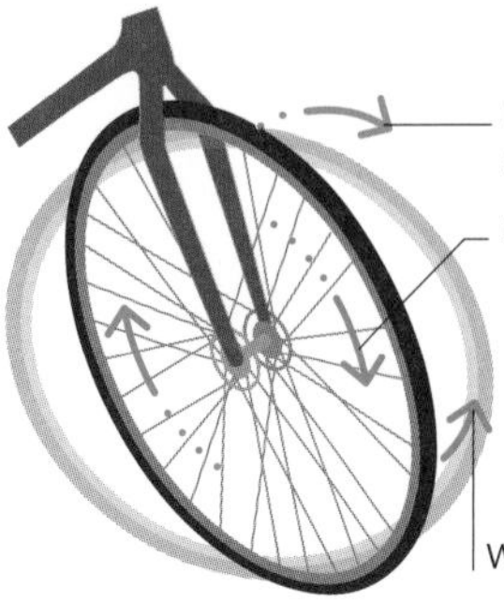

THE GYROSCOPIC EFFECT
The front wheel acts like a gyroscope. If the bicycle falls to one side, the gyroscopic effect turns the wheel to the same side, keeping the bicycle upright.

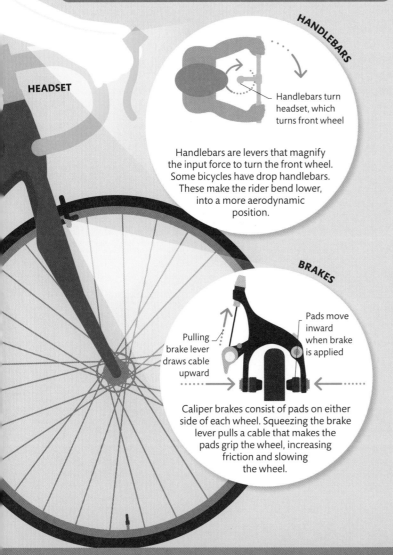

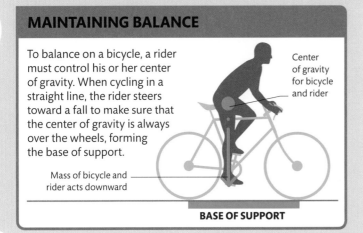

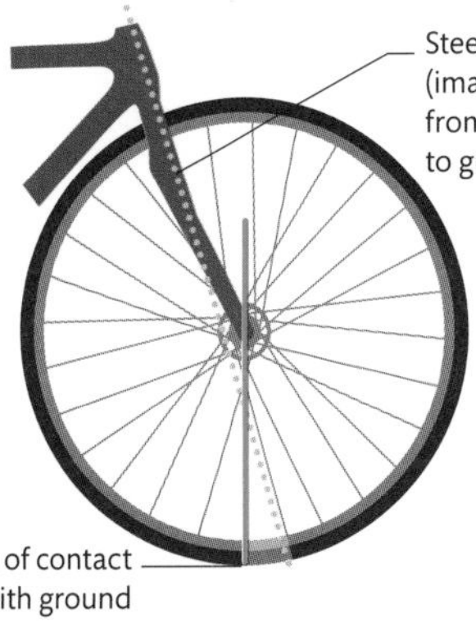

THE CASTER EFFECT
The point at which the front wheel meets the ground trails behind the steering axis, like a caster on a trolley. This means that the wheel always turns in the same direction as the bicycle is traveling.

Internal combustion engines

Many machines, from cars to power tools, use an internal combustion engine to generate power. A car's engine converts chemical energy from fuel into heat energy and then into kinetic energy to drive the wheels.

HOW DOES A DIESEL ENGINE WORK?

A diesel engine works in a similar way to a gas engine, but a diesel engine uses hot, pressurized air, rather than a spark plug, to ignite the fuel.

Four-stroke engines

An internal combustion engine burns a mixture of fuel (usually gasoline or diesel) and air inside a cylinder. A four-stroke engine creates power by repeating four stages, or strokes: intake, compression, power, and exhaust. A heated fuel-air mixture, ignited by a spark plug, pushes a piston down inside the cylinder, causing the connected crankshaft to rotate. This rotation is transferred to the wheels via the car's transmission. Multiple cylinders firing at different times produce a smooth power output.

ONE OF **RUDOLF DIESEL'S** EARLY **ENGINES** RAN ON **PEANUT OIL**

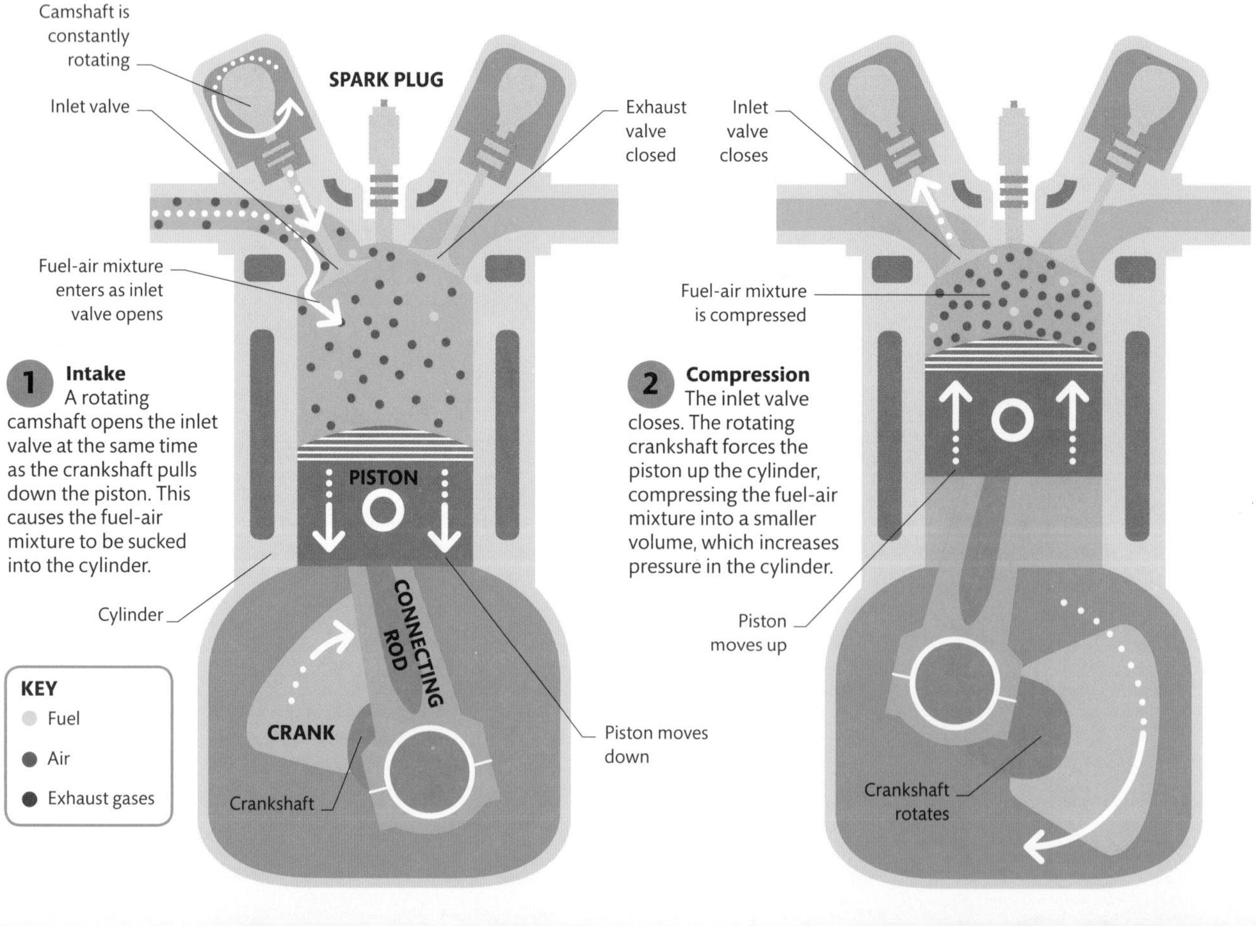
Camshaft is constantly rotating
SPARK PLUG
Inlet valve
Exhaust valve closed
Fuel-air mixture enters as inlet valve opens
1 Intake
A rotating camshaft opens the inlet valve at the same time as the crankshaft pulls down the piston. This causes the fuel-air mixture to be sucked into the cylinder.
PISTON
CONNECTING ROD
Cylinder
CRANK
Crankshaft
Piston moves down
KEY
Fuel
Air
Exhaust gases
Inlet valve closes
Fuel-air mixture is compressed
2 Compression
The inlet valve closes. The rotating crankshaft forces the piston up the cylinder, compressing the fuel-air mixture into a smaller volume, which increases pressure in the cylinder.
Piston moves up
Crankshaft rotates

Two-stroke engines

Four-stroke engines are heavy so are impractical for many uses, such as for powering chain saws and lawn mowers. These use a smaller two-stroke engine instead, which fires the spark plug once every revolution of the crankshaft, rather than once every second revolution.

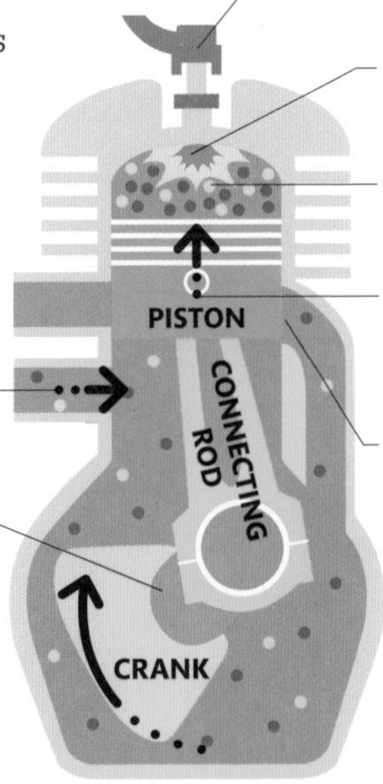
Spark plug
Spark ignites fuel-air mixture
Fuel-air mixture is compressed
Piston moves up
PISTON
CONNECTING ROD
Inlet port opens and fuel-air mixture enters
Transfer port blocked by piston
Crankshaft
CRANK

1 Upstroke
The piston moves up, compressing the fuel-air mixture, which is then ignited by the spark plug. The piston creates a partial vacuum behind it, drawing in more fuel and air through an inlet port.

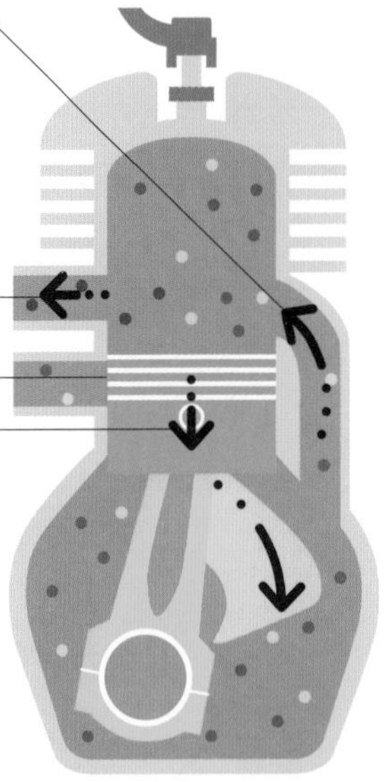
Transfer port opens and fuel-air mixture moves upward
Exhaust gases escape
Inlet port closed
Piston moves down, precompressing fuel-air mixture

2 Downstroke
The ignited fuel pushes the piston down, turning the crankshaft. The transfer port is uncovered as the piston moves, forcing more fuel to the top of the cylinder.

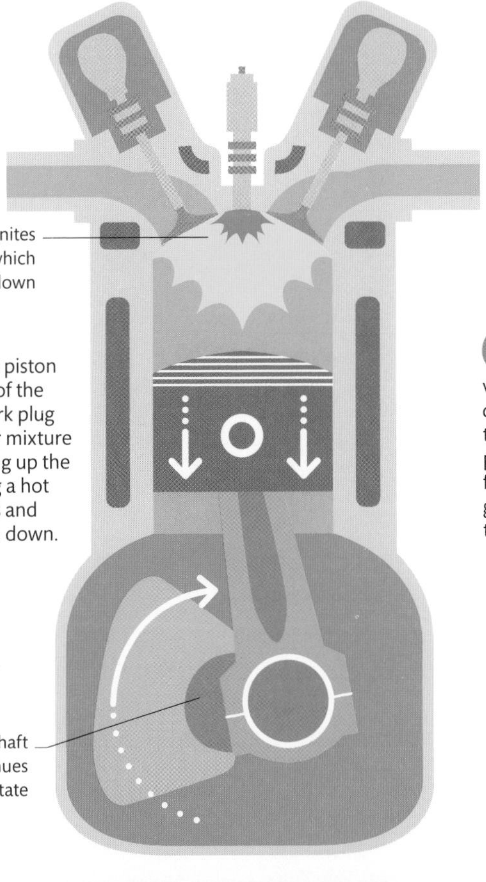
Spark plug ignites fuel-air mixture, which forces piston down
Crankshaft continues to rotate

3 Power
When the piston reaches the top of the cylinder, the spark plug fires. The fuel-air mixture explodes, burning up the fuel and creating a hot gas that expands and forces the piston down.

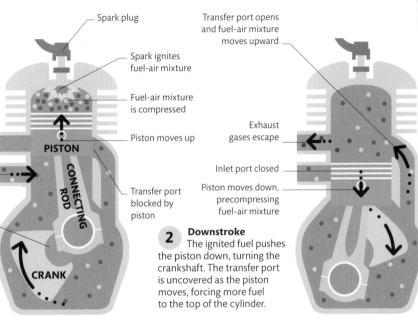

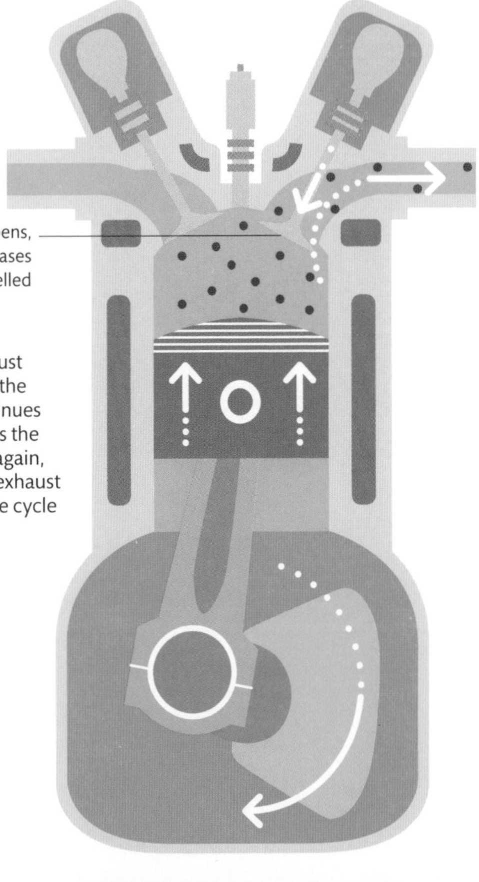
Exhaust valve opens, and exhaust gases are expelled

4 Exhaust
The exhaust valve opens. As the crankshaft continues to turn, it pushes the piston back up again, forcing out the exhaust gases. The whole cycle then repeats.

How a car works

A car is a collection of systems that generate power in an engine and transmit it to the wheels. Other systems allow the driver to control the car by turning the wheels to change direction and by applying brake force to either slow down or stop.

WHAT WAS THE FIRST MASS-PRODUCED CAR WITH AN AUTOMATIC TRANSMISSION?

The first fully automatic transmission was an optional extra on Oldsmobile cars in the US from 1940.

Transmitting power

A car's engine is linked to its wheels by a system of shafts and gears, collectively known as the drivetrain, which make use of the engine power in the most efficient way. Most cars have a two-wheel-drive arrangement, in which either the two front wheels or two back wheels are driven by the engine. Off-road vehicles, which require more grip on unstable surfaces, have four-wheel drive, meaning all four wheels are driven directly by the engine.

Inside the car
The heaviest parts of a car are its engine and driveshaft, including its transmission. They are mounted low down in the car to improve stability, especially when cornering.

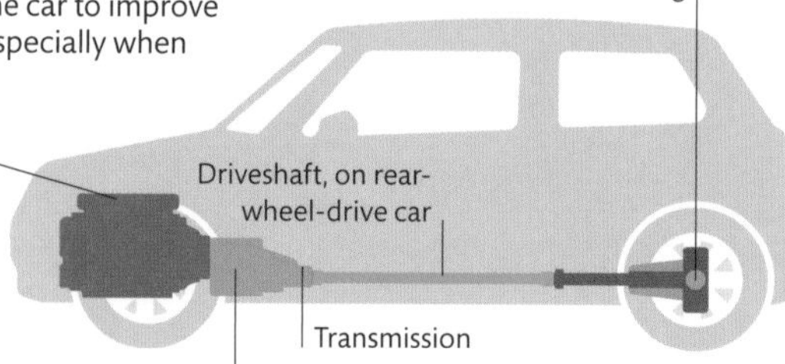

Differential gear aids cornering
Engine
Driveshaft, on rear-wheel-drive car
Transmission
Clutch engages and disengages driveshaft from transmission

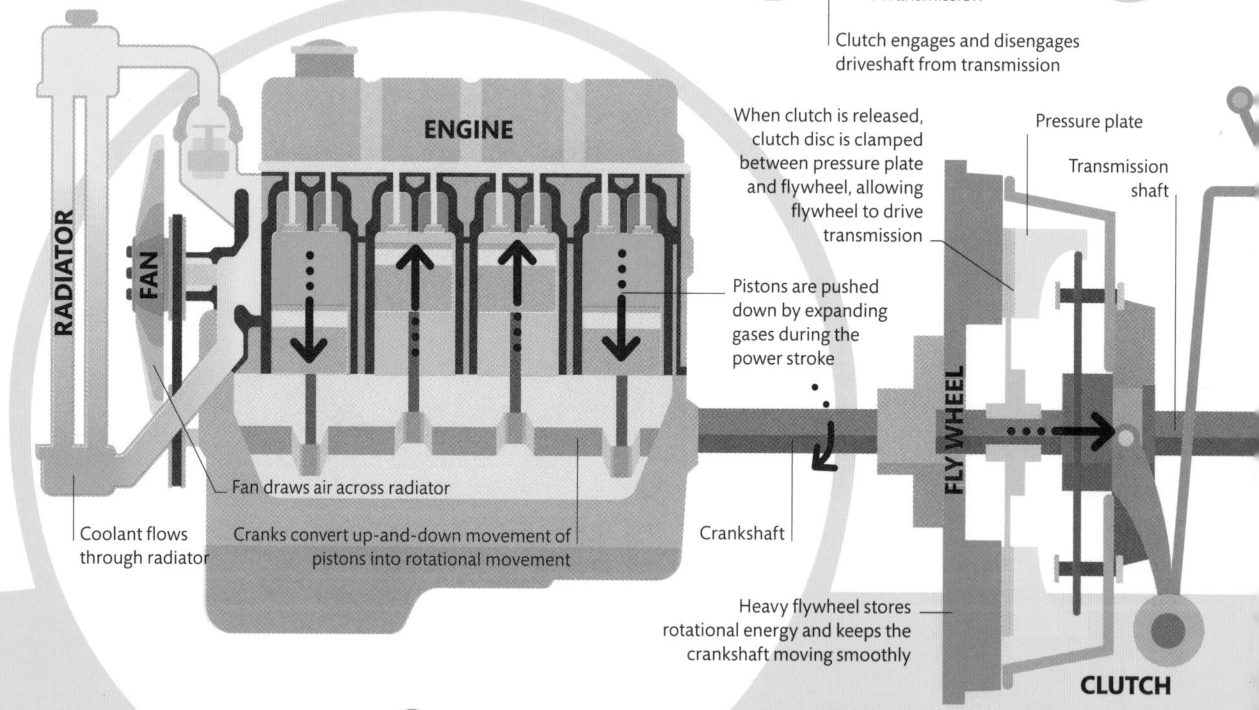

RADIATOR
FAN
ENGINE
Fan draws air across radiator
Coolant flows through radiator
Cranks convert up-and-down movement of pistons into rotational movement
Crankshaft
Pistons are pushed down by expanding gases during the power stroke
When clutch is released, clutch disc is clamped between pressure plate and flywheel, allowing flywheel to drive transmission
Pressure plate
Transmission shaft
FLY WHEEL
Heavy flywheel stores rotational energy and keeps the crankshaft moving smoothly
CLUTCH

Driving off
A car is set in motion by a series of operations that generate power and transfer it to the driven wheels in a controlled way. Turning the ignition key or pressing the start button switches on a small battery-powered electric motor, which starts the car's piston engine.

1 Engine
A car's motion begins with its engine. Starting the engine ignites fuel and releases energy (see pp.42–43). This moves the pistons, which turn the engine's crankshaft. A flywheel attached to the crankshaft smooths out the power provided by the pistons.

2 Clutch
In a car with a manual transmission, when the car first starts, the driver must push the clutch pedal in to disconnect the engine from the wheels so that the car does not lurch forward. The driver then releases the clutch pedal, allowing the engine to turn the wheels.

Steering

The simplest steering systems in cars rely on a type of gear mechanism called a rack and pinion. Turning a car's steering wheel rotates a pinion—a small, round gear. Its teeth engage the teeth on a flat bar called a rack. When the pinion turns, it moves the rack sideways and turns the wheels. In a car with power steering, high-pressure oil or electric motors help to move the rack.

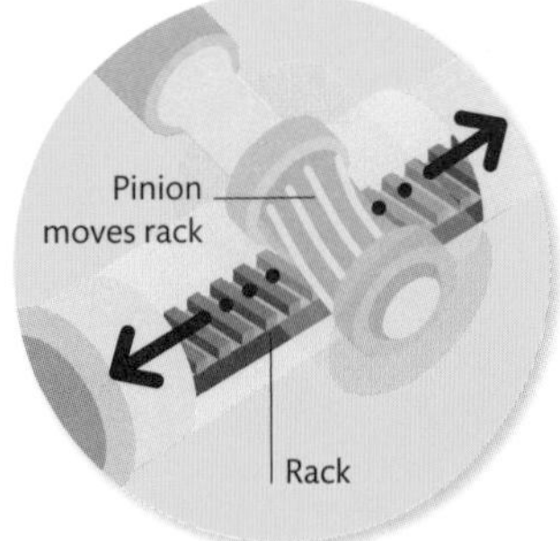
Pinion moves rack
Rack

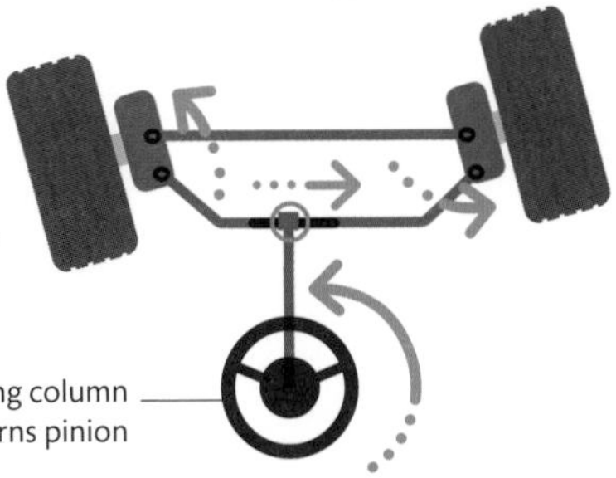
Steering column turns pinion

Braking

Most cars have disc brakes. A disc is fixed to each wheel, and when the wheel spins, so does the disc. When the driver presses the brake pedal, hydraulic fluid forces brake pads, mounted on callipers, to push against the disc to slow the wheel down.

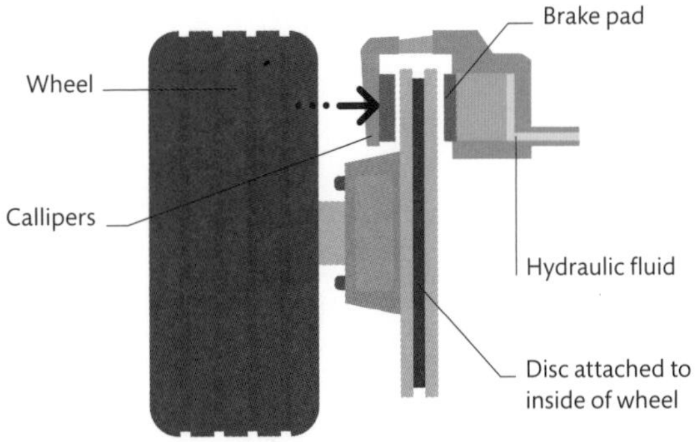
Brake pad
Wheel
Callipers
Hydraulic fluid
Disc attached to inside of wheel

5 Differential
During a turn, the wheel on the outside has to travel farther than the inside wheel, and so the outer wheel has to turn faster. A differential, which is attached midway between the two wheels, allows this to happen.

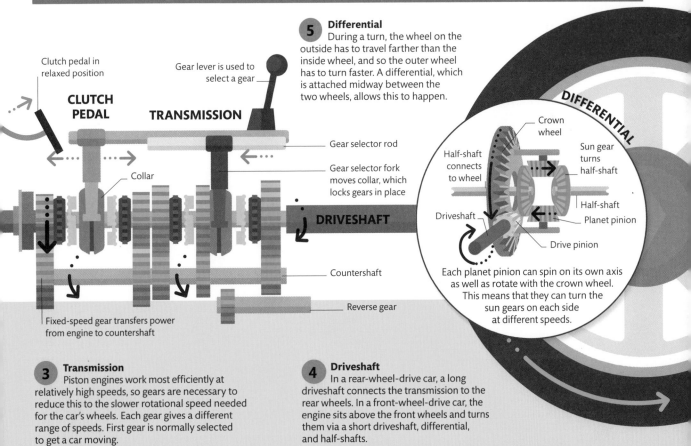

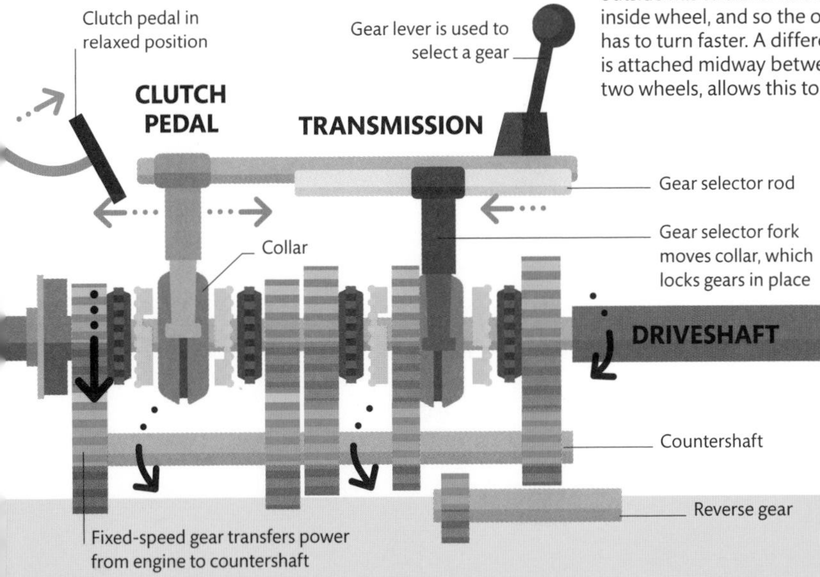
Clutch pedal in relaxed position
Gear lever is used to select a gear
CLUTCH PEDAL
TRANSMISSION
Gear selector rod
Gear selector fork moves collar, which locks gears in place
Collar
DRIVESHAFT
Countershaft
Reverse gear
Fixed-speed gear transfers power from engine to countershaft

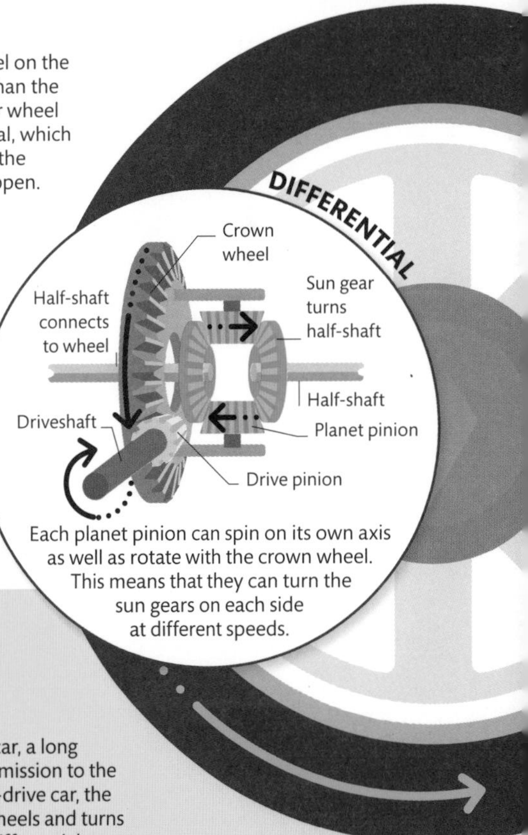
DIFFERENTIAL
Crown wheel
Half-shaft connects to wheel
Sun gear turns half-shaft
Half-shaft
Driveshaft
Planet pinion
Drive pinion
Each planet pinion can spin on its own axis as well as rotate with the crown wheel. This means that they can turn the sun gears on each side at different speeds.

3 Transmission
Piston engines work most efficiently at relatively high speeds, so gears are necessary to reduce this to the slower rotational speed needed for the car's wheels. Each gear gives a different range of speeds. First gear is normally selected to get a car moving.

4 Driveshaft
In a rear-wheel-drive car, a long driveshaft connects the transmission to the rear wheels. In a front-wheel-drive car, the engine sits above the front wheels and turns them via a short driveshaft, differential, and half-shafts.

Electric and hybrid cars

Most cars are powered by an internal combustion engine burning gasoline or diesel fuel. However, concerns about the harmful air pollution produced by these engines has led to the development of less-polluting electric and hybrid cars.

WHEN WAS THE FIRST HYBRID CAR BUILT?

Engineer Ferdinand Porsche built the world's first hybrid car in 1900. He named it the Lohner-Porsche Semper Vivus ("always alive").

Electric cars

An electric car is powered by one or more electric motors. The motor is connected to a rechargeable battery pack. Electric cars are simpler than conventional piston-engine cars because they do not need a fuel system, ignition system, water cooling system, or oil lubrication system. A transmission is not necessary either as, unlike internal combustion engines, electric motors deliver the maximum turning force (torque) across their whole speed range.

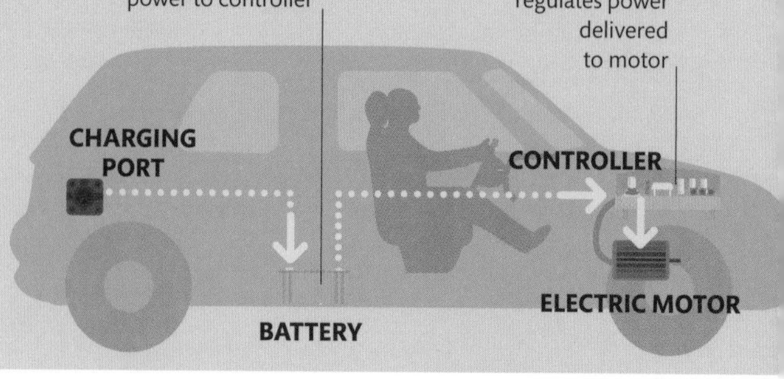

Hybrid cars

A hybrid car has two or more different power sources driving the wheels—an internal combustion engine and at least one electric motor. There are two main types of hybrid cars. A series hybrid is always powered by its electric motor. The role of its internal combustion engine is to run a generator that produces electricity to power the electric motor and charge its battery. The second type of hybrid car, the more popular parallel hybrid (see right), can be powered by either one of its power sources, or both can be used together when maximum power or acceleration is needed.

KEY

- Electric power
- Power from internal combustion engine

Pulling away
Most hybrid cars start moving by using only their battery-powered electric motor. The internal combustion engine is not needed. For short trips at low speeds, electric power alone may be used for the whole trip.

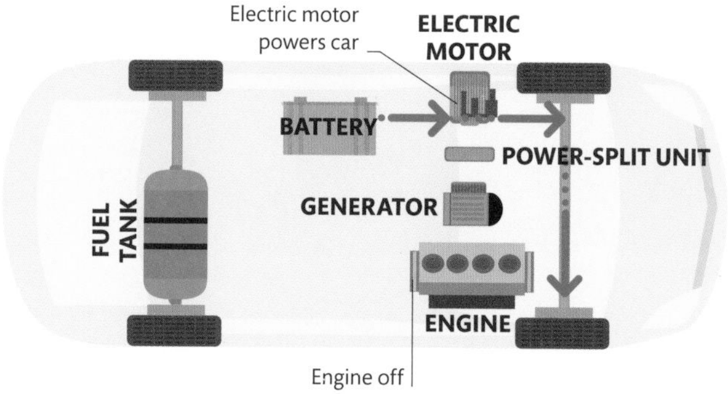

Accelerating
If rapid acceleration is needed, the internal combustion engine starts. The car's wheels are driven by the power of both the engine and the electric motor. The engine also runs a generator that recharges the electric motor's battery.

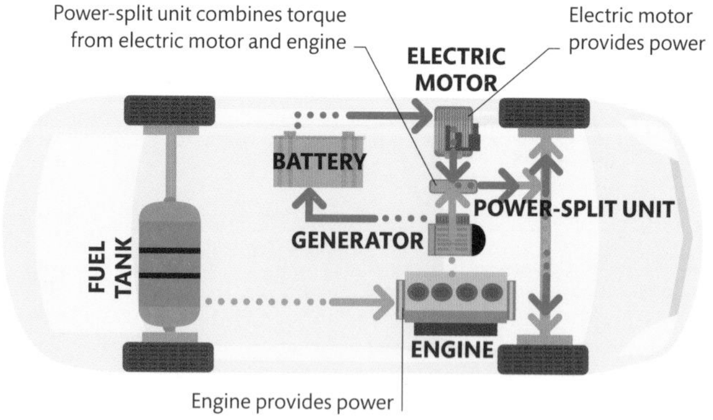

Regenerative braking

Most cars brake by using friction pads (see p.45), which transform the wheels' kinetic energy into wasted heat energy. Electric and hybrid cars convert the wheels' energy into electrical energy instead to recharge the battery.

THE FIRST ELECTRIC CAR WAS BUILT BY INVENTOR ROBERT ANDERSON IN THE 1830s

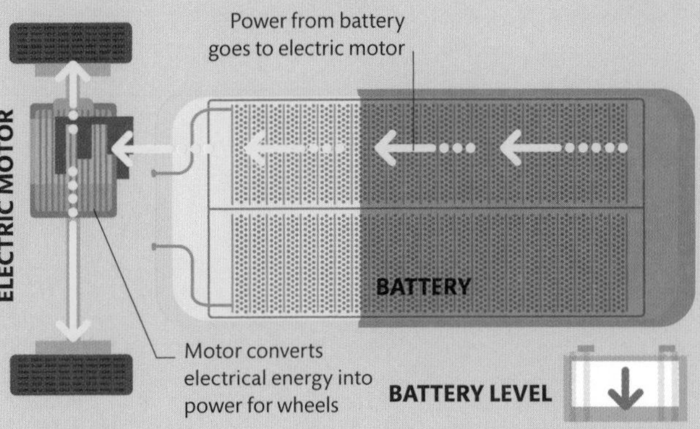
Power from battery goes to electric motor
ELECTRIC MOTOR
BATTERY
Motor converts electrical energy into power for wheels
BATTERY LEVEL

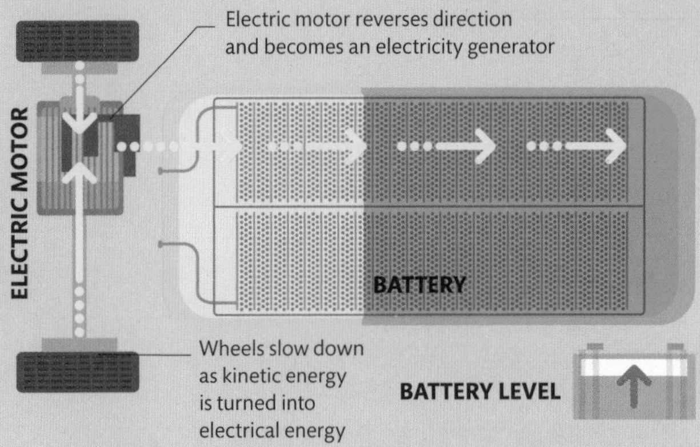
Electric motor reverses direction and becomes an electricity generator
ELECTRIC MOTOR
BATTERY
Wheels slow down as kinetic energy is turned into electrical energy
BATTERY LEVEL

1 Acceleration
When an electric or hybrid car accelerates, its motor draws the energy it needs from the battery. The motor converts the battery's electrical energy into the car's kinetic energy. The level of charge in the battery falls as it is progressively drained of energy.

2 Braking
When the driver applies the brakes, the electric motor becomes a generator. Instead of drawing energy from the battery, it transforms the kinetic energy of the car's spinning wheels into electrical energy, which is returned to the battery to be reused.

Cruising
While the car is cruising at high speed on a long trip, the internal combustion engine operates on its own. The electric motor is not needed.

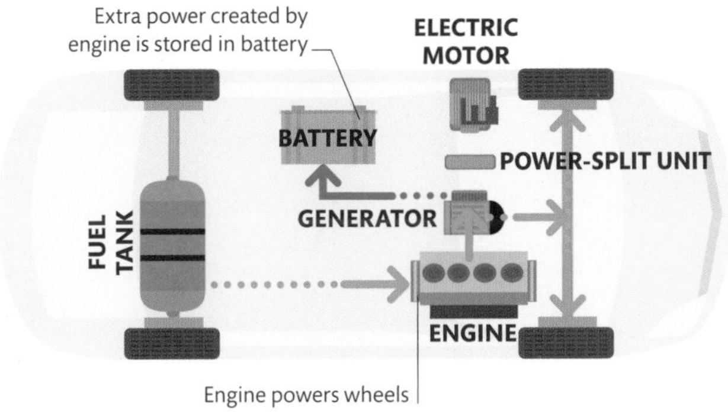
Extra power created by engine is stored in battery
ELECTRIC MOTOR
BATTERY
POWER-SPLIT UNIT
GENERATOR
FUEL TANK
ENGINE
Engine powers wheels

Braking
When the car begins to slow down, the internal combustion engine and electric motor switch off. During braking, the car's excess energy is converted into electricity to charge the battery.

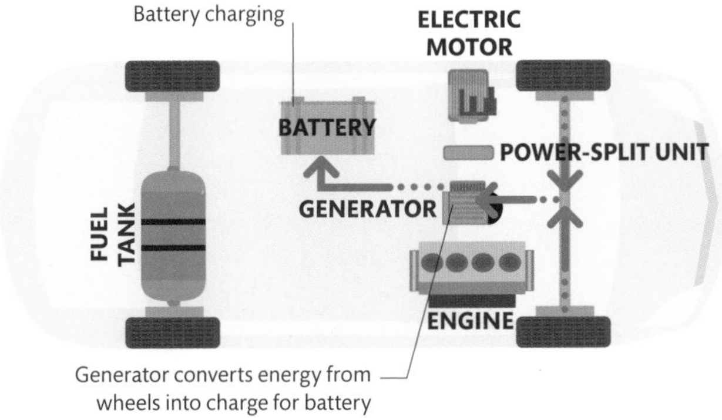
Battery charging
ELECTRIC MOTOR
BATTERY
POWER-SPLIT UNIT
GENERATOR
FUEL TANK
ENGINE
Generator converts energy from wheels into charge for battery

DRIVERLESS CARS

A driverless car has various cameras, lasers, and radar that create a real-time 3-D image of the car's surroundings. Together with computers, satellite navigation, and artificial intelligence (AI), these enable the car to drive itself.

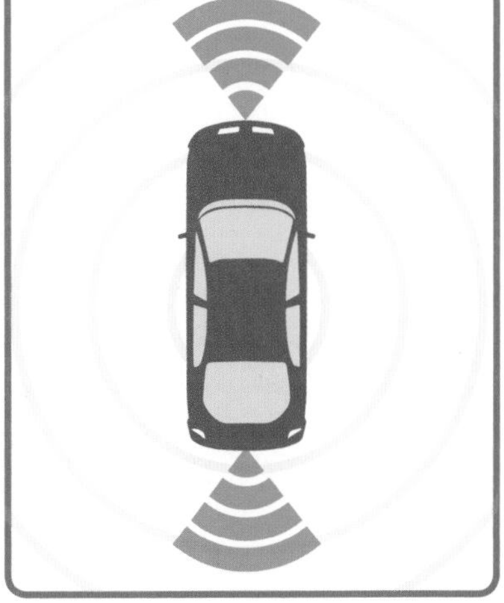

Radar

Radar is used to locate distant objects by sending out high-frequency radio waves (see pp.180–181) and detecting any waves that are reflected back. Radar is vital to air traffic control systems and is used to track aircraft in flight and control their movements safely.

2 Waves bounce back
Large metal objects such as aircraft reflect radio waves. Some of these reflected waves return to the antenna. The distance to the aircraft is calculated from the time taken for radar pulses to travel out to the aircraft and bounce back.

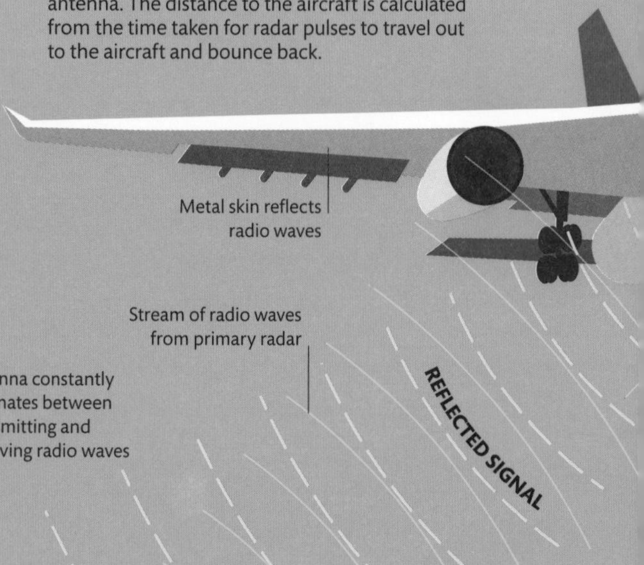
Metal skin reflects radio waves
Stream of radio waves from primary radar
REFLECTED SIGNAL

Air traffic control radar

Two types of radar are used in air traffic control—primary radar and secondary radar. Primary radar transmits radio waves that are reflected back by an aircraft, revealing its position. Secondary radar relies on an aircraft actively sending signals using a device called a transponder to add information about the aircraft, such as its identity and altitude.

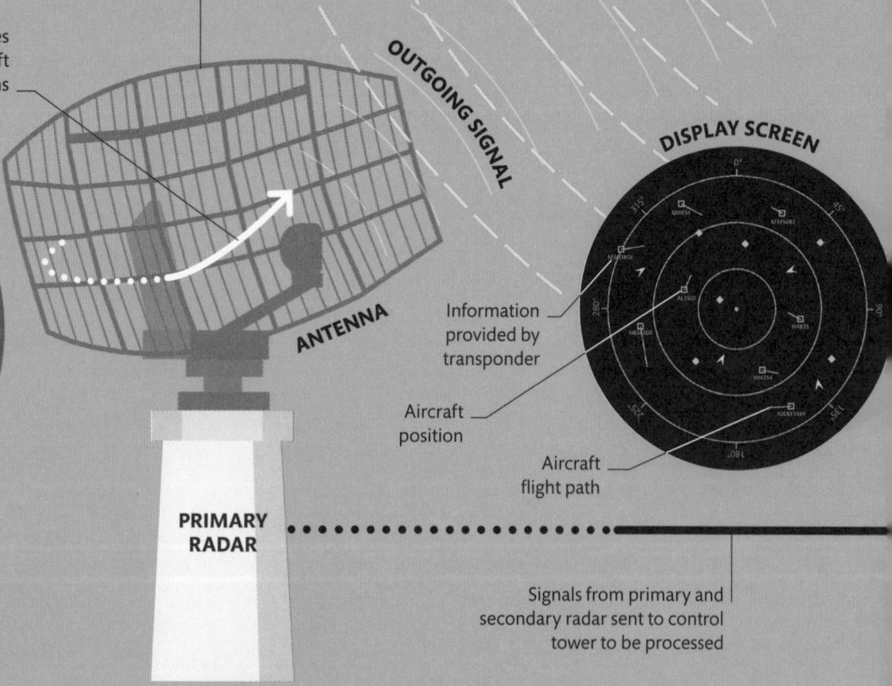
Antenna constantly alternates between transmitting and receiving radio waves
Antenna rotates 360° to scan for aircraft in all directions
OUTGOING SIGNAL
ANTENNA
DISPLAY SCREEN
Information provided by transponder
Aircraft position
Aircraft flight path
PRIMARY RADAR
Signals from primary and secondary radar sent to control tower to be processed

WHAT OTHER TECHNOLOGIES USE RADAR?

Radar has several other uses, including in ocean and geological surveys, mapping, astronomy, and in intruder alarms and cameras.

1 Primary radar
A rotating antenna sends out pulses of radio waves in all directions. They travel in a straight line at the speed of light. The antenna can both transmit and receive radio waves.

RADAR WAS USED TO **MAP THE SURFACES** OF THE PLANETS **MERCURY** AND **VENUS**

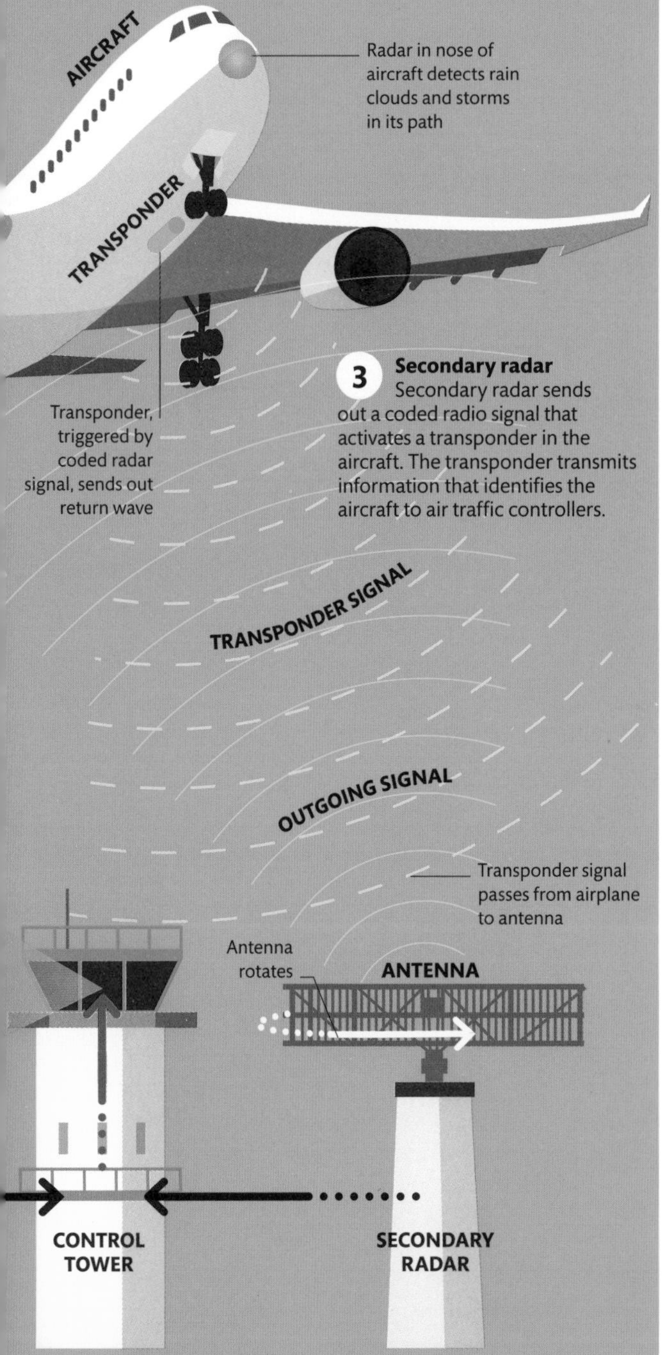
AIRCRAFT
Radar in nose of aircraft detects rain clouds and storms in its path
TRANSPONDER
Transponder, triggered by coded radar signal, sends out return wave
TRANSPONDER SIGNAL
OUTGOING SIGNAL
Transponder signal passes from airplane to antenna
Antenna rotates
ANTENNA
CONTROL TOWER
SECONDARY RADAR

3 Secondary radar
Secondary radar sends out a coded radio signal that activates a transponder in the aircraft. The transponder transmits information that identifies the aircraft to air traffic controllers.

4 Control tower
A signal processor inside the control tower analyzes the information from both radars and then sends it to a display screen. Aircraft appear as a spot or a line.

EVADING RADAR

Some military aircraft, like the B-2 bomber, are designed to evade enemy radar. The airplane's shape reflects radio waves away from their source. The airplane is also covered with radar-absorbing materials that reduce reflections and make it harder to detect. This is known as stealth technology.

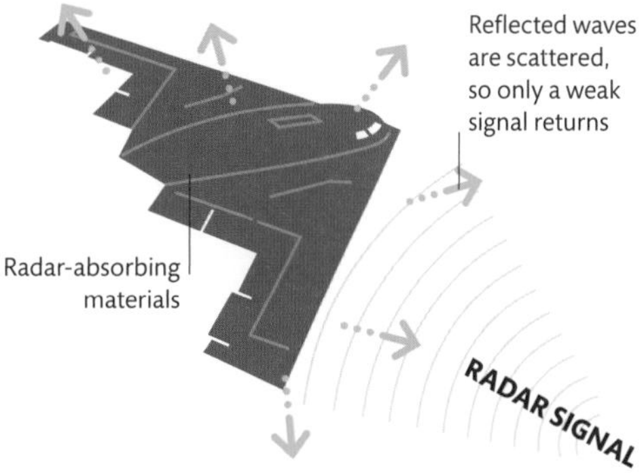
Reflected waves are scattered, so only a weak signal returns
Radar-absorbing materials
RADAR SIGNAL

Ground-penetrating radar

Radar can also reveal what is below the ground. Radio waves bounce off any objects or soil disturbances they encounter, and these reflections are processed by a computer to produce a map. Ground-penetrating radar (GPR) is used in a variety of fields, including archaeology, engineering, and military activities.

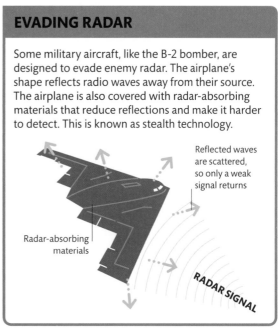

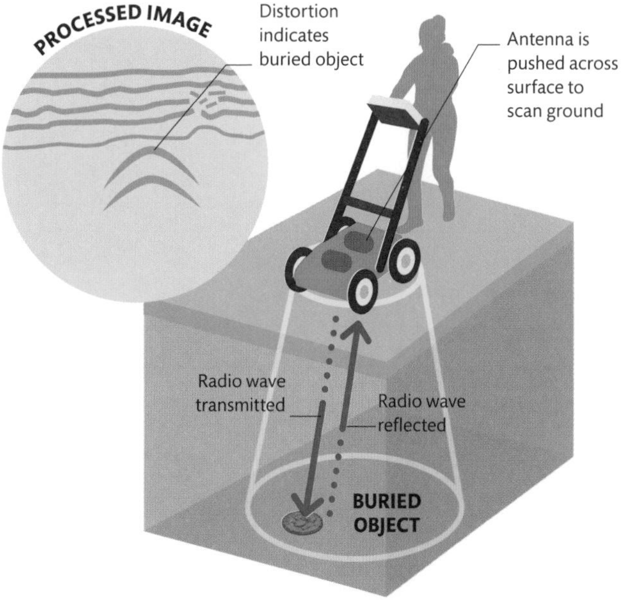
PROCESSED IMAGE
Distortion indicates buried object
Antenna is pushed across surface to scan ground
Radio wave transmitted
Radio wave reflected
BURIED OBJECT

Speed cameras

A REVIEW OF 35 INTERNATIONAL STUDIES FOUND THAT SPEED CAMERAS REDUCE AVERAGE SPEEDS BY UP TO 15 PERCENT

Many types of speed cameras use radar (see pp.48–49) to measure the speed of a vehicle. They transmit radio waves at a vehicle and use the waves reflected back to calculate its speed.

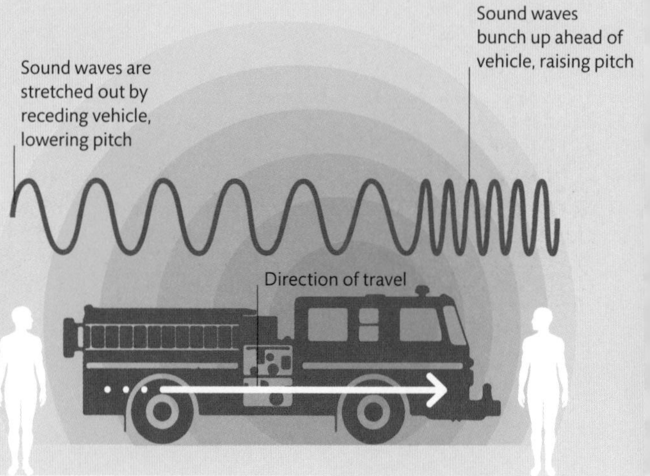
Sound waves bunch up ahead of vehicle, raising pitch
Sound waves are stretched out by receding vehicle, lowering pitch
Direction of travel

The Doppler effect

When radio waves strike a vehicle that is moving toward or away from a transmitter, such as a speed camera, the vehicle's motion changes the wavelength of the reflected waves. This change is called the Doppler effect. The same effect makes an emergency vehicle's siren rise in pitch as the vehicle approaches and fall in pitch as it moves farther away.

How a speed camera works

A speed camera sends out bursts of radio waves and then detects the waves that are reflected back from a moving vehicle. It uses differences between the transmitted and reflected waves, caused by the Doppler effect, to determine the vehicle's speed. The very short radio waves emitted by a speed camera are called microwaves. They are about a centimeter long and travel at the speed of light.

1 Transmission
The camera's radar unit transmits a beam of microwaves, which fans out across the road. Less than a microsecond (one millionth of a second) later, the waves reach the back of the passing vehicle.

2 Reflection
The microwaves bounce off the vehicle's bodywork like light bouncing off a mirror. The curved shape of the vehicle sends the reflected waves away in all directions.

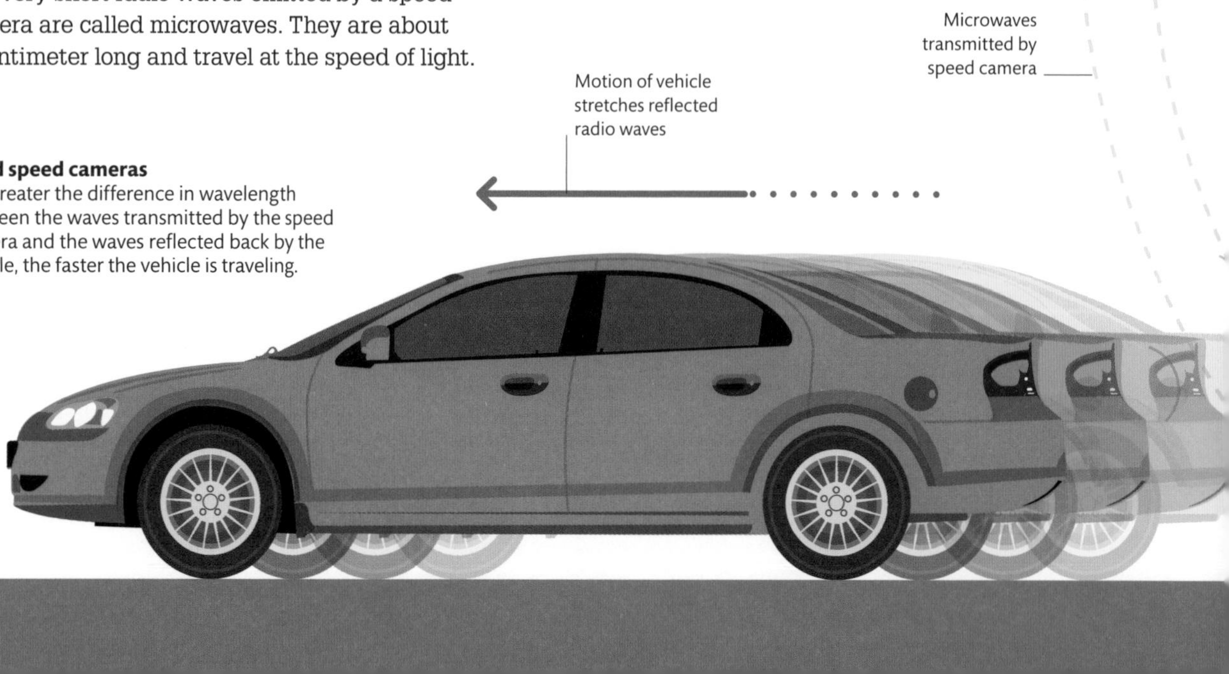
Microwaves transmitted by speed camera
Motion of vehicle stretches reflected radio waves

Fixed speed cameras
The greater the difference in wavelength between the waves transmitted by the speed camera and the waves reflected back by the vehicle, the faster the vehicle is traveling.

INSIDE A SPEED CAMERA

A speed camera houses a radar unit, camera, power supply, and control unit. It usually points at the backs of vehicles so that the flash of the camera does not dazzle drivers.

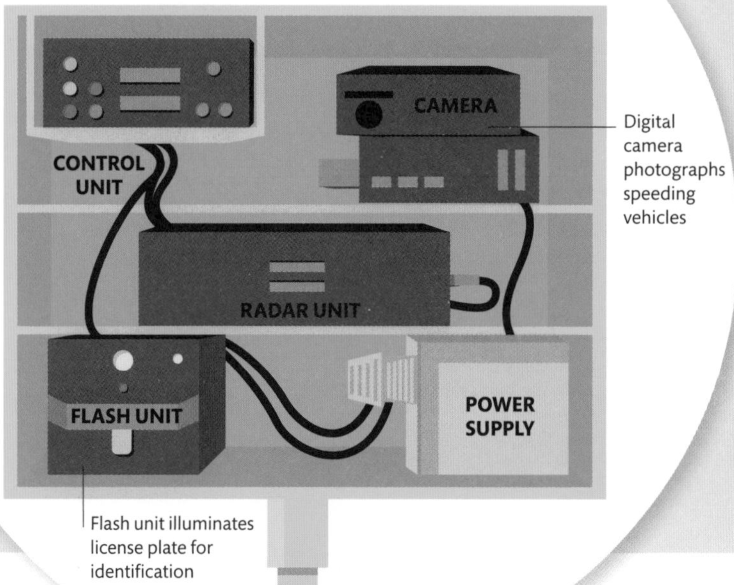

LIDAR

Some handheld speed detectors fire a series of laser pulses at vehicles and measure the return time of reflected pulses to calculate a vehicle's distance and speed. This technique is known as LiDAR (Light Detection and Ranging).

3 Reception
The radar unit receives some of the reflected microwaves. If their longer wavelength indicates a speed above the speed limit, a digital camera is activated to photograph the car.

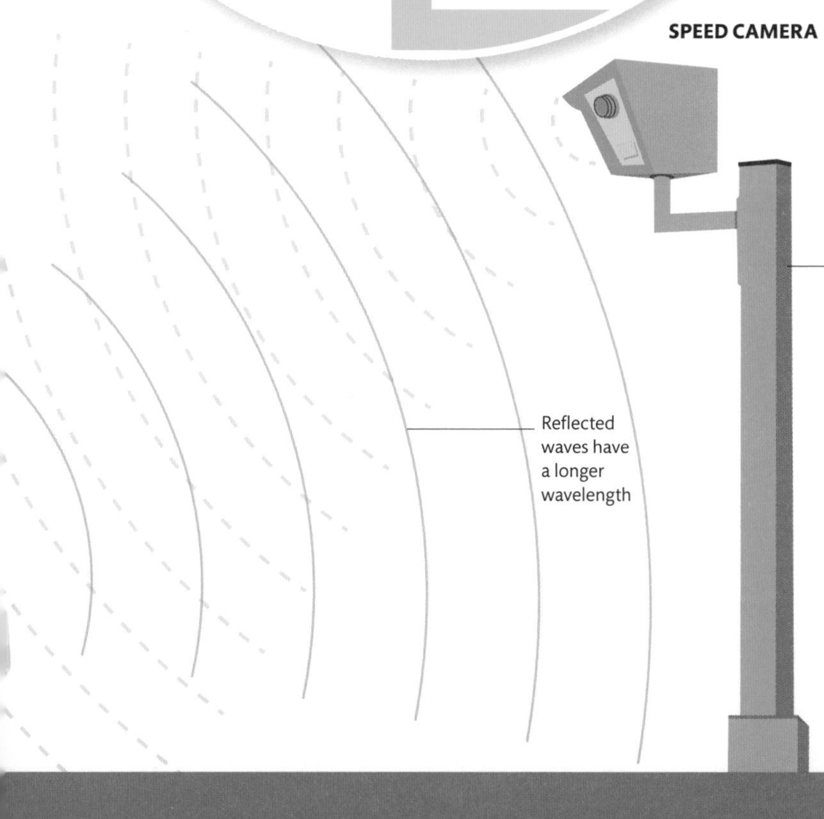

Mounting pole holds camera at required height and angle

WHEN WERE SPEED CAMERAS INVENTED?

Although the idea of developing speed cameras dates back to at least the early 1900s, the first radar speed cameras were made in the US for military use during World War II.

Trains

Trains provide one of the most time-efficient transportation solutions for traveling long distances. Most modern trains are powered by a diesel engine or an external source of electricity.

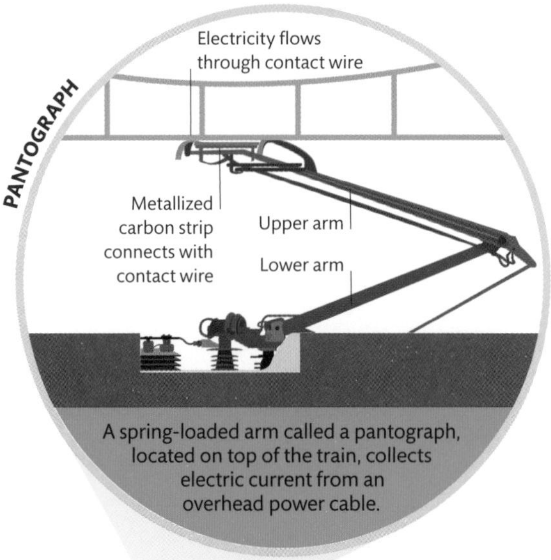
PANTOGRAPH
Electricity flows through contact wire
Metallized carbon strip connects with contact wire
Upper arm
Lower arm

A spring-loaded arm called a pantograph, located on top of the train, collects electric current from an overhead power cable.

Electric trains

Electric trains are powered by electricity supplied either from overhead cables or by a third rail in the track. Since they do not have to carry their own power-generating equipment, electric locomotives are lighter than diesel equivalents and are therefore capable of faster acceleration.

WHO BUILT THE FIRST RAILROAD LOCOMOTIVE?

In 1804, English engineer Richard Trevithick built the first railroad locomotive. It was used to haul iron from Penydarren Ironworks in Wales.

Current conversion

Many modern electric locomotives convert the high-voltage alternating current (AC) supply to the lower-voltage AC needed by the electric motors that turn the train's wheels.

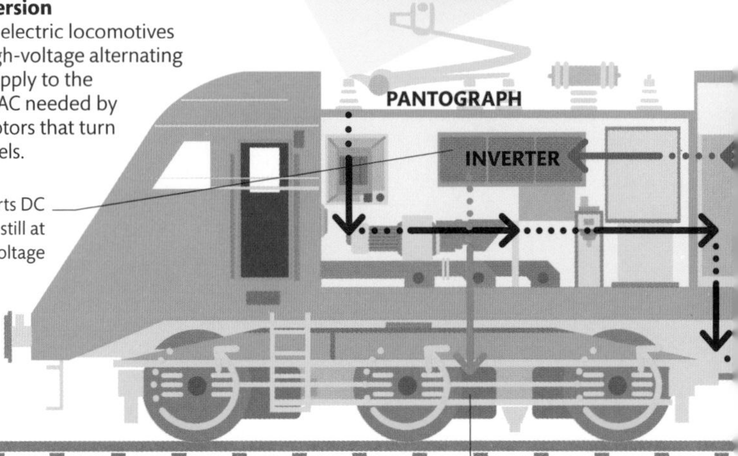
Inverter converts DC back to AC but still at lower voltage
PANTOGRAPH
INVERTER
Traction motor, powered by AC, turns wheels

KEY

- High-voltage AC
- Lower-voltage DC
- Lower-voltage AC
- Fuel

Diesel-electric trains

Most modern diesel trains employ a diesel-electric powerplant housed inside the locomotive. Rather than power the wheels directly, the diesel engine drives a generator or alternator (see pp.16–17) to produce electricity, which operates the train's electrical systems and traction motors. Since diesel trains need no external power supply, they are used on rail lines where electrification is uneconomical.

Engine power

The alternating current (AC) from the engine-driven alternator is converted to direct current (DC) by a rectifier. An inverter converts this into AC to supply the motors.

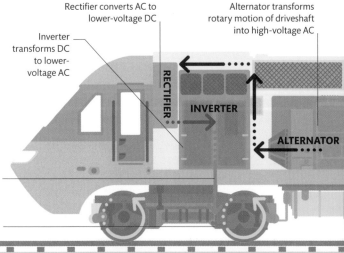

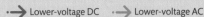

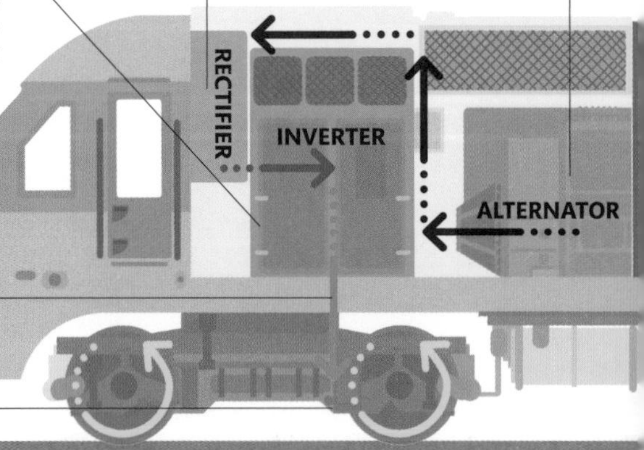
Rectifier converts AC to lower-voltage DC
Alternator transforms rotary motion of driveshaft into high-voltage AC
Inverter transforms DC to lower-voltage AC
RECTIFIER
INVERTER
ALTERNATOR
Lower-voltage AC powers traction motor
Traction motor powers train using current generated by alternator

HYPERLOOP

A Hyperloop is an experimental train designed to travel faster than a jet airliner. Passenger pods travel inside a tube that is a near-vacuum. Air is removed to reduce the piston effect (a buildup of air in front of the train) and to enable the pods to travel faster by reducing friction. Electromagnets beneath the train and on the track repel or attract each other to generate lift and thrust.

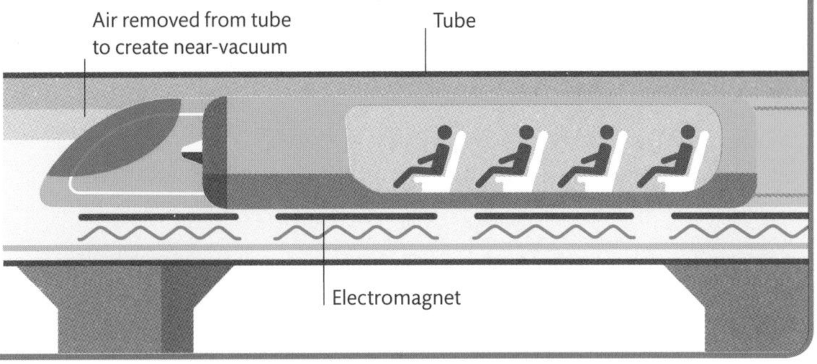
Air removed from tube to create near-vacuum
Tube
Electromagnet

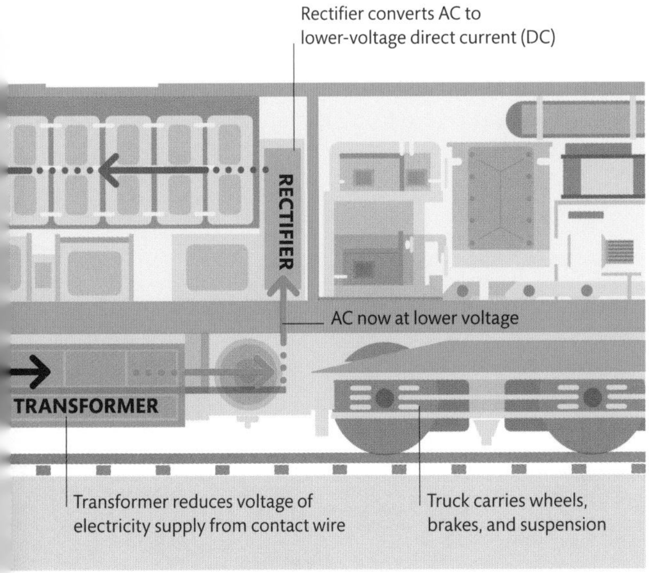
Rectifier converts AC to lower-voltage direct current (DC)
RECTIFIER
AC now at lower voltage
TRANSFORMER
Transformer reduces voltage of electricity supply from contact wire
Truck carries wheels, brakes, and suspension

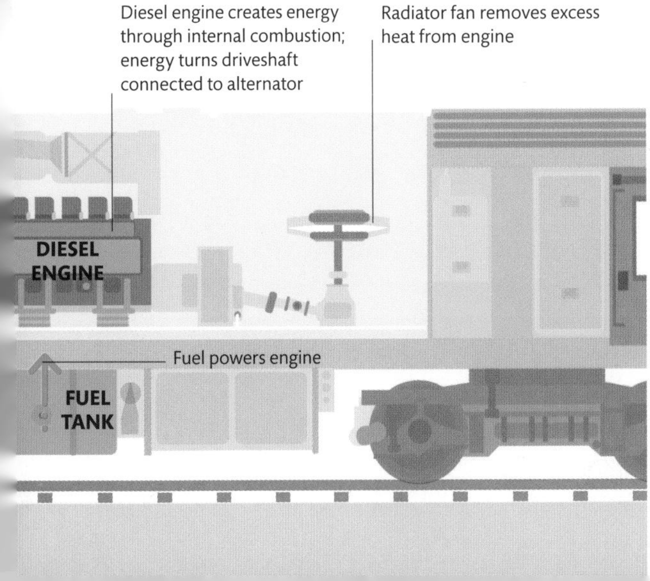
Diesel engine creates energy through internal combustion; energy turns driveshaft connected to alternator
Radiator fan removes excess heat from engine
DIESEL ENGINE
Fuel powers engine
FUEL TANK

Railroad trucks and wheels

Every section of a train is supported at each end by a frame system, or railroad truck, in which the wheel sets (axles and wheels) are mounted. Some trucks can turn to follow bends in the track. The wheels are made of solid steel and run on steel rails to minimize rolling friction. Each wheel has a projecting rim, or flange, on one side that helps to hold it on the rails.

Smoothing the ride
A railroad truck has a built-in suspension system. It uses coil springs, dampers, and air bags to soak up bumps and vibrations caused by uneven tracks. The wheels stay in contact with the rails, while the locomotive and coaches above move along smoothly.

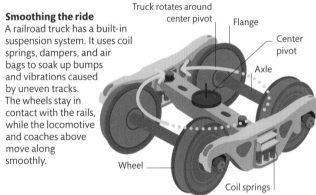

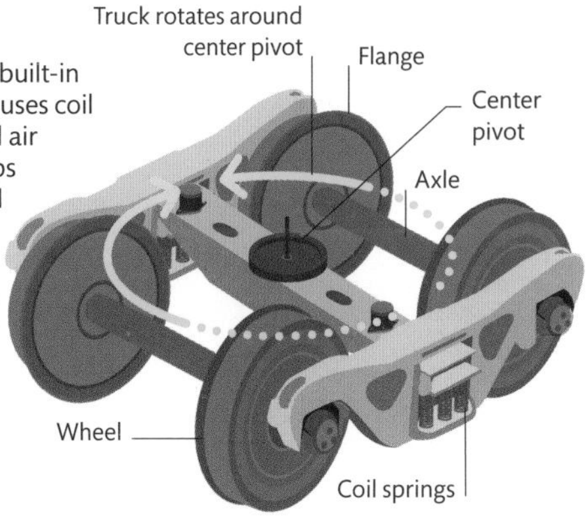
Truck rotates around center pivot
Flange
Center pivot
Axle
Wheel
Coil springs

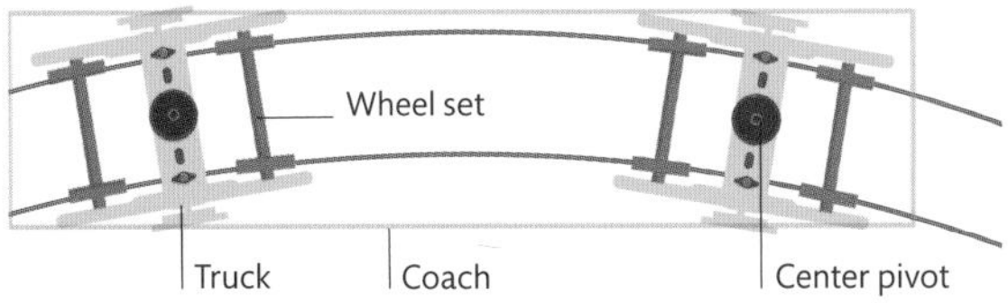
Wheel set
Truck
Coach
Center pivot

Turning bends
A long train with steel wheel sets traveling along steel rails is inherently rigid. To allow trains to follow bends, some modern railroad trucks have a built-in steering mechanism, with a steering beam and levers hinged around a center pivot, which allows the wheel sets to turn.

Sailboats

Sailboats use a combination of wind and water movement to propel a boat without the need for an engine. Once used for commerce and warfare, they are now used mainly for sport and leisure.

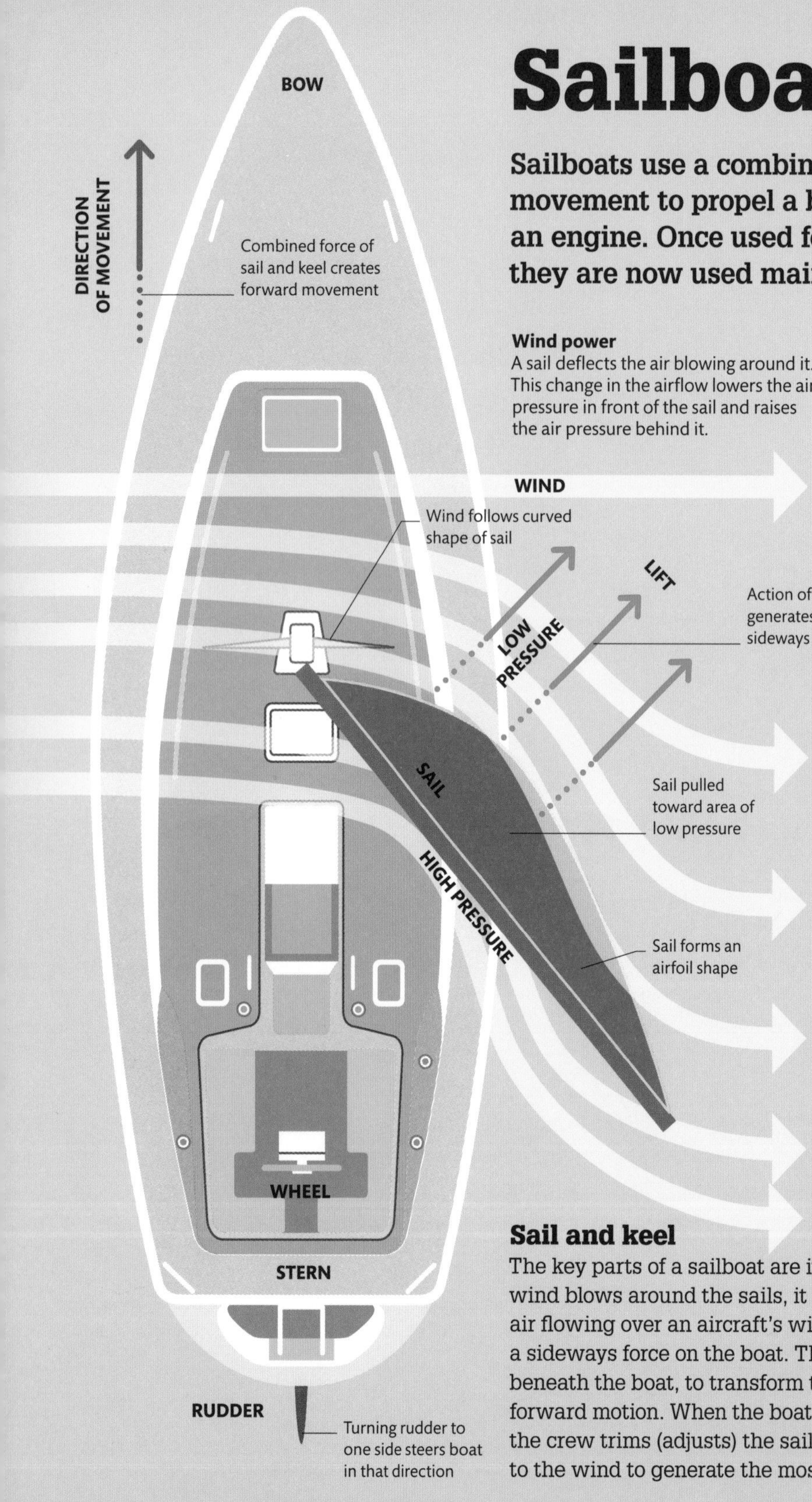

BOW
DIRECTION OF MOVEMENT
Combined force of sail and keel creates forward movement
WIND
Wind follows curved shape of sail
LIFT
LOW PRESSURE
Action of wind on sails generates forward and sideways lift
SAIL
Sail pulled toward area of low pressure
HIGH PRESSURE
Sail forms an airfoil shape
WHEEL
STERN
RUDDER
Turning rudder to one side steers boat in that direction

Wind power
A sail deflects the air blowing around it. This change in the airflow lowers the air pressure in front of the sail and raises the air pressure behind it.

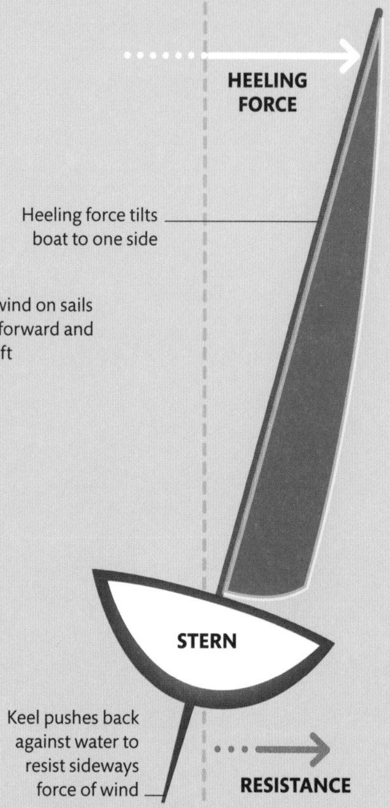

HEELING FORCE
Heeling force tilts boat to one side
STERN
Keel pushes back against water to resist sideways force of wind
RESISTANCE

Going forward
A keel generates forward movement by pushing against the water to cancel out the heeling force (the sideways motion created by the wind). The sideways force is not completely eliminated.

Sail and keel

The key parts of a sailboat are its sail, or sails, and its keel. When wind blows around the sails, it creates lift in the same way as air flowing over an aircraft's wing (see p.62). The wind creates a sideways force on the boat. The sailboat uses its keel, located beneath the boat, to transform the sideways force of the wind into forward motion. When the boat, or the wind, changes direction, the crew trims (adjusts) the sails so that they are at the best angle to the wind to generate the most lift.

Buoyancy and stability

Any boat displaces its own volume of water. Its weight is balanced by an upward force, called buoyancy force or upthrust. As long as the boat's density is equal to or less than that of the water, the upthrust will be enough to make the boat float. To float upright in the water, the boat's center of gravity—the midpoint of its mass—must be directly above its center of buoyancy, the point at which all buoyancy force is considered to act. When a boat heels over (see left), its center of gravity remains the same, but its center of buoyancy moves in the direction of the tilt. The two points must be brought back into alignment to bring the boat upright again.

Buoyancy
The density of an object is found by dividing its mass by its volume. The boat and the steel weight shown here both weigh the same. However, the steel weight will sink because it is more dense than water, while the boat will float because it is less dense.

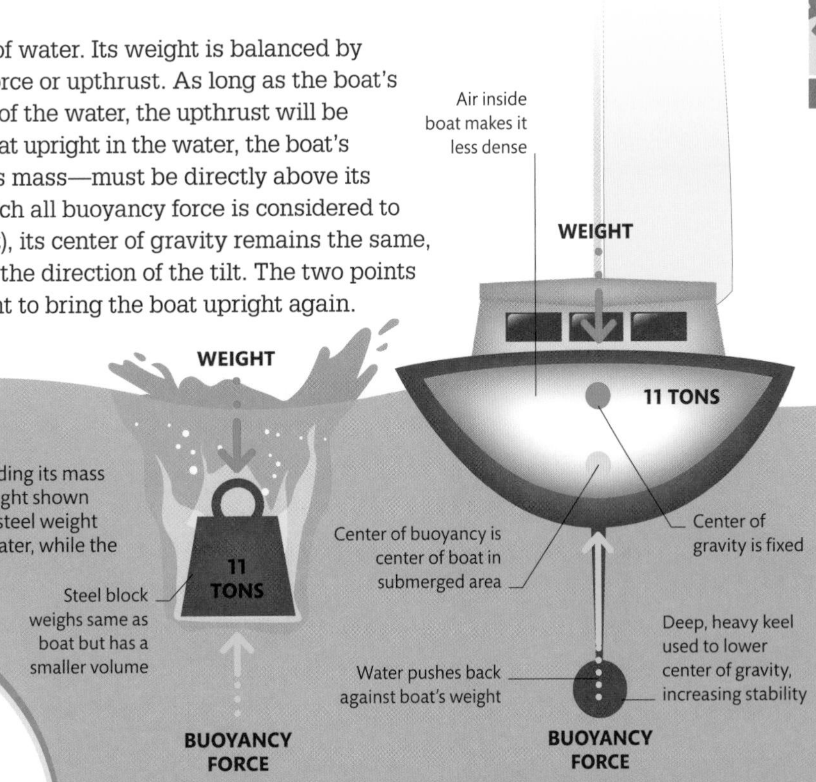
Air inside boat makes it less dense
WEIGHT
11 TONS
WEIGHT
11 TONS
Steel block weighs same as boat but has a smaller volume
Center of buoyancy is center of boat in submerged area
Center of gravity is fixed
Deep, heavy keel used to lower center of gravity, increasing stability
Water pushes back against boat's weight
BUOYANCY FORCE
BUOYANCY FORCE

WHICH IS THE FASTEST SAILBOAT?

The Vestas Sailrocket 2 holds the outright world sailing speed record with a speed of 75.2 miles (121.1 km) per hour.

40 DAYS, 23 HOURS, AND 30 MINUTES—THE RECORD-BREAKING TIME TAKEN TO SAIL AROUND THE WORLD

TYPES OF HULLS

The hull is the main body of a vessel. Sailboats can have one hull (monohulls) or have several hulls (multihulls). Multihulls are often used for racing because they are lighter than monohulls, since they do not require a heavy keel to keep them stable. The most popular multihulls are catamarans and trimarans. Catamarans have two hulls, and trimarans have three.

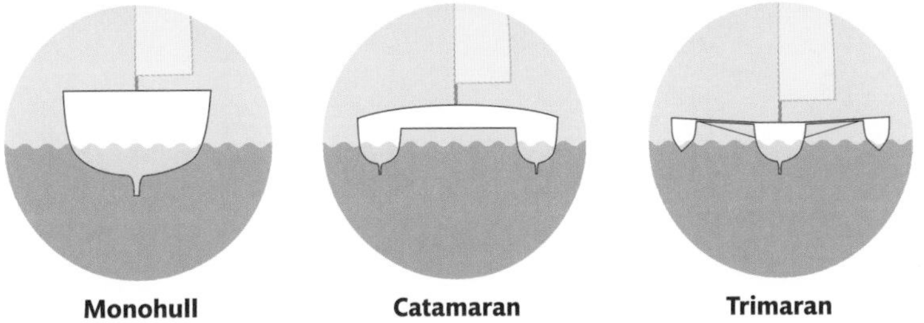

Monohull
A monohull has a spacious single hull below deck.

Catamaran
Catamarans are wider and so more stable than monohulls.

Trimaran
Trimarans have a main hull with two small outrigger hulls.

Propellers

A motor vessel's engine power is usually converted into the motion of the vessel through water by one or more propellers. When a propeller spins, its angled blades force water backward. The water pushes back against the propeller blades, generating thrust, which moves the boat. Water rushes in to fill the space that has been created behind the moving blade. This creates a pressure difference on either side of the blade, with low pressure at the front of the blade and high pressure behind. This pulls the front surface of the blade forward. Propellers are also called screws, because of their screwlike motion through water.

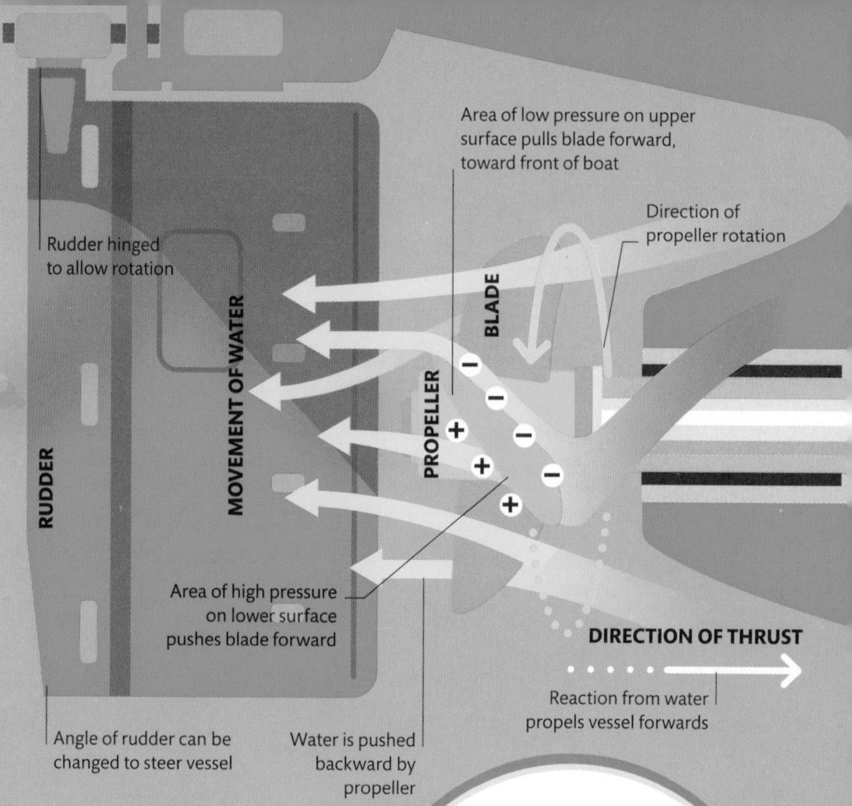

Area of low pressure on upper surface pulls blade forward, toward front of boat
Direction of propeller rotation
Rudder hinged to allow rotation
BLADE
MOVEMENT OF WATER
PROPELLER
RUDDER
Area of high pressure on lower surface pushes blade forward
DIRECTION OF THRUST
Reaction from water propels vessel forwards
Angle of rudder can be changed to steer vessel
Water is pushed backward by propeller

Motor vessels

The power provided by an engine frees motor vessels from the limitations of wind and sail. It also enables boats to generate electrical and hydraulic power to operate additional equipment.

WHAT IS THE FASTEST MOTORBOAT?

In 1978, Australian motorboat racer Ken Warby set a speed record of 317 miles (511 km) per hour in his jet-propelled powerboat.

Engines

A motor vessel can be powered in a number of different ways. Many use a diesel engine (see pp.42–43) to turn a shaft connected to a propeller. Other ships, including ocean liners, are powered by steam turbines. Warships often have gas-turbine engines similar to jet engines (see pp.60–61), and a handful of the biggest ships are nuclear-powered. On smaller boats, the engine is often mounted on the outside of the boat, while larger vessels usually have inboard motors.

Stability

A motor vessel's inboard engine may be used to power more than one propeller, as well as bow thrusters to help with steering (see opposite). The engine and heavy equipment are positioned low down in the hull to improve stability.

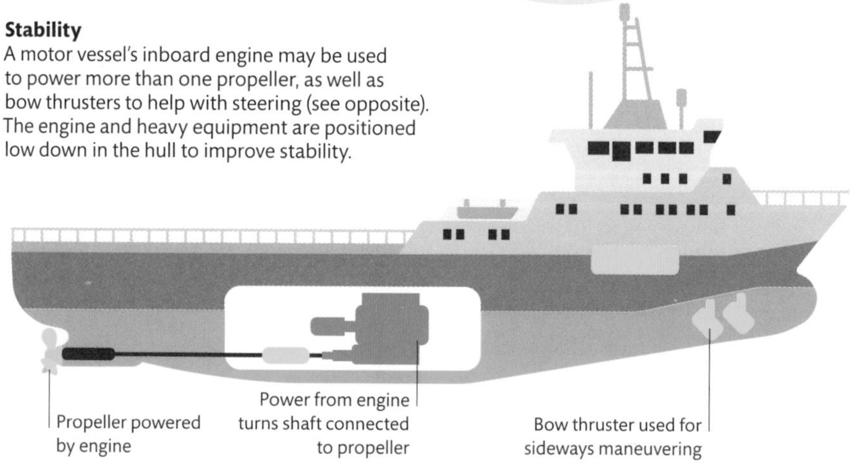

Propeller powered by engine
Power from engine turns shaft connected to propeller
Bow thruster used for sideways maneuvering

THE **FIRST SUCCESSFUL** MARINE **PROPELLERS** WERE DEVELOPED IN THE **1830s**

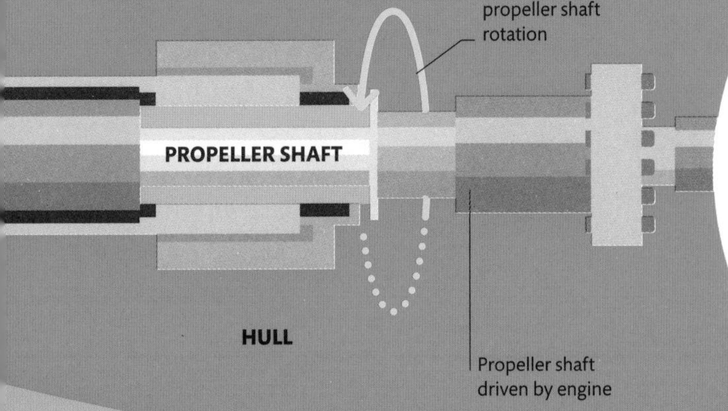
Direction of propeller shaft rotation
PROPELLER SHAFT
HULL
Propeller shaft driven by engine

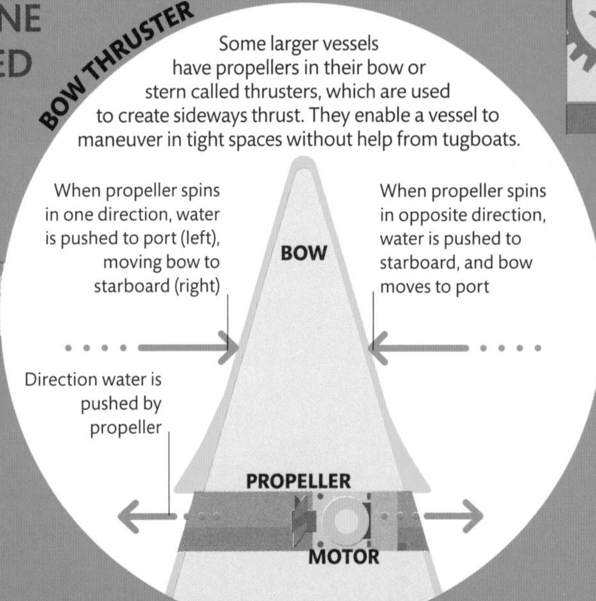
BOW THRUSTER
Some larger vessels have propellers in their bow or stern called thrusters, which are used to create sideways thrust. They enable a vessel to maneuver in tight spaces without help from tugboats.
When propeller spins in one direction, water is pushed to port (left), moving bow to starboard (right)
BOW
When propeller spins in opposite direction, water is pushed to starboard, and bow moves to port
Direction water is pushed by propeller
PROPELLER
MOTOR

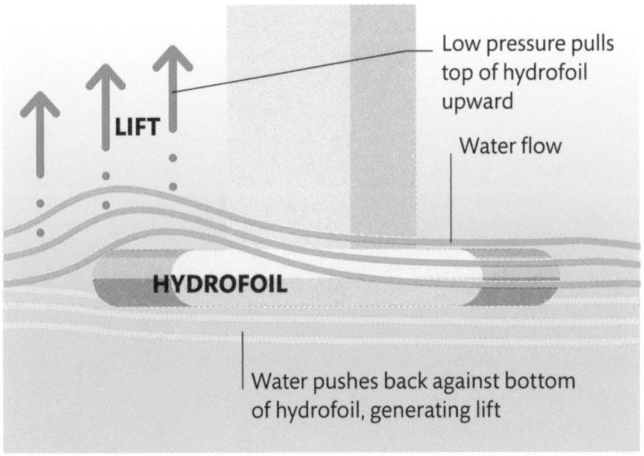
Low pressure pulls top of hydrofoil upward
LIFT
Water flow
HYDROFOIL
Water pushes back against bottom of hydrofoil, generating lift

Hydrofoils

Water pressing against a vessel's hull causes drag, which slows the vessel down since its engine has to work harder to overcome this resistance. Hydrofoil boats minimize drag by using underwater wings called hydrofoils, or foils, which work in the same way as an airplane's wing (see p.62) to lift the whole hull out of the water. Because water is denser than air, compared with an aircraft's wing, a hydrofoil can create more lift at a lower speed.

Types of hydrofoils
Surface-piercing foils break through the surface of the water, while fully submerged foils stay underwater.

SURFACE-PIERCING HYDROFOIL

FULLY SUBMERGED HYDROFOIL

POD PROPULSION

It is common today for large ships to be propelled and steered by devices called azimuth thrusters. These contain an electric motor turning a propeller. The whole pod can be rotated to provide thrust in any direction.

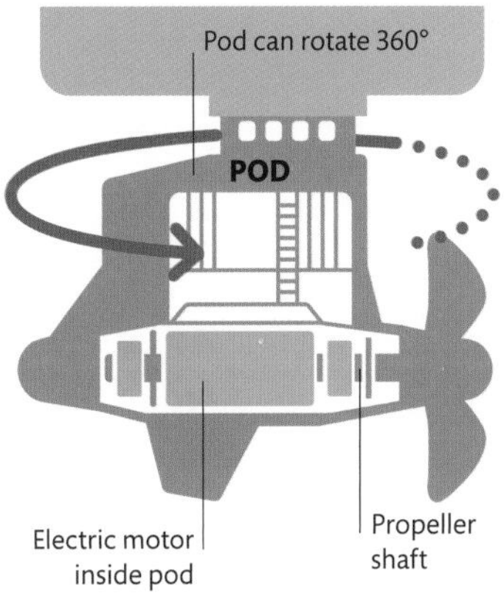
Pod can rotate 360°
POD
Electric motor inside pod
Propeller shaft

Submarines

A submarine is a vessel designed to be used underwater, often for military use. Ballast tanks allow submarines to sink or float. Usually powered by a nuclear reactor or a diesel-electric engine, submarines contain high-tech navigation and communication systems and can remain hidden for months at a time.

Moving through water

As a submarine moves through the ocean, propelled by its powerful engine, the crew steer it by moving three types of control surfaces—bow planes, stern planes, and rudders. They tilt the bow planes to make the submarine climb higher or dive deeper in the water. They adjust the stern planes to keep the submarine level. They use rudders to steer the submarine to port (left) or starboard (right).

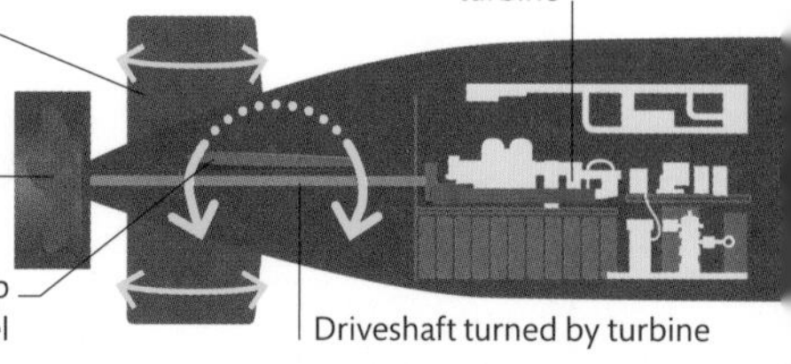

Steam-driven turbine
Top and bottom rudders steer submarine left and right
Shroud reduces propeller noise
Stern planes keep submarine level
Driveshaft turned by turbine

1 Floating

When a submarine's ballast tanks are full of air, it floats at the surface. All of the ballast tank valves are closed to stop water from rushing in.

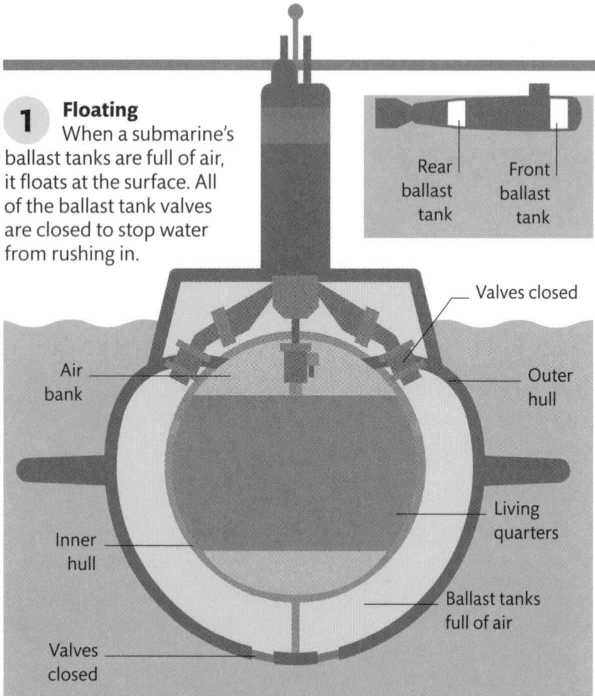

Rear ballast tank
Front ballast tank
Valves closed
Air bank
Outer hull
Living quarters
Inner hull
Ballast tanks full of air
Valves closed

How a submarine dives and rises

Submarines are able to dive to great depths and return to the surface again because they can change their density relative to the water around them. If a submarine's density is greater than the surrounding water, it sinks. Reducing the submarine's density makes it more buoyant, so it floats up toward the surface. The crew changes the submarine's density by filling its ballast tanks, located between the inner and outer hulls, with seawater or compressed air.

2 Diving

A submarine dives by opening its ballast tank valves to let seawater flood inside. The submarine, now heavier than the same volume of water, sinks. Taking in more water makes it sink to a greater depth.

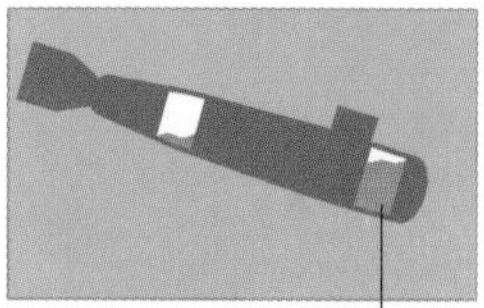

Front tank is filled first to lower bow

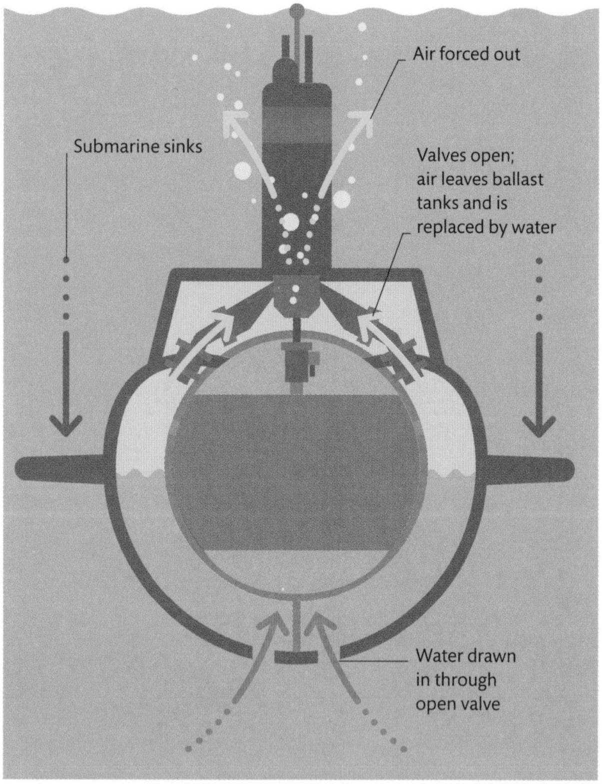

Air forced out
Submarine sinks
Valves open; air leaves ballast tanks and is replaced by water
Water drawn in through open valve

Naval submarine
To avoid detection, the machinery in a naval submarine is isolated from the hull to stop vibration from being transmitted into the water. The submarine's propeller is often enclosed in a shroud to reduce noise generated.

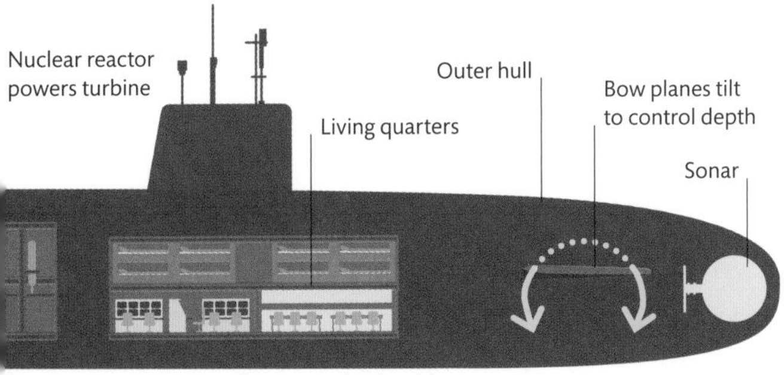

3 Surfacing
To rise, compressed air is pumped into the ballast tanks, and water is gradually forced out. The air supply for the air banks is replenished at the surface.

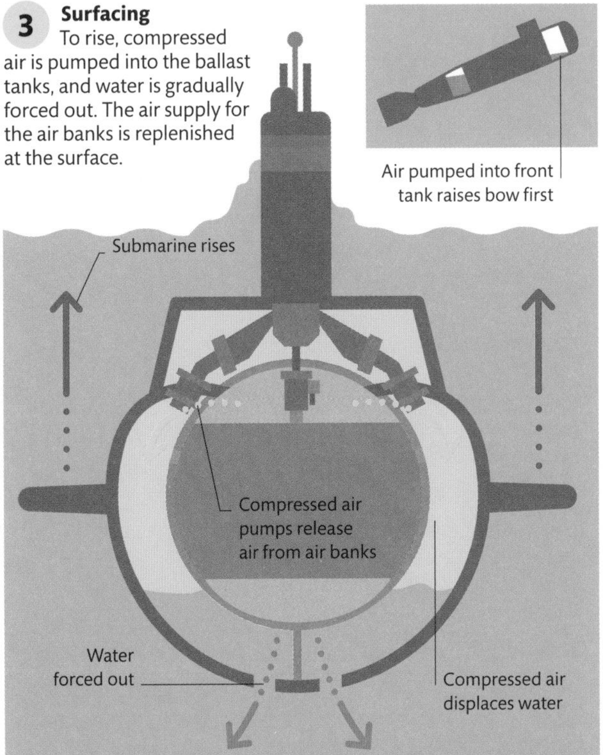

THE **FIRST WORKING SUBMARINE** WAS BUILT BY **CORNELIS DREBBEL** IN **1620**

Submersibles
Submersibles are manned or unmanned diving craft that are smaller than submarines. While submarines operate independently, submersibles are carried to their dive location by ship. To withstand huge water pressures at great depths, submersibles have a very strong spherical compartment for the crew. Submersibles use electric-powered thrusters to maneuver.

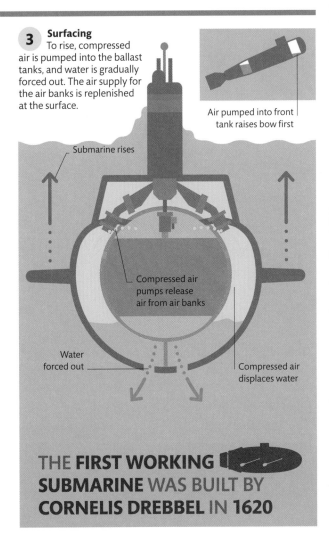

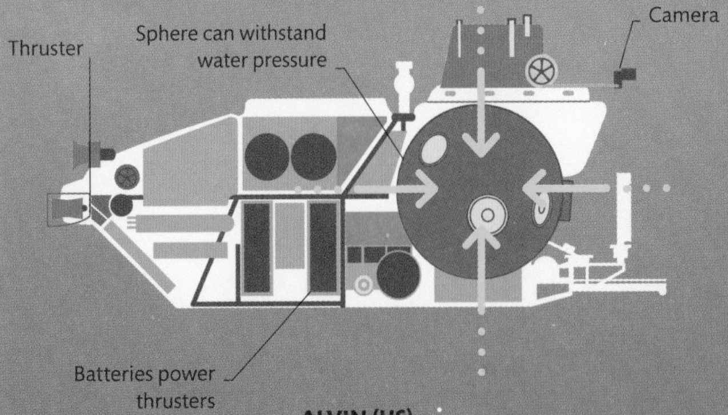

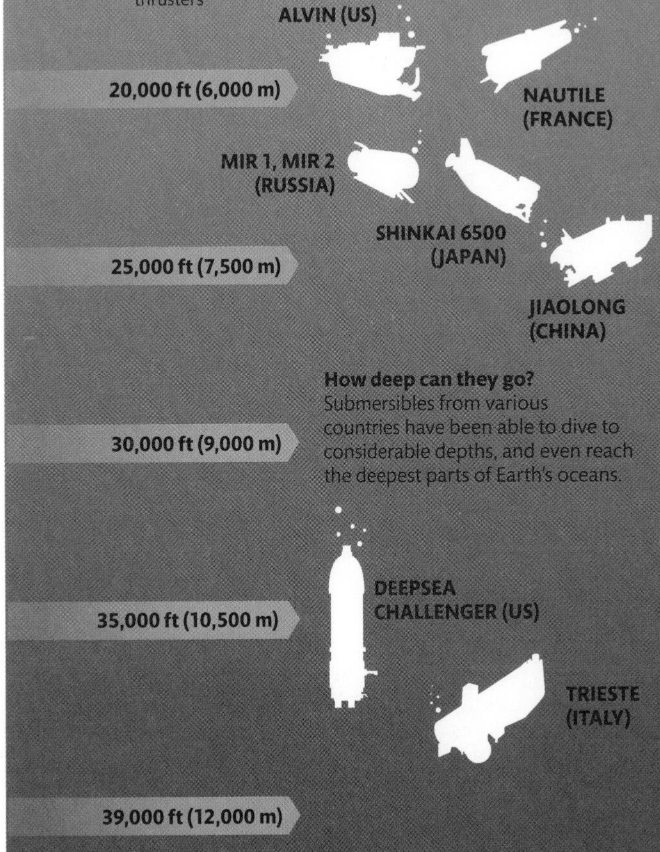

How deep can they go?
Submersibles from various countries have been able to dive to considerable depths, and even reach the deepest parts of Earth's oceans.

Jet engines and rockets

Jet engines and rockets are both types of reaction engines that use thrust to propel them forward or upward. The rapid expulsion of gas in one direction generates thrust in the opposite direction.

Turbofan engine
The most common type of jet engine used by passenger airliners is called a turbofan, named for the large fan at its front. In this type of engine, the main source of thrust is air that bypasses the central core.

Aircraft engines

Jet engines revolutionized aviation by enabling aircraft to become faster and more fuel efficient than their propeller-driven predecessors. Most modern commercial airliners and military fighter aircraft are jet powered. Although there are different types, all jet engines work by the same principle. They take in air, add fuel, then burn the mixture. The resulting explosive exhaust gases produce the jet propulsion.

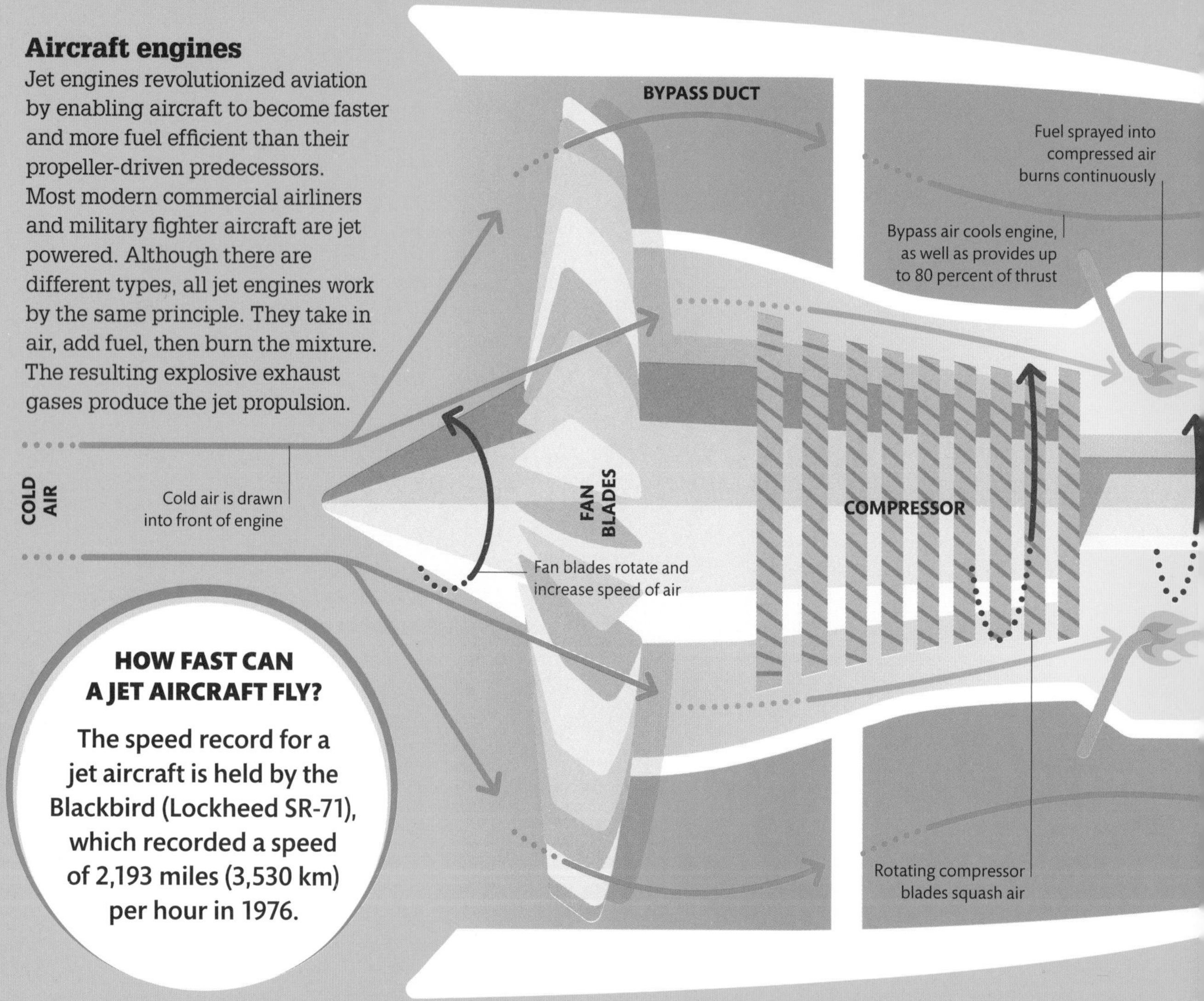

BYPASS DUCT
Fuel sprayed into compressed air burns continuously
Bypass air cools engine, as well as provides up to 80 percent of thrust
COLD AIR
Cold air is drawn into front of engine
FAN BLADES
COMPRESSOR
Fan blades rotate and increase speed of air
Rotating compressor blades squash air

HOW FAST CAN A JET AIRCRAFT FLY?

The speed record for a jet aircraft is held by the Blackbird (Lockheed SR-71), which recorded a speed of 2,193 miles (3,530 km) per hour in 1976.

1 Air intake
Fan blades at the front of the engine draw in cold air. Most of the air is propelled through bypass ducts to the back of the engine. The rest travels into the engine's core.

2 Compressor
The air enters the compressor, which contains a series of fan blades. This compresses the air, raising its temperature and pressure dramatically.

3 Combustion chamber
A steady stream of compressed air passes through to the combustion chamber. Here, fuel is sprayed in through nozzles, and the mixture burns at very high temperatures.

THE SOUND BARRIER

Planes flying faster than the speed of sound compress the air in front of them so much that they form a high-pressure shock wave. This spreads out and is heard on the ground as a loud sonic boom.

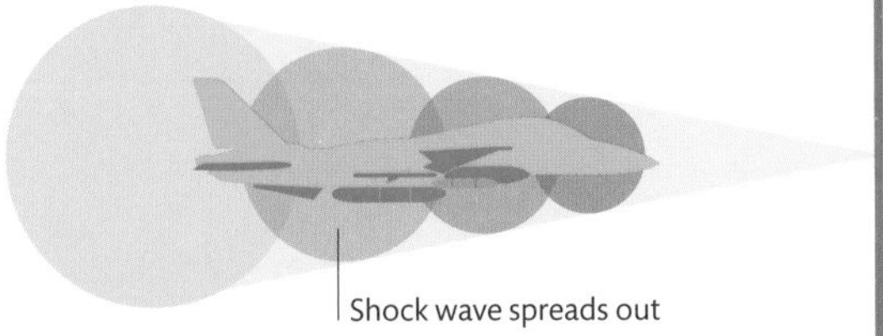
Shock wave spreads out

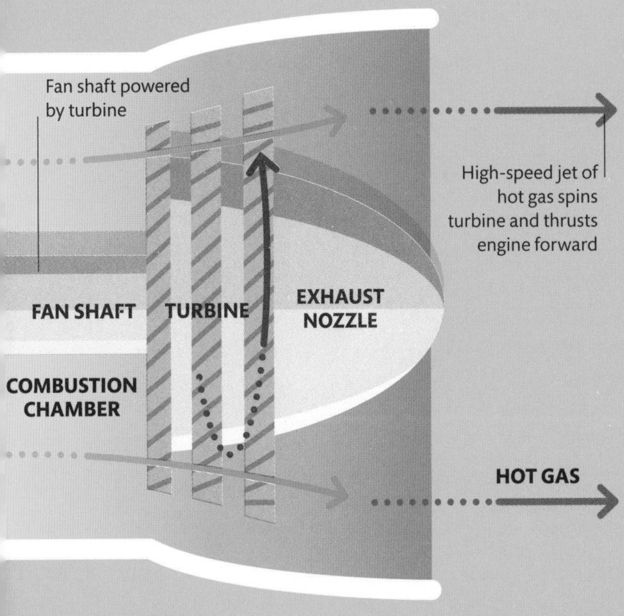
Fan shaft powered by turbine
High-speed jet of hot gas spins turbine and thrusts engine forward
FAN SHAFT
TURBINE
EXHAUST NOZZLE
COMBUSTION CHAMBER
HOT GAS

4 Turbine The hot gas expands explosively and rushes out of the engine, spinning the blades of a turbine. The spinning turbine powers the fan and compressor.

5 Exhaust nozzle The jet of hot exhaust gas leaves the engine, along with the cold bypass air, pushing back against the engine, and generating thrust.

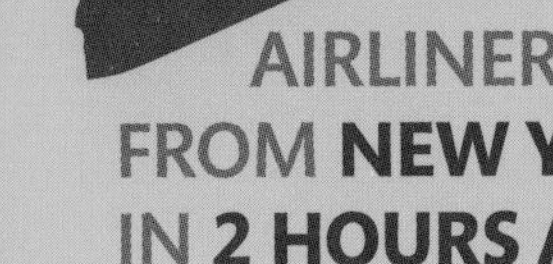

THE SUPERSONIC AIRLINER **CONCORDE** FLEW FROM **NEW YORK TO LONDON** IN **2 HOURS AND 52 MINUTES**

Rocket engines

Unlike jet engines, which use oxygen from the atmosphere to burn their fuel, rockets carry their own oxygen supply, which means that they can operate in the vacuum of space. The oxygen supply, or oxidizer, can take the form of pure liquid oxygen or an oxygen-rich chemical compound.

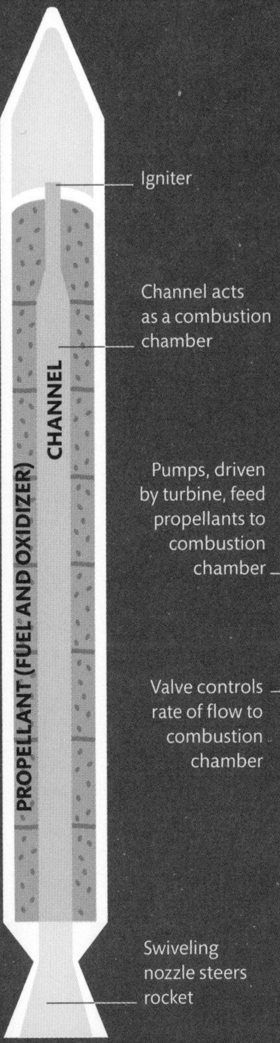
Igniter
Channel acts as a combustion chamber
CHANNEL
PROPELLANT (FUEL AND OXIDIZER)
Swiveling nozzle steers rocket

Solid-fueled rocket
The fuel and oxidizer are mixed together as a solid compound with an open channel in the middle. When the igniter fires, the fuel burns along the channel until there is none left.

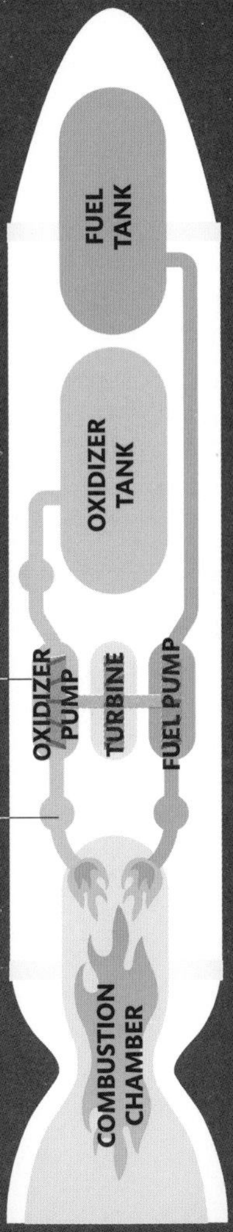
FUEL TANK
OXIDIZER TANK
Pumps, driven by turbine, feed propellants to combustion chamber
OXIDIZER PUMP
TURBINE
FUEL PUMP
Valve controls rate of flow to combustion chamber
COMBUSTION CHAMBER

Liquid-fueled rocket
The fuel and oxidizer are stored as liquids. Unlike a solid-fueled rocket, a liquid-fueled rocket can be restarted. It can also be throttled by varying the flows of fuel and oxidizer.

Airplanes

Airplanes come in a wide variety of shapes and sizes, but they all fly according to the same principles. Power generated by an engine or propeller thrusts the plane forward, while wings generate lift.

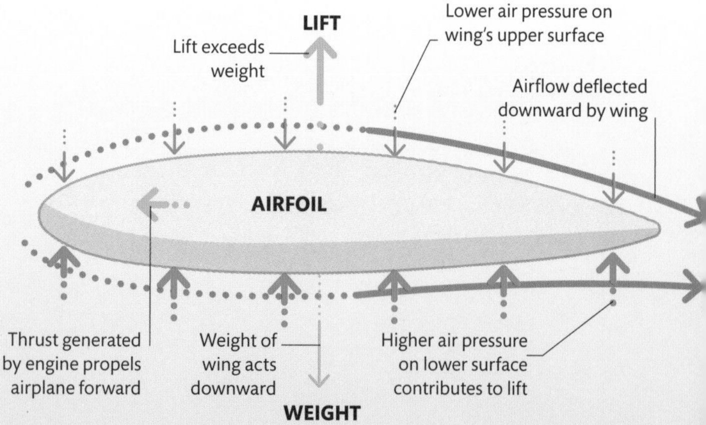
LIFT
Lift exceeds weight
Lower air pressure on wing's upper surface
Airflow deflected downward by wing
AIRFOIL
Thrust generated by engine propels airplane forward
Weight of wing acts downward
Higher air pressure on lower surface contributes to lift
WEIGHT

How an airplane flies

When an airplane is propelled forward by its engines (see pp.60–61), its wings slice through the air. A wing's shape, called an airfoil, deflects the air downward. When the wing pushes air down, the air acts in accordance with Isaac Newton's third law of motion by pushing back and producing an upward reaction force known as lift. The air pressure above the wing falls, and the pressure below it rises, contributing to lift generated.

Angle of attack

The angle between a wing and the oncoming air is called the angle of attack. By increasing this angle, more lift is created. If the angle is too big, the airflow separates from the wing, which loses lift or stalls.

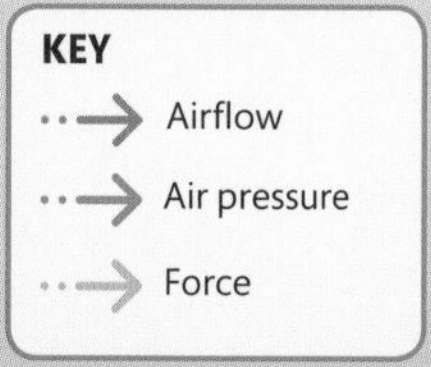
KEY
Airflow
Air pressure
Force

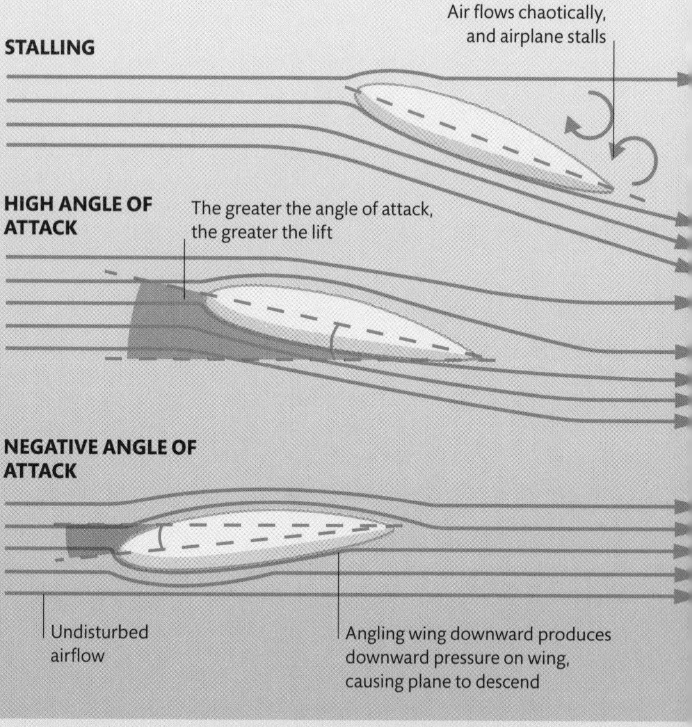
STALLING
Air flows chaotically, and airplane stalls
HIGH ANGLE OF ATTACK
The greater the angle of attack, the greater the lift
NEGATIVE ANGLE OF ATTACK
Undisturbed airflow
Angling wing downward produces downward pressure on wing, causing plane to descend

THE AIRBUS A380, THE WORLD'S LARGEST AIRLINER, HAS 4 MILLION PARTS

Controlling an airplane

An airplane is steered by moving panels in the wings and tail called control surfaces. There are three types—elevators, ailerons, and a rudder. When the pilot moves the flight controls, the control surfaces move out into the air flowing past the plane, which rotates the plane in three ways—pitch, roll, and yaw.

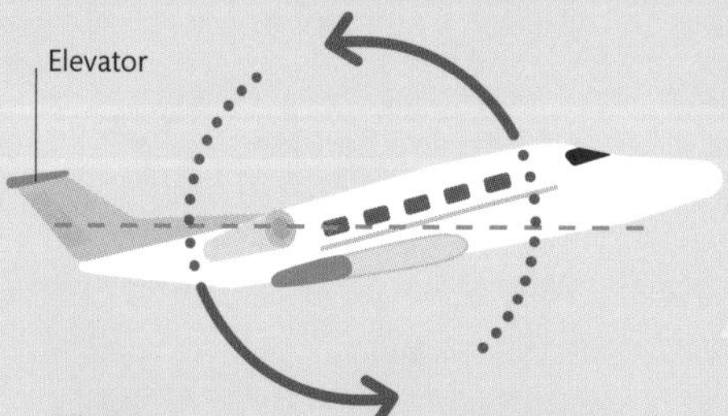
Elevator

Pitch
Elevators in the tail's horizontal stabilizers tilt up and down. Tilting them up pushes the tail down, and the plane climbs. Tilting them down makes the plane dive.

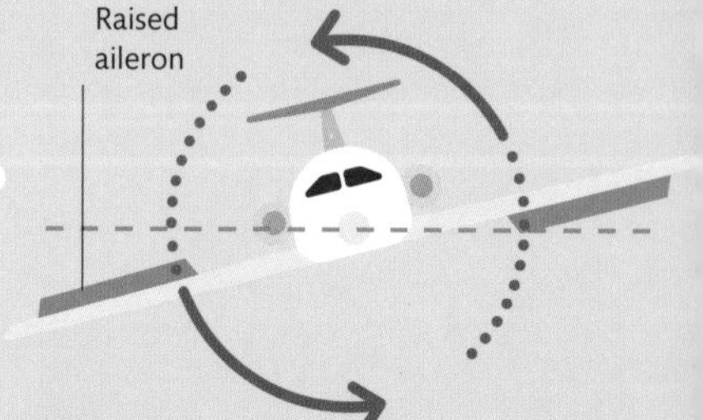
Raised aileron

Roll
The aileron in one wing is raised, while the aileron in the other wing is lowered. This makes the first wing fall and the second rise, making the plane roll.

AIR PRESSURE

The air pressure experienced on the ground is caused by the weight of the atmosphere above pressing down. At ground level, the pressure inside and outside an airplane is the same. As it climbs to its cruising altitude, the air pressure outside the airplane falls. The air pressure inside the cabin is maintained at a higher level by a system that pumps air from the engines into the cabin. This ensures that there is enough oxygen for people to breathe.

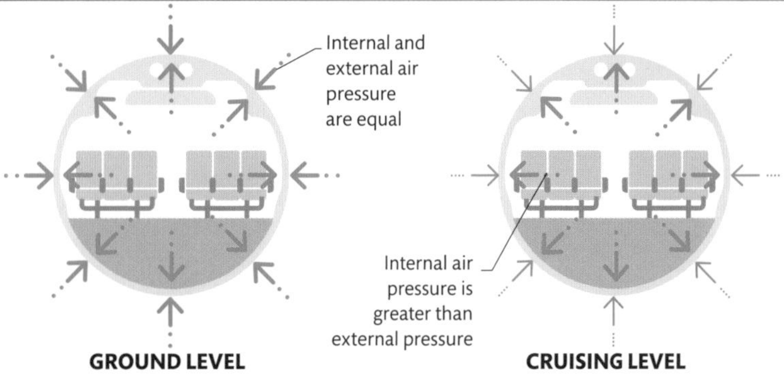
Internal and external air pressure are equal
GROUND LEVEL
Internal air pressure is greater than external pressure
CRUISING LEVEL

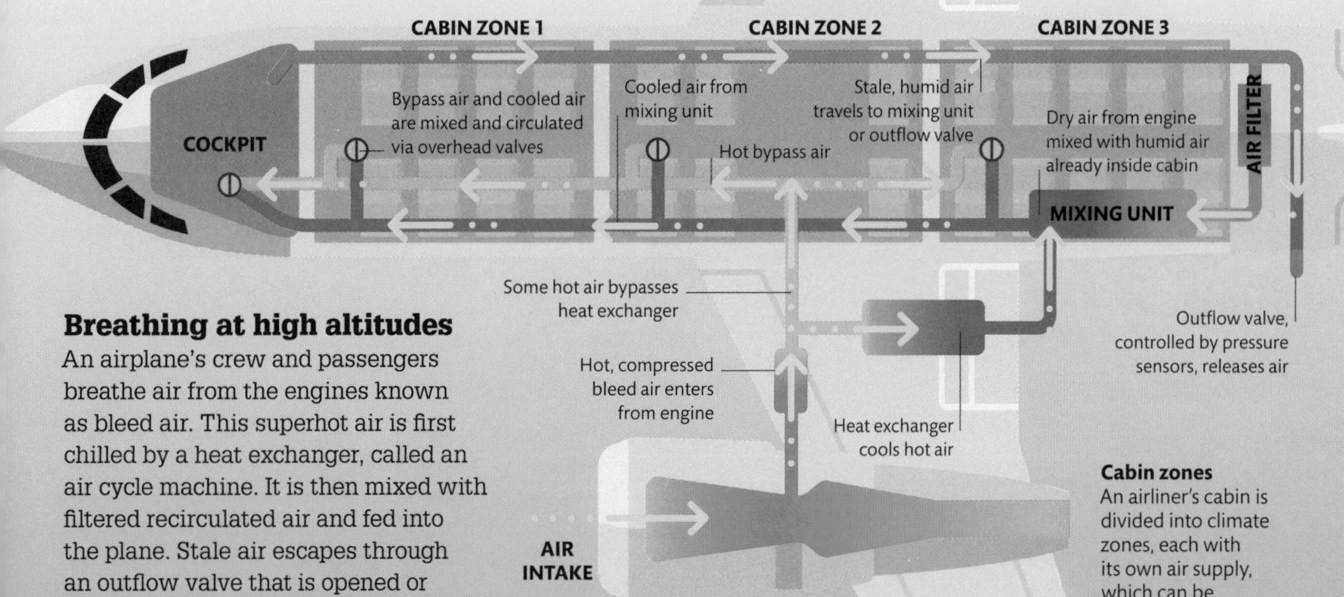
CABIN ZONE 1
CABIN ZONE 2
CABIN ZONE 3
COCKPIT
Bypass air and cooled air are mixed and circulated via overhead valves
Cooled air from mixing unit
Hot bypass air
Stale, humid air travels to mixing unit or outflow valve
Dry air from engine mixed with humid air already inside cabin
AIR FILTER
MIXING UNIT
Some hot air bypasses heat exchanger
Hot, compressed bleed air enters from engine
Heat exchanger cools hot air
Outflow valve, controlled by pressure sensors, releases air
AIR INTAKE

Breathing at high altitudes

An airplane's crew and passengers breathe air from the engines known as bleed air. This superhot air is first chilled by a heat exchanger, called an air cycle machine. It is then mixed with filtered recirculated air and fed into the plane. Stale air escapes through an outflow valve that is opened or closed to regulate the pressure.

Cabin zones
An airliner's cabin is divided into climate zones, each with its own air supply, which can be controlled separately.

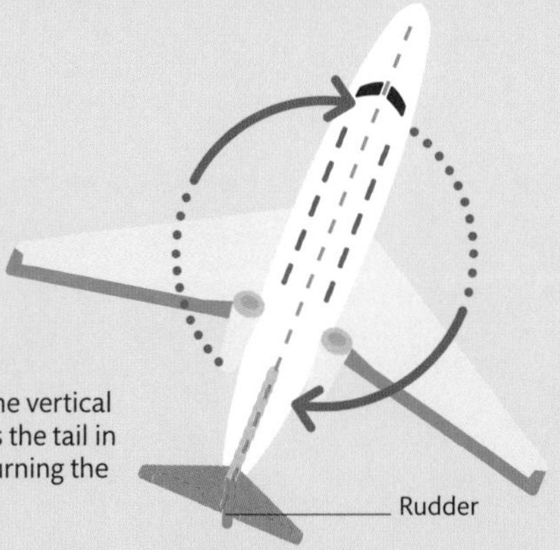
Rudder

Yaw
Swiveling the rudder in the vertical tail fin to one side pushes the tail in the opposite direction, turning the plane's nose left or right.

WHICH IS THE LONGEST SCHEDULED FLIGHT?

A nonstop flight from Singapore to New York covers 9,532 miles (15,341 km) in 17 hours and 25 minutes.

Helicopters

The long, thin blades of a helicopter create lift using the same principle as an airplane's wing (see p.62). The blades rotate many hundreds of times a minute, generating enough lift for a helicopter to take off vertically.

The parts of a helicopter

A helicopter consists of a fuselage made of lightweight aluminum or composite materials, one or more overhead rotors, one or more engines, and a transmission system. Most helicopters have a small rotor at the end of the tail. Without it, the body of the helicopter would spin in the opposite direction to the main rotor blades.

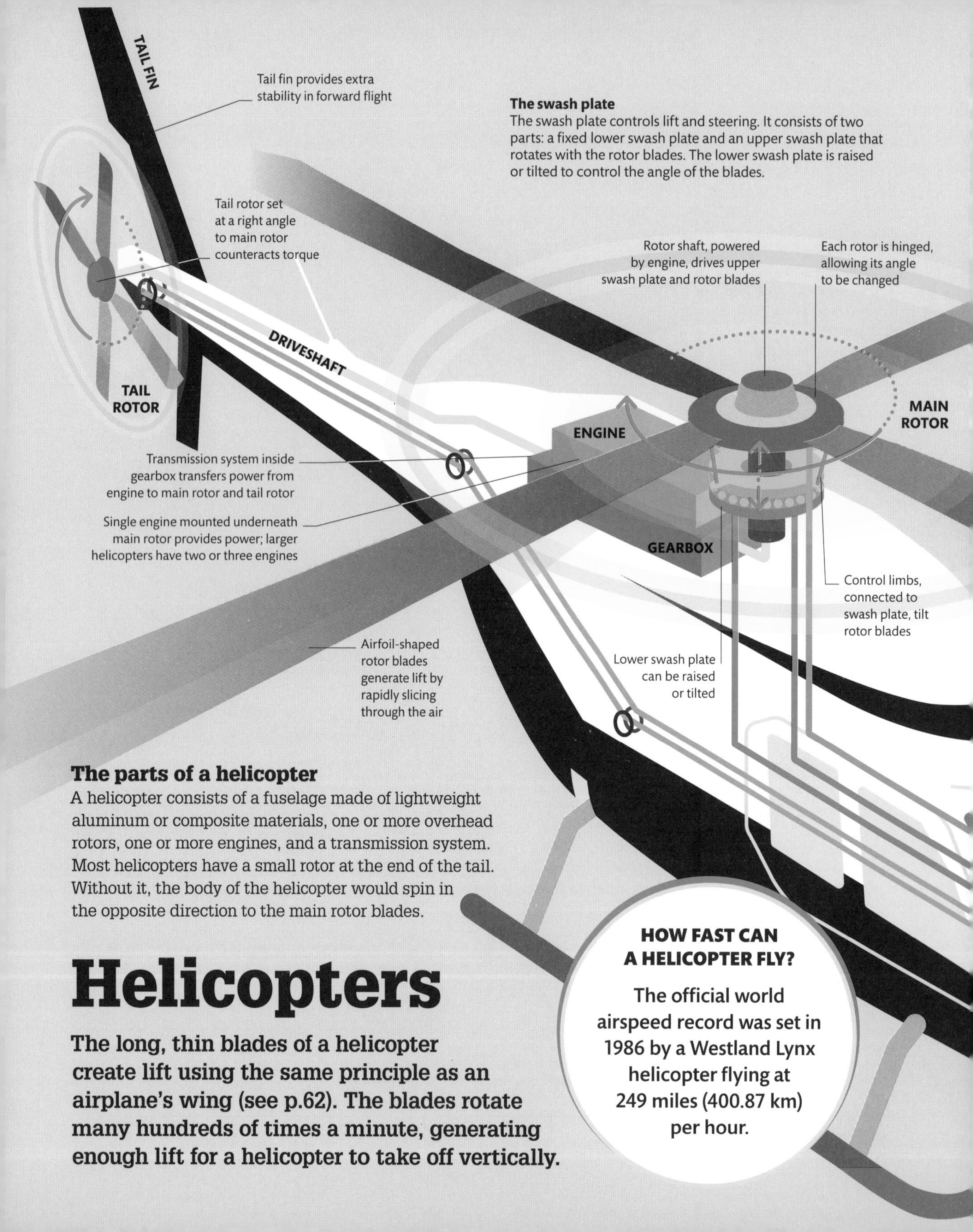

TAIL FIN
Tail fin provides extra stability in forward flight
Tail rotor set at a right angle to main rotor counteracts torque
TAIL ROTOR
DRIVESHAFT
Transmission system inside gearbox transfers power from engine to main rotor and tail rotor
Single engine mounted underneath main rotor provides power; larger helicopters have two or three engines
ENGINE
GEARBOX
Rotor shaft, powered by engine, drives upper swash plate and rotor blades
Each rotor is hinged, allowing its angle to be changed
MAIN ROTOR
Control limbs, connected to swash plate, tilt rotor blades
Lower swash plate can be raised or tilted
Airfoil-shaped rotor blades generate lift by rapidly slicing through the air

The swash plate

The swash plate controls lift and steering. It consists of two parts: a fixed lower swash plate and an upper swash plate that rotates with the rotor blades. The lower swash plate is raised or tilted to control the angle of the blades.

HOW FAST CAN A HELICOPTER FLY?

The official world airspeed record was set in 1986 by a Westland Lynx helicopter flying at 249 miles (400.87 km) per hour.

The collective and cyclic controls

To generate lift and change direction, the pilot uses the collective and cyclic controls. To increase or decrease lift, the collective lever raises or lowers the swash plate, changing the angle, or pitch, of all the blades at the same time. To change direction, the cyclic stick is used to tilt the swash plate, giving the blades unequal pitch depending on whether they are in front of or behind the rotor shaft.

KEY
Lift
Weight

Takeoff
To take off, the pilot increases the engine speed and raises the collective lever to produce more lift.

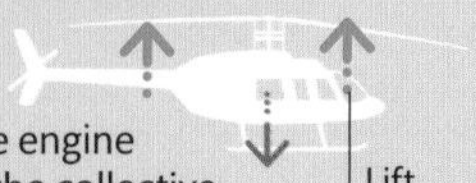
Lift exceeds weight

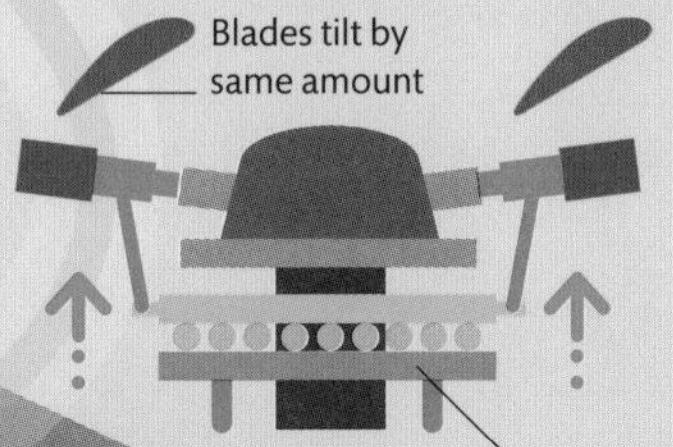
Blades tilt by same amount
Swash plate raised

Hovering
To hover in one spot, the rotor blades produce just enough lift to balance the helicopter's weight.

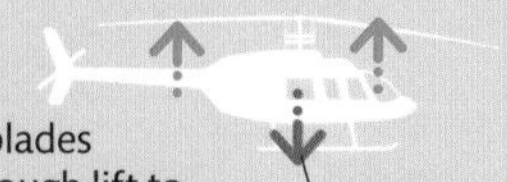
Lift and weight are in balance

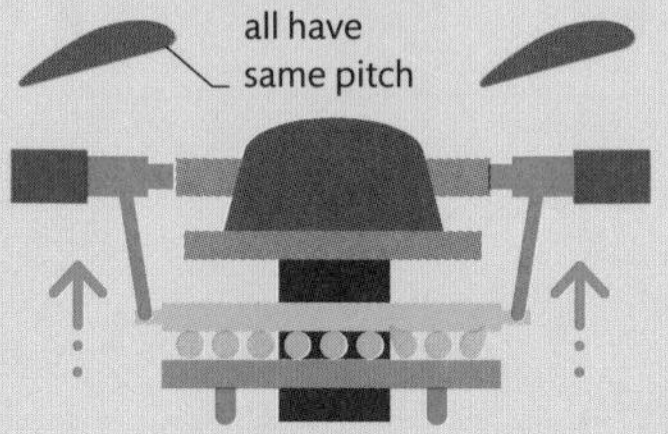
Rotor blades all have same pitch

Forward flight
To go forward, the cyclic stick is pushed forward, which tilts the rotor up at the back.

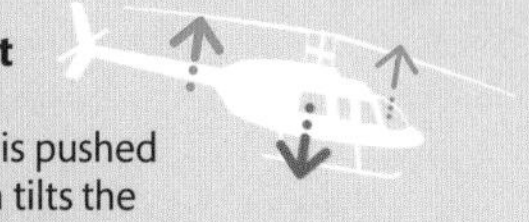

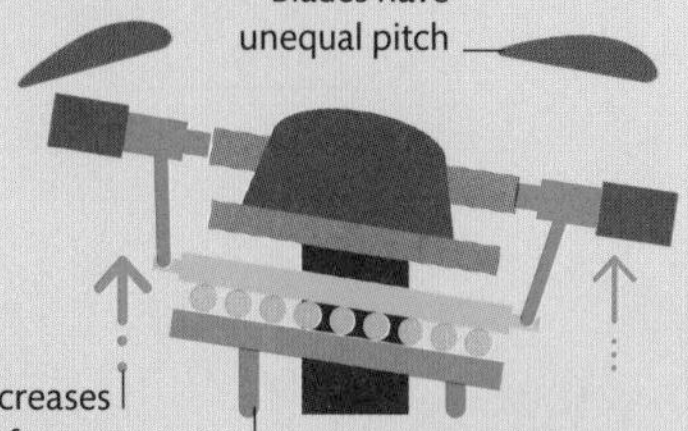
Blades have unequal pitch
Lift increases at back of rotor, causing helicopter to tilt forward
Swash plate tilted by cyclic controls

IN 1480, LEONARDO DA VINCI SKETCHED AN IDEA FOR AN AIRCRAFT THAT COULD FLY VERTICALLY

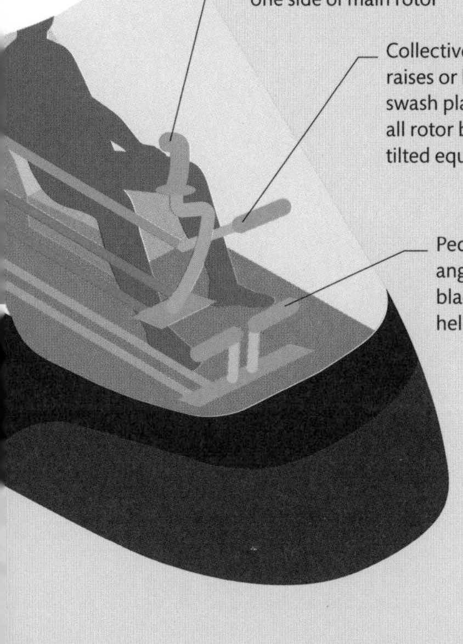
Cyclic stick allows pilot to tilt swash plate, which increases lift on one side of main rotor
Collective lever raises or lowers swash plate, meaning all rotor blades are tilted equally
Pedal changes angle of tail rotor blades, allowing helicopter to turn

TANDEM ROTOR BLADES

Instead of using a tail rotor to counteract torque, some helicopters have two overhead rotors that spin in opposite directions. The helicopter is steered by tilting the front rotor in one direction and the rear rotor in the opposite direction.

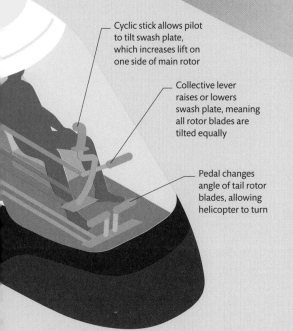

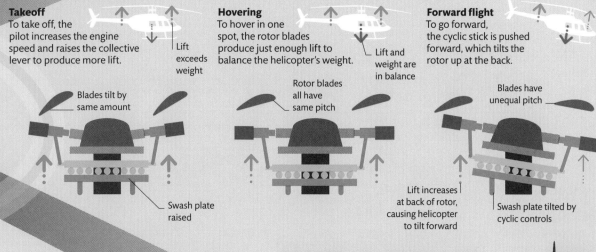

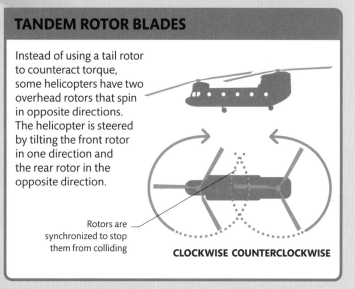

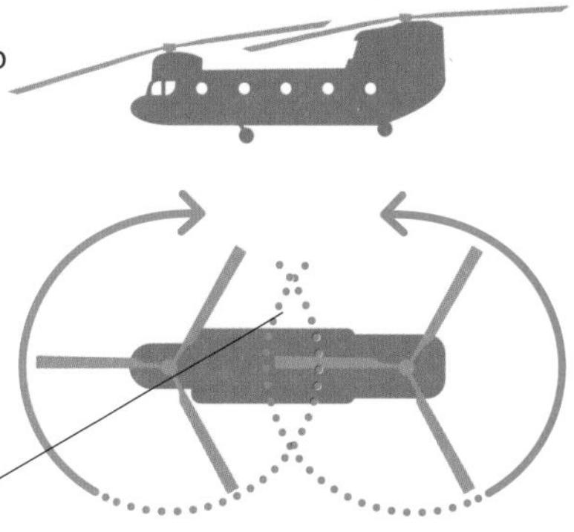
Rotors are synchronized to stop them from colliding
CLOCKWISE
COUNTERCLOCKWISE

Drones

A drone is a kind of flying robot. Drones are often flown for recreation, but they also serve commercial and military purposes as well as have other important uses.

What is a drone?

A drone is an unmanned aerial vehicle (UAV). Most drones are flown by remote control, but some can be programmed to operate autonomously. To reduce weight, drones are made of lightweight materials such as plastic, composites, and aluminum. Since they are often used for photography and filming, many have a camera attached.

How drones fly

Drones are propelled by rotors driven by electric motors. They move in a similar way to helicopters (see pp.64–65) but usually have several propellers to produce both lift and thrust. Four-propeller "quadcopters" are the most common.

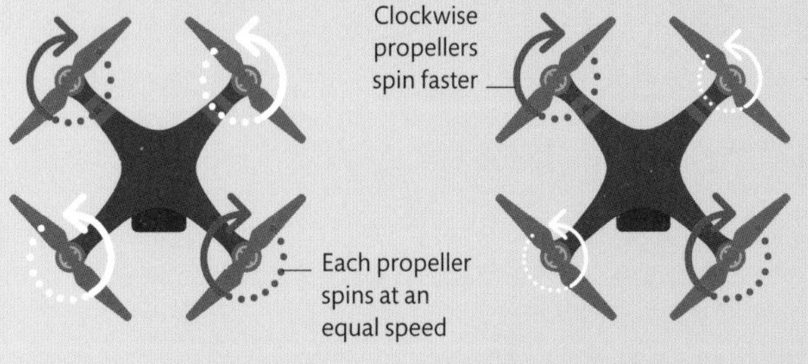
Clockwise propellers spin faster
Each propeller spins at an equal speed

Hovering
Quadcopters have two propellers spinning clockwise and two counterclockwise. This balances their torque (turning force). To hover, all four spin at the same speed.

Turning to the left
To make the drone turn (yaw) left, the clockwise propellers spin faster. To turn right, more power is applied to the counterclockwise propellers.

IN 2014, A **DRONE FILMED ITSELF** BEING **CAUGHT** IN MIDAIR **BY A HAWK**

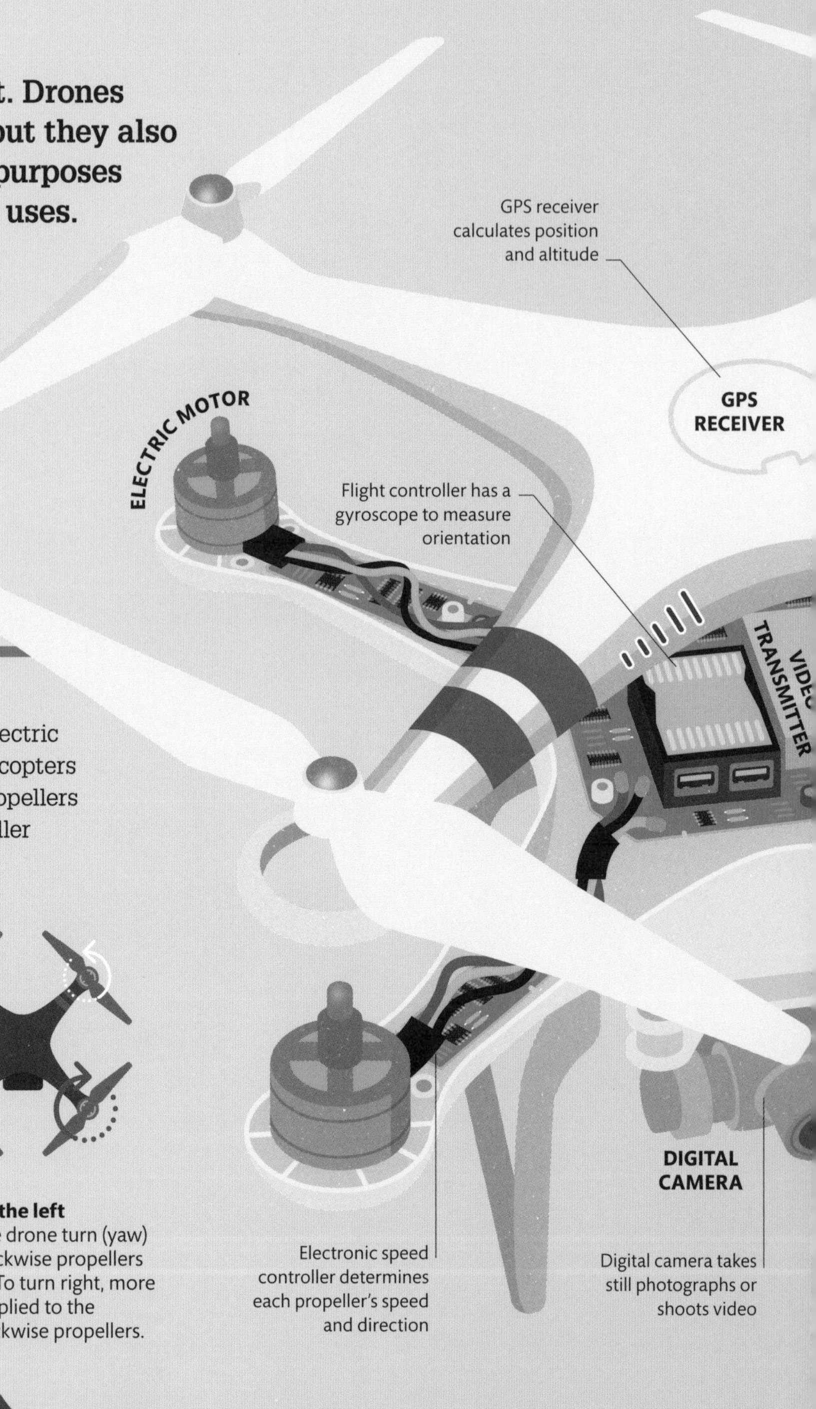
GPS receiver calculates position and altitude
GPS RECEIVER
ELECTRIC MOTOR
Flight controller has a gyroscope to measure orientation
VIDEO TRANSMITTER
DIGITAL CAMERA
Electronic speed controller determines each propeller's speed and direction
Digital camera takes still photographs or shoots video

Quadcopter
A quadcopter is usually equipped with a global positioning system (GPS), a flight controller, speed controllers, and a transmitter/receiver system to receive commands and send back data.

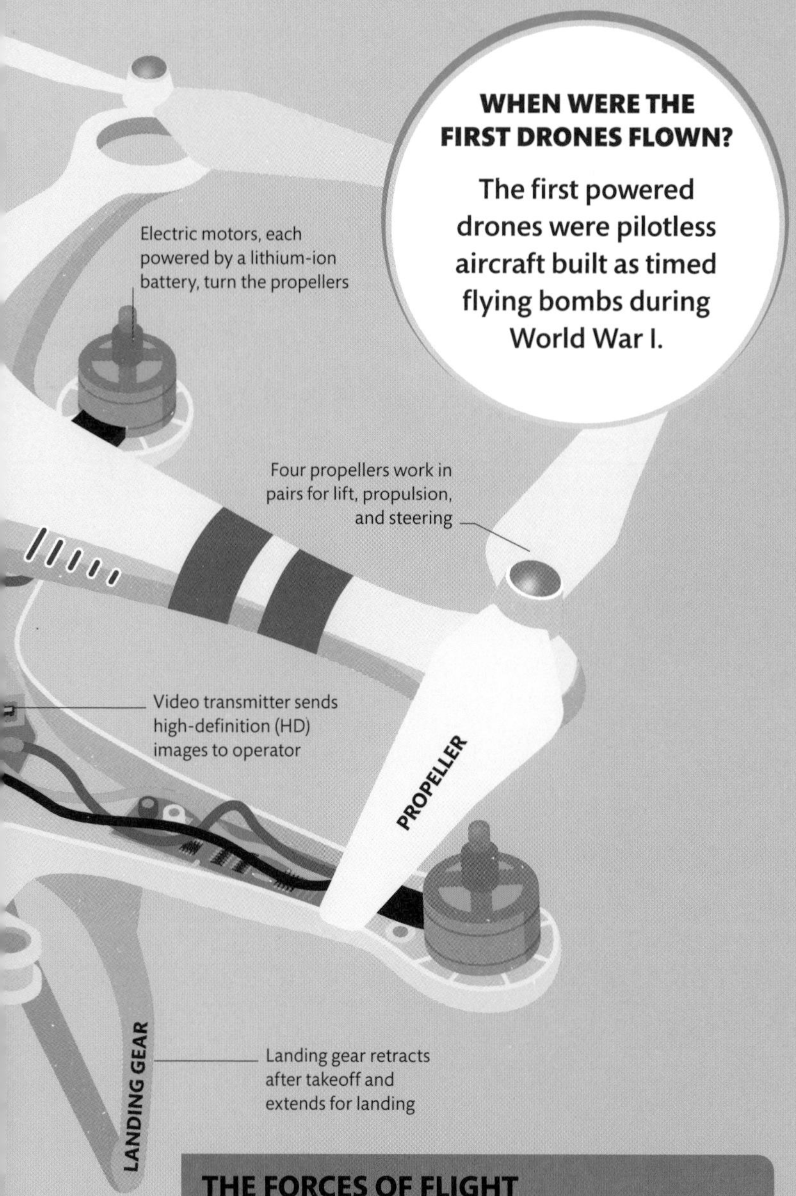

Electric motors, each powered by a lithium-ion battery, turn the propellers
Four propellers work in pairs for lift, propulsion, and steering
Video transmitter sends high-definition (HD) images to operator
PROPELLER
LANDING GEAR
Landing gear retracts after takeoff and extends for landing

WHEN WERE THE FIRST DRONES FLOWN?

The first powered drones were pilotless aircraft built as timed flying bombs during World War I.

THE FORCES OF FLIGHT

Drones achieve a balance between the four forces of flight (see p.38). Lift and thrust are produced by propellers. They work against gravity and drag respectively to produce vertical and horizontal movement.

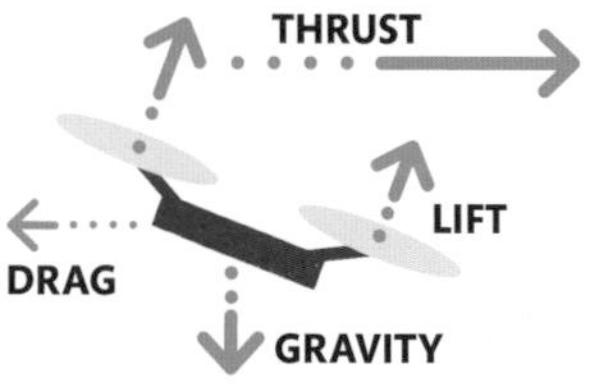

THRUST
LIFT
DRAG
GRAVITY

Uses of drones

A drone's ability to take off and land almost anywhere, as well as hover at a fixed point over the ground, makes it ideal for a wide range of applications, including surveillance, aerial photography, scientific research, mapmaking, and filming. Broadcasters use drones to capture aerial views of events; farmers use them to assess the health of their crops (see p.220); archaeologists use drones to monitor, map, and protect their sites; and wildlife organizations use them to help protect animals from poachers.

Surveying
Drones can take aerial photographs to map sites faster than traditional methods on the ground.

Military use
Drones flown from very remote distances are used for surveillance, intelligence, and attack without risk to a pilot.

Disaster relief
Medical equipment and medicines can be delivered by drones when land transportation is impractical.

Search and rescue
Some drones are used in search-and-rescue missions. They can deliver equipment to inaccessible places.

Delivery
Courier companies have begun to trial drones for delivering packages weighing up to 4½ lbs (2 kg).

Underwater exploration
Most drones are aircraft, but the term is also used for unmanned underwater craft used for research purposes.

Space probes

Space probes are unmanned spacecraft sent to explore the solar system. They have visited every planet and a few smaller bodies, such as comets and moons, taking images and sending back data.

WHICH SPACE PROBE IS FARTHEST FROM EARTH?

Launched in 1977, Voyager 1 is the most distant human-made object, having traveled more than 13 billion miles (21 billion km) across space.

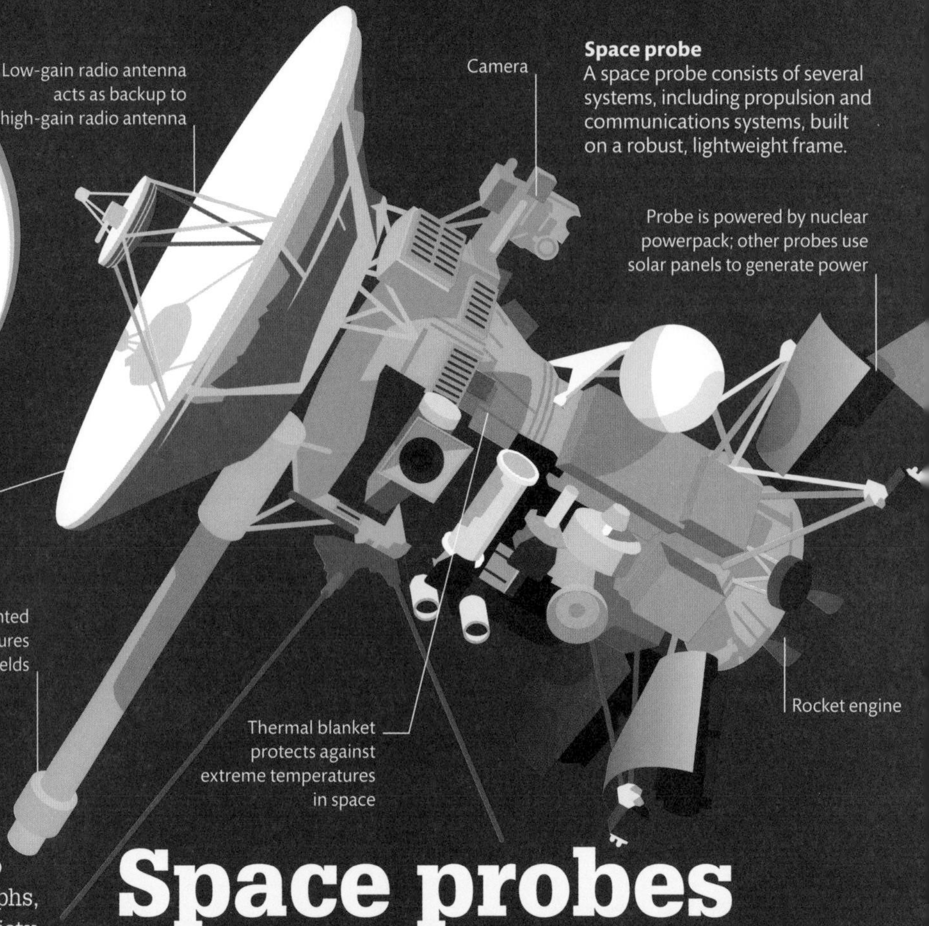
Low-gain radio antenna acts as backup to high-gain radio antenna
Camera
Space probe
A space probe consists of several systems, including propulsion and communications systems, built on a robust, lightweight frame.
Probe is powered by nuclear powerpack; other probes use solar panels to generate power
High-gain radio antenna sends and receives radio waves to and from Earth
Boom-mounted magnetometer measures magnetic fields
Thermal blanket protects against extreme temperatures in space
Rocket engine

Exploring space

A space probe's primary role is to transport scientific instruments to remote parts of the solar system. On arriving at its target, a probe may go into orbit. Its cameras take photographs, and its instruments record a variety of measurements, including magnetic field strength, radiation and dust levels, and temperature. Data is returned to Earth by using radio waves (see pp.180–181).

Types of space probes

There are several types of space probes. Flybys pass planetary bodies and study them from a distance, while orbiters travel in circles around them. Some probes send mini-probes into an object's atmosphere; others send vehicles to land on the surface. Probes can also carry rovers, which are able to move across an object's surface.

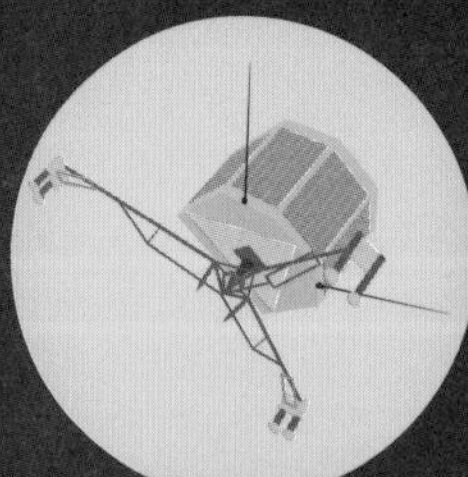

Lander
A lander is designed to descend from a space probe and reach the surface of a planetary object. It remains stationary and sends information back to Earth.

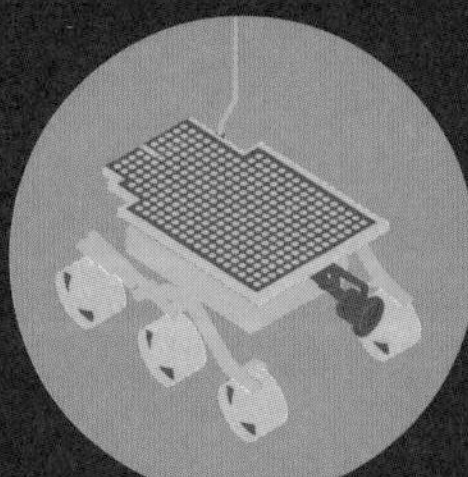

Rover
Unlike landers, rovers are built to travel across the surface of a planetary object. They can be autonomous or semiautonomous.

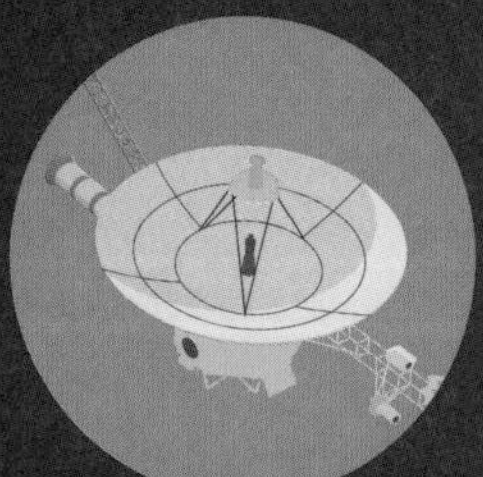

Flyby
Flyby probes fly close to a planet or other body and collect data. They remain far enough away that they are not captured by the gravitational pull.

RADIOISOTOPE THERMOELECTRIC GENERATOR

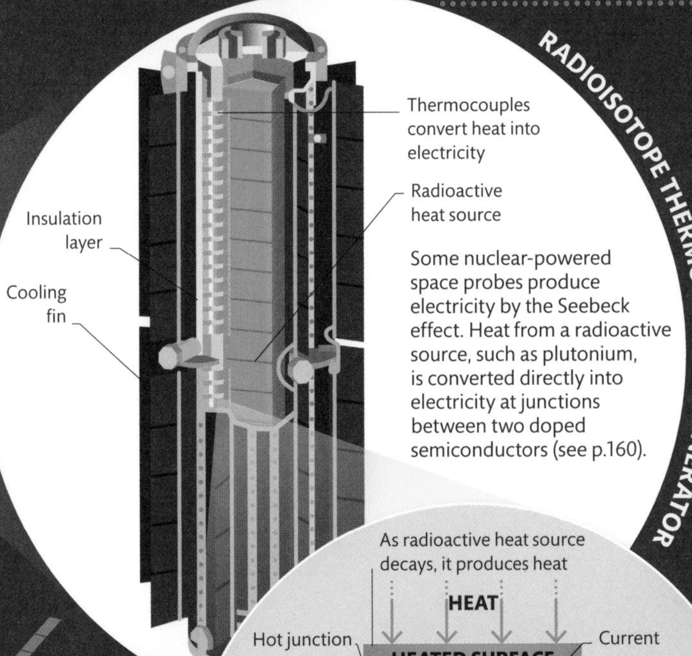
Thermocouples convert heat into electricity
Radioactive heat source
Insulation layer
Cooling fin

Some nuclear-powered space probes produce electricity by the Seebeck effect. Heat from a radioactive source, such as plutonium, is converted directly into electricity at junctions between two doped semiconductors (see p.160).

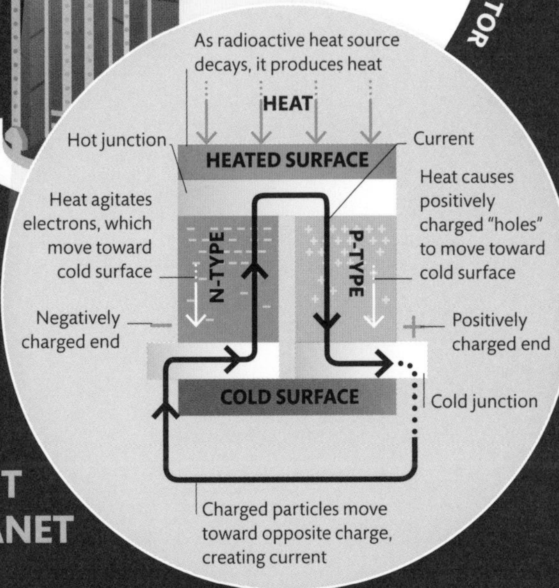
As radioactive heat source decays, it produces heat
HEAT
Hot junction
HEATED SURFACE
Current
Heat agitates electrons, which move toward cold surface
Heat causes positively charged "holes" to move toward cold surface
N-TYPE
P-TYPE
Negatively charged end
Positively charged end
COLD SURFACE
Cold junction
Charged particles move toward opposite charge, creating current

IN 1962, **MARINER 2** BECAME THE **FIRST SPACE PROBE** TO **VISIT ANOTHER PLANET**

SPACECRAFT PROPULSION SYSTEMS
Chemical rockets Rockets burning chemical propellants (see p.61) provide the huge thrust needed to launch probes and make directional corrections and changes in orbits. Gas thrusters are fired to make smaller positional changes.
Ion drive An ion drive, also known as an ion thruster, uses electricity to accelerate small amounts of electrically charged particles (called ions) into space, generating thrust. Ion drives require fuel to generate electricity.
Photon sail A photon sail, or solar sail, requires no fuel. It uses the radiation pressure of sunlight acting on a giant, mirrorlike sail to propel a spacecraft. The photons in sunlight bounce off the sail, pushing it in the opposite direction.

Landing a probe

Landers use a variety of methods to land on a planet or other object. Typically, parachutes slow the craft as it descends through the atmosphere. Retrorockets slow the descent even more, and then inflated bags may also cushion the landing.

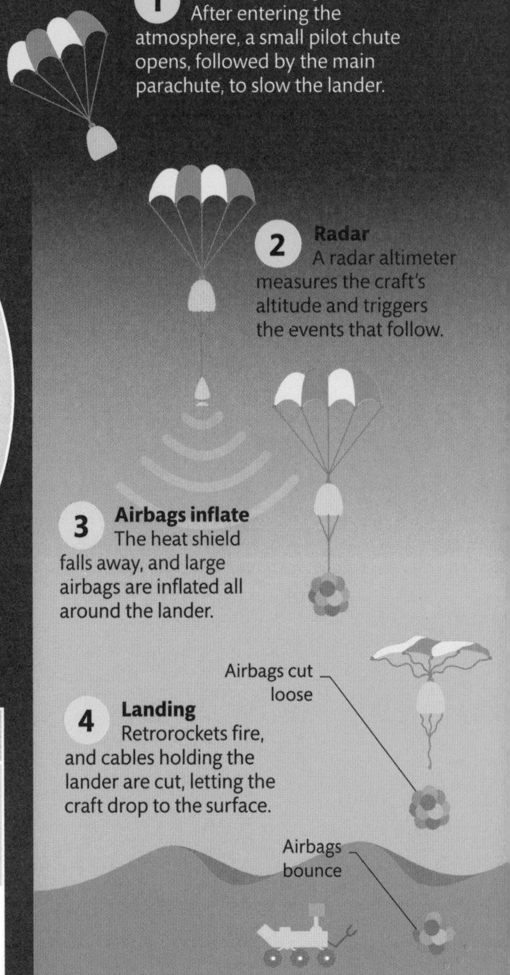
Airbags cut loose
Airbags bounce

1 Enters atmosphere
After entering the atmosphere, a small pilot chute opens, followed by the main parachute, to slow the lander.

2 Radar
A radar altimeter measures the craft's altitude and triggers the events that follow.

3 Airbags inflate
The heat shield falls away, and large airbags are inflated all around the lander.

4 Landing
Retrorockets fire, and cables holding the lander are cut, letting the craft drop to the surface.

5 On the surface
The lander touches down and bounces across the surface. When it comes to rest, the airbags deflate, and the lander turns itself upright. From entry to landing takes just a few minutes.

MATERIALS AND CONSTRUCTION TECHNOLOGY

Metals

We have been using metals, either in the form of pure elements or combined with other elements as alloys, for thousands of years to make all sorts of useful objects, from jewelry and cutlery to bridges and spacecraft.

The properties of metals

Metals tend to be strong but malleable, are good conductors of heat and electricity, and have high melting points. However, pure metals tend to be too soft or brittle to be useful. Their properties can often be improved by combining them with other elements to form alloys. Most metals in everyday use are alloys, one of the most common being steel.

Shiny
Metals have many electrons on their surface that absorb and then reemit light, giving metals a shiny appearance.

Good heat conductors
Electrons in metals can move freely, so when they gain heat energy, they can pass it on quickly.

Strong
The atoms in metals are arranged in a regular pattern and strongly bonded together. This makes metals strong.

Good electrical conductors
Because electrons in metals can carry electrical charge and move freely, electric currents flow through them easily.

High melting point
The strong bonds between atoms in a metal mean that it takes a lot of heat to free the atoms and melt the metal.

Malleable
The molecular structure of metals allows layers of atoms to slide so that the metal is malleable and can be easily shaped.

Steelmaking

Basic steel is an alloy of iron and a small amount of carbon (if the carbon content is more than about 2 percent, the alloy is known as cast iron). There are two main processes for making steel. The primary method uses a basic oxygen furnace (BOF) to produce steel from iron made in a blast furnace. The other method uses an electric arc furnace (EAF), which utilizes scrap steel. The basic steel may subsequently be refined into higher grades by adding alloying elements.

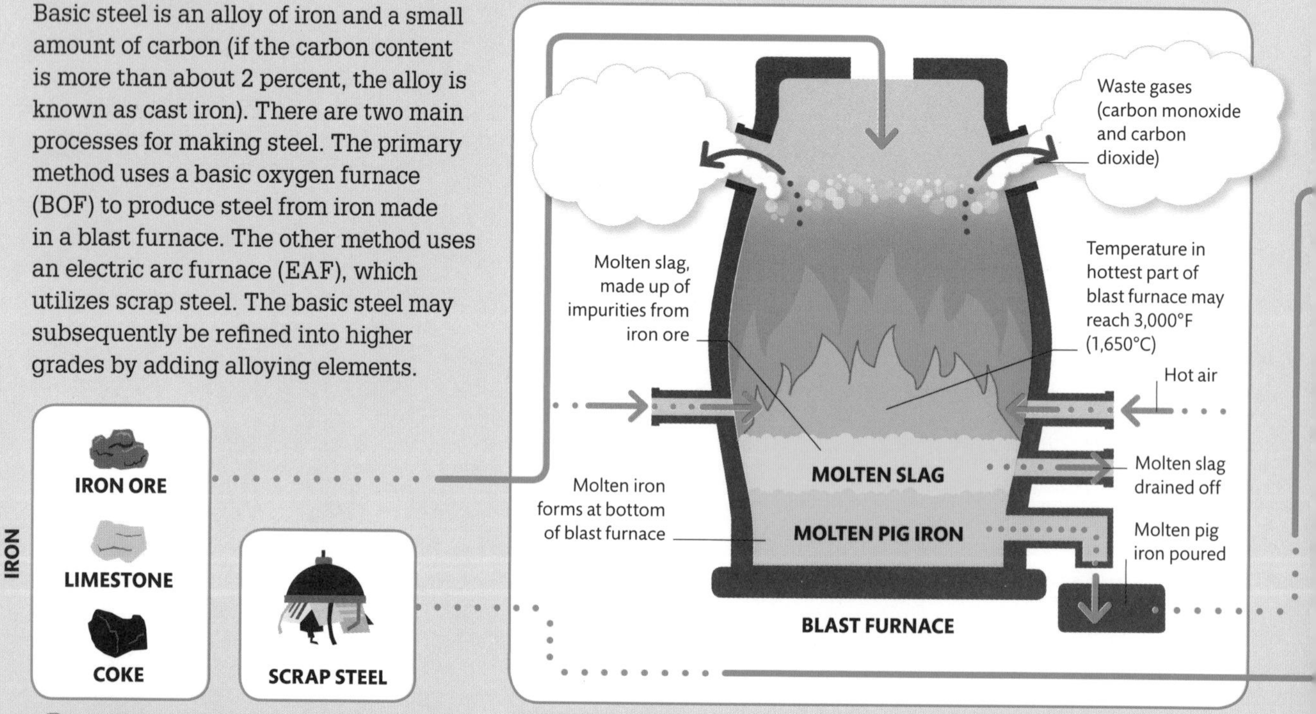

1 Raw materials
The raw materials for making iron are iron ore (iron oxide plus impurities), limestone (calcium carbonate), and coke (carbon). Steel is produced using iron from a blast furnace, sometimes with scrap steel added, or directly from scrap steel alone.

2 Making iron
In a blast furnace, coke reacts with hot air to produce carbon monoxide, which then reacts with the iron ore to produce pig iron (iron with a high carbon content). The limestone removes most of the impurities from the iron ore. The impurities form a molten slag on top of the molten pig iron.

COMMON ALLOYS

Bronze
The first man-made alloy, bronze was produced around 5,000 years ago by smelting copper and tin together. Bronze is resistant to atmospheric corrosion and is extremely strong.

Sterling silver
Sterling silver is an alloy consisting of 92.5 percent silver and 7.5 percent other metals, most commonly copper. These other metals make sterling silver harder and stronger than pure silver.

Solder
Traditionally, solder was an alloy of tin and lead, but modern solder usually consists of tin, copper, and silver and typically melts between about 355°F (180°C) and 375°F (190°C).

Cast iron
Cast iron is an alloy of iron and carbon, with a carbon content greater than about 2 percent. It is easy to cast and has good corrosion resistance and excellent compressive strength.

Brass
An alloy of copper and zinc, brass has a relatively low melting point (about 1,650°F/ 900°C), which makes it easy to cast. Brass is durable, more malleable than bronze, and has a bright, goldlike finish.

Stainless steel
Stainless steel varies in composition but commonly consists of 74 percent iron, 18 percent chromium, and 8 percent nickel. The chromium makes the alloy resistant to corrosion.

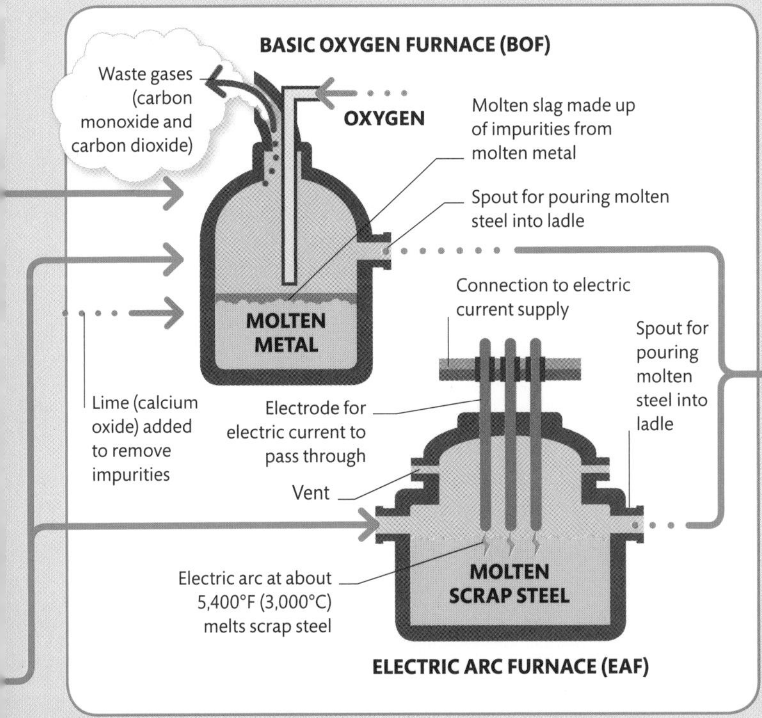

PURE IRON IS JUST SOFT ENOUGH TO CUT WITH A SHARP KNIFE

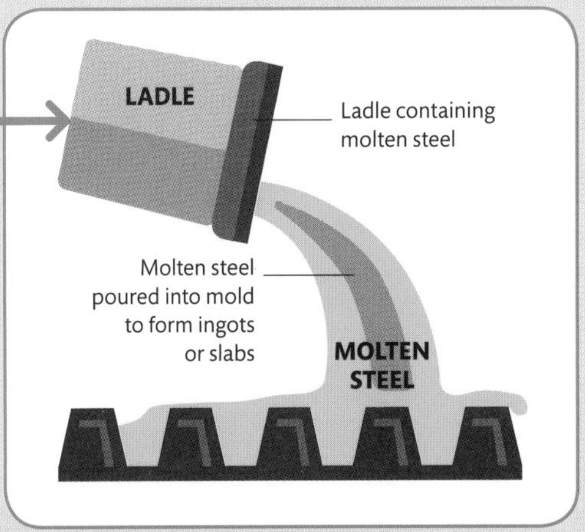

3 Making molten steel
In a basic oxygen furnace, oxygen is blown into molten pig iron, which reduces the iron's carbon content and produces steel. Lime is also added to remove impurities, which form a molten layer of slag. Sometimes, scrap steel may also be added. In an electric arc furnace, scrap steel is simply melted.

4 Casting or rolling molten steel
The molten steel may be poured into a ladle and then into a mold or put through rollers to shape it. This basic steel may be made into finished products, or it may be reprocessed by adding alloying elements to produce high-grade or special steels.

Working with metals

Most metals are produced as simple ingots, sheets, or bars, which usually require shaping or joining to other items to make finished products. Metals may also need to be treated to improve their properties, for example, to make them easier to shape or more corrosion resistant.

Shaping metals

Metals have a crystalline structure that breaks down when heat is applied. The metal softens and then becomes molten, allowing it to be easily shaped. As the metal cools, it recrystallizes and becomes hard once more. Processes that take advantage of these transformations to shape metals are known as hot working and include casting, extruding, forging, and rolling. Metals can also be worked without the application of heat, by processes known as cold working. In these processes, change in the metal is brought about through mechanical stresses rather than heat.

HOT METHODS

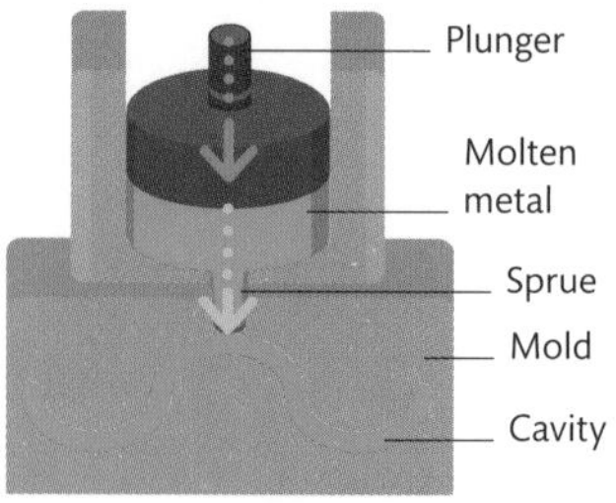

Casting
Molten metal is poured through a channel (called a sprue) into a mold. Once the metal has cooled, it can be extracted. Casting is typically used for complex 3-D shapes.

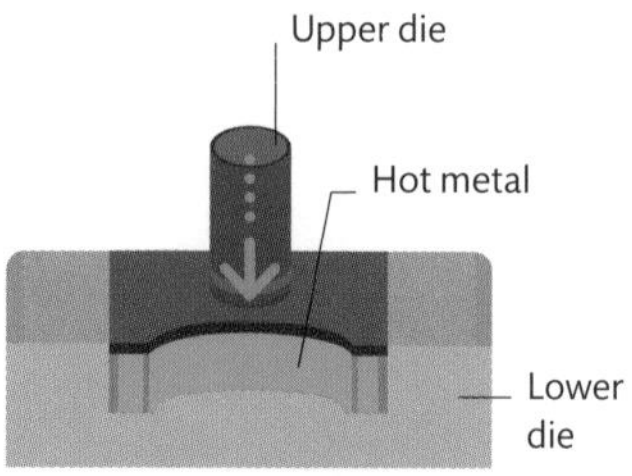

Forging
Forging replaces the blacksmith's hammer and anvil with modern machinery. The hot metal is compressed to the desired shape between two shaped dies—one die is fixed, and one is movable.

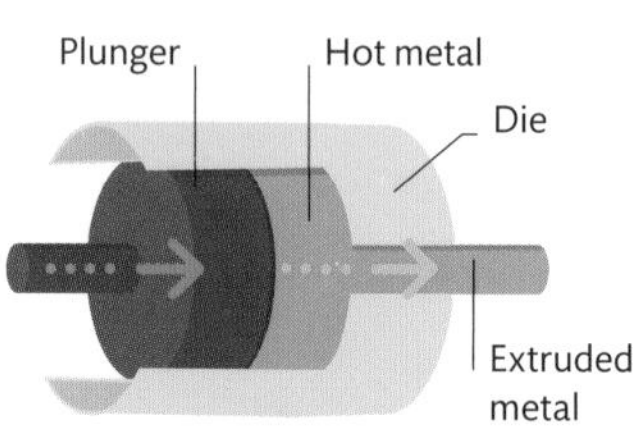

Extruding
The metal is softened with heat and then pushed through a die. Extrusion is used to create uniform cross sections, typically simple shapes such as rods or pipes.

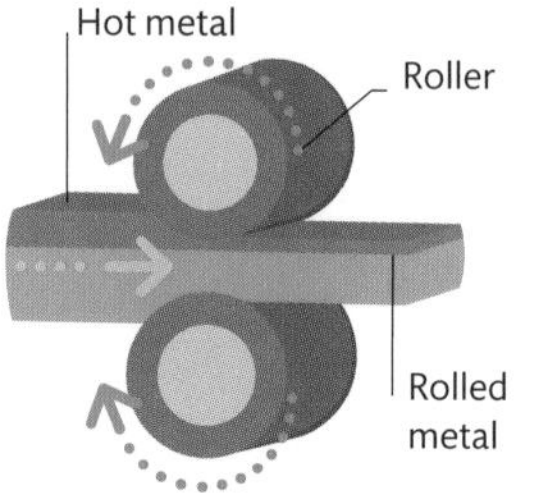

Rolling
In this process, a slab of hot metal is fed through rollers to reduce its thickness. Rolling is used to create sheet metal and other structural components.

Joining metals

The main methods for joining metals are soldering, welding, and riveting. Soldering and welding rely on the principle that metals become molten and pliable when heated and then return to a hardened state on cooling. Soldering forms the weakest bond, because it uses a soft metal with a lower melting point to act as the "glue." In welding, the two metals to be joined are melted and fused together, creating an exceptionally strong bond. Riveting also creates a strong bond and has a higher tolerance for thermal expansion and contraction. It is also cheaper than welding. However, riveting is less aesthetically pleasing than welding and so is usually used on internal or industrial structures.

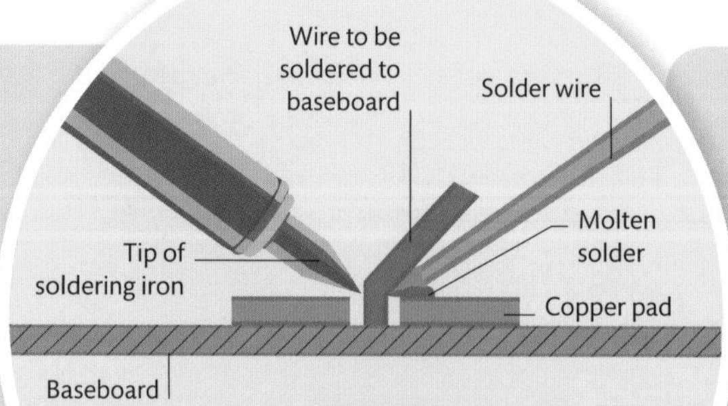

Soldering
Soldering is commonly used to create bonds in electronic equipment. Soft solder is melted so that it flows into the space between two pieces of metal, joining them together when it cools.

COLD METHODS

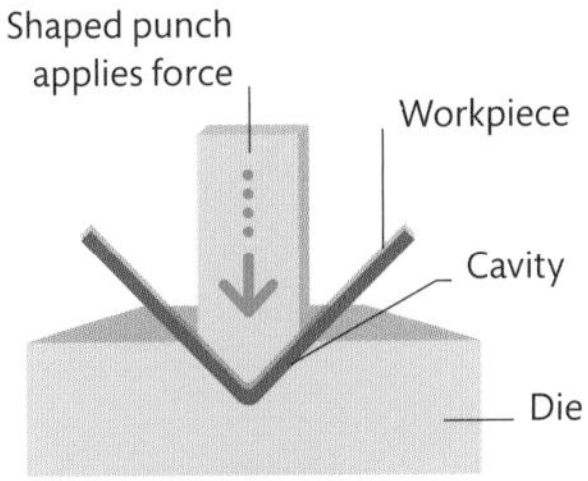

Bending
Many products are made by a process that is also known as cold forging, in which pressure is applied to force a metal workpiece into a cavity to achieve the desired shape.

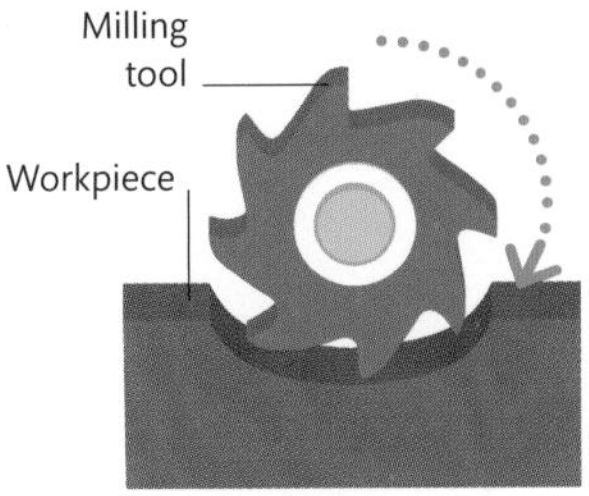

Milling
A milling machine shapes the workpiece by wearing away the excess parts with a milling bit. During the process, the machine sprays both bit and metal with a coolant.

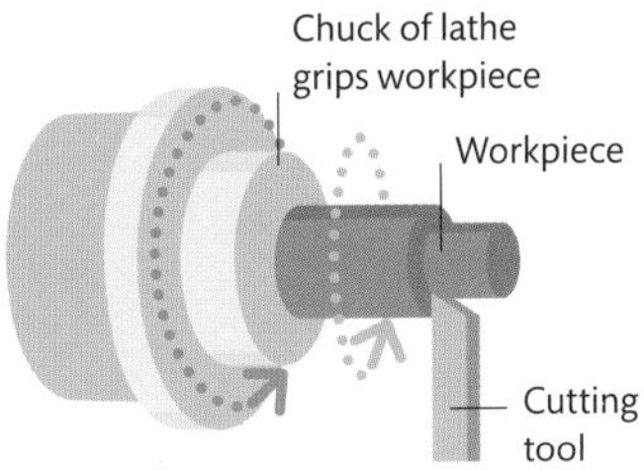

Turning
The workpiece is shaped by a fixed cutting tool while being rotated by a lathe. Turning can produce only objects that are symmetrical around the axis of rotation.

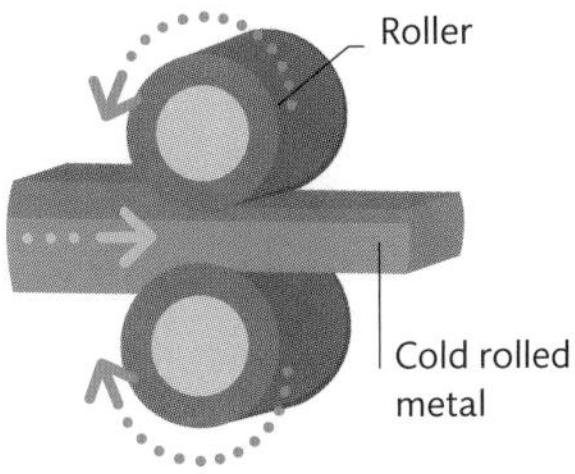

Rolling
The metal is shaped by rollers. Sheets, strips, bars, and rods are cold rolled to obtain products that have smooth surfaces and accurate dimensions.

5,700°F

(3,150°C)—THE TEMPERATURE IN THE **FLAMES** OF SOME **OXYACETYLENE TORCHES**

TREATING METALS

Metals can be treated in different ways to adapt their properties. Some common treatments aim to make the metal less brittle, while others prevent rust and corrosion.

Tempering
The metal is heated to a specific temperature and allowed to gradually cool. The process removes hardness and increases the toughness.

Anodizing
Metal is submerged into an electrolytic solution through which a current is passed. This forms a metal oxide film that increases corrosion resistance.

Galvanizing
The metal is submerged into a bath of molten zinc, resulting in a protective zinc coating that prevents rusting.

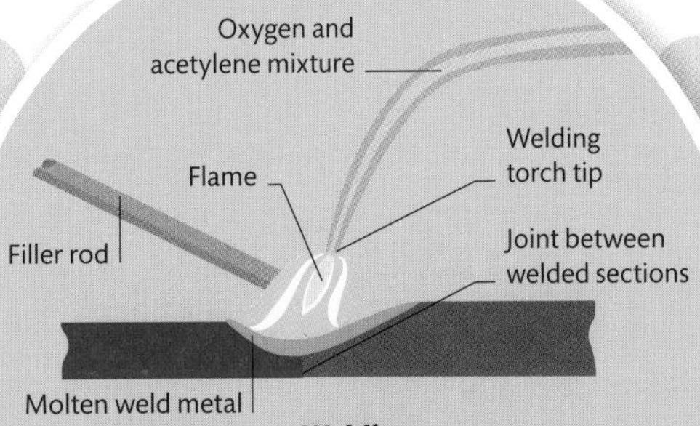

Welding
In welding, the two pieces of metal are heated by a localized heat source and melted along the joint. A filler with a similar melting point can be used to strengthen the joint.

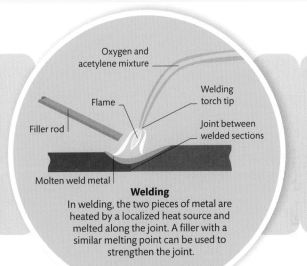

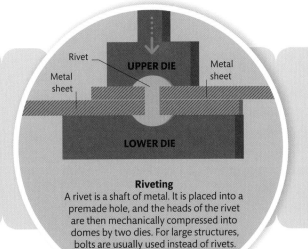

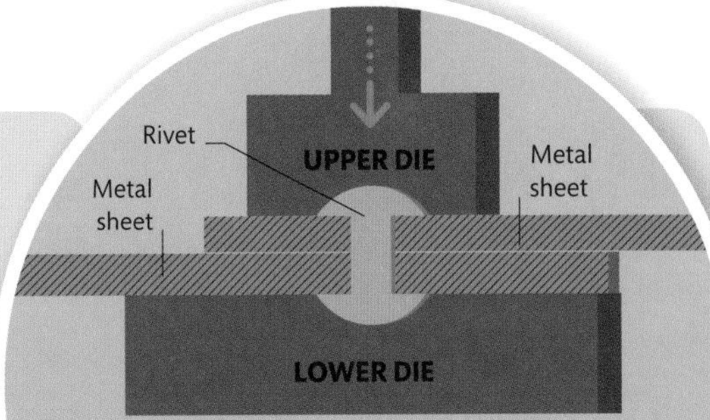

Riveting
A rivet is a shaft of metal. It is placed into a premade hole, and the heads of the rivet are then mechanically compressed into domes by two dies. For large structures, bolts are usually used instead of rivets.

Concrete

Essentially a type of artificial stone, concrete is one of the most versatile and commonly used building materials. It is relatively inexpensive and easy to produce, and its properties are useful for construction. Concrete is strong (especially under compression), durable, resistant to fire, corrosion, and decay, requires relatively little maintenance, and can be molded or cast into almost any shape.

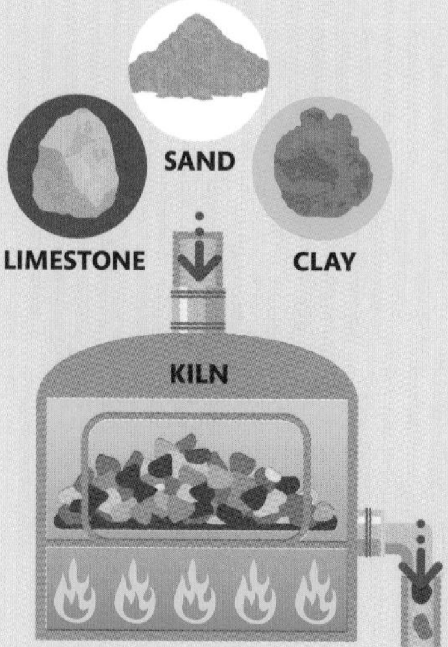

1 Raw materials of cement
Cement is one of the two key ingredients of concrete. A fine, powdery substance, it is made from limestone, sand, and clay.

2 Raw materials heated
The raw materials are heated in a kiln at about 2,600–2,900°F (1,400–1,600°C) to form lumps of a rock-hard substance known as clinker.

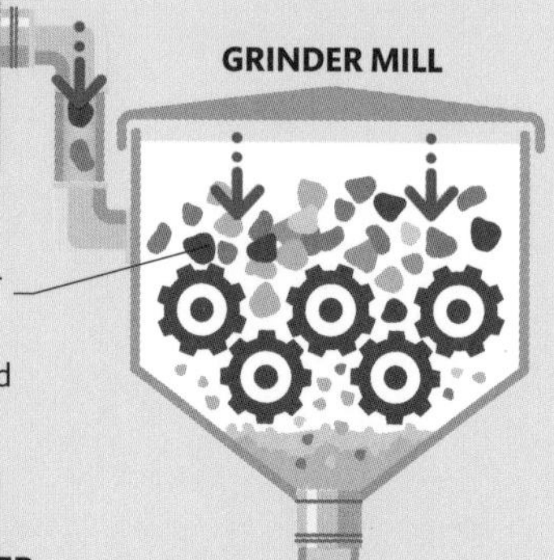

3 Cement produced
The clinker is cooled and then ground in a mill until it has turned into a fine powder; this powder is dry cement.

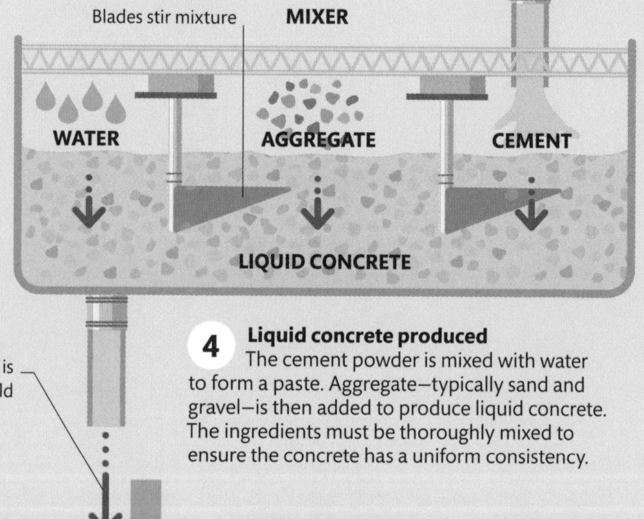

4 Liquid concrete produced
The cement powder is mixed with water to form a paste. Aggregate—typically sand and gravel—is then added to produce liquid concrete. The ingredients must be thoroughly mixed to ensure the concrete has a uniform consistency.

Making concrete

Concrete is a composite material consisting of a binder and a filler. The binder is a paste made of cement and water; the filler consists of aggregate—hard, particulate matter, such as sand, gravel, slag from steelmaking (see pp.72–73), or recycled glass. Typically, concrete consists of about 60–75 percent aggregate, 7–15 percent cement, 14–21 percent water, and up to 8 percent air.

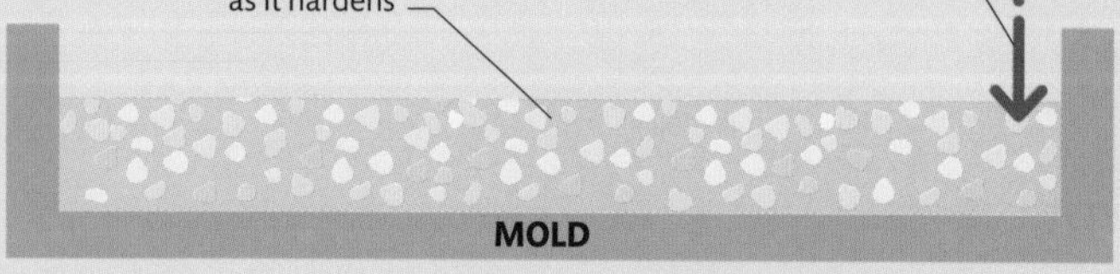

5 Concrete molded
The liquid concrete is poured into a mold, vibrated to remove air bubbles, and left to cure (harden). Curing is a chemical reaction between the cement and water rather than a drying-out process. The reaction generates heat, and the concrete becomes stronger as curing progresses.

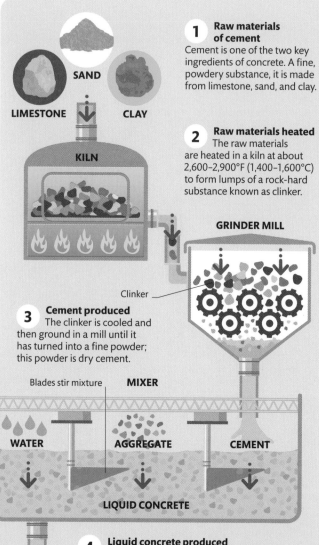

Strengthening concrete

Large concrete structures often use concrete reinforced with steel mesh or bars (called rebars) to increase the strength of the concrete. Concrete can be made even stronger by prestressing—putting the rebars under tension while the concrete is hardening.

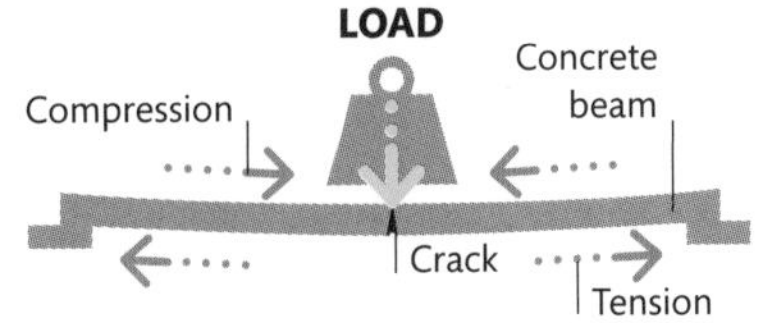

Unreinforced concrete
Concrete is strong under compression but relatively weak under tension. A heavy load can cause concrete to bend and crack.

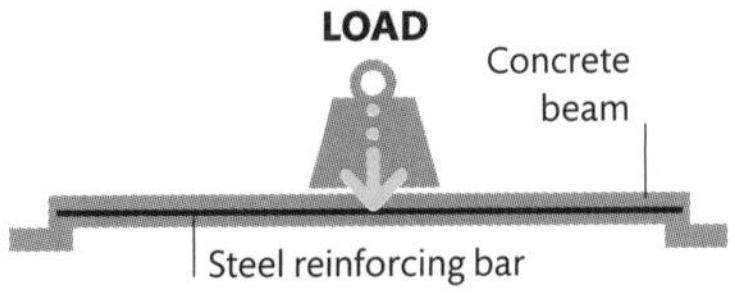

Steel-reinforced concrete
Placing a steel bar inside concrete helps to prevent it from bending and cracking under heavy loads.

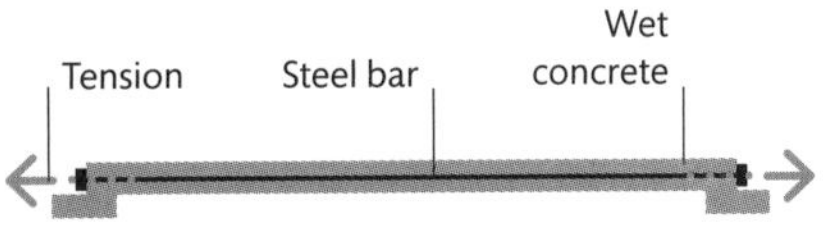

Forming prestressed concrete
Concrete is poured around a steel bar that is under tension. As the concrete sets, it bonds to the bar.

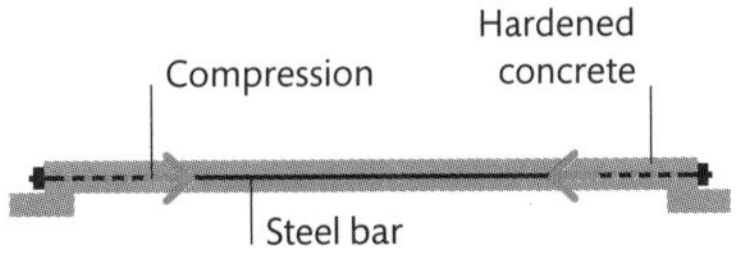

Hardened prestressed concrete
When the concrete has set, tension on the bar is released. The bar compresses the concrete, making it stronger.

WHAT IS CONCRETE CANCER?

Concrete cancer is staining, cracking, and eventual breaking of reinforced concrete when rust expands the steel inside the concrete, destroying the concrete from within.

THE **ANCIENT ROMANS** USED **VOLCANIC ASH,** CALLED **POZZOLANA,** TO MAKE **CONCRETE**

MASSIVE CONCRETE STRUCTURES

Many of the world's largest man-made structures are made of concrete. The most massive is the Three Gorges Dam in China, made of more than 72 million tons (65 million metric tons) of concrete. The Petronas Twin Towers is the most massive concrete building.

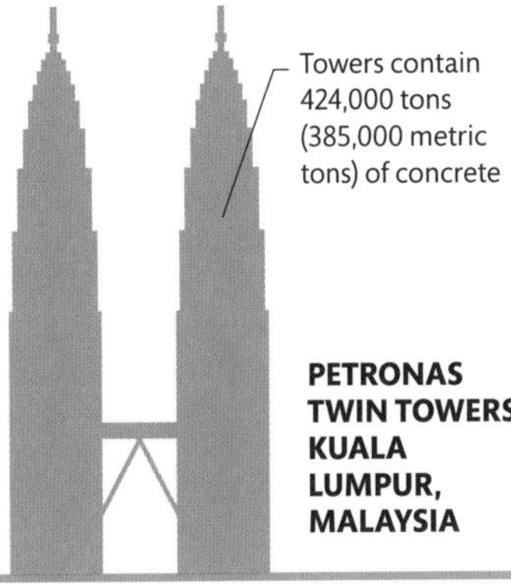

PETRONAS TWIN TOWERS, KUALA LUMPUR, MALAYSIA

TYPES OF CONCRETE	
Type	**Characteristics**
Precast concrete	Unlike standard concrete, which is cast and cured on-site, precast concrete is cast and cured elsewhere then transported to the construction site and lifted into place.
Heavyweight concrete	Using special aggregates, such as iron, lead, or barium sulphate, heavyweight concrete is much denser than normal concrete and is mainly used for radiation shielding.
Shotcrete	Shotcrete is concrete that is applied by high-pressure spraying, usually onto a steel mesh framework. It is often used for artificial rock walls, tunnel linings, and pools.
Pervious concrete	Pervious concrete is made with coarse particles of aggregate, which makes the concrete porous, allowing water to drain through it.
Rapid-strength concrete	This type of concrete contains additives, such as calcium chloride, to speed up curing so that the concrete becomes strong and hard enough to bear loads within a few hours.
Glass concrete	Glass concrete uses recycled glass as the aggregate. It is stronger and provides better thermal insulation than standard concrete and resembles marble in appearance.

Plastics

Plastics are synthetic materials made of polymers—long-chain molecules consisting of repeating units called monomers. Due to their low cost, ease of manufacture, and versatility, plastics are one of the most widely used types of materials in the modern world.

Types of plastic

There are two main types of plastic. Thermoplastics are easy to melt and recycle. Examples include polyethylene, polystyrene, and polyvinyl chloride (PVC). Thermosetting plastics are hardened by heat and cannot be remelted. Less commonly used than thermoplastics, thermosetting plastics include polyurethane, melamine, and epoxy resins.

Thermoplastics
In a thermoplastic, long polymer molecules are joined to one another by weak bonds that easily break apart when the plastic is heated and quickly reform again as it cools.

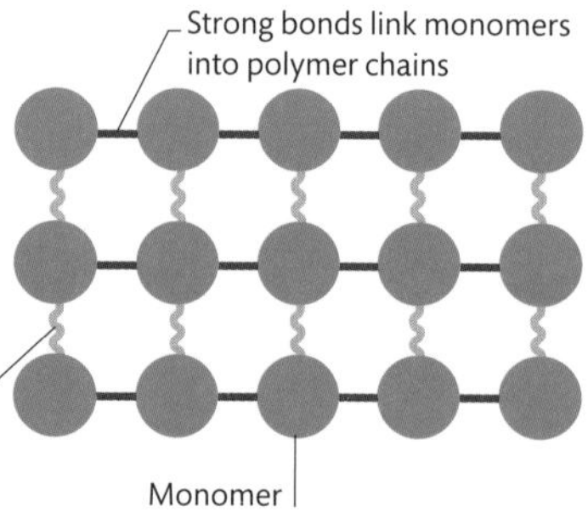

Strong bonds link monomers into polymer chains
Weak attractive force between monomers
Monomer

Thermosetting plastics
Thermosetting plastics have strong cross-link bonds binding the polymer chains. The plastics are soft at low temperatures and then permanently set (hardened) by the application of heat.

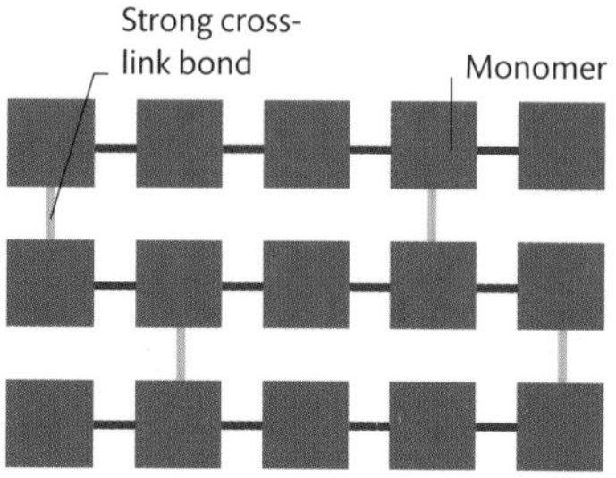

Strong cross-link bond
Monomer

Making polyethylene
Polyethylene is made by the polymerization of ethylene, a colorless hydrocarbon that is gaseous at room temperature and which is obtained from oil. Polyethylene is manufactured in two main forms: low-density polyethylene (LDPE), used for plastic bags and sheets, and high-density polyethylene (HDPE), used to produce harder plastics. The process shown here, known as the slurry process, is used to produce high-density polyethylene.

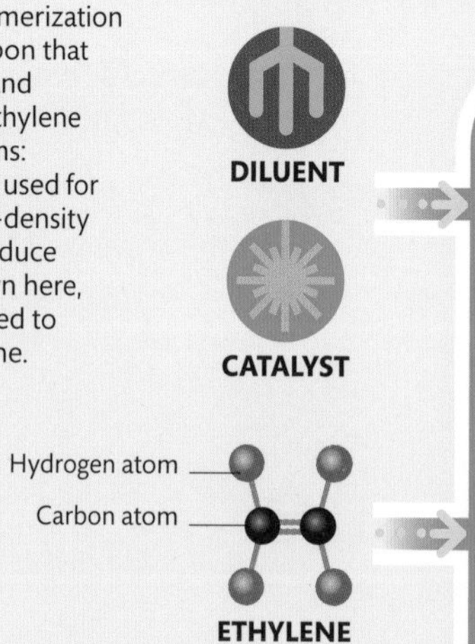

DILUENT
CATALYST
Hydrogen atom
Carbon atom
ETHYLENE

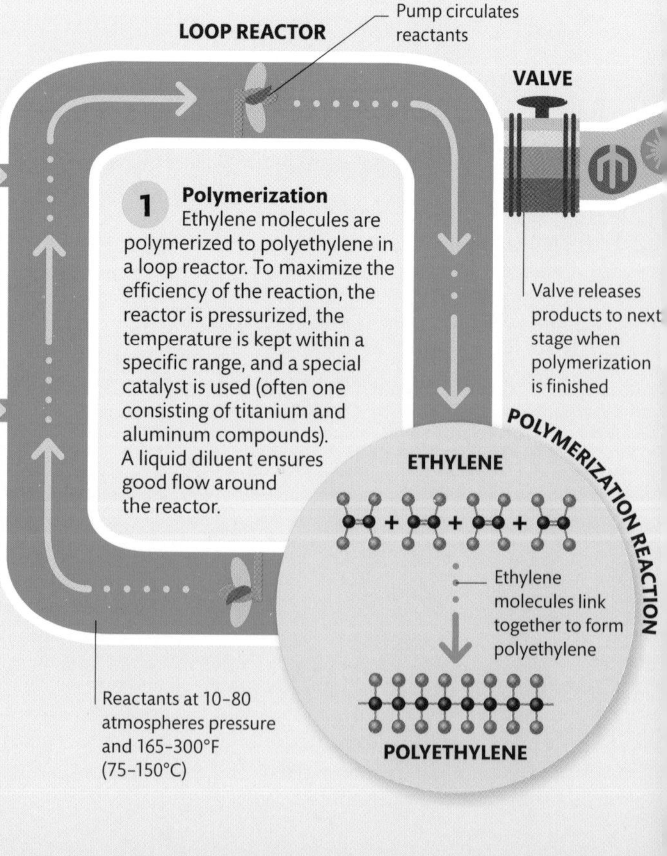

LOOP REACTOR
Pump circulates reactants
VALVE
1 Polymerization
Ethylene molecules are polymerized to polyethylene in a loop reactor. To maximize the efficiency of the reaction, the reactor is pressurized, the temperature is kept within a specific range, and a special catalyst is used (often one consisting of titanium and aluminum compounds). A liquid diluent ensures good flow around the reactor.
Valve releases products to next stage when polymerization is finished
POLYMERIZATION REACTION
ETHYLENE
Ethylene molecules link together to form polyethylene
POLYETHYLENE
Reactants at 10–80 atmospheres pressure and 165–300°F (75–150°C)

Making plastics

Most plastics are made from petrochemicals obtained from crude oil by fractional distillation (see pp.14–15). These petrochemicals are processed to make monomers such as ethylene (also known as ethene), which are then polymerized. In polymerization, monomers react together to form long polymer chains. Other chemicals can be added to the polymers to change their properties. The process results in polymer resins, which can then be shaped into various products.

WHAT WAS THE FIRST PLASTIC?

The first plastic was Parkesine, invented in 1856 and named after its creator, Alexander Parkes. Now better known as celluloid, Parkesine was initially used to make billiard balls.

500 BILLION

THE **NUMBER OF PLASTIC BAGS** USED WORLDWIDE **EVERY YEAR**

COMMON TYPES OF PLASTIC	
Name	**Characteristics**
PET Polyethylene terephthalate	The most common type of plastic, PET comes in soft forms, used to make fibers for clothing, and harder forms used to make items such as beverage bottles.
PVC Polyvinyl chloride	Rigid and strong, PVC is used to make credit cards and in construction for pipes and door and window frames. In a softer form, it is a substitute for leather and rubber.
PP Polypropylene	Similar to PET but harder and more heat resistant, polypropylene is the second most widely used plastic, often used in packaging, including microwavable meal trays and bottle caps.
PC Polycarbonate	Polycarbonates are tough, and some grades are transparent. It is used for CDs and DVDs, sunglasses and safety goggles, and in construction for dome lights and flat or curved glazing.
PS Polystyrene	Polystyrene can be clear, hard, and brittle, often used for cases of small items. It can also be filled with tiny gas bubbles to make the light foam used for egg cartons and single-use cups.

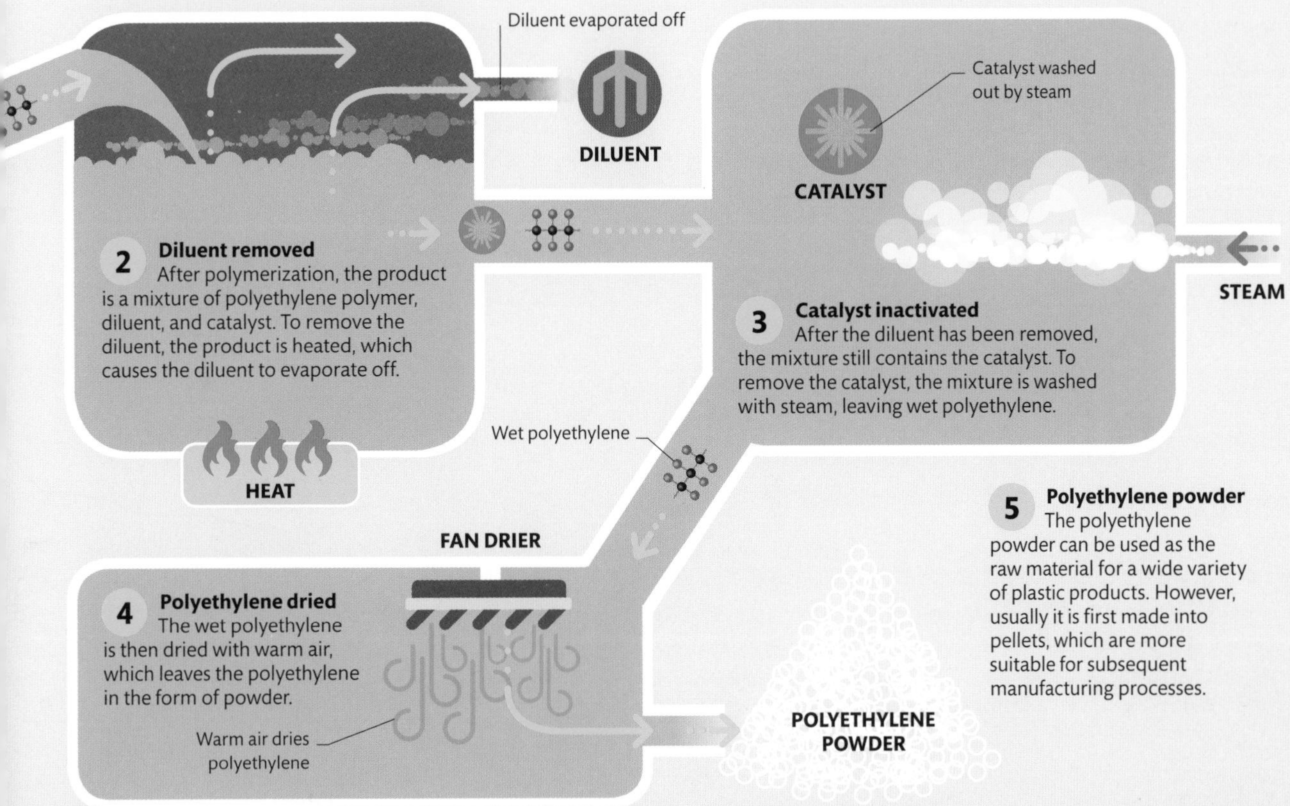

2 Diluent removed
After polymerization, the product is a mixture of polyethylene polymer, diluent, and catalyst. To remove the diluent, the product is heated, which causes the diluent to evaporate off.

3 Catalyst inactivated
After the diluent has been removed, the mixture still contains the catalyst. To remove the catalyst, the mixture is washed with steam, leaving wet polyethylene.

4 Polyethylene dried
The wet polyethylene is then dried with warm air, which leaves the polyethylene in the form of powder.

5 Polyethylene powder
The polyethylene powder can be used as the raw material for a wide variety of plastic products. However, usually it is first made into pellets, which are more suitable for subsequent manufacturing processes.

Composites

A composite material comprises two or more materials that, when combined, have superior qualities to the originals. Many modern synthetic composites are made to be strong but light.

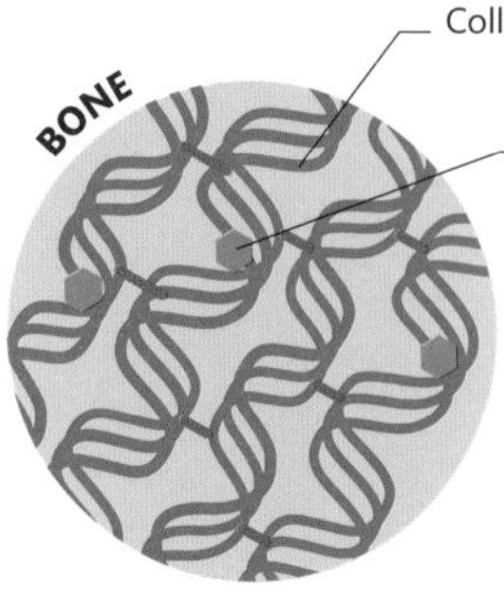

Structure of bone
Bone is made from a hard but brittle material called hydroxyapatite (which is mainly calcium phosphate) and a soft, flexible material called collagen (which is a protein).

Natural composites

Almost all the materials we see around us are composites, including many natural materials, such as wood and rock. Our bodies contain composite materials, most notably our bones and teeth, which both feature a combination of a hard outer layer and a soft inner layer. Mud bricks and the wattle-and-daub wall construction method are examples of simple combinations of basic natural materials forming composites used for their superior strength.

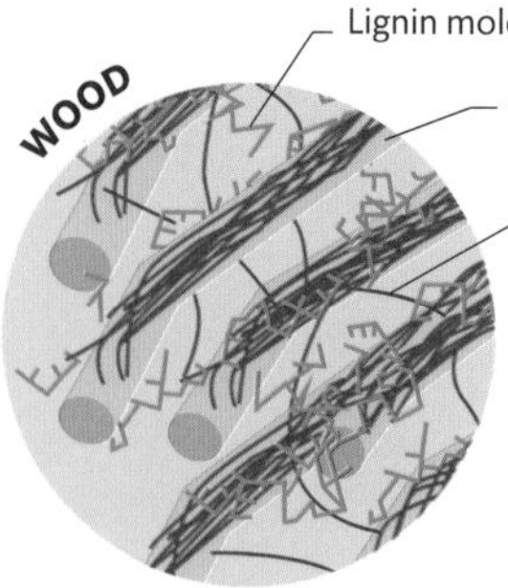

Structure of wood
Wood is made from long cellulose fibers held together by other materials. When combined, these weak materials form a strong composite.

Synthetic composites

One of the first modern composites was fiberglass. It combines fine threads of glass with plastic. Advanced composites are now made out of carbon fibers instead of glass. These fibers are narrower than the width of a human hair. They are twisted together to form a yarn that is woven into cloth then molded together with a resin. The resulting composite material is extremely strong and light.

Making carbon fiber polymer
The part-chemical, part-mechanical process of making carbon fiber polymer involves a variety of gases and liquids. Exact compositions vary and are often considered trade secrets.

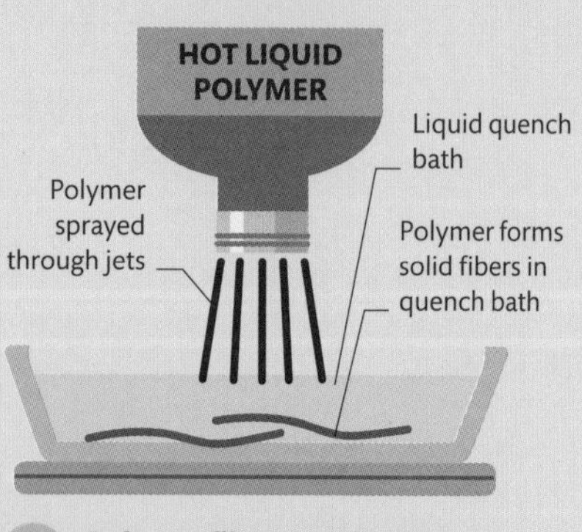

1 Polymer fibers produced
The raw material used to make carbon fiber is a polymer. About 90 percent of all carbon fiber is made from a polymer called polyacrylonitrile (PAN). In the first stage, the PAN is formed into long fibers.

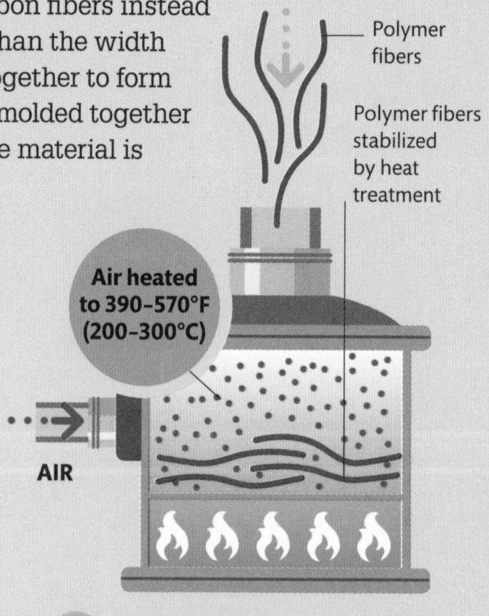

2 Fibers stabilized
Heat chemically alters the fibers, converting their atomic bonds to a form that is more thermally stable. Oxygen molecules in the introduced air facilitate this process.

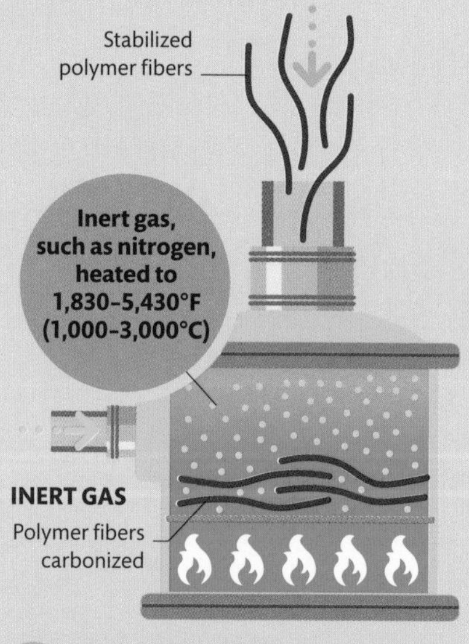

3 Fibers carbonized
The fibers are then heated to a far higher temperature in a furnace filled with an oxygen-free, inert gas to prevent the fibers from burning. As a result, the fibers lose their noncarbon atoms and become carbonized.

USES OF SYNTHETIC COMPOSITES

Breathable fabric
Traditional waterproof clothing traps sweat inside. Composite versions mix nylon with polytetrafluoroethylene (PTFE), which does not allow rainwater to pass through but lets water molecules from sweat escape.

Disc brakes
Some high-performance cars and heavy vehicles use disc brakes made from carbon-fiber-reinforced ceramics. Not only is the material lightweight and strong, but it also has exceptionally high heat tolerance.

Bicycle frames
The frames of most racing bicycles are made from a variety of carbon fibers of different types, each used in different places for very specific purposes. Carbon fiber is also used to make some other components, such as wheels and handlebars.

Boat hulls
Fiberglass has been widely used in boat-hull construction since the 1950s. At the leading edge of boat building, composites using aramid fibers—a particularly strong fiber used in aerospace—are used to reinforce key areas.

Kevlar
Kevlar is a composite fiber that is about five times stronger than steel. It can be woven into cloth to create bulletproof garments, used to create mooring lines, or added to polymers to form racing sails or linings for bicycle tires.

Reinforced concrete
One of the oldest and most common synthetic composites, concrete is a mix of cement, water, sand, and gravel (see pp.76–77). Its poor tensile strength can be improved by embedding steel bars within the concrete.

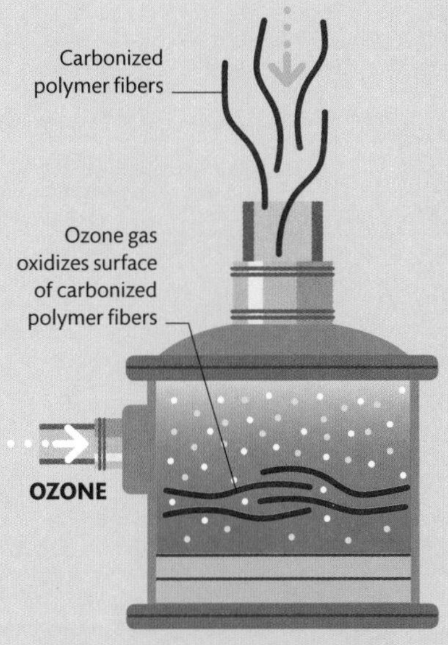
Carbonized polymer fibers
Ozone gas oxidizes surface of carbonized polymer fibers
OZONE

4 Surface of fibers oxidized
After carbonizing, the fibers have a surface that does not bond well. To improve the bonding properties, their surface is slightly oxidized by the addition of oxygen atoms from ozone.

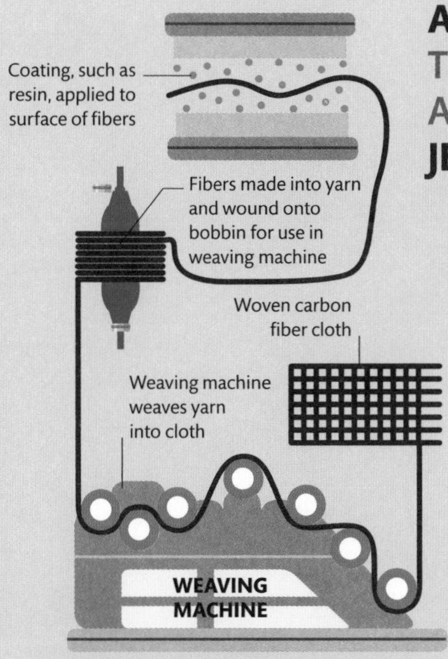
Coating, such as resin, applied to surface of fibers
Fibers made into yarn and wound onto bobbin for use in weaving machine
Woven carbon fiber cloth
Weaving machine weaves yarn into cloth
WEAVING MACHINE

5 Fibers coated and woven
After the surface treatment, the fibers are coated for protection and twisted together to form yarn. The yarn is wound onto bobbins, which are loaded onto weaving machines to produce cloth.

AROUND HALF OF THE MATERIALS IN A MODERN PASSENGER JET ARE COMPOSITES

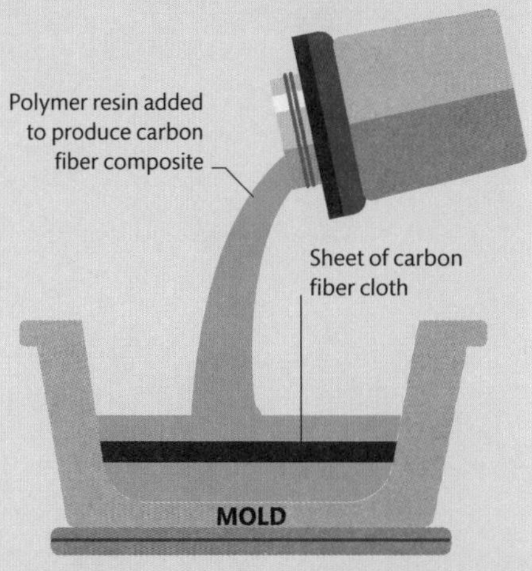
Polymer resin added to produce carbon fiber composite
Sheet of carbon fiber cloth
MOLD

6 Carbon fiber polymer produced
The carbon fiber cloth is delivered to manufacturers to complete the process to their individual requirements. This involves placing it into a mold and adding polymer resin to form the composite.

1 Manual sorting Often, mixed waste is sorted by hand to remove any nonrecyclables. These usually go to landfill or, if burnable, may be incinerated.

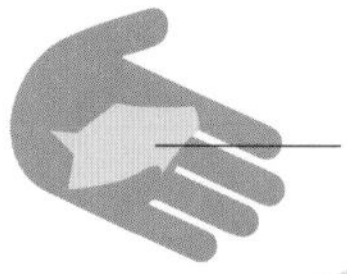
Large items and items consisting of a mixture of materials tend not to be suitable for recycling

NONRECYCLABLES

2 Paper and cardboard recovery Paper and cardboard are recovered by a screen system that separates them from heavier materials. They are sent to specialist plants for reprocessing into new items.

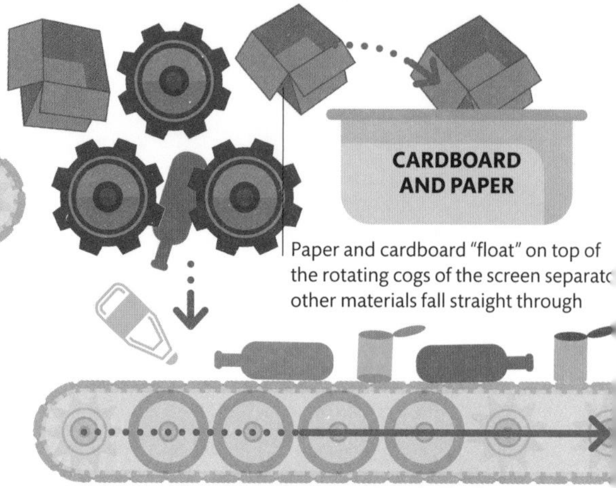
CARDBOARD AND PAPER
Paper and cardboard "float" on top of the rotating cogs of the screen separatc
other materials fall straight through

Materials' recovery

The sorting and cleaning of recyclables happens at a materials recovery facility (MRF). Through a combination of systems and processes that varies between MRFs, materials are recovered and forwarded to specialist plants for processing. Recyclable materials include paper and cardboard, which are made into new paper and card products, and glass, which is turned into new bottles and jars. Some items, such as electronics, that are complex and contain many different components are processed at specialist recycling facilities.

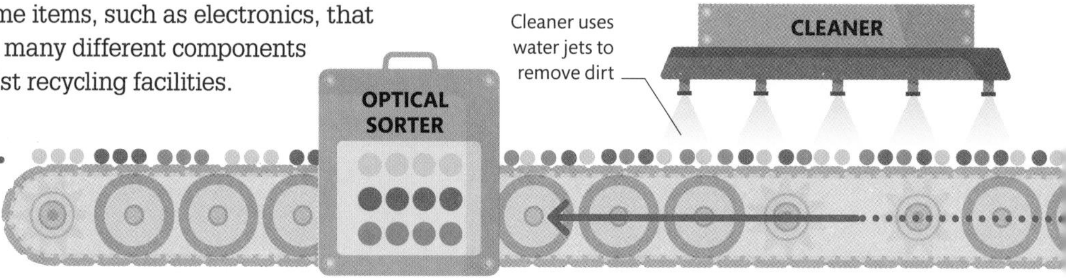
Cleaner uses water jets to remove dirt
CLEANER
OPTICAL SORTER

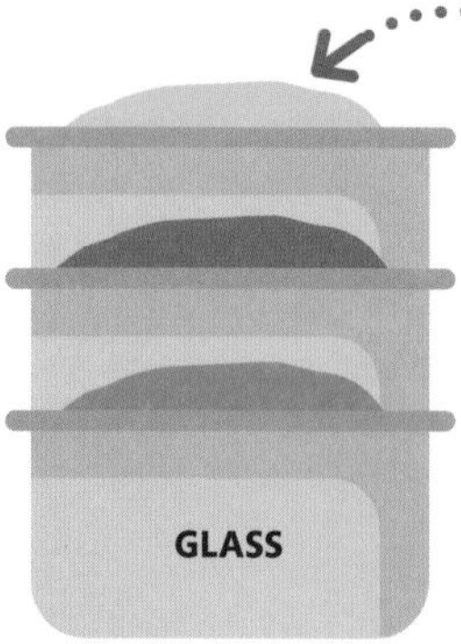
GLASS

9 Glass recovery Sorted glass may go on to be melted down and remolded into new bottles and jars or other glass items of uniform color.

8 Glass sorted Some glass recycling plants use advanced optical scanners to sort the glass fragments by color.

7 Glass cleaned The crushed glass is cleaned to remove any impurities. The cleaned glass may be sorted by color or used in products such as roadbed.

Recycling

Recycling is the process of collecting waste items and breaking them down into materials that can be turned into new products. A key part of the process is the sorting of items into their different materials, such as glass or plastic, so that they can be sent to the appropriate reprocessing facility.

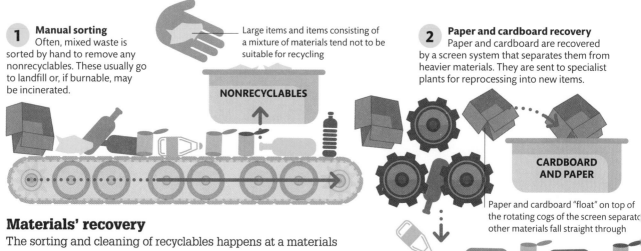

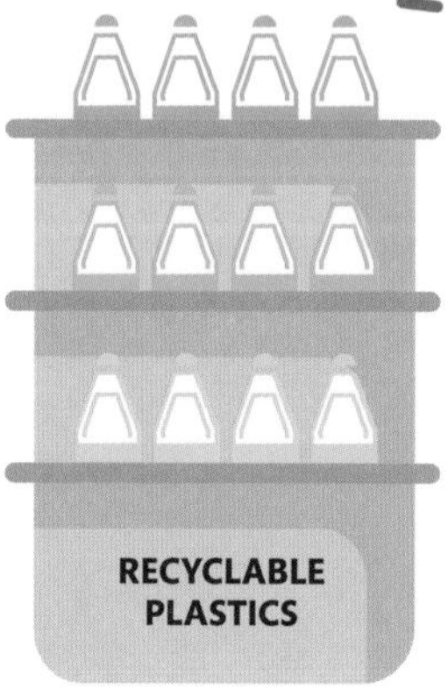
RECYCLABLE PLASTICS

11 Plastics' recovery Plastics such as polyethylene terephthalate (PET)—used in some plastic bottles—can be melted and reformed. Others must be mixed with other materials for reuse.

3 Ferrous metal recovery
Metal items that are high in iron (or ferrous), such as steel cans, are pulled out by powerful magnets and then sent to smelting plants, where they are melted into ingots.

How an eddy current separator works
An eddy current separator consists of several rotating magnets. These induce an electric current in nonferrous metals that pass over the separator, producing a temporary magnetic field in the metals. This magnetic field interacts with that of the separator, causing the metals to be repelled and thrown outward.

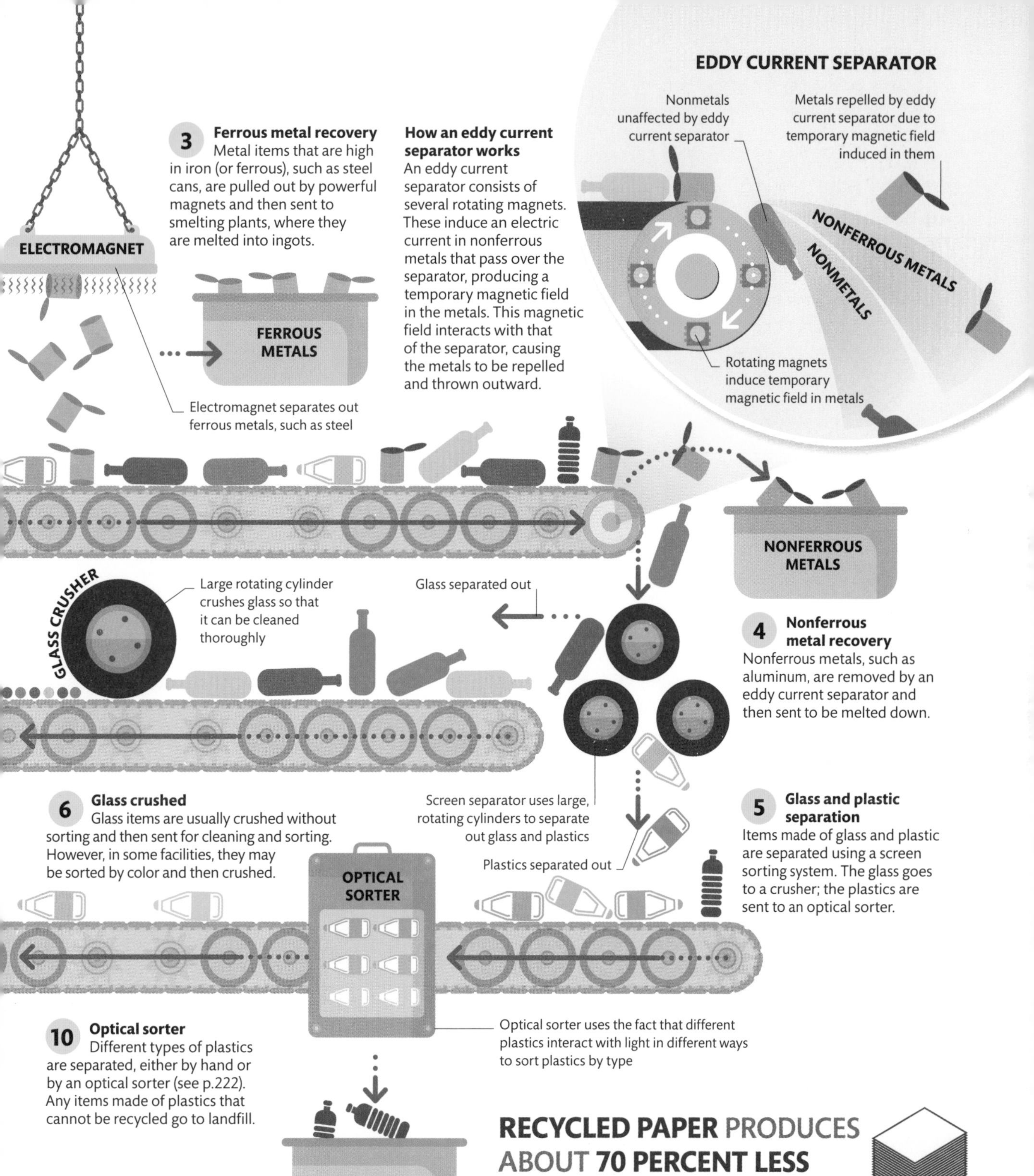
EDDY CURRENT SEPARATOR
Nonmetals unaffected by eddy current separator
Metals repelled by eddy current separator due to temporary magnetic field induced in them
NONFERROUS METALS
NONMETALS
Rotating magnets induce temporary magnetic field in metals
ELECTROMAGNET
FERROUS METALS
Electromagnet separates out ferrous metals, such as steel
NONFERROUS METALS
GLASS CRUSHER
Large rotating cylinder crushes glass so that it can be cleaned thoroughly
Glass separated out
Screen separator uses large, rotating cylinders to separate out glass and plastics
Plastics separated out
OPTICAL SORTER
Optical sorter uses the fact that different plastics interact with light in different ways to sort plastics by type
NONRECYCLABLE PLASTICS
Some types of plastics, such as certain thermosetting plastics, cannot be recycled

4 Nonferrous metal recovery
Nonferrous metals, such as aluminum, are removed by an eddy current separator and then sent to be melted down.

5 Glass and plastic separation
Items made of glass and plastic are separated using a screen sorting system. The glass goes to a crusher; the plastics are sent to an optical sorter.

6 Glass crushed
Glass items are usually crushed without sorting and then sent for cleaning and sorting. However, in some facilities, they may be sorted by color and then crushed.

10 Optical sorter
Different types of plastics are separated, either by hand or by an optical sorter (see p.222). Any items made of plastics that cannot be recycled go to landfill.

RECYCLED PAPER PRODUCES ABOUT **70 PERCENT LESS** AIR **POLLUTION** THAN PAPER MADE FROM RAW MATERIALS

Nanotechnology

Nanotechnology is the area of technology that involves creating and manipulating matter and objects at an extremely small scale, known as the nanoscale.

The nanoscale

Nanoscale objects measure between 1 and 100 nanometers (nm), where 1 nm is a billionth of a meter. Some molecules, such as glucose, antibodies (large protein molecules), and viruses are nanoscale objects.

Nanomaterials

A nanomaterial is any material or object with at least one dimension (length, width, or height) smaller than 100 nm. Some nanomaterials occur naturally—such as smoke particles, spider silk, and some butterfly wing scales—while others are deliberately created with unique properties. For example, gold nanoparticles can be engineered so that they emit a burst of heat when illuminated with light, a property that can be used to destroy cancer cells.

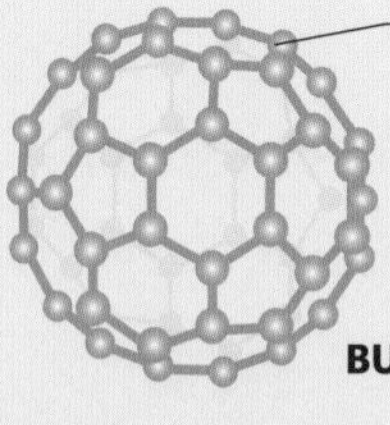
Hexagonal and pentagonal rings of carbon atoms arranged in a sphere
BUCKYBALL

Nanoparticles

A nanoparticle is an object with all three dimensions on the nanoscale. Many nanoparticles have unusual properties due to their size or shape; for example, the hollow structure of buckyballs means they could carry other molecules inside.

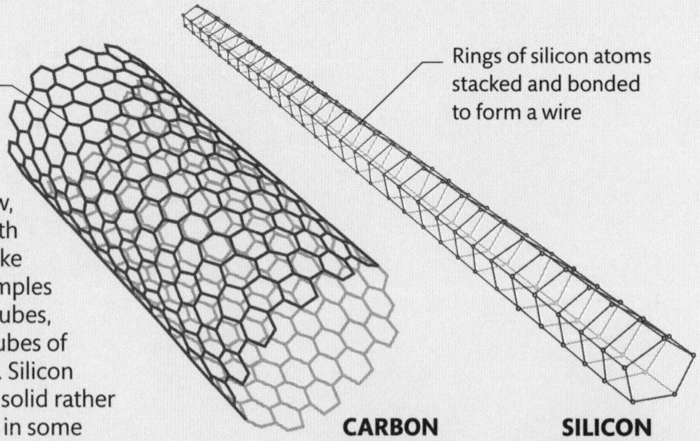
Hexagonal rings of carbon atoms rolled into a tube shape
Rings of silicon atoms stacked and bonded to form a wire
CARBON NANOTUBE
SILICON NANOWIRE

Nanotubes and nanowires

Nanotubes are narrow, tubelike structures with walls made of sheet-like lattices of atoms. Examples include carbon nanotubes, which are rolled-up tubes of graphene (see below). Silicon nanowires, which are solid rather than hollow, are used in some types of battery.

GRAPHENE

Graphene is a one-atom-thick layer of carbon atoms, arranged in a hexagonal (honeycomb) lattice. It is very stiff in all directions and is the strongest material ever tested. Graphene is also an excellent conductor of heat and electricity.

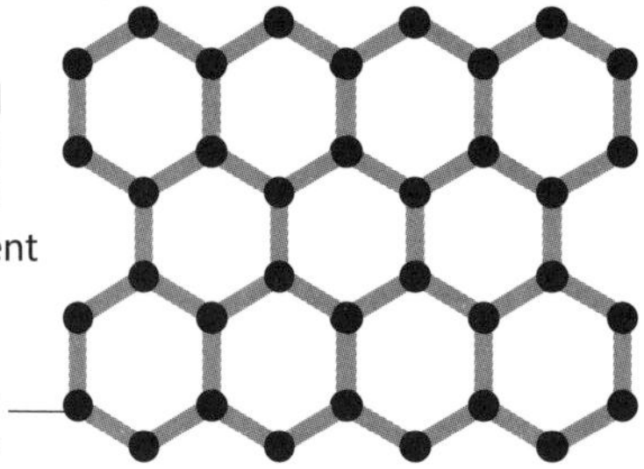
Graphene sheet, formed of a single layer of carbon atoms

Quantum dot television

Some television screens use nanoparticles in the form of quantum dots to achieve a brighter, sharper, more colorful image. In these displays, an array of quantum dots is positioned on top of the LED and liquid crystal layers. When the differently sized dots are stimulated with blue light from the LEDs, they emit pure red and green light. The combination of red, green, and blue light from each pixel of the screen is perceived as a single color.

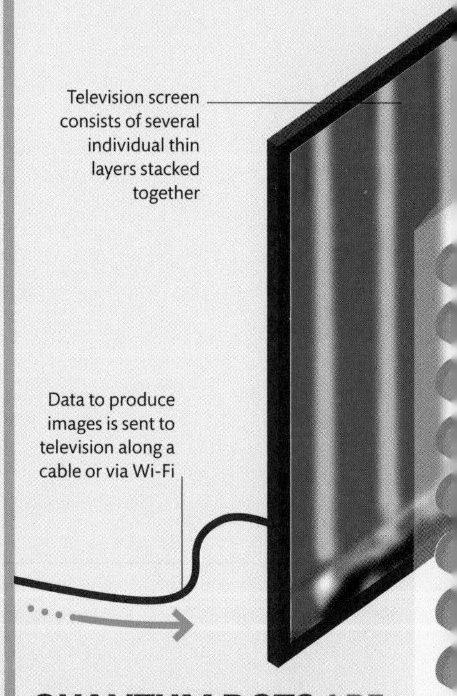
Television screen consists of several individual thin layers stacked together
Data to produce images is sent to television along a cable or via Wi-Fi

QUANTUM DOTS ARE ABOUT 10,000 TIMES NARROWER THAN A HUMAN HAIR

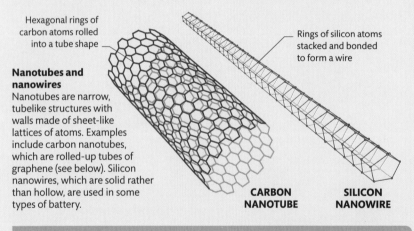

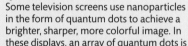

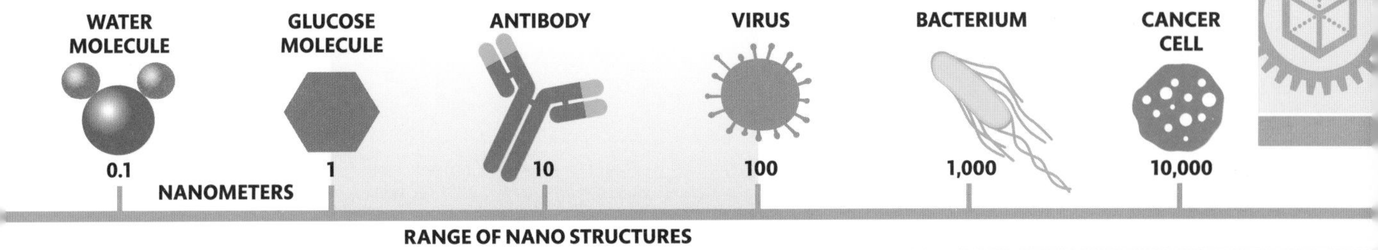
WATER MOLECULE
GLUCOSE MOLECULE
ANTIBODY
VIRUS
BACTERIUM
CANCER CELL
0.1
1
10
100
1,000
10,000
NANOMETERS
RANGE OF NANO STRUCTURES

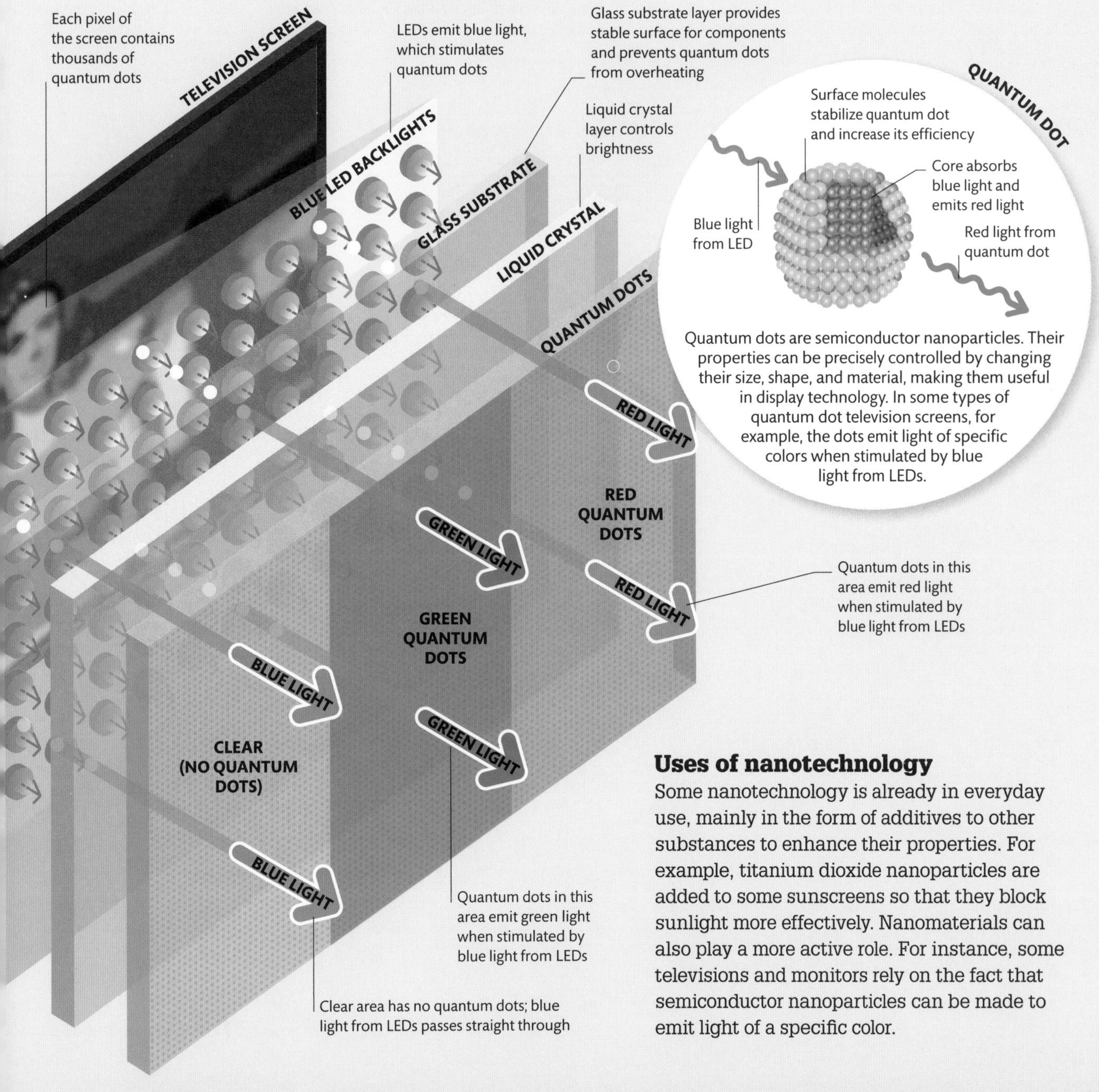
Each pixel of the screen contains thousands of quantum dots
TELEVISION SCREEN
LEDs emit blue light, which stimulates quantum dots
BLUE LED BACKLIGHTS
Glass substrate layer provides stable surface for components and prevents quantum dots from overheating
GLASS SUBSTRATE
Liquid crystal layer controls brightness
LIQUID CRYSTAL
QUANTUM DOTS
QUANTUM DOT
Surface molecules stabilize quantum dot and increase its efficiency
Core absorbs blue light and emits red light
Blue light from LED
Red light from quantum dot
Quantum dots are semiconductor nanoparticles. Their properties can be precisely controlled by changing their size, shape, and material, making them useful in display technology. In some types of quantum dot television screens, for example, the dots emit light of specific colors when stimulated by blue light from LEDs.
RED LIGHT
RED QUANTUM DOTS
GREEN LIGHT
RED LIGHT
Quantum dots in this area emit red light when stimulated by blue light from LEDs
GREEN QUANTUM DOTS
BLUE LIGHT
GREEN LIGHT
CLEAR (NO QUANTUM DOTS)
BLUE LIGHT
Quantum dots in this area emit green light when stimulated by blue light from LEDs
Clear area has no quantum dots; blue light from LEDs passes straight through

Uses of nanotechnology

Some nanotechnology is already in everyday use, mainly in the form of additives to other substances to enhance their properties. For example, titanium dioxide nanoparticles are added to some sunscreens so that they block sunlight more effectively. Nanomaterials can also play a more active role. For instance, some televisions and monitors rely on the fact that semiconductor nanoparticles can be made to emit light of a specific color.

3-D printing

Most items we use involve complex manufacturing processes. 3-D printing offers the prospect of being able to make a wide variety of objects simply by printing them out from digital files.

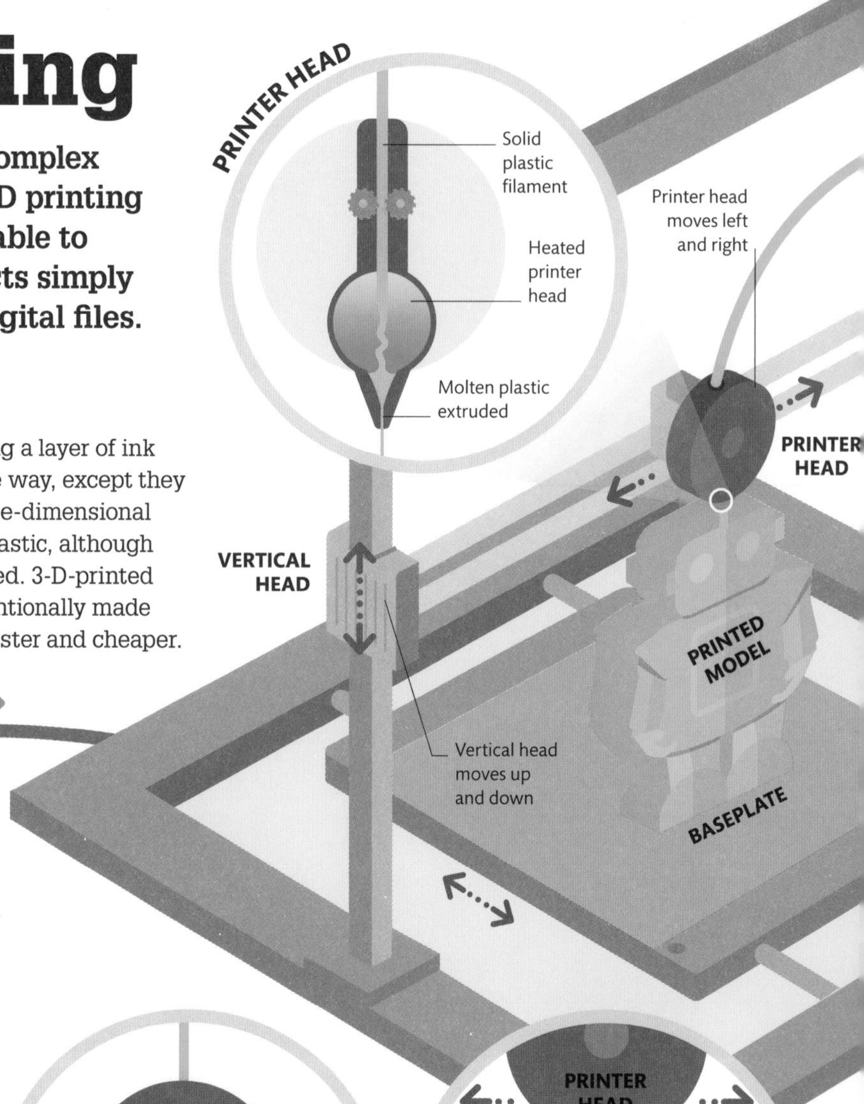

How 3-D printing works

Traditional printing works by depositing a layer of ink on paper. 3-D printers work in the same way, except they build up multiple layers to create a three-dimensional object. Instead of ink, they often use plastic, although various other materials may also be used. 3-D-printed items are not as well finished as conventionally made versions, but they can often be made faster and cheaper.

MANY **3-D PRINTERS** USE **PLASTIC** MADE FROM **CORNSTARCH**

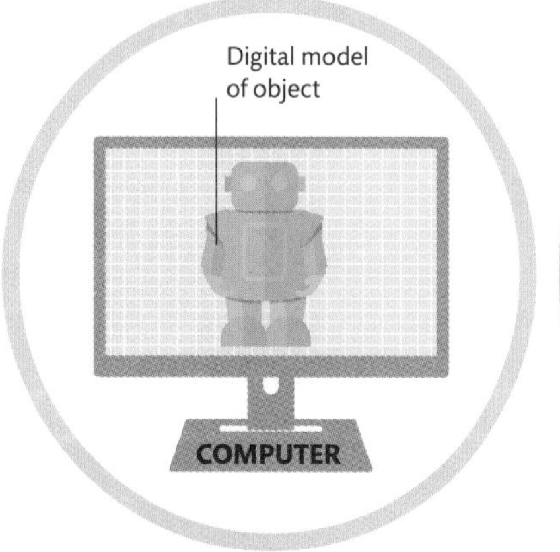

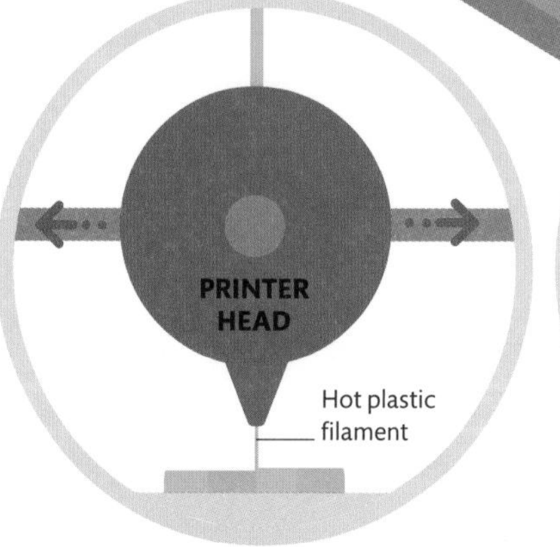

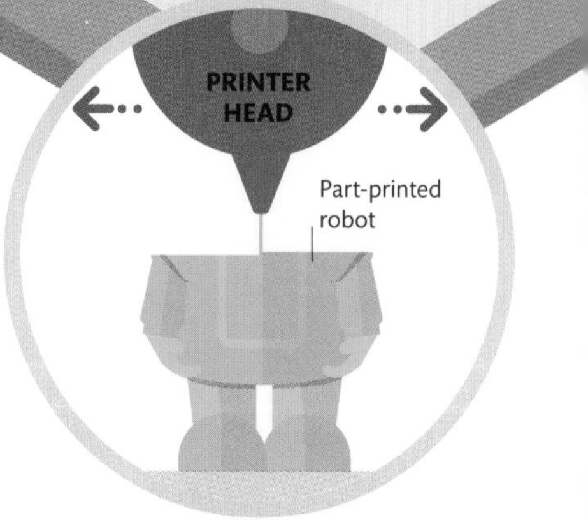

1 Computer design
3-D printing starts with the creation of a 3-D digital model in a computer. The model may be produced by special software or by scanning an object with a laser and then digitizing and processing the scanning data.

2 Start printing
A plastic filament is fed through the printer head, which contains a heating element that melts the plastic. Data from the computer moves the printer head from side to side, the vertical head up and down, and the baseplate back and forth.

3 Building up layers
The printed object is gradually built up layer by layer from the bottom upward. As each layer is added, the molten plastic cools and solidifies. Depending on the size and complexity of the object, it may take up to several hours to print.

4 Finishing
Because of the layer-by-layer nature of the printing process, 3-D printed objects have a rough surface. It is usually necessary to treat them with chemicals or to smooth them mechanically to achieve a clean finish. They may also be painted.

Uses of 3-D printers

3-D printing is still a young technology and is not yet commonly used to manufacture mass-produced consumer items. It is mainly utilized to produce specialized or custom-made items, such as pills and prosthetic body parts in medicine, musical instruments, and prototypes of potential new products.

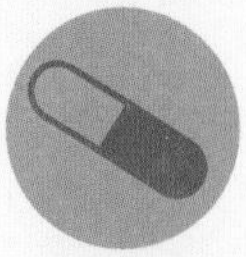

Pills
3-D printing allows pharmaceutical manufacturers to better fine-tune the composition of pills compared with traditional pill-making methods. It also makes possible production of pills that dissolve almost instantly.

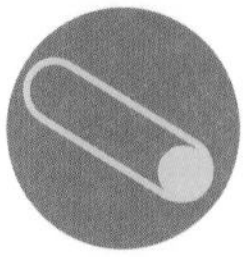

Synthetic blood vessels
Scientists have 3-D-printed blood vessels incorporating living cells. These vessels have been successfully implanted into mice and could in the future be used to replace damaged blood vessels in humans.

Sports shoes
Several sportswear companies have produced 3-D-printed sports shoes. They have been worn by athletes competing in international events but are available only in limited numbers.

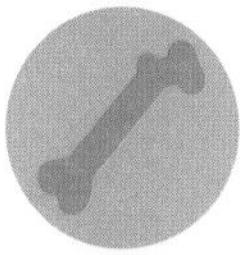

Prosthetic bones
Some patients who have had a section of bone removed (to treat cancer, for example) have received 3-D-printed implants made of titanium or synthetic bone that are an exact match for the area of bone removed.

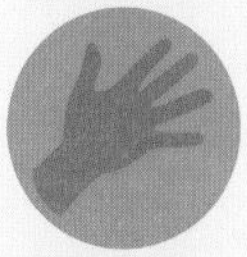

Prosthetic limbs
The use of 3-D printing to make prosthetic limbs has led to more lightweight designs than conventional prostheses. 3-D-printed limbs are also cheaper to produce and easier to customize for each individual.

Musical instruments
A wide variety of musical instruments have been experimentally 3-D printed, and many are available commercially, including some wind and string instruments, such as flutes, guitars, and violins.

MADE IN SPACE

In 2014, astronauts on the International Space Station printed a ratchet wrench with a design file transmitted from the ground. 3-D printing could avoid the need to carry items that might never be used or to supply spare parts over large distances at great expense.

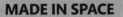

Arches and domes

For many traditionally constructed buildings, arches and domes are commonly used to span openings and big spaces, because they enable a large area to be covered with the least amount of supporting structure.

Arches

The simplest way to create an opening in a wall is to use two pillars (also called posts) with a flat beam across (the lintel) to carry the load above. However, this design is unable to support large loads and so does not allow for large openings. An arch can span wider openings, because the downward force from the weight of masonry pushes the individual blocks of the arch together, thereby utilizing the natural compressive strength of materials such as brick and stone. While an arch is under construction, it has to be supported by scaffolding until the keystone is in place to lock the structure securely.

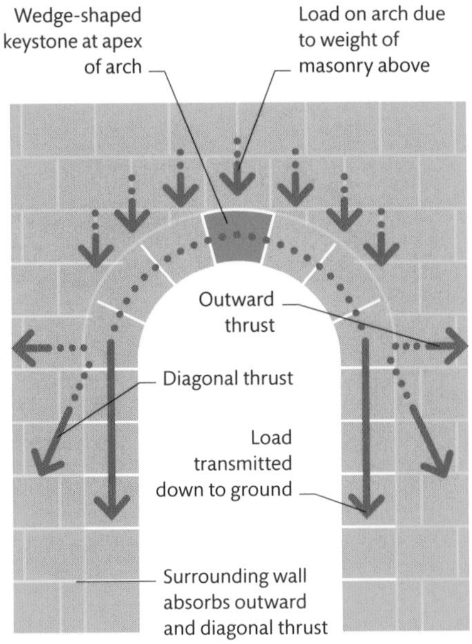

Forces in an arch
The load on an arch is carried along the curve and down. The load also produces outward and diagonal thrusts, which are countered by the surrounding wall or by buttresses.

Domes

A dome is like an arch rotated in a circle to form a three-dimensional shape. Like an arch, a dome is self-supporting, with all the weight transferred down to the structure on which it rests. However, unlike an arch, a dome does not require a keystone to lock it in place, and domes are stable during construction as each level is a complete and self-supporting ring. The weight of the dome creates forces that thrust outward. To counter this outward force, tension rings, which act like hoops on a barrel, are wrapped around the dome.

THE **WORLD'S FIRST GEODESIC DOME** OPENED IN **1926** IN GERMANY. IT WAS **82 FT (25 M) IN DIAMETER**

ROME'S PANTHEON

Almost 2,000 years after it was built, the dome of the Pantheon is still the world's largest unreinforced concrete dome, with an inner diameter of about 142 ft (43.3 m) and a weight of 5,000 tons (4,535 metric tons). To minimize the dome's weight, the concrete is thinner at the top and thicker at the base. The weight is further reduced by indentations in the dome, called coffers, and by a 26 ft (8 m) diameter hole at the apex, known as an oculus.

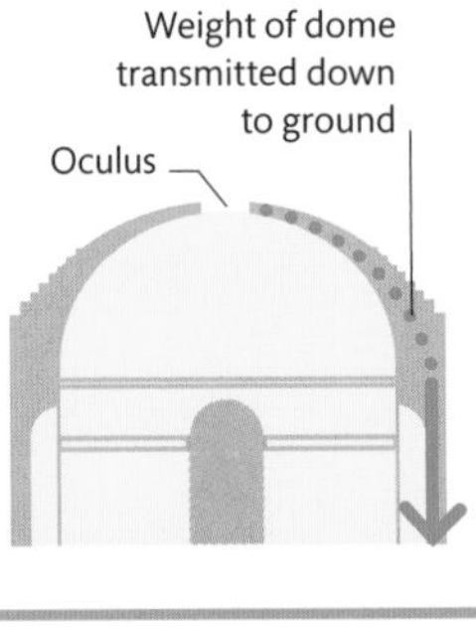

Brunelleschi's dome
The dome of Florence cathedral, commonly known as Brunelleschi's dome after its designer, is the largest masonry dome ever built, at about 148 ft (45 m) across and rising to 376 ft (114.5 m) above ground level. It consists of two concentric octagonal domes, or shells: an inner shell visible from inside the cathedral and a larger external shell.

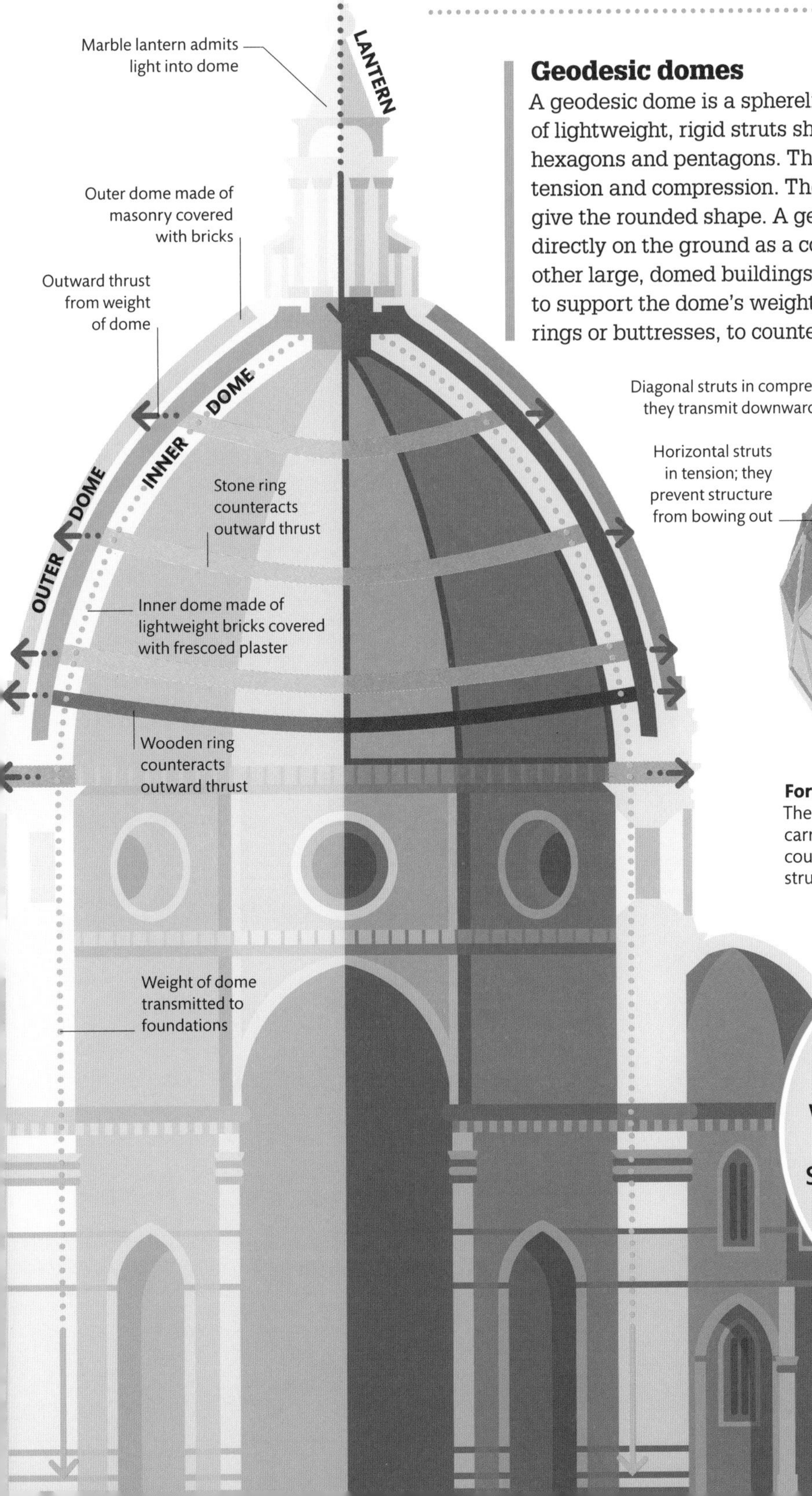
Marble lantern admits light into dome
LANTERN
Outer dome made of masonry covered with bricks
Outward thrust from weight of dome
INNER DOME
OUTER DOME
Stone ring counteracts outward thrust
Inner dome made of lightweight bricks covered with frescoed plaster
Wooden ring counteracts outward thrust
Weight of dome transmitted to foundations

Geodesic domes

A geodesic dome is a spherelike structure constructed of lightweight, rigid struts shaped into triangles within hexagons and pentagons. The triangles are strong under tension and compression. The hexagons and pentagons give the rounded shape. A geodesic dome can be set directly on the ground as a complete structure, unlike other large, domed buildings, which require foundations to support the dome's weight and structures, such as rings or buttresses, to counteract the outward thrust.

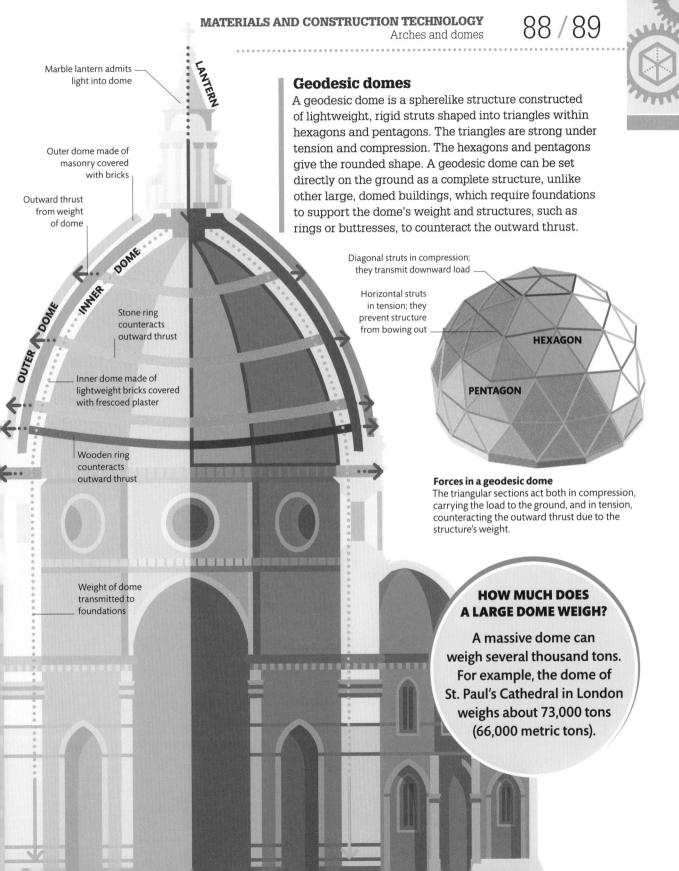

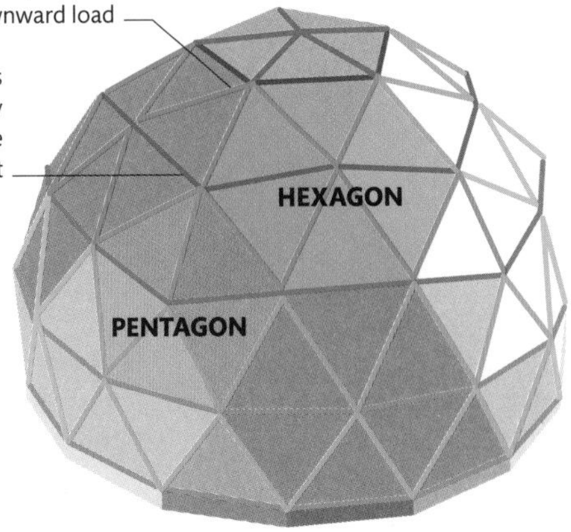
Diagonal struts in compression; they transmit downward load
Horizontal struts in tension; they prevent structure from bowing out
HEXAGON
PENTAGON

Forces in a geodesic dome
The triangular sections act both in compression, carrying the load to the ground, and in tension, counteracting the outward thrust due to the structure's weight.

HOW MUCH DOES A LARGE DOME WEIGH?

A massive dome can weigh several thousand tons. For example, the dome of St. Paul's Cathedral in London weighs about 73,000 tons (66,000 metric tons).

Drilling

Drilling holes deep below Earth's surface makes it possible to access natural resources such as water, oil, and gas. Holes are also drilled for scientific purposes, for example to remove ice core samples that can be analyzed to provide information about past environmental conditions.

Drilling for oil

Oil is a naturally occurring organic substance that forms liquid deposits underground. An oil drilling rig consists of a drill supported by a structure called a derrick. As the drill moves down through the ground, sections of steel casing are placed around the hole. A mixture of fluid known as mud is also pumped into the hole to make the cutting tool, or drill bit, work more effectively. Once the drill reaches the oil, the derrick and drilling equipment are removed and replaced by a pump.

DRILLING ICE CORES

Ice forms from the gradual buildup of snow, so lower layers are older than upper ones, and analyzing ice cores can provide information about past climatic conditions. Ice cores are drilled with a hollow pipe, and some can be 2 miles (3 km) deep.

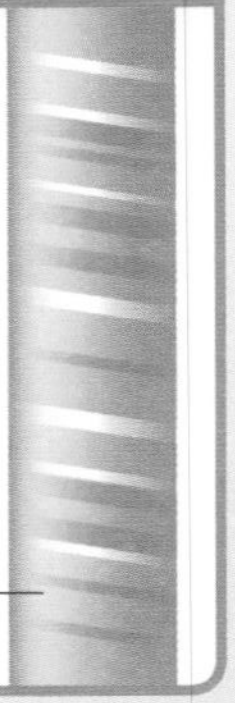

Layers of ice build up year by year

Derrick supports drilling equipment

Standpipe carries mud to drill bit

Offshore drilling

To access oil deposits below the seabed, oil companies use specialized mobile offshore drilling units (MODUs). Once an oil deposit has been found, some MODUs can be converted into oil production rigs. Usually, however, a MODU is replaced with a more permanent oil production platform after oil has been struck.

Jackup
A jackup is a MODU with legs that can be extended down to rest on the seabed. This keeps the rig safe from tidal motions and waves.

Semisubmersible
Semisubmersibles float on the sea surface on top of submerged pontoons. Some can convert to production rigs once oil is found.

Drillship
These are specialist ships with a drilling rig on the top deck. The drill operates through a hole in the hull. Drillships can operate in deep water.

Drilling barge
A drilling barge is a small vessel fitted with a rig raised off its deck. Drilling barges are suitable for use only in calm, shallow water.

Onshore drilling rig
Land-based oil rigs vary in height, depending on the depth of hole to be drilled. The drill bit is rotated by a rotary drive unit on the main platform and is raised and lowered by a pulley system powered by the motorized draw works.

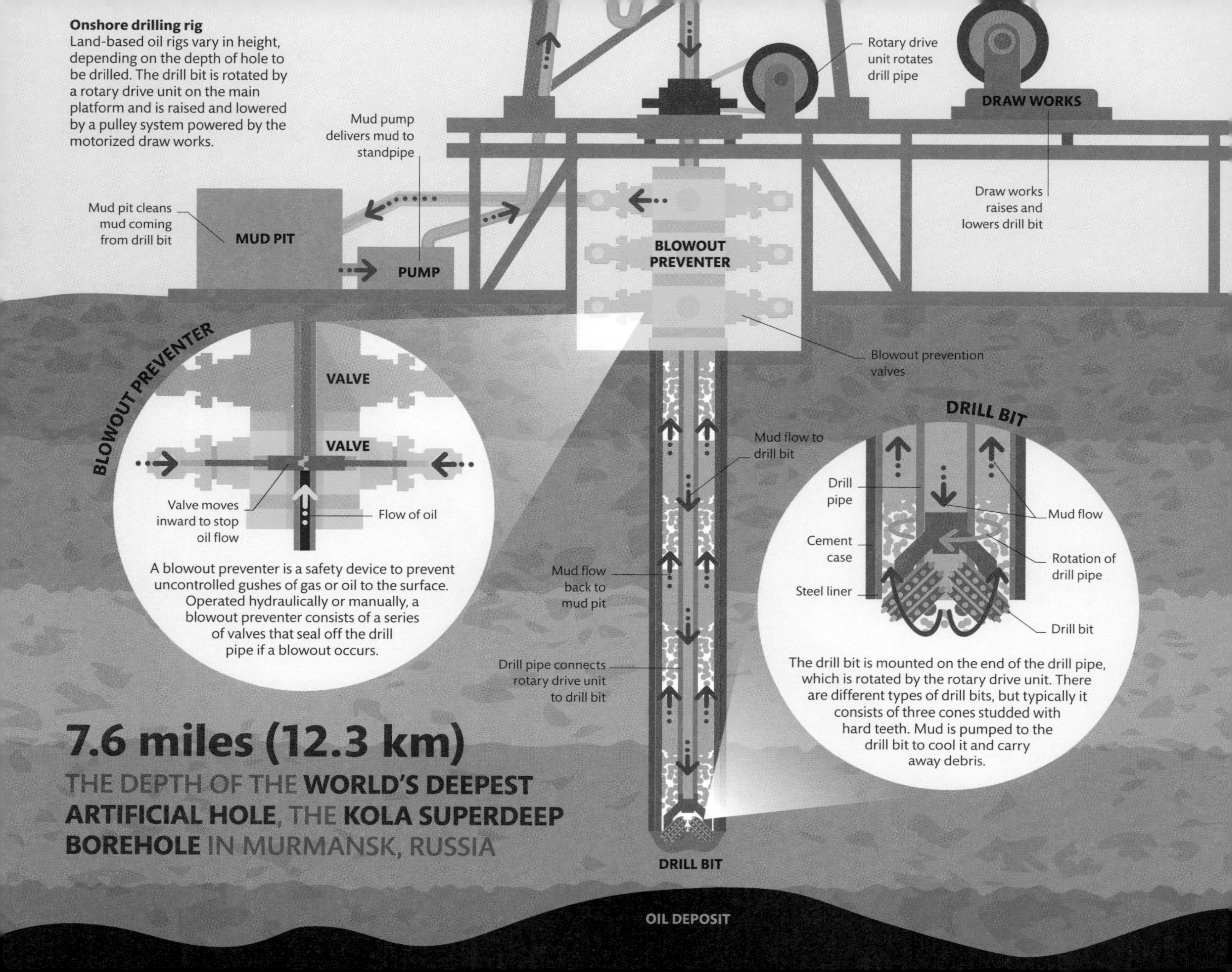
Mud pump delivers mud to standpipe
Rotary drive unit rotates drill pipe
DRAW WORKS
Draw works raises and lowers drill bit
Mud pit cleans mud coming from drill bit
MUD PIT
PUMP
BLOWOUT PREVENTER
Blowout prevention valves
BLOWOUT PREVENTER
VALVE
VALVE
Valve moves inward to stop oil flow
Flow of oil
A blowout preventer is a safety device to prevent uncontrolled gushes of gas or oil to the surface. Operated hydraulically or manually, a blowout preventer consists of a series of valves that seal off the drill pipe if a blowout occurs.
Mud flow to drill bit
Mud flow back to mud pit
Drill pipe connects rotary drive unit to drill bit
DRILL BIT
Drill pipe
Cement case
Steel liner
Mud flow
Rotation of drill pipe
Drill bit
The drill bit is mounted on the end of the drill pipe, which is rotated by the rotary drive unit. There are different types of drill bits, but typically it consists of three cones studded with hard teeth. Mud is pumped to the drill bit to cool it and carry away debris.
DRILL BIT
OIL DEPOSIT

7.6 miles (12.3 km)

THE DEPTH OF THE **WORLD'S DEEPEST ARTIFICIAL HOLE**, THE **KOLA SUPERDEEP BOREHOLE** IN MURMANSK, RUSSIA

Earth movers

Earth moving is a key part of the construction process. It includes digging and removing material, leveling, and filling. Earth-moving machines operate by using levers and hydraulics.

How an excavator works
An excavator's caterpillar track is driven by a diesel engine housed in the engine compartment. The engine also drives a pump, housed in the same compartment, that powers the hydraulic systems that move the excavator's arm and bucket.

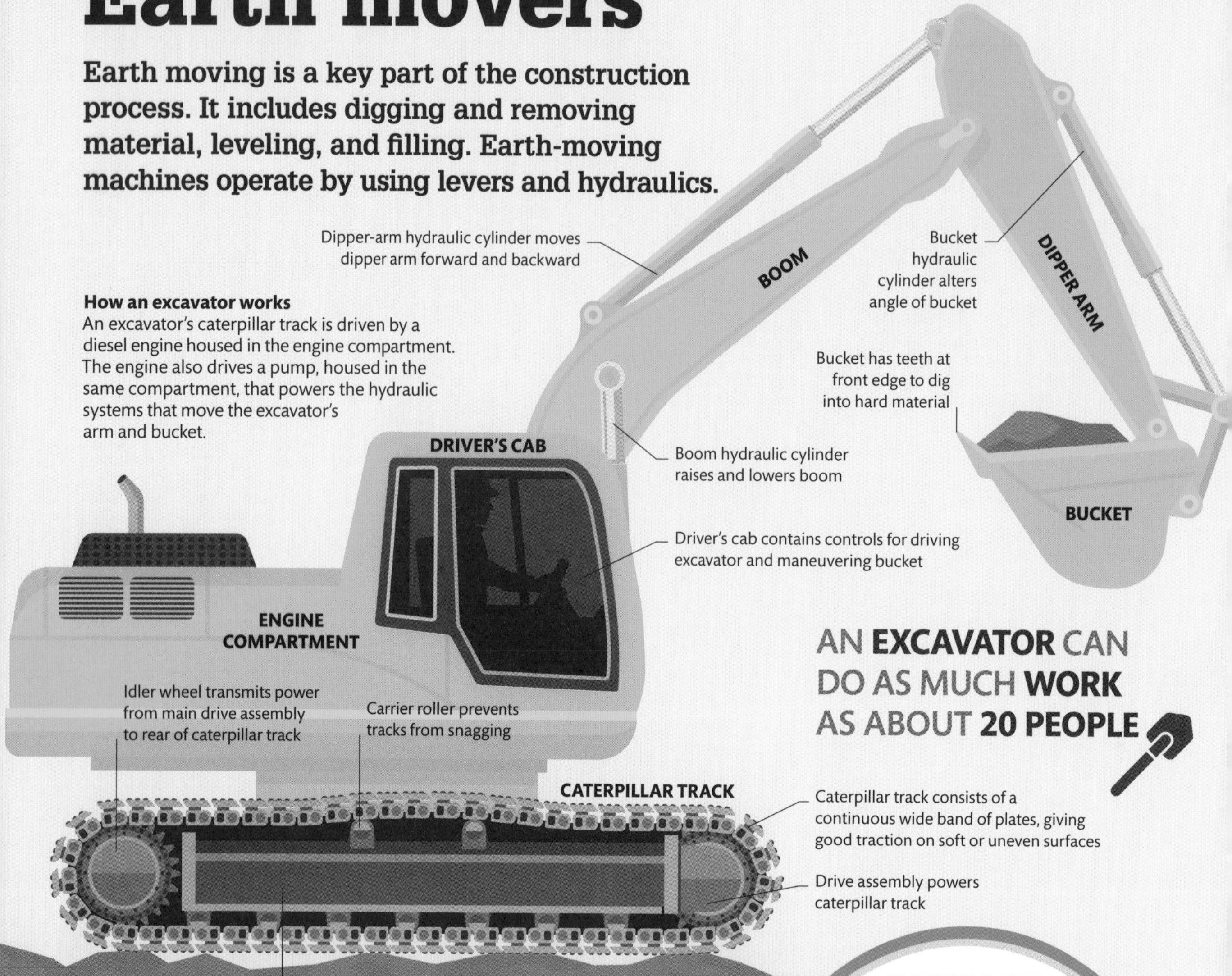
Dipper-arm hydraulic cylinder moves dipper arm forward and backward
BOOM
Bucket hydraulic cylinder alters angle of bucket
DIPPER ARM
Bucket has teeth at front edge to dig into hard material
DRIVER'S CAB
Boom hydraulic cylinder raises and lowers boom
BUCKET
Driver's cab contains controls for driving excavator and maneuvering bucket
ENGINE COMPARTMENT
Idler wheel transmits power from main drive assembly to rear of caterpillar track
Carrier roller prevents tracks from snagging
CATERPILLAR TRACK
Caterpillar track consists of a continuous wide band of plates, giving good traction on soft or uneven surfaces
Drive assembly powers caterpillar track
Track adjuster alters tension of caterpillar track

AN **EXCAVATOR** CAN DO AS MUCH **WORK** AS ABOUT **20 PEOPLE**

Earth-moving machinery

An excavator, or digger, digs into and scoops up material before depositing it elsewhere. It is one of many types of heavy earth-moving machinery used on construction sites. A bulldozer is a multipurpose earth mover that shunts material with a large, hydraulically operated front blade. A front loader is a type of tractor with a wide, front-mounted bucket used for scooping and lifting; the bucket is raised and lowered by hydraulics. A backhoe loader is a combination of a front loader and an excavator.

HOW BIG ARE THE LARGEST EARTH MOVERS?

The largest excavator is the Bucyrus RH400 hydraulic shovel, which is three stories tall, weighs 1,080 tons (980 metric tons), and can hold 1,590 cubic feet (45 cubic meters) of rock in a single scoop.

Hydraulics

Liquids cannot be compressed (unlike gases), which means that any force or pressure applied to a liquid is transferred through it. In a basic hydraulic system, when pressure is applied to one end of a liquid inside a closed pipe or cylinder, the force is passed all the way to the other end. A small force can be multiplied by changing the width of one piston and cylinder column relative to the other.

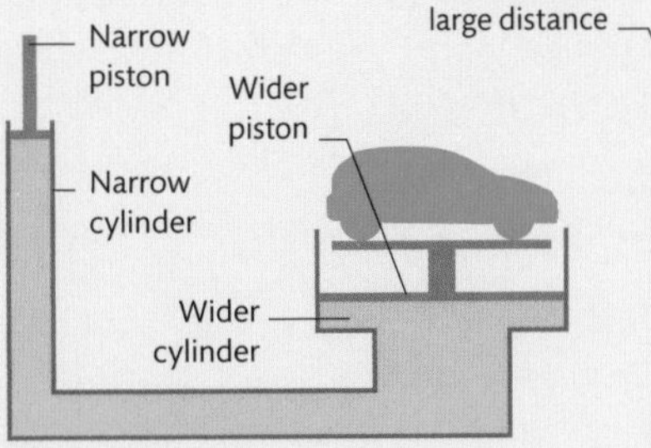

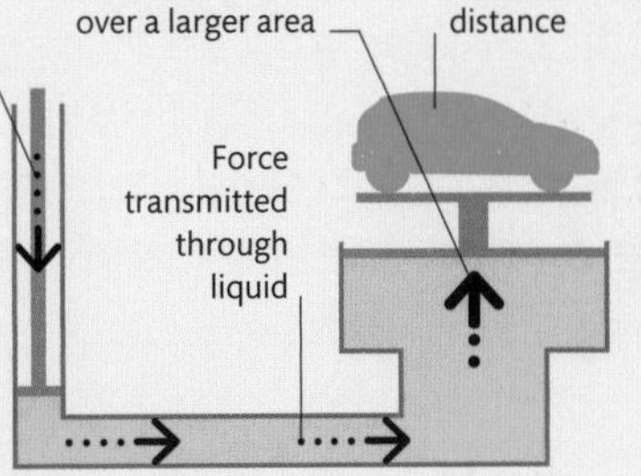

1 Multiplying a force
A force applied by a piston in a narrow cylinder is multiplied into a larger force by a wider piston at the other end, although the liquid's pressure remains the same.

2 Double the force, half the distance
If the larger piston has twice the area of the small piston, the force exerted will double. The cost is that this greater force operates through half of the distance.

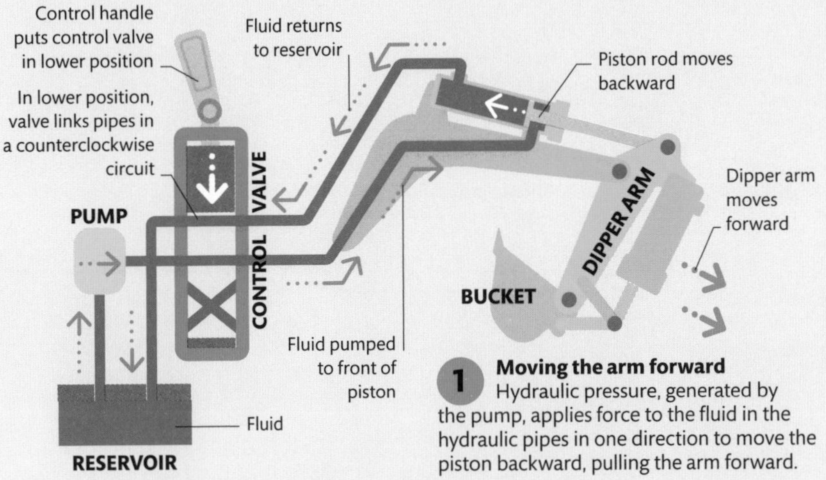

1 Moving the arm forward
Hydraulic pressure, generated by the pump, applies force to the fluid in the hydraulic pipes in one direction to move the piston backward, pulling the arm forward.

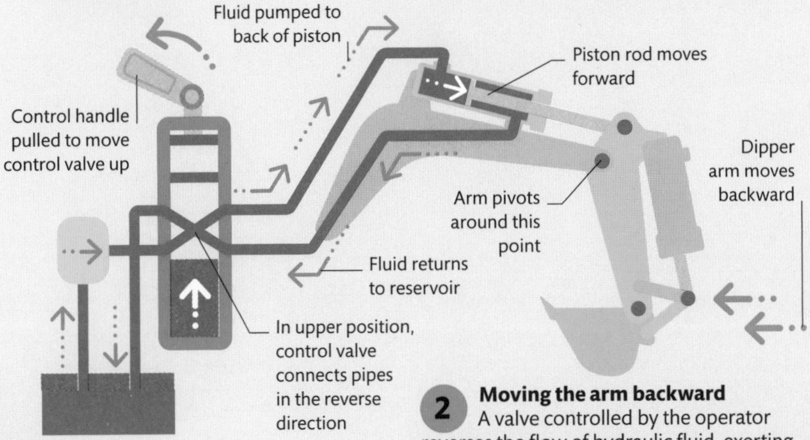

2 Moving the arm backward
A valve controlled by the operator reverses the flow of hydraulic fluid, exerting pressure on the other side of the piston and pushing the arm in the opposite direction.

LEVERS

There are three classes of levers, defined by where the effort and output forces are located relative to the fulcrum. They can be used to increase either power or movement in different directions.

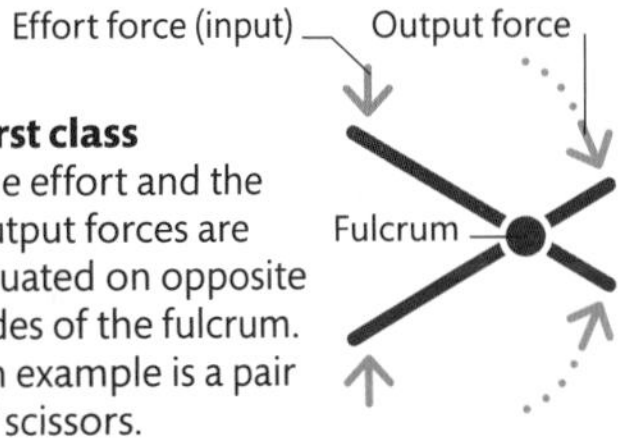

First class
The effort and the output forces are situated on opposite sides of the fulcrum. An example is a pair of scissors.

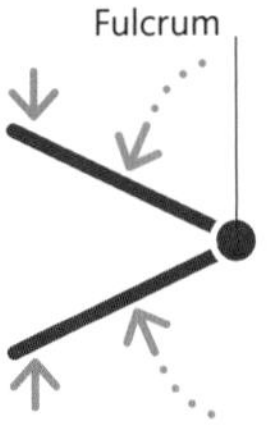

Second class
The output force is located between the fulcrum and the effort force. An example is a pair of nutcrackers.

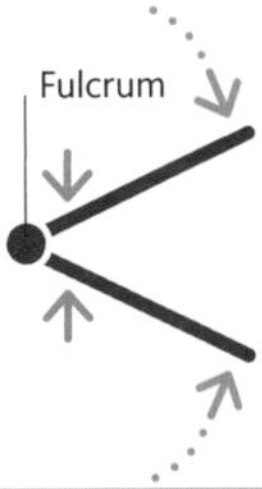

Third class
The effort force is applied between the fulcrum and the output force. An example is a pair of tongs or tweezers.

Bridges

Whether crossing a small gap or spanning more than 60 miles, a bridge must be able to withstand and transfer the forces of tension and compression from the bridge's weight and its load.

Types of bridges

While bridges come in all shapes and sizes, nearly all are variations on a few basic types. Beam and truss bridges are the simplest forms. Similar to laying a plank of wood between two banks, they can be used for only relatively short spans. The arch bridge is also best suited to shorter spans, unless multiple arches are joined together. Cable-stayed and, particularly, suspension designs offer the greatest scope for lengthy spans.

Beam bridge
In a beam bridge, piers or posts on either end support a flat deck. The deck consists of beams, such as hollow steel box girders.

Arch bridge
An arch constructed below the bridge supports the deck, transferring the compression forces to the piers.

Truss bridge
In a truss bridge, the deck has added support from a girder framework with diagonal posts to counter compressive forces.

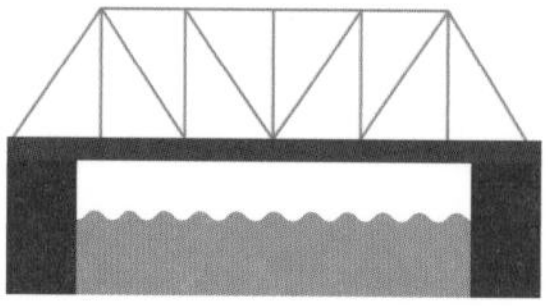

Cantilever bridge
This incorporates two "seesaws" whose ends meet in the middle. The ends are anchored on both sides.

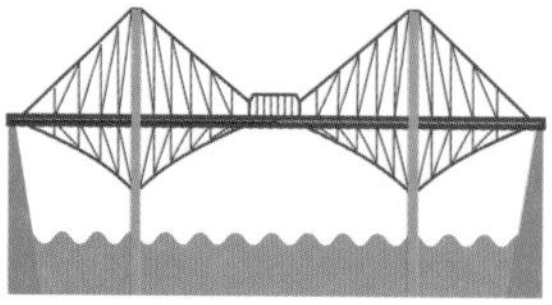

Cable-stayed bridge
The deck is supported by multiple cables that are directly connected to one or more vertical towers.

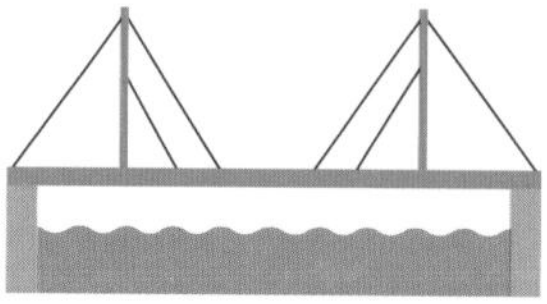

Suspension bridges

In a cable-stayed bridge (see left), cables connect the deck directly to vertical towers. In a suspension bridge, the main cables connect the tops of the towers to anchor blocks embedded in banks at the ends of the bridge. The deck is supported by vertical suspension cables that hang from the main cables. This is a system that allows for very large spans.

Suspension bridge structure
The weight of the bridge's deck, and any extra load applied, is transmitted through suspension cables to the main cables, putting the suspension cables and main cables under tension. The main cables transmit the load to the fixed anchor blocks and to the towers. This produces a compressive force in the towers. The towers ultimately pass the load down to the foundations.

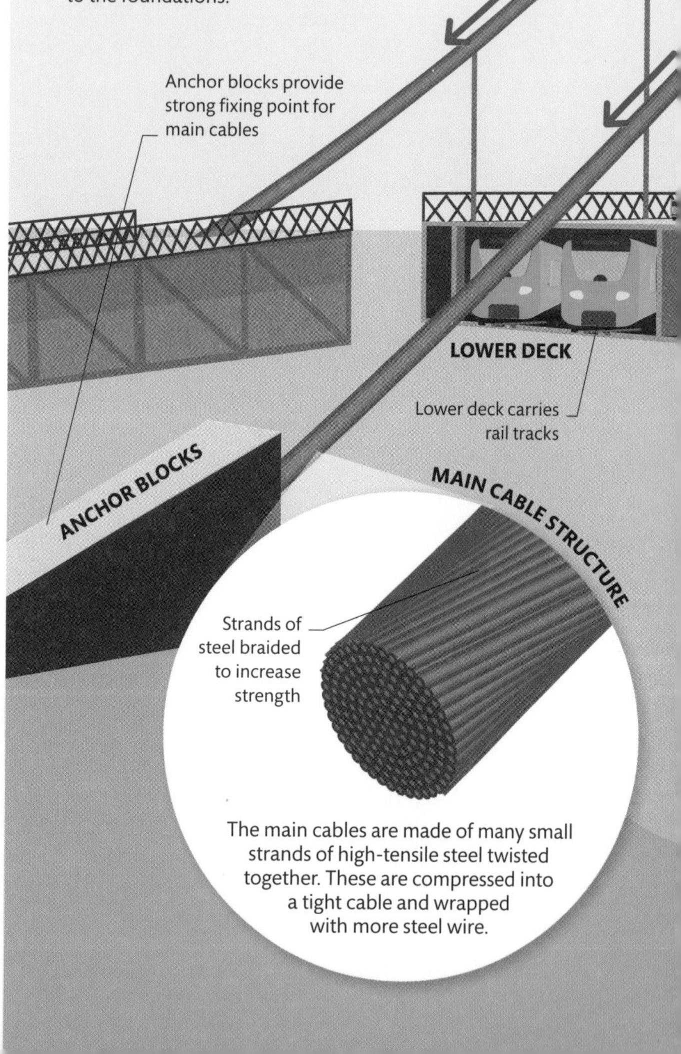
Strong tension forces transferred by main cable to anchor blocks and tower
Anchor blocks provide strong fixing point for main cables
LOWER DECK
Lower deck carries rail tracks
ANCHOR BLOCKS
MAIN CABLE STRUCTURE
Strands of steel braided to increase strength
The main cables are made of many small strands of high-tensile steel twisted together. These are compressed into a tight cable and wrapped with more steel wire.

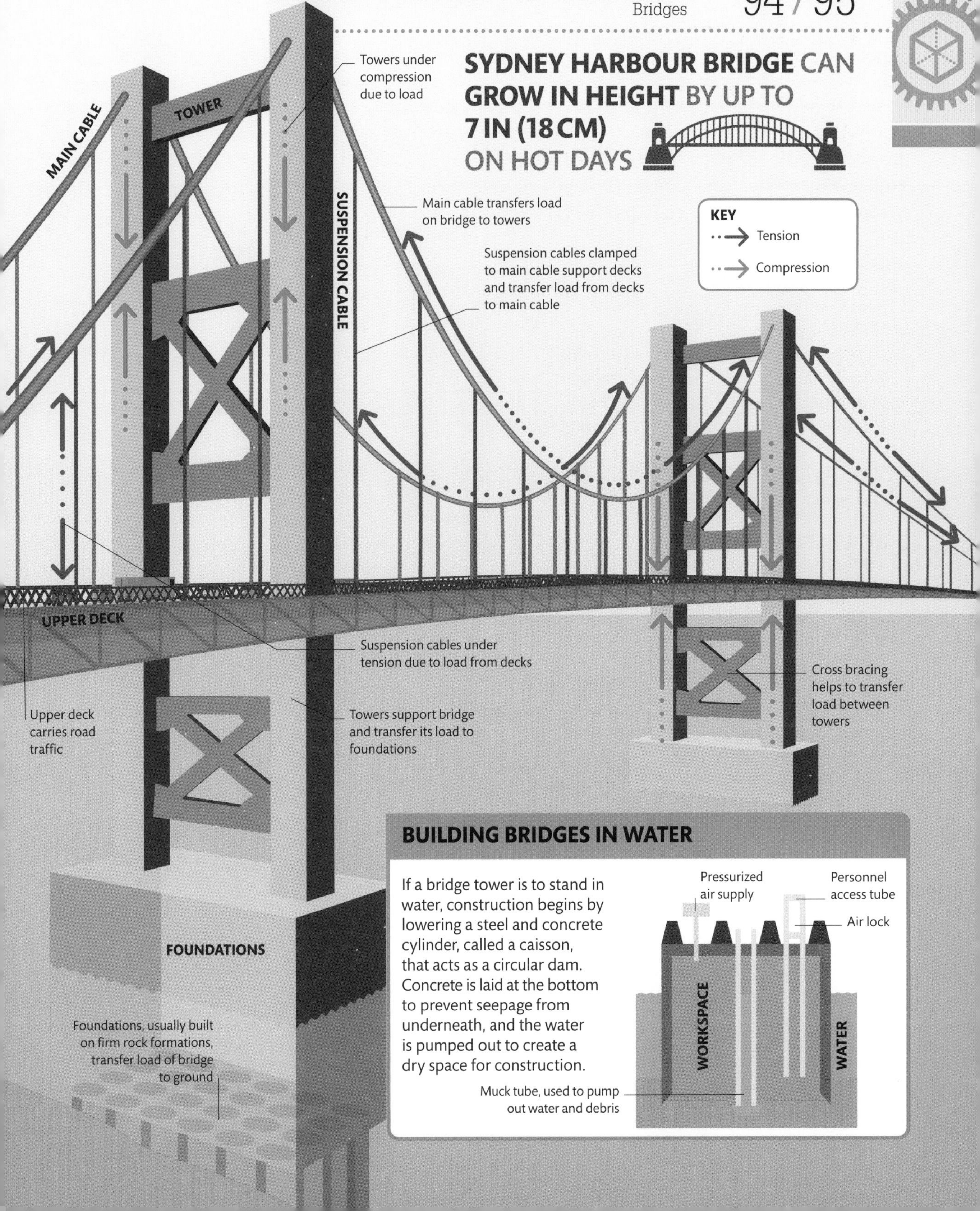
Towers under compression due to load
TOWER
MAIN CABLE
SUSPENSION CABLE
Main cable transfers load on bridge to towers
KEY
Tension
Compression
Suspension cables clamped to main cable support decks and transfer load from decks to main cable
UPPER DECK
Suspension cables under tension due to load from decks
Cross bracing helps to transfer load between towers
Upper deck carries road traffic
Towers support bridge and transfer its load to foundations
FOUNDATIONS
Foundations, usually built on firm rock formations, transfer load of bridge to ground

SYDNEY HARBOUR BRIDGE CAN **GROW IN HEIGHT** BY UP TO **7 IN (18 CM)** ON HOT DAYS

BUILDING BRIDGES IN WATER

If a bridge tower is to stand in water, construction begins by lowering a steel and concrete cylinder, called a caisson, that acts as a circular dam. Concrete is laid at the bottom to prevent seepage from underneath, and the water is pumped out to create a dry space for construction.

Pressurized air supply
Personnel access tube
Air lock
WORKSPACE
WATER
Muck tube, used to pump out water and debris

Tunnels

A tunnel is basically a large tube through earth or rock that has been reinforced to prevent collapse. Building tunnels usually requires specialized machinery.

Underwater tunnels

Tunnels can be bored under bodies of water using a tunnel boring machine (see below). One example of a bored underwater tunnel is the Channel Tunnel, which connects Britain and France. However, it is often faster and more cost-effective to build tunnels under water using the immersed tube method.

Immersed tube tunnel
The immersed tube method involves making a tunnel in sections on land and then bringing the sections to the construction site, where they are submerged and joined to each other.

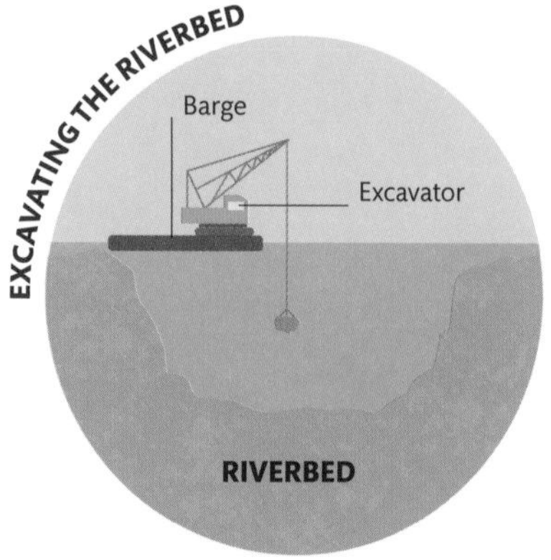
EXCAVATING THE RIVERBED
Barge
Excavator
RIVERBED

1 To reduce the risk of the tunnel interfering with shipping, a trench for the tunnel is dug into the bed of the river, lake, or sea using an excavator mounted on a barge.

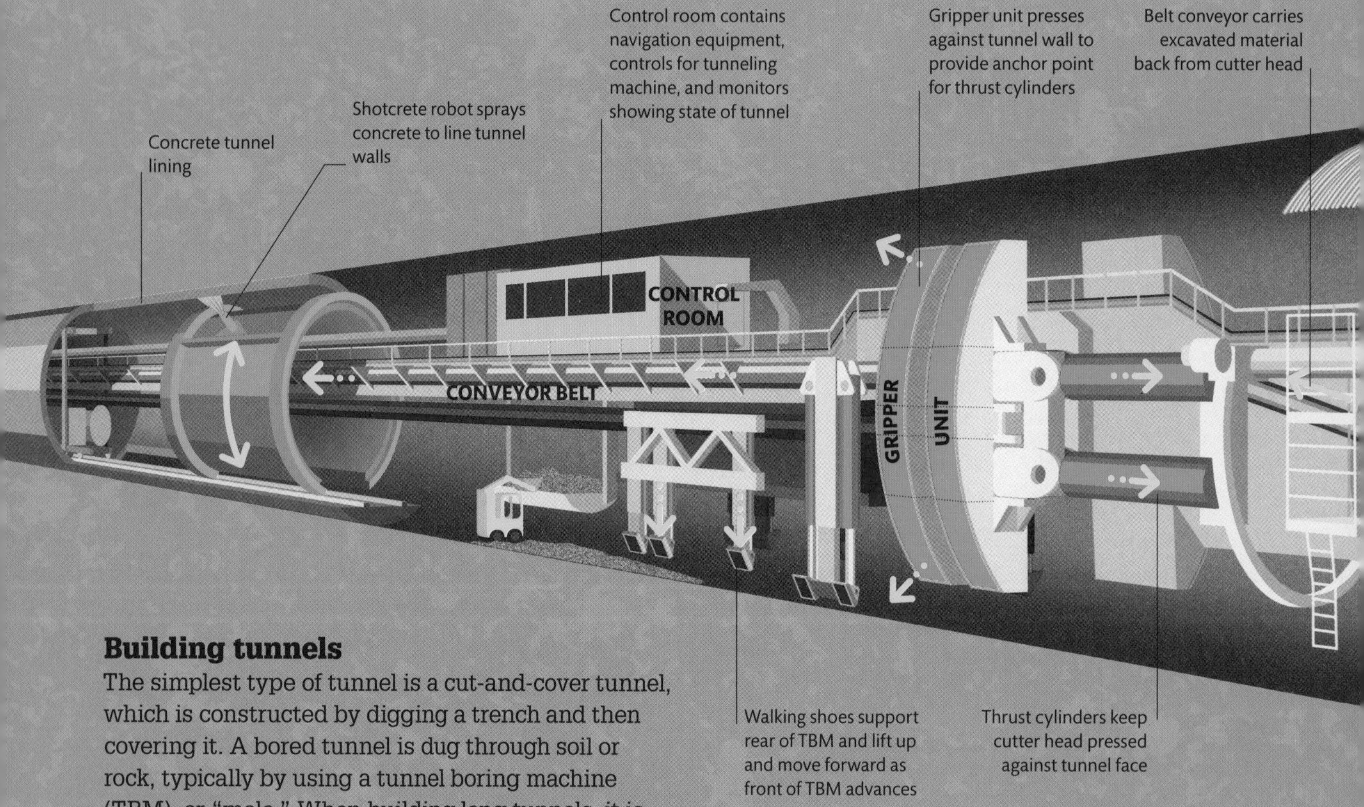
Control room contains navigation equipment, controls for tunneling machine, and monitors showing state of tunnel
Gripper unit presses against tunnel wall to provide anchor point for thrust cylinders
Belt conveyor carries excavated material back from cutter head
Shotcrete robot sprays concrete to line tunnel walls
Concrete tunnel lining
CONTROL ROOM
CONVEYOR BELT
GRIPPER
UNIT
Walking shoes support rear of TBM and lift up and move forward as front of TBM advances
Thrust cylinders keep cutter head pressed against tunnel face

Building tunnels

The simplest type of tunnel is a cut-and-cover tunnel, which is constructed by digging a trench and then covering it. A bored tunnel is dug through soil or rock, typically by using a tunnel boring machine (TBM), or "mole." When building long tunnels, it is often necessary to bore additional shafts to provide emergency escape routes and remove toxic fumes.

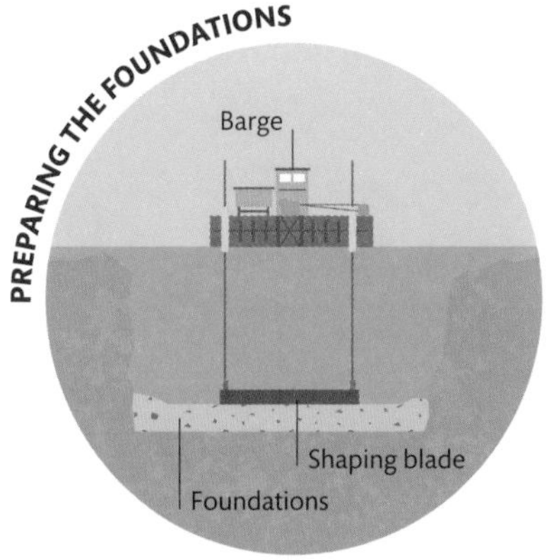
PREPARING THE FOUNDATIONS
Barge
Shaping blade
Foundations

2 The bottom of the trench is prepared by laying a foundation of aggregate and sand. The foundation is smoothed with a shaping blade to ensure a level base for the tunnel sections.

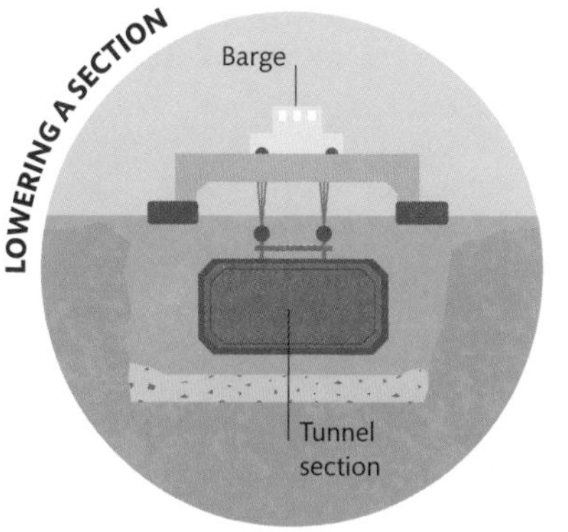
LOWERING A SECTION
Barge
Tunnel section

3 Prefabricated cast-concrete sections are floated to the installation site and lowered to the bottom. A hydraulic arm pulls each new section close up to the adjacent section to form a watertight seal.

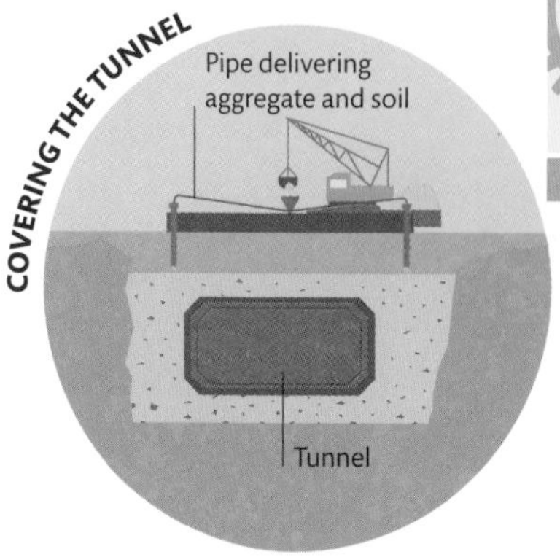
COVERING THE TUNNEL
Pipe delivering aggregate and soil
Tunnel

4 Pipes running from a barge deliver more aggregate and soil to cover the completed tunnel. The top of the tunnel may also be covered by a layer of large stones to protect it from damage by ship anchors.

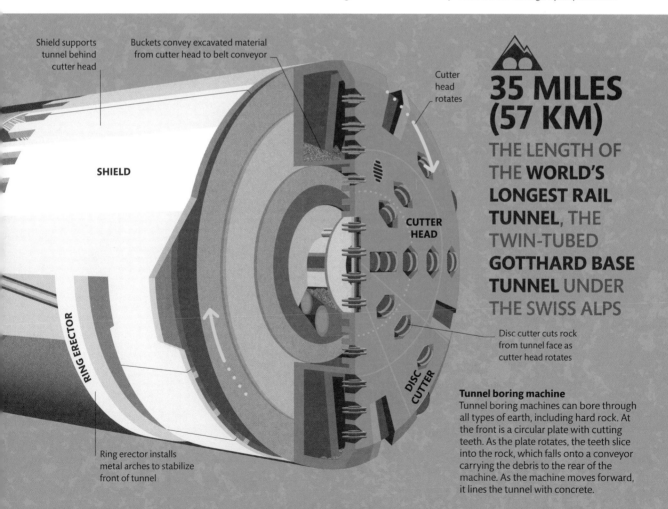

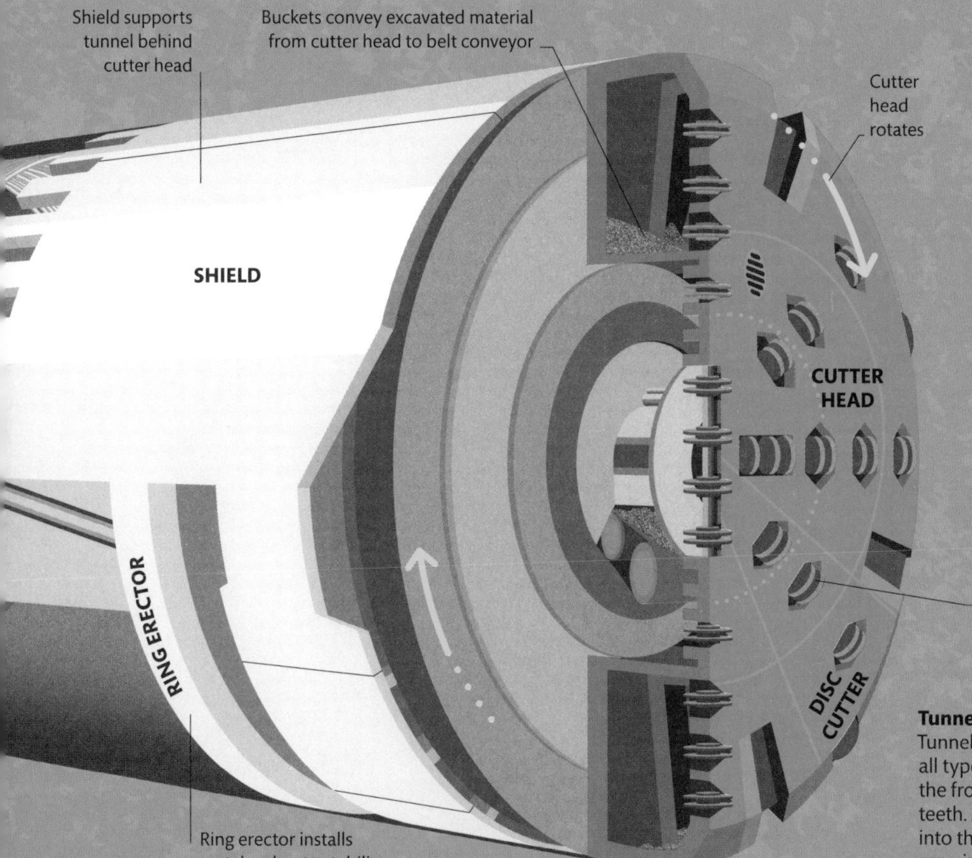
Shield supports tunnel behind cutter head
Buckets convey excavated material from cutter head to belt conveyor
Cutter head rotates
SHIELD
CUTTER HEAD
RING ERECTOR
DISC CUTTER
Disc cutter cuts rock from tunnel face as cutter head rotates
Ring erector installs metal arches to stabilize front of tunnel

35 MILES (57 KM)

THE LENGTH OF THE **WORLD'S LONGEST RAIL TUNNEL**, THE TWIN-TUBED **GOTTHARD BASE TUNNEL** UNDER THE SWISS ALPS

Tunnel boring machine
Tunnel boring machines can bore through all types of earth, including hard rock. At the front is a circular plate with cutting teeth. As the plate rotates, the teeth slice into the rock, which falls onto a conveyor carrying the debris to the rear of the machine. As the machine moves forward, it lines the tunnel with concrete.

WOODEN SKYSCRAPERS

New types of engineered wood have made it possible to construct high-rise buildings with wooden frames. Made from thin layers of wood crisscrossed and stuck together with glue, this engineered wood—known as glulam—is as strong as steel. Existing wooden-framed skyscrapers include the 18-story Brock Commons student residence in Vancouver.

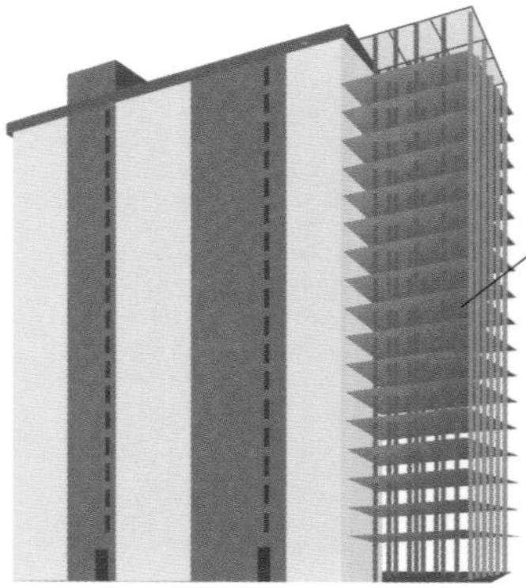

Engineered-wood frame

BROCK COMMONS, VANCOUVER

CURTAIN WALL

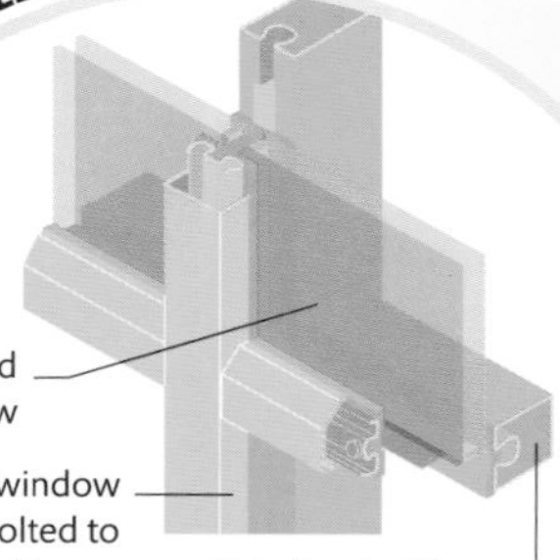

Double-glazed window
Outer window frame bolted to tubular steel frame
Tubular steel frame

The exterior walls—called curtain walls—are attached to the building's framework. The walls do not bear the weight of the building and need to support only their own weight.

Anatomy of a skyscraper
A typical skyscraper consists of a steel framework around a central concrete core, which contains elevator shafts and various services, such as the water supply. Wrapped around the steel frame is the non-load-bearing exterior curtain wall.

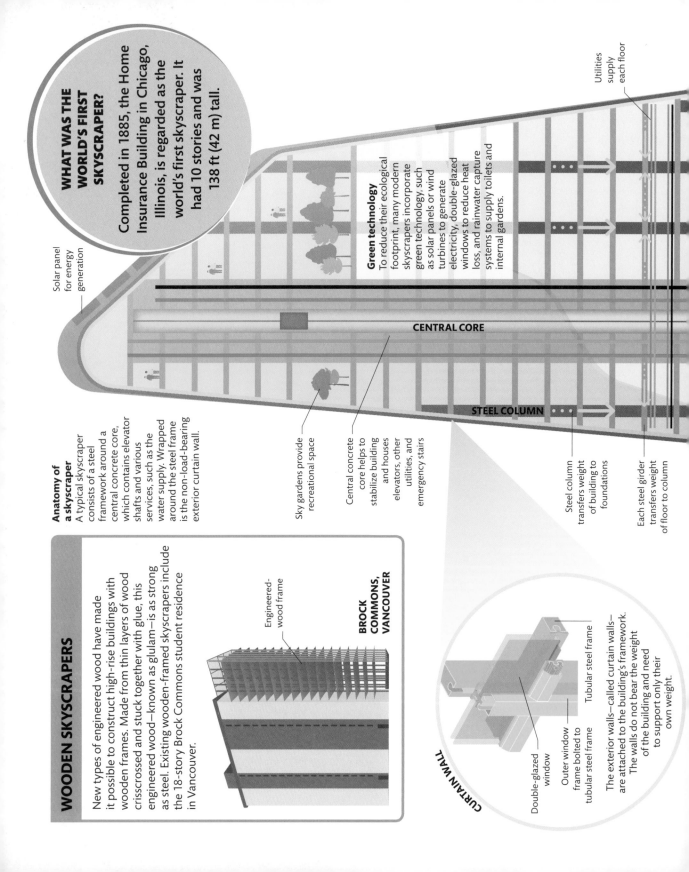

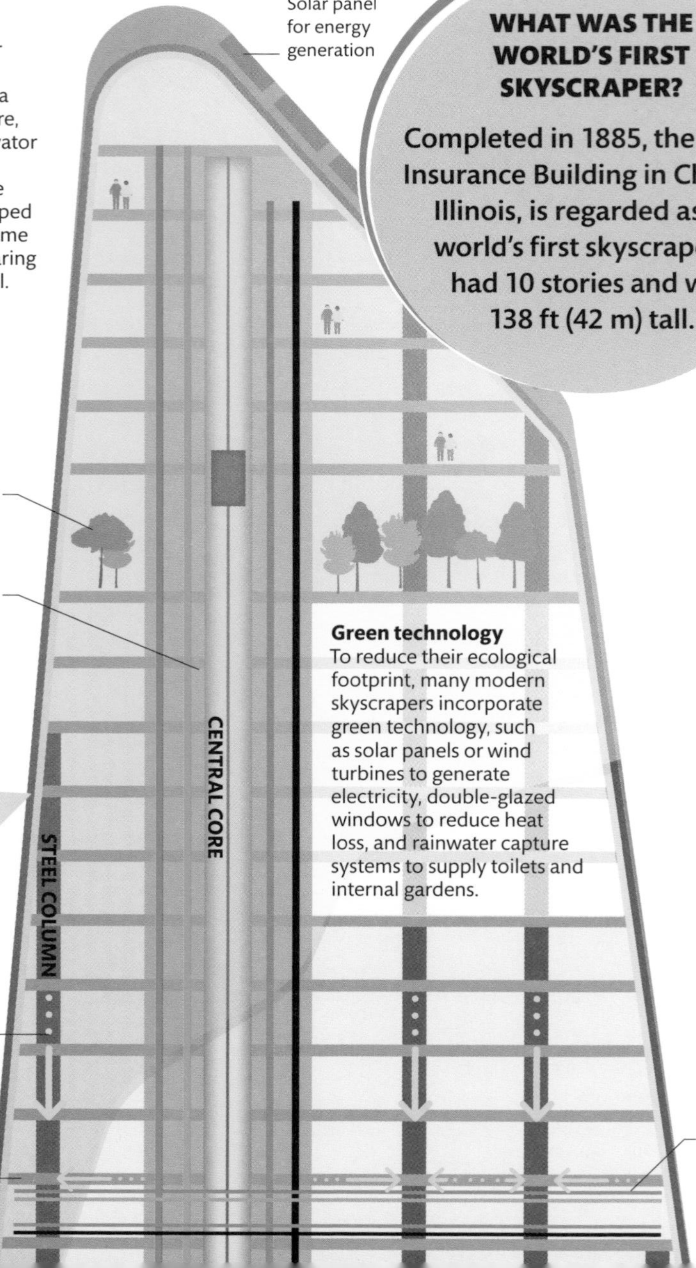

Solar panel for energy generation
Sky gardens provide recreational space
Central concrete core helps to stabilize building and houses elevators, other utilities, and emergency stairs
CENTRAL CORE
STEEL COLUMN
Steel column transfers weight of building to foundations
Each steel girder transfers weight of floor to column
Utilities supply each floor

Green technology
To reduce their ecological footprint, many modern skyscrapers incorporate green technology, such as solar panels or wind turbines to generate electricity, double-glazed windows to reduce heat loss, and rainwater capture systems to supply toilets and internal gardens.

WHAT WAS THE WORLD'S FIRST SKYSCRAPER?

Completed in 1885, the Home Insurance Building in Chicago, Illinois, is regarded as the world's first skyscraper. It had 10 stories and was 138 ft (42 m) tall.

Skyscrapers

High-rise buildings dominate the skylines of many cities, because they offer large accommodation space but occupy a small area of land. As construction technology has improved, ever-taller skyscrapers have been built, and buildings with more than 160 floors are now feasible.

Skyscraper structures

Buildings made of brick or stone require thick, heavy walls, which makes it impractical to have more than five or six floors. Skyscrapers can be built much higher because they have lightweight steel frames and walls. However, they must be able to resist high-altitude winds that would make them sway and must also have elevators to move people efficiently up and down the building (see pp.100–101).

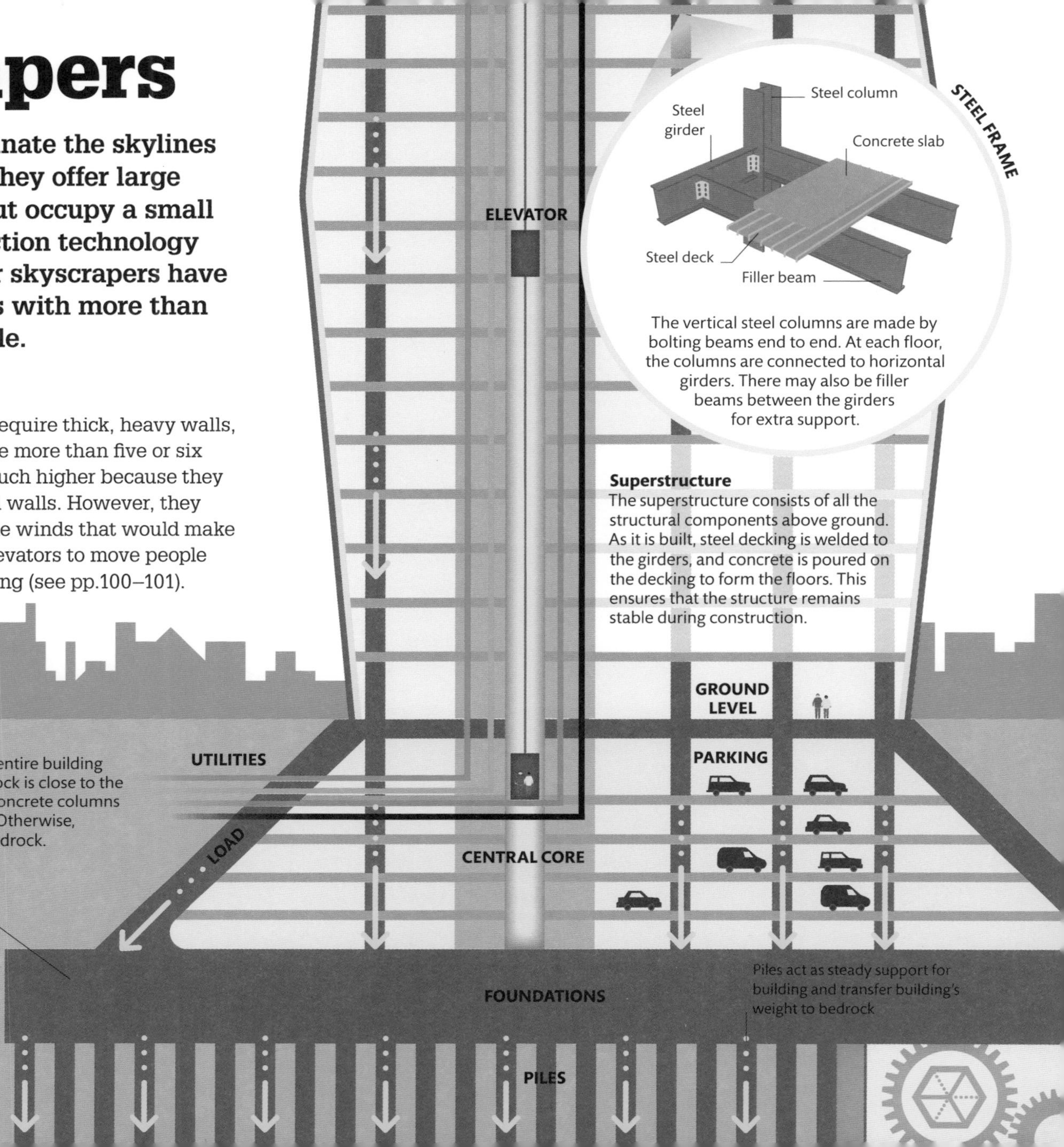
STEEL FRAME
Steel column
Steel girder
Concrete slab
Steel deck
Filler beam
The vertical steel columns are made by bolting beams end to end. At each floor, the columns are connected to horizontal girders. There may also be filler beams between the girders for extra support.
ELEVATOR
GROUND LEVEL
PARKING
UTILITIES
LOAD
CENTRAL CORE
FOUNDATIONS
PILES
Piles act as steady support for building and transfer building's weight to bedrock
Foundations help to distribute weight of building over a large area and also help to transfer weight of building to piles
KEY
Heating and cooling
Water
Electricity
Sewage

Superstructure
The superstructure consists of all the structural components above ground. As it is built, steel decking is welded to the girders, and concrete is poured on the decking to form the floors. This ensures that the structure remains stable during construction.

Substructure
The substructure carries the weight of the entire building and transfers it to the bedrock. If the bedrock is close to the surface, the building's steel or reinforced concrete columns are placed in holes drilled in the bedrock. Otherwise, supporting piles are driven down to the bedrock.

Elevators

Elevators make use of motors, counterweights, and strong cables to move elevator cars carrying passengers or freight up and down. In the 1800s, the invention of the safety elevator and steel-framed buildings made skyscrapers practical (see pp.98–99).

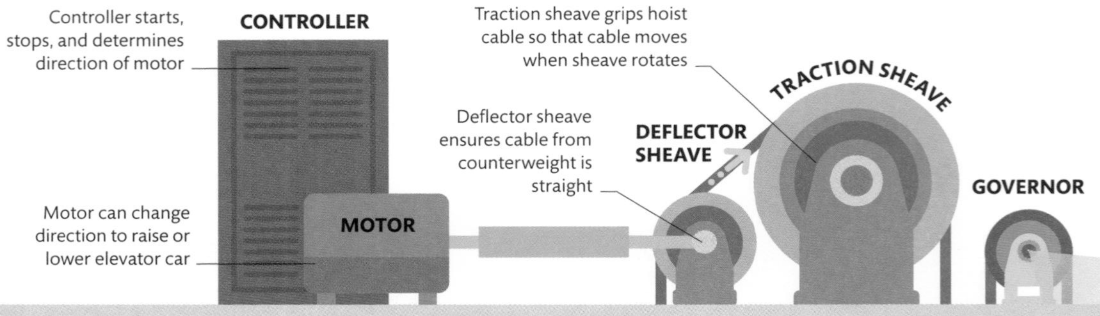

Controller starts, stops, and determines direction of motor
CONTROLLER
Motor can change direction to raise or lower elevator car
MOTOR
Traction sheave grips hoist cable so that cable moves when sheave rotates
TRACTION SHEAVE
Deflector sheave ensures cable from counterweight is straight
DEFLECTOR SHEAVE
GOVERNOR

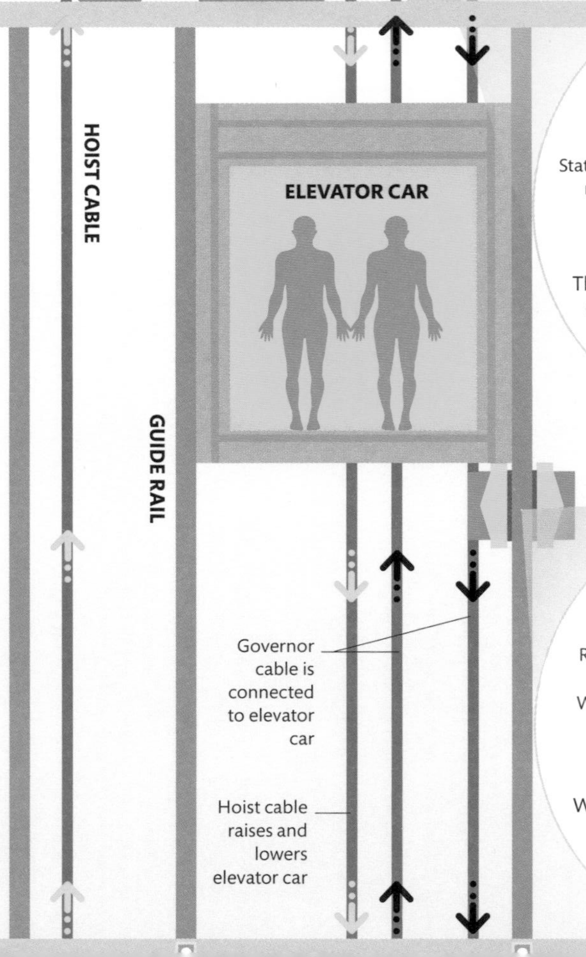

HOIST CABLE
GUIDE RAIL
ELEVATOR CAR
Governor cable is connected to elevator car
Hoist cable raises and lowers elevator car

How elevators work

Most elevators are raised and lowered by metal hoist cables that pass over a pulley called a traction sheave. The sheave is connected to an electric motor that powers the elevator. At one end of the cables is the elevator car and at the other end is a counterweight. The car runs along guide rails, which prevent it from swaying sideways. In an emergency, safety brakes clamp against the guide rail to stop the elevator car. The controller and power systems are usually housed in a machine room above the elevator shaft.

Safety systems

All elevators have safety features that mean it is almost impossible for a car to plummet down a shaft. These include multiple cables, each of which can support the weight of a loaded car on its own, as well as speed control and safety brakes.

GOVERNOR

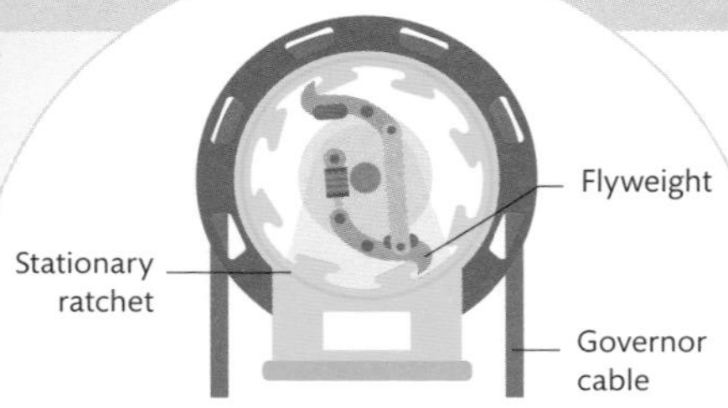

Flyweight
Stationary ratchet
Governor cable

The governor limits the speed of the elevator car. If the governor cable runs too fast, the flyweight engages with the ratchet, which stops the governor from rotating, triggering the safety brakes.

SAFETY BRAKES

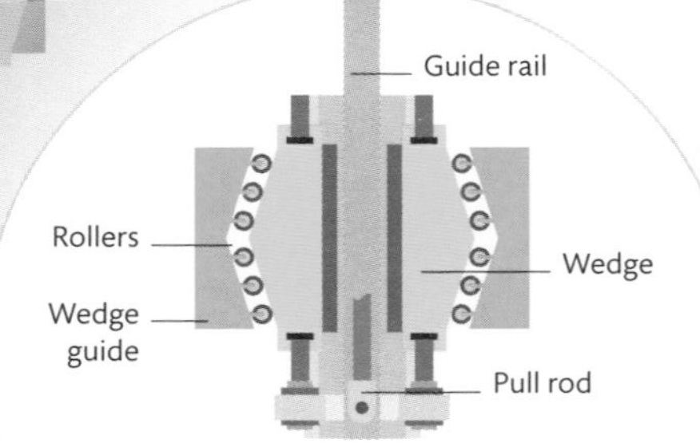

Guide rail
Rollers
Wedge
Wedge guide
Pull rod

When the governor stops rotating, it jerks on the pull rod. This, in turn, causes wedges to press against the guide rail, thereby bringing the elevator car to a stop by friction.

HOIST CABLE

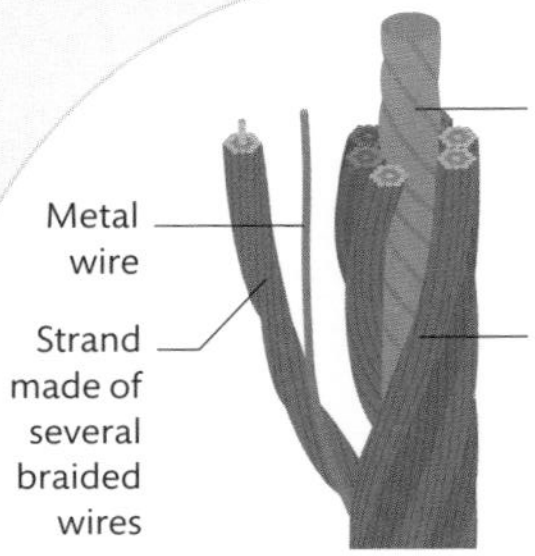
Metal wire
Strand made of several braided wires
Metal or synthetic core
Rope made of strands wrapped in a helical pattern around core

Each cable is made from many thin wires braided together. One cable can support the weight of the elevator car on its own, but most elevators have between four and eight cables.

ELEVATORS ARE THE SAFEST FORM OF TRAVEL AND ARE 50 TIMES SAFER THAN STAIRS

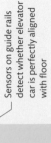

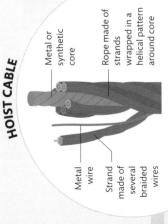

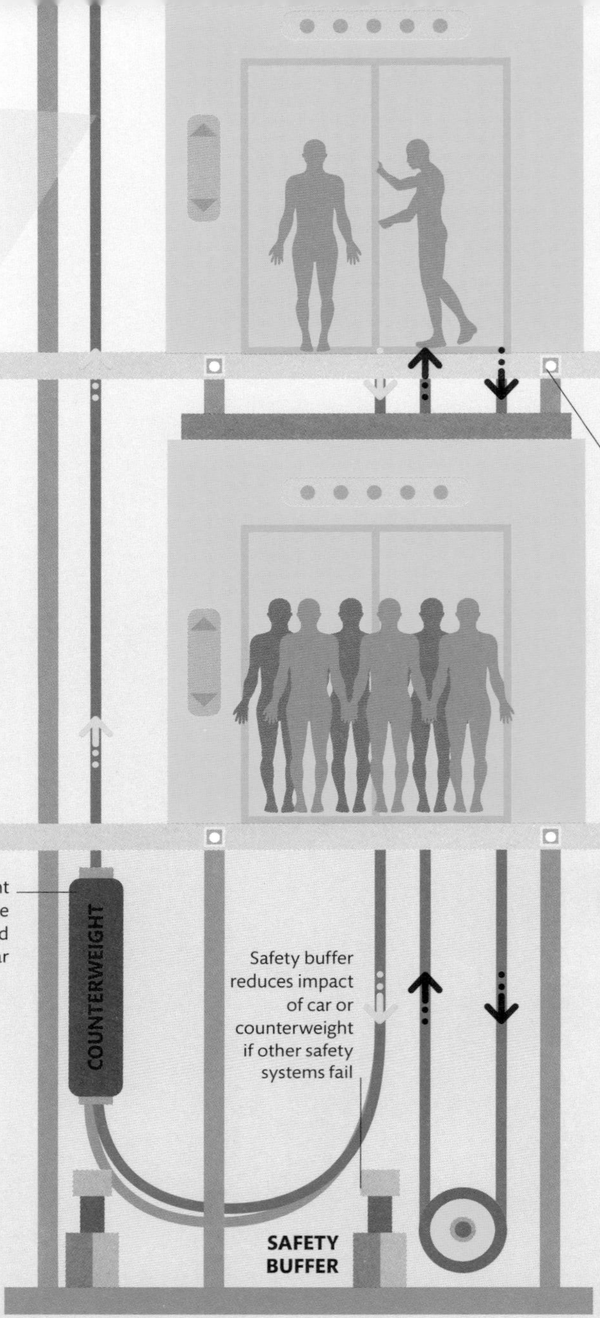
Sensors on guide rails detect whether elevator car is perfectly aligned with floor
Counterweight reduces the energy needed to raise car
COUNTERWEIGHT
Safety buffer reduces impact of car or counterweight if other safety systems fail
SAFETY BUFFER

Safety doors
Elevators have inner and outer doors. The inner doors are part of the elevator car, whereas the outer doors are part of the elevator shaft. The cars have a mechanism that unlocks the outer doors and pulls them open. In this way, the outer doors will open only if there is a car at that floor.

Weight limit
All elevators have a maximum weight limit, which varies according to the size of elevator and its machinery. If the elevator's sensors detect overloading, they will prevent the doors from closing. Freight elevators are designed to carry much heavier loads than passenger elevators.

HOW FAST CAN AN ELEVATOR TRAVEL?

The fastest elevators can travel upward at about 67 ft (20.5 m) per second. The maximum down speed is about 33 ft (10 m) per second for most elevators.

ELEVATOR PROGRAMMING

Elevators are controlled by computers that are programmed with efficient strategies for running the elevator cars. Usually, a car on its way up will not answer a "down" call until it has fulfilled all its "up" calls, and vice versa. Advanced systems take passenger traffic patterns into account and direct the elevator cars according to demand.

Anatomy of a tower crane

The tower rises from a heavy base set into the ground. Attached to the top of the tower is the slewing ring that rotates the jib. The counterweight helps to support the load. A system of wires and pulleys lifts the load and moves it along the jib.

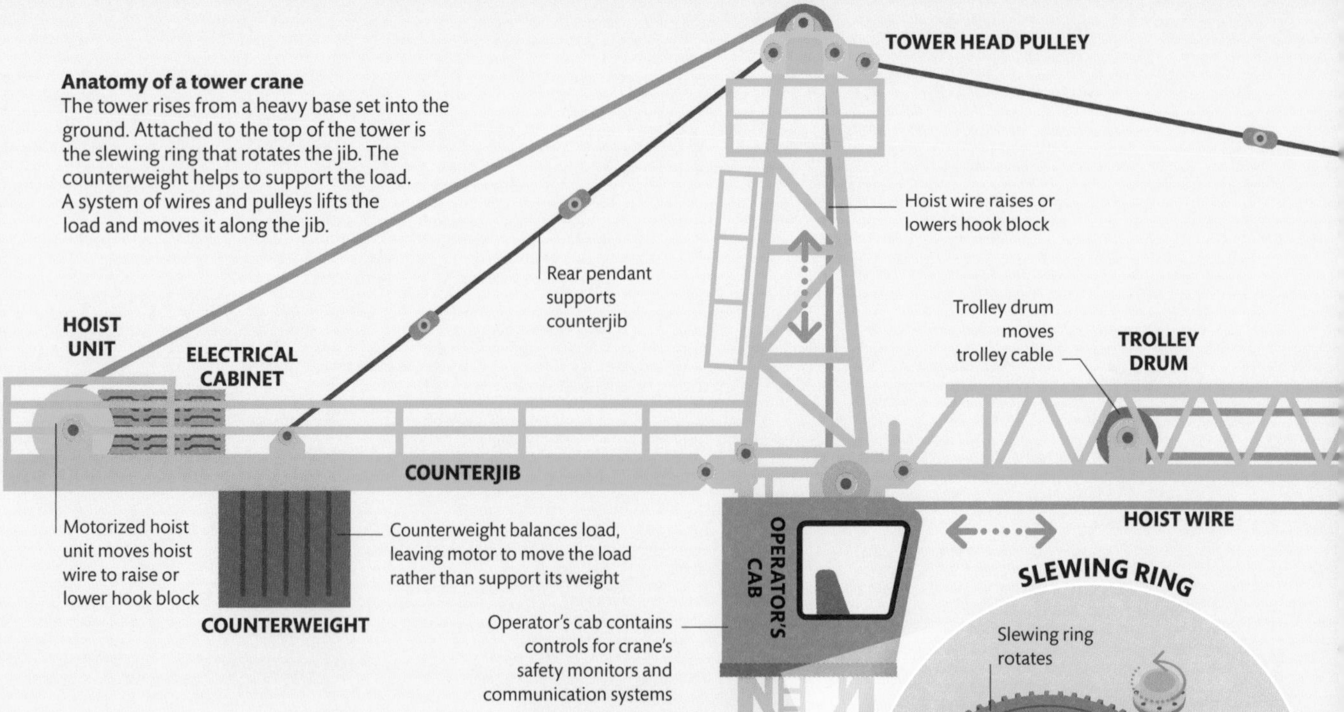
TOWER HEAD PULLEY
Hoist wire raises or lowers hook block
Rear pendant supports counterjib
HOIST UNIT
ELECTRICAL CABINET
Trolley drum moves trolley cable
TROLLEY DRUM
COUNTERJIB
HOIST WIRE
Motorized hoist unit moves hoist wire to raise or lower hook block
Counterweight balances load, leaving motor to move the load rather than support its weight
COUNTERWEIGHT
OPERATOR'S CAB
Operator's cab contains controls for crane's safety monitors and communication systems
Tower climbing unit lifts upper parts of crane, enabling a new section to be added

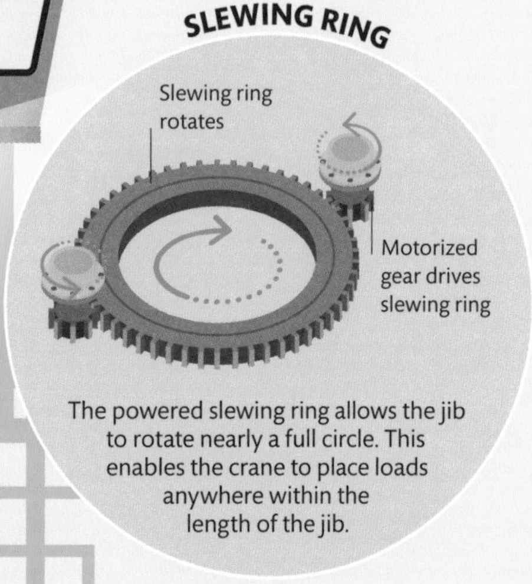
SLEWING RING
Slewing ring rotates
Motorized gear drives slewing ring

The powered slewing ring allows the jib to rotate nearly a full circle. This enables the crane to place loads anywhere within the length of the jib.

Cranes

Heavy loads require equally heavyweight equipment to move them into place. In most instances, this requires a crane. The masses of cranes on so many city skylines demonstrate how vital these machines are in shaping our world.

Tower cranes

Tower cranes consist of a mast (or tower) and a horizontal main arm (or jib). They can soar up to about 260 ft (80 m) high—or even higher if tethered to a building as it goes up. The jib can extend out 250 ft (75 m). It carries a pulley with a trolley that moves along the jib. Attached to the trolley is the hook block, which supports the load. A shorter arm—the counterjib—extends outward in the opposite direction. The counterjib carries a concrete counterweight, winding gear, and motors.

TOWER

WHY DON'T TOWER CRANES FALL OVER?

Tower cranes are bolted to concrete pads set into the ground that weigh about 200 tons (180 tonnes). Tall cranes may also be secured to the building with a metal tie-in.

SOME **SPECIALIST CRANES** CAN LIFT **1,800 TONS (1,600 TONNES)**– NEARLY THE WEIGHT OF **400 ELEPHANTS**

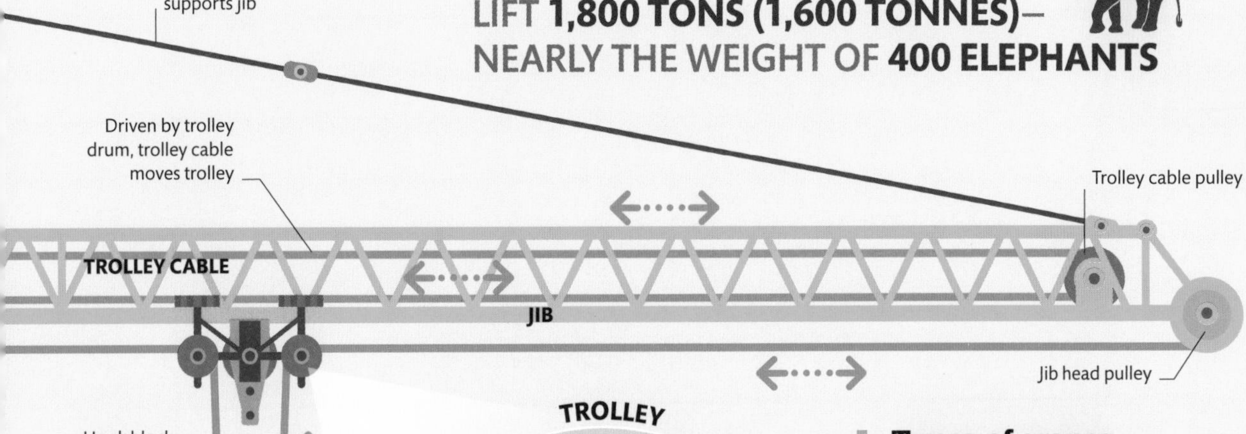

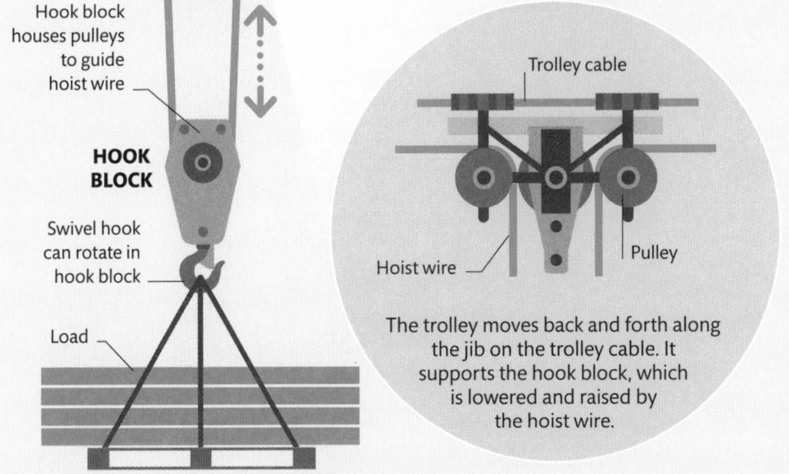
Front pendant supports jib
Driven by trolley drum, trolley cable moves trolley
Trolley cable pulley
TROLLEY CABLE
JIB
Jib head pulley
Hook block houses pulleys to guide hoist wire
HOOK BLOCK
Swivel hook can rotate in hook block
Load
TROLLEY
Trolley cable
Hoist wire
Pulley

The trolley moves back and forth along the jib on the trolley cable. It supports the hook block, which is lowered and raised by the hoist wire.

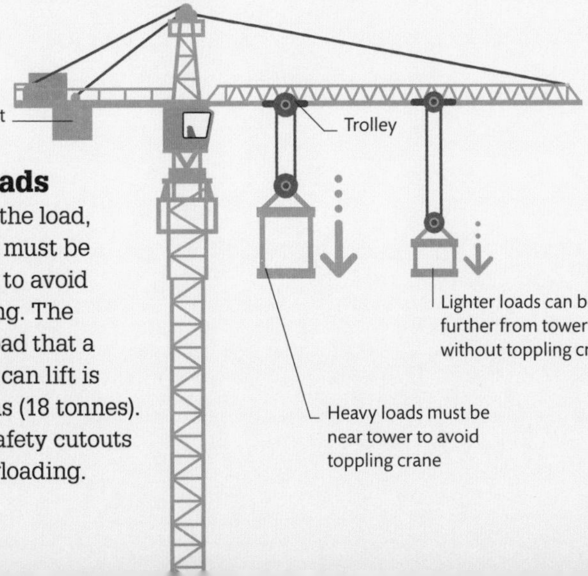
Counterweight
Trolley
Lighter loads can be further from tower without toppling crane
Heavy loads must be near tower to avoid toppling crane

Lifting loads

The heavier the load, the nearer it must be to the tower to avoid overbalancing. The maximum load that a tower crane can lift is about 20 tons (18 tonnes). Automatic safety cutouts prevent overloading.

Types of cranes

Most land-based cranes fall into four main types: tower cranes, including cantilever cranes; overhead cranes, such as gantry cranes; level-luffing cranes; and mobile cranes, which often use hydraulics for lifting loads.

Mobile crane
A mobile crane is mounted on a truck chassis with the crane mounted on a turntable. The crane is powered by hydraulics.

Level-luffing crane
In this crane, the hook stays at the same level while the jib moves up and down, moving the hook inward and outward.

Gantry crane
This type of crane is on a fixed structure straddling an object or workspace. Gantry cranes are often used in shipyards or container depots.

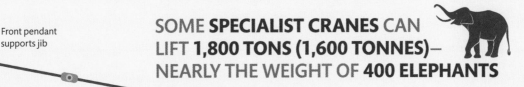

Cantilever crane
Cantilever cranes are the low-level forerunners of tower cranes. They have a steel tower with a rotating, counterbalanced jib.

TECHNOLOGY IN THE HOME

Hot-air heating
In hot-air heating systems, cool air is drawn out of rooms and travels through a return duct to a heating unit. There, it is warmed by a heat exchanger powered by a furnace, typically burning oil or gas. The warm air rises and is channeled around the property through supply ducts.

Central heating
Water is heated, often using an oil- or gas-fired boiler, and then circulated in a closed system of pipes and radiators to warm rooms (see pp.108–109). Thermostats monitor temperature to ensure the desired heat level is maintained.

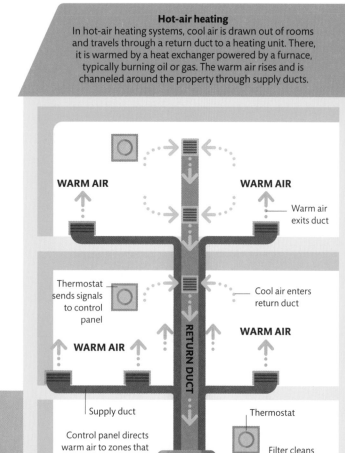

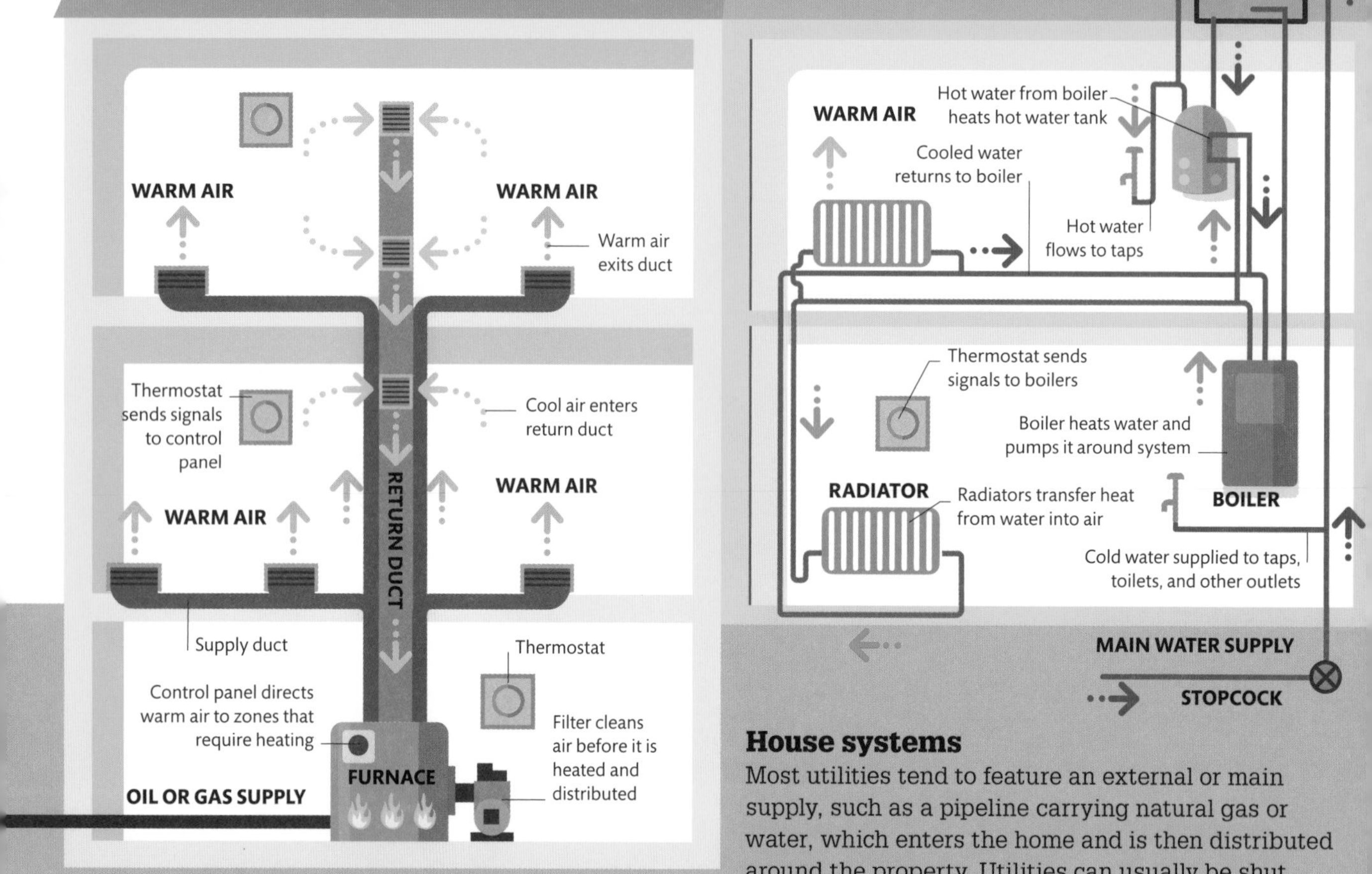

HEATING SYSTEM FEED TANK
WARM AIR
WARM AIR
Warm air exits duct
Thermostat sends signals to control panel
Cool air enters return duct
WARM AIR
WARM AIR
RETURN DUCT
Supply duct
Thermostat
Control panel directs warm air to zones that require heating
Filter cleans air before it is heated and distributed
FURNACE
OIL OR GAS SUPPLY
WARM AIR
Hot water from boiler heats hot water tank
Cooled water returns to boiler
Hot water flows to taps
Thermostat sends signals to boilers
Boiler heats water and pumps it around system
RADIATOR
Radiators transfer heat from water into air
BOILER
Cold water supplied to taps, toilets, and other outlets
MAIN WATER SUPPLY
STOPCOCK

House systems

Most utilities tend to feature an external or main supply, such as a pipeline carrying natural gas or water, which enters the home and is then distributed around the property. Utilities can usually be shut off or disconnected easily in the event of a problem or if a property is lying empty.

Utilities in the home

Utilities include power, heating, water, and communications services supplied to homes. They are usually provided by external companies, although some properties may have an independent water supply or heating source, such as a wood-burning fire.

Electricity supply

Electricity is metered and distributed around a home by a consumer unit. Plug sockets and other power outlets are commonly found on ring circuits with both ends of the circuit connected to the consumer unit. Radial circuits, often used for lighting, branch off from one central point.

Water supply

A water pipe carries clean, fresh water under pressure into a home, where it may be routed to a cistern or storage tanks, or be available on demand when a tap is turned on. Wastewater is carried away via further pipes, usually to a sewage treatment plant.

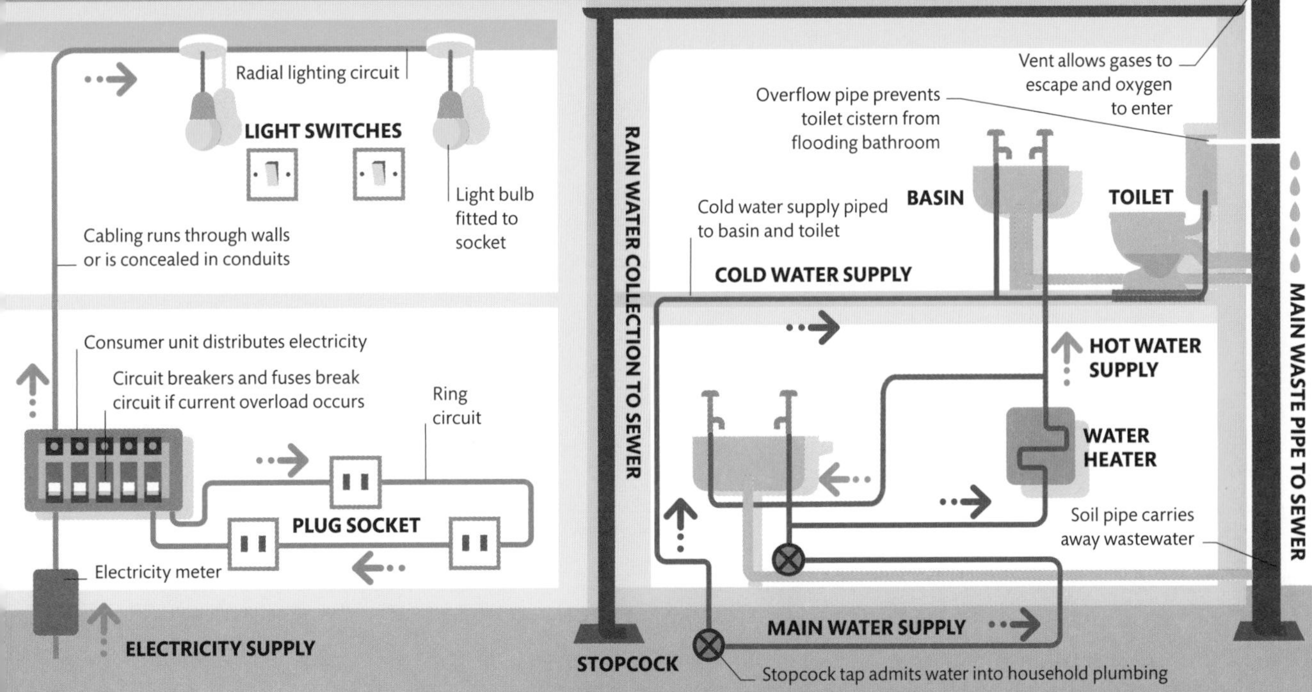
VENT
Radial lighting circuit
LIGHT SWITCHES
Light bulb fitted to socket
Cabling runs through walls or is concealed in conduits
Consumer unit distributes electricity
Circuit breakers and fuses break circuit if current overload occurs
Ring circuit
PLUG SOCKET
Electricity meter
ELECTRICITY SUPPLY
RAIN WATER COLLECTION TO SEWER
Vent allows gases to escape and oxygen to enter
Overflow pipe prevents toilet cistern from flooding bathroom
Cold water supply piped to basin and toilet
BASIN
TOILET
COLD WATER SUPPLY
HOT WATER SUPPLY
WATER HEATER
Soil pipe carries away wastewater
MAIN WASTE PIPE TO SEWER
MAIN WATER SUPPLY
STOPCOCK
Stopcock tap admits water into household plumbing

KEY

- Warm air
- Cold air
- Warm water
- Cold water
- Electricity

HOW CAN WE SMELL ODORLESS NATURAL GAS?

Methane and propane have no smell. Suppliers add an odorant such as ethyl mercaptan, which smells of rotten eggs, so that gas leaks can be detected by smell.

MAGNETIC CIRCUIT BREAKERS

These safety switches protect against circuit overloads. Electric current flows through the circuit breaker and its two contacts, which complete the circuit. If the current exceeds a limit, an electromagnet attracts a metal lever toward it, pulling the contacts apart to break the circuit.

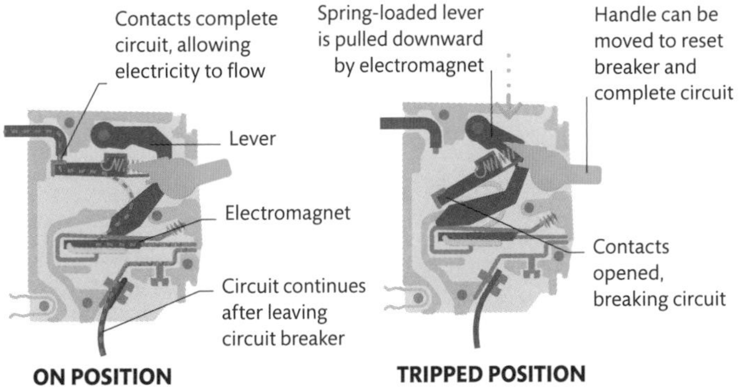
Contacts complete circuit, allowing electricity to flow
Lever
Electromagnet
Circuit continues after leaving circuit breaker
ON POSITION
Spring-loaded lever is pulled downward by electromagnet
Handle can be moved to reset breaker and complete circuit
Contacts opened, breaking circuit
TRIPPED POSITION

Heating

Heating systems are one of the main consumers of energy in most homes. Depending on location and available utilities, many different devices—from space heaters to entire central heating systems—are used to warm homes.

3 Water is heated
Heat is transferred to cold water running through pipes looping around the heat exchanger.

2 Combustion
Gas and air enter the combustion chamber and are ignited. Their burning heats up the heat exchanger.

Hot water on demand

Some home heating systems heat water, which is stored in a tank and used when needed. Other systems heat cold water only when a user demands it, such as by turning on a hot tap. Combination (or "combi") boilers provide hot water on demand but also make use of two heat exchangers to send hot water around a closed-loop system of pipes and radiators to centrally heat a home.

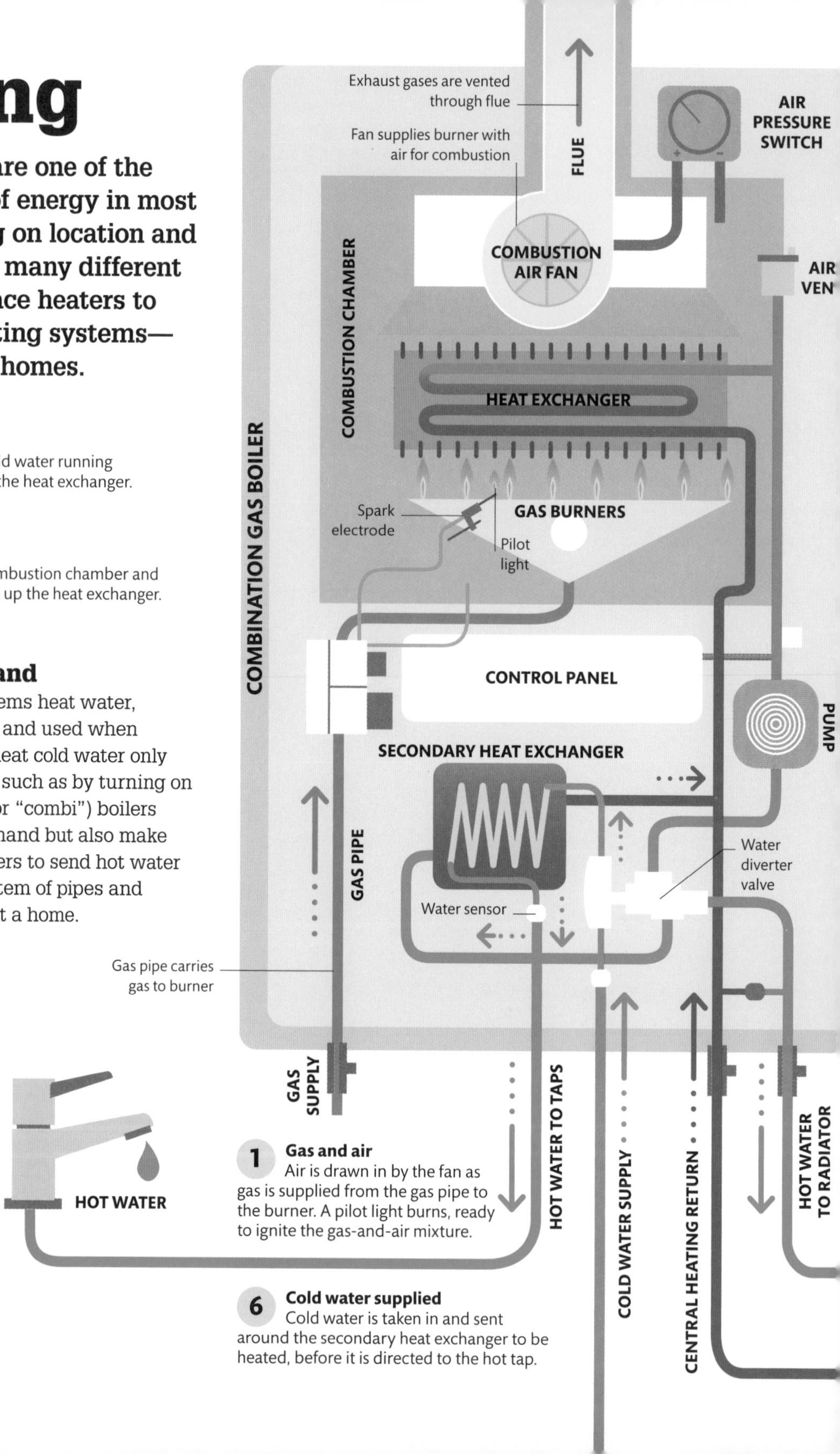

7 Hot water reaches tap
Hot water flows from the tap. When the tap is turned off, the diverter valve closes to allow the central heating to continue.

1 Gas and air
Air is drawn in by the fan as gas is supplied from the gas pipe to the burner. A pilot light burns, ready to ignite the gas-and-air mixture.

5 Hot water tap turned on
Turning on a hot water tap makes the boiler's water diverter valve reroute some hot water to the secondary heat exchanger.

6 Cold water supplied
Cold water is taken in and sent around the secondary heat exchanger to be heated, before it is directed to the hot tap.

Thermostats

Used to maintain temperatures in a home, thermostats may be local—fitted to a room—or global. When the temperature drops below the temperature set by the user, the thermostat completes a circuit that sends a signal instructing the boiler to fire and generate more heat.

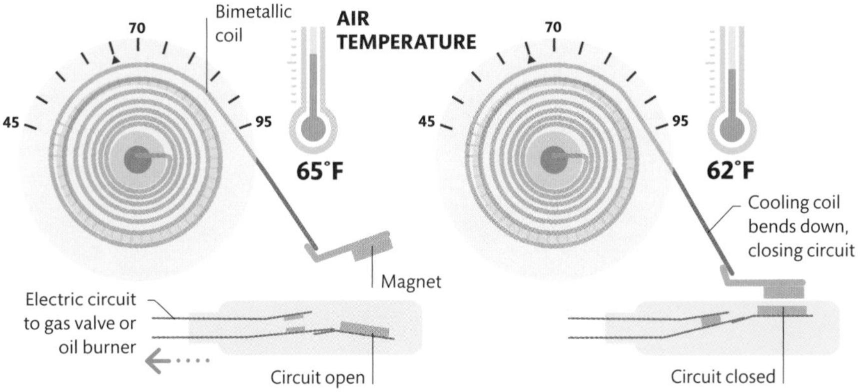
Bimetallic coil
AIR TEMPERATURE
70
45
95
65°F
Magnet
Electric circuit to gas valve or oil burner
Circuit open
70
45
95
62°F
Cooling coil bends down, closing circuit
Circuit closed

1 Warm enough When the temperature is higher than the desired setting (64°F/18°C in this example), the coil warms up and straightens, pulling the magnet away from its contact and breaking the circuit. The boiler stops firing.

2 Cooler than desired As it cools, the coil bends and the magnet moves toward the contact. The contact closes and completes the electric circuit, which sends a signal to the boiler to fire and heat water.

Central heating

Hot water is pumped from the boiler through pipes or channels inside radiators, heating the radiator's outer panels, which warm the surrounding air. The lockshield valve adjusts the speed of the water's flow through the radiator—a slower flow makes radiators become hotter.

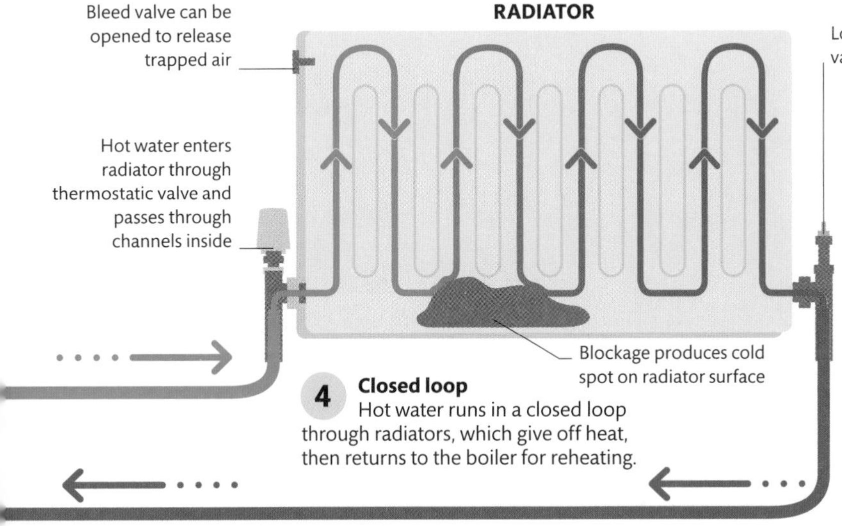
RADIATOR
Bleed valve can be opened to release trapped air
Hot water enters radiator through thermostatic valve and passes through channels inside
Lockshield valve
Blockage produces cold spot on radiator surface

4 Closed loop Hot water runs in a closed loop through radiators, which give off heat, then returns to the boiler for reheating.

UNDERFLOOR HEATING

There are two main types of underfloor heating systems. A wet system uses networks of pipes or tubes through which heated water is pumped. Dry systems feature heated coils powered by electricity. Both can be expensive to install and run, but they can let heat radiate up through the floor to warm an entire room evenly without cold spots.

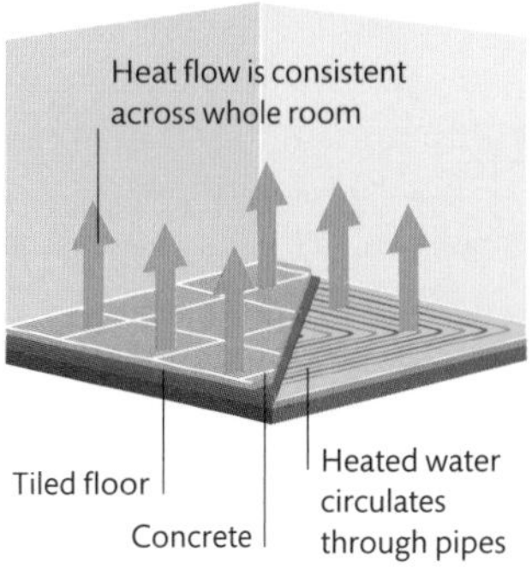
Heat flow is consistent across whole room
Tiled floor
Concrete
Heated water circulates through pipes

WATER-BASED (WET) UNDERFLOOR HEATING

DOES TURNING UP A THERMOSTAT HEAT A HOUSE FASTER?

No. When a thermostat is set, the boiler runs at its maximum until the desired temperature is reached. It does not operate faster to reach higher temperatures.

WHY DO I NEED TO PIERCE THE FILM WHEN HEATING READY MEALS?

As microwaves heat water molecules in food, they expand to become steam. Piercing the film lets steam escape the container, which might otherwise explode.

4 Heating food
Microwaves reflect off the oven's metal interior but pass through containers made of plastic, glass, or ceramics into the food, heating it up.

3 Microwave delivery
A wave guide channels the microwaves from the magnetron to the sealed cooking chamber of the oven. The microwaves bounce around the interior of the cooking chamber.

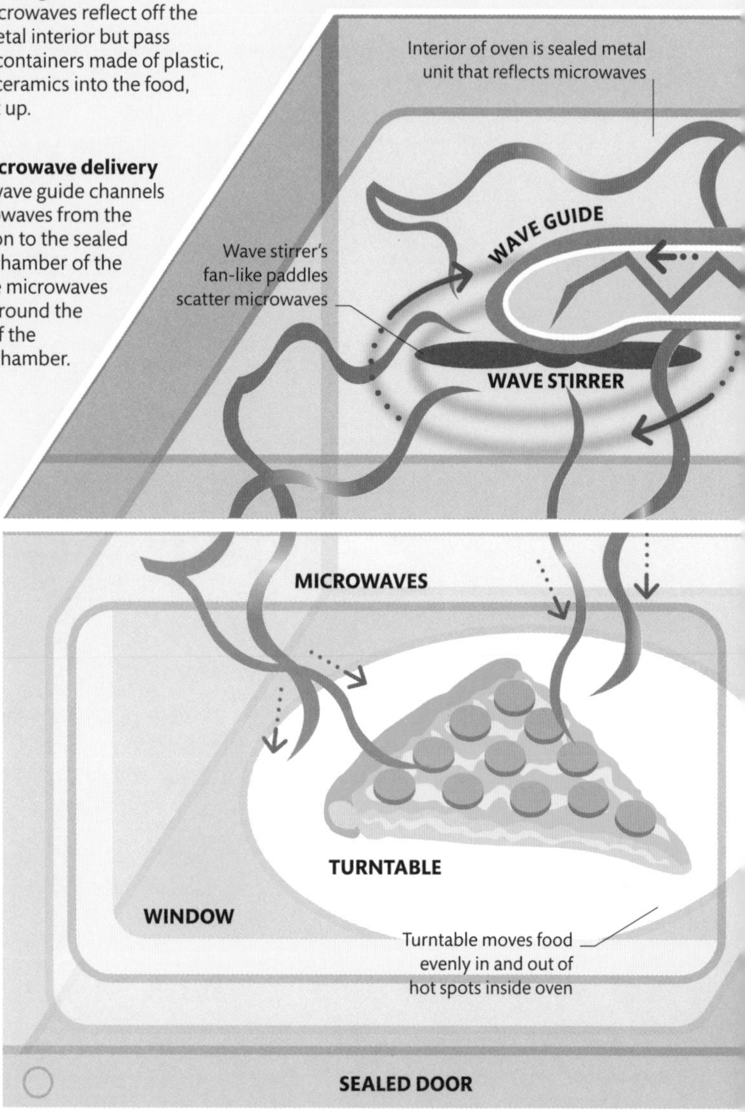
Interior of oven is sealed metal unit that reflects microwaves
WAVE GUIDE
Wave stirrer's fan-like paddles scatter microwaves
WAVE STIRRER
MICROWAVES
TURNTABLE
WINDOW
Turntable moves food evenly in and out of hot spots inside oven
SEALED DOOR

How a microwave works

A home microwave oven uses AC power to power a magnetron. This device uses interacting electric and magnetic fields to produce microwaves, which oscillate and reverse their electric field several billion times a second. The microwaves are directed into the oven's cooking chamber—a sealed metal box—where they bounce around, striking and exciting molecules in food, which heat up as a result.

2 Microwave creation
The magnetron generates microwaves oscillating at a frequency of 2.45GHz (2.45 billion times per second).

1 Control settings
The user selects power and time settings, often using a touchscreen panel. Safety switches in the door cut power if the door is opened while the oven is working.

Microwave ovens

Microwaves are a type of energy found on the electromagnetic spectrum between infrared and radio waves (see pp.136–137). They pass through many, but not all, materials and can penetrate food to agitate water and fat molecules, generating heat so that foods are cooked evenly and more quickly than in a conventional oven.

THE **FIRST COMMERCIAL MICROWAVE** OVEN WAS **5½ FT (1.7 M) TALL**

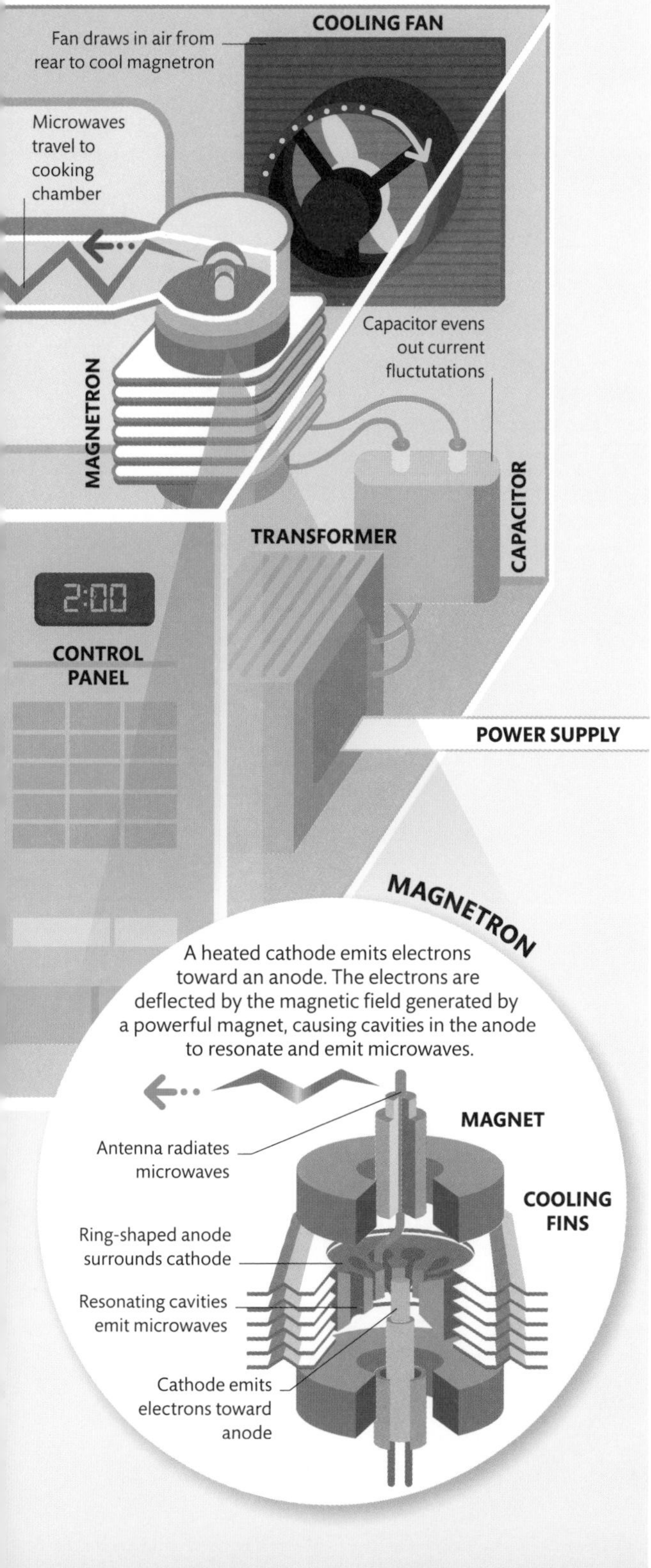
COOLING FAN
Fan draws in air from rear to cool magnetron
Microwaves travel to cooking chamber
MAGNETRON
Capacitor evens out current fluctutations
CAPACITOR
TRANSFORMER
2:00
CONTROL PANEL
POWER SUPPLY
MAGNETRON
A heated cathode emits electrons toward an anode. The electrons are deflected by the magnetic field generated by a powerful magnet, causing cavities in the anode to resonate and emit microwaves.
MAGNET
Antenna radiates microwaves
COOLING FINS
Ring-shaped anode surrounds cathode
Resonating cavities emit microwaves
Cathode emits electrons toward anode

Moving molecules

Water molecules comprise a negatively charged oxygen atom and two positively charged hydrogen atoms. The molecules turn to align with the polarity of the microwaves' electric field. This field changes its polarity billions of times each second, causing the molecules to constantly flip back and forth.

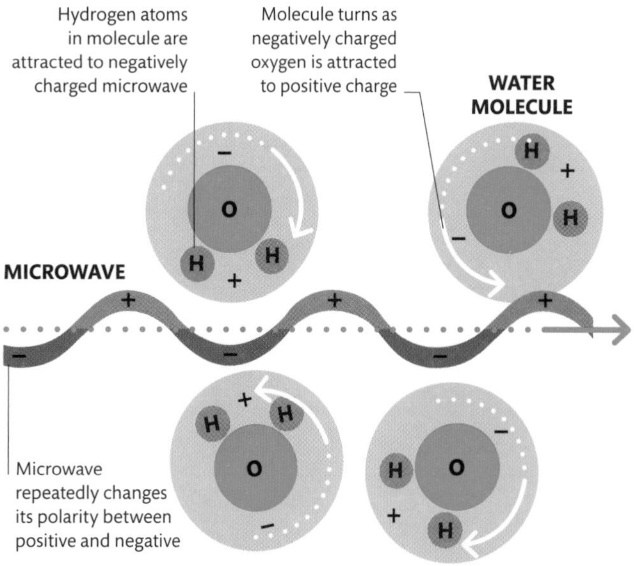
Hydrogen atoms in molecule are attracted to negatively charged microwave
Molecule turns as negatively charged oxygen is attracted to positive charge
WATER MOLECULE
MICROWAVE
O
H
+
–
Microwave repeatedly changes its polarity between positive and negative

Generating heat
As the water molecules rotate back and forth to align with the changing electric field, they rub against each other, generating heat through friction.

INDUSTRIAL MICROWAVE OVENS

Large microwave ovens are used in industry to dry and cure plastic reinforced with carbon fiber, to remove moisture to create dried foodstuffs, and, in some instances, to vulcanize rubber.

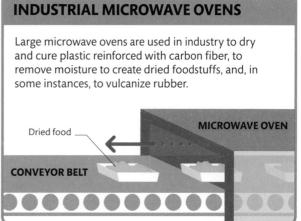

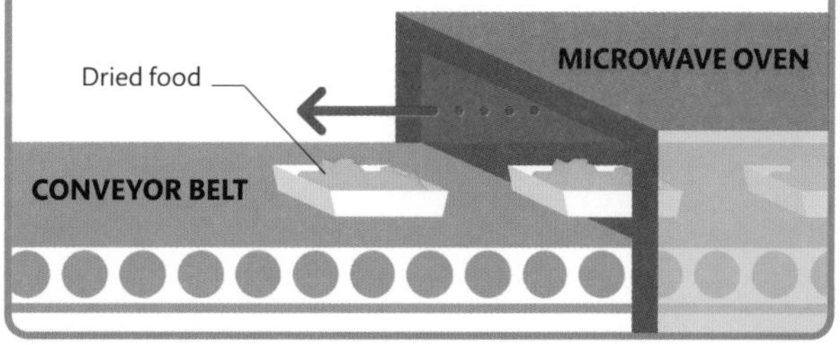
MICROWAVE OVEN
Dried food
CONVEYOR BELT

Kettles and toasters

When an electric current flows through a wire, electrical energy is converted into heat energy. This principle is used in the heating elements found in several kitchen appliances.

Kettles

At the bottom of an electric kettle's watertight chamber is a large heating element. When the kettle is switched on, electricity passes through the element, heating it rapidly. The heat is transferred to the water until the thermostat detects that boiling temperature has been reached. The thermostat breaks an electrical contact, and power to the element is stopped.

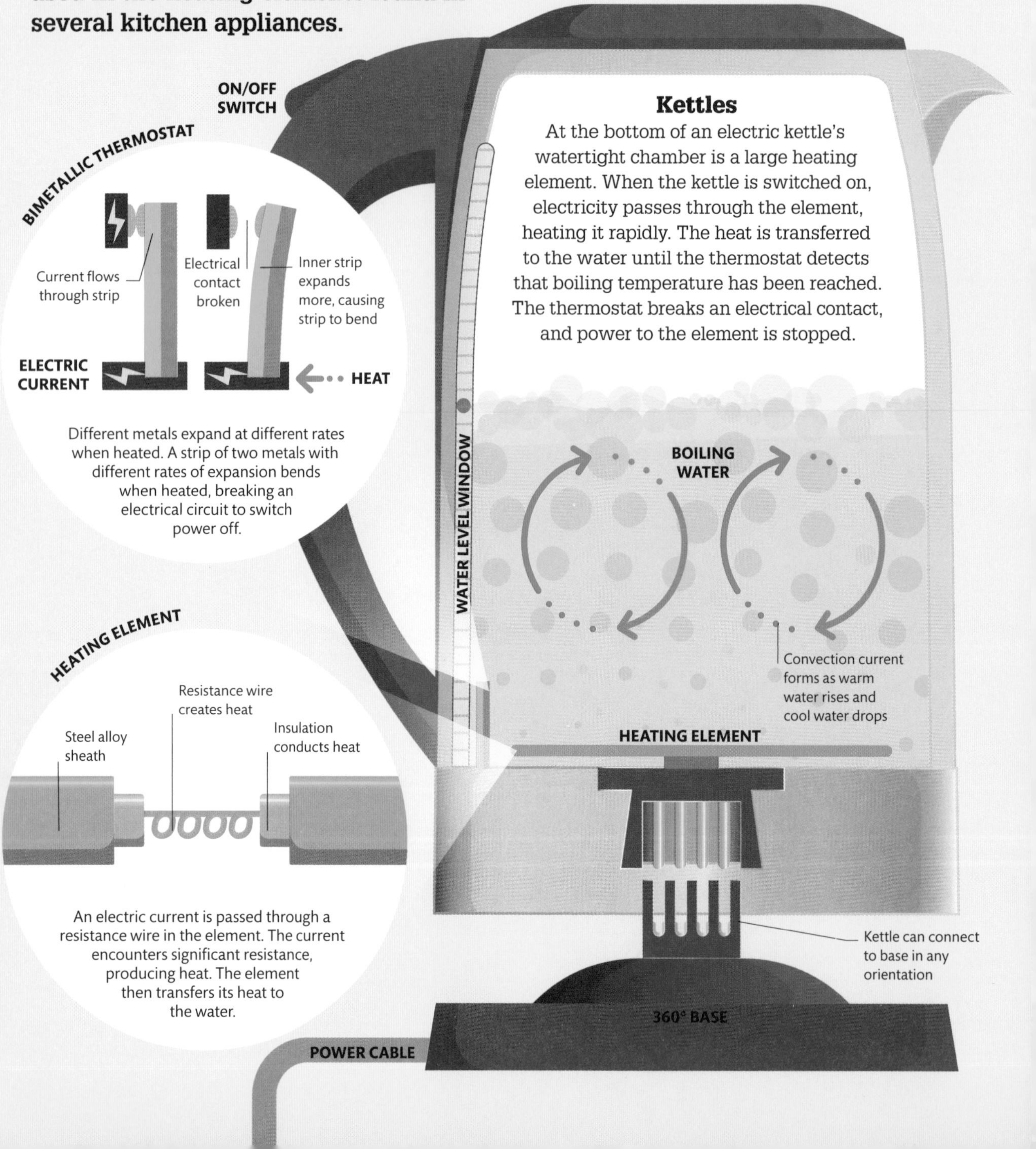

ON/OFF SWITCH
BIMETALLIC THERMOSTAT
Current flows through strip
Electrical contact broken
Inner strip expands more, causing strip to bend
ELECTRIC CURRENT
HEAT
Different metals expand at different rates when heated. A strip of two metals with different rates of expansion bends when heated, breaking an electrical circuit to switch power off.
HEATING ELEMENT
Resistance wire creates heat
Steel alloy sheath
Insulation conducts heat
An electric current is passed through a resistance wire in the element. The current encounters significant resistance, producing heat. The element then transfers its heat to the water.
WATER LEVEL WINDOW
BOILING WATER
Convection current forms as warm water rises and cool water drops
HEATING ELEMENT
Kettle can connect to base in any orientation
360° BASE
POWER CABLE

Toasters

Thin wires made of nichrome, an alloy of nickel and chromium, glow red hot when electricity passes through them. These wires form the heating elements that caramelize the starch and sugars in bread to produce toast. An electric circuit is completed when the bread tray is pressed down, allowing current to flow through the elements, and broken by an adjustable timing mechanism.

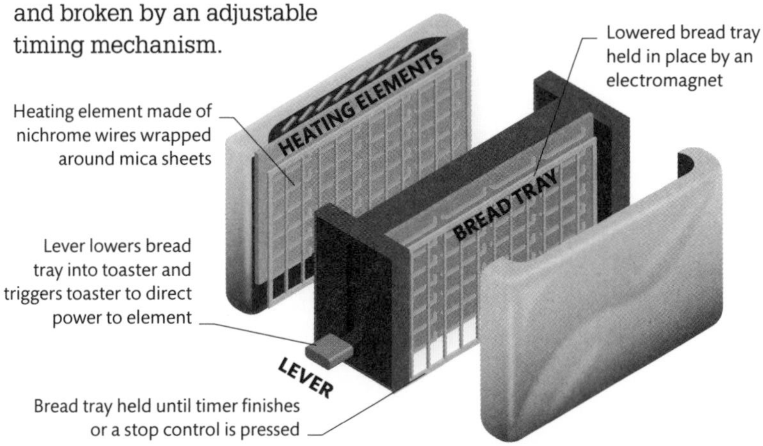
Lowered bread tray held in place by an electromagnet
Heating element made of nichrome wires wrapped around mica sheets
HEATING ELEMENTS
BREAD TRAY
Lever lowers bread tray into toaster and triggers toaster to direct power to element
LEVER
Bread tray held until timer finishes or a stop control is pressed

MOKA POTS

Heated on a stove top, pressure builds in a moka pot's water chamber. Pressure forces the water up a funnel, bubbling through the coffee grounds and finally into the upper chamber as ready-to-drink coffee.

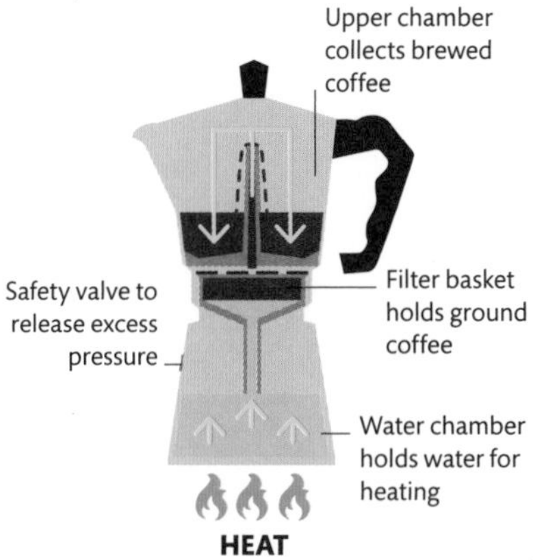
Upper chamber collects brewed coffee
Safety valve to release excess pressure
Filter basket holds ground coffee
Water chamber holds water for heating
HEAT

Espresso machines

The machine's element heats a large reservoir of water, creating steam. The steam travels through a heat exchanger, where it flash-heats fresh, cold water, which is pumped in under pressure. The heated, pressurized water flows slowly through a portafilter containing tamped (compressed) ground coffee to make an espresso.

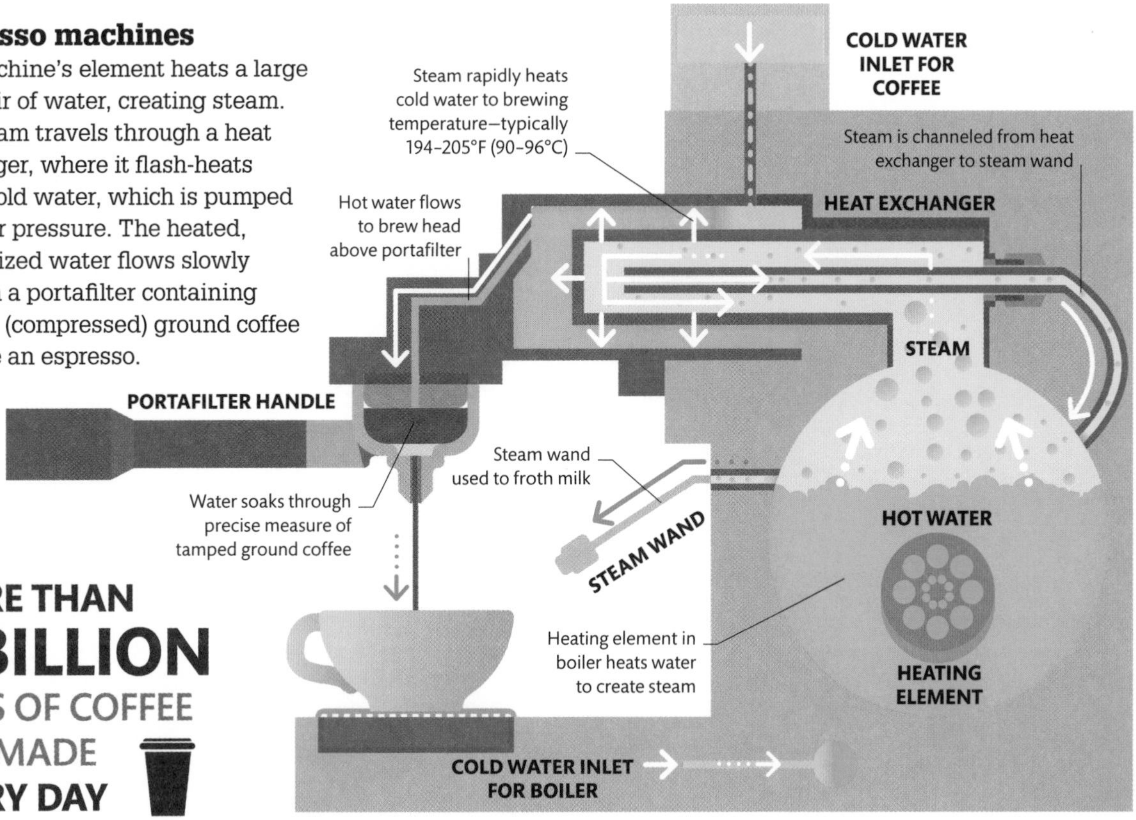
COLD WATER INLET FOR COFFEE
Steam rapidly heats cold water to brewing temperature—typically 194–205°F (90–96°C)
Steam is channeled from heat exchanger to steam wand
Hot water flows to brew head above portafilter
HEAT EXCHANGER
STEAM
PORTAFILTER HANDLE
Steam wand used to froth milk
Water soaks through precise measure of tamped ground coffee
STEAM WAND
HOT WATER
Heating element in boiler heats water to create steam
HEATING ELEMENT
COLD WATER INLET FOR BOILER

MORE THAN
2 BILLION
CUPS OF COFFEE ARE MADE EVERY DAY

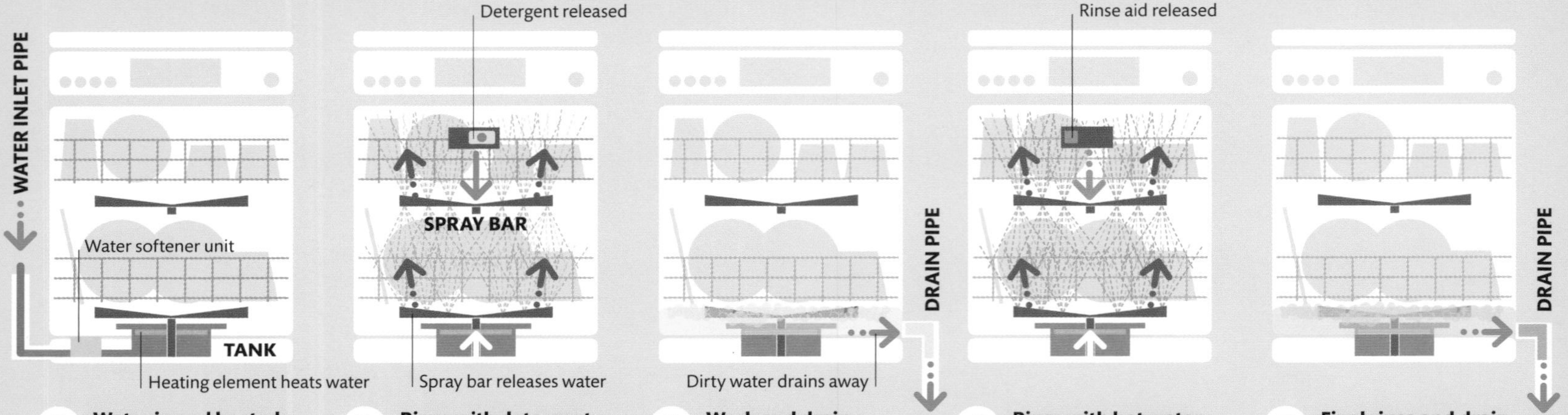
WATER INLET PIPE
Water softener unit
TANK
Heating element heats water
Detergent released
SPRAY BAR
Spray bar releases water
DRAIN PIPE
Dirty water drains away
Rinse aid released
DRAIN PIPE

1 Water in and heated A pump draws water from the water supply, usually through a water softener unit. An electric element in the base heats the water.

2 Rinse with detergent As water is pumped through the spray bars, detergent is released. Pressure causes the bars to spin and spray water out in all directions.

3 Wash and drain Hot water and detergent are repeatedly pumped through the spray bars to clean the load. The cycle ends with dirty water draining away.

4 Rinse with hot water Clean water is pumped in and mixed with rinse aid, which lowers surface tension so that water runs off quickly without streaking cleaned objects.

5 Final rinse and drain A final rinse with clean water is an option with some dishwasher programs. The water is drained, and heat inside the machine assists drying.

Dishwashers

A dishwasher combines pumps, electric heating elements, high-pressure water sprays, and cleaning chemicals—all coordinated by a microprocessor—to wash, rinse, and dry kitchenware in a sequence of stages.

How a dishwasher works

Dishwashers heat and apply water under pressure to dirty dishes, cutlery, and kitchenware held in baskets and racks. Small, powerful jets of spray combined with dissolved detergent dislodge debris and stains. The cleaning process is aided by the high temperature of the water, which helps cut through grease and fatty deposits. To finish, a load is rinsed with water and rinse aid and then, in some machines, dried using heated air.

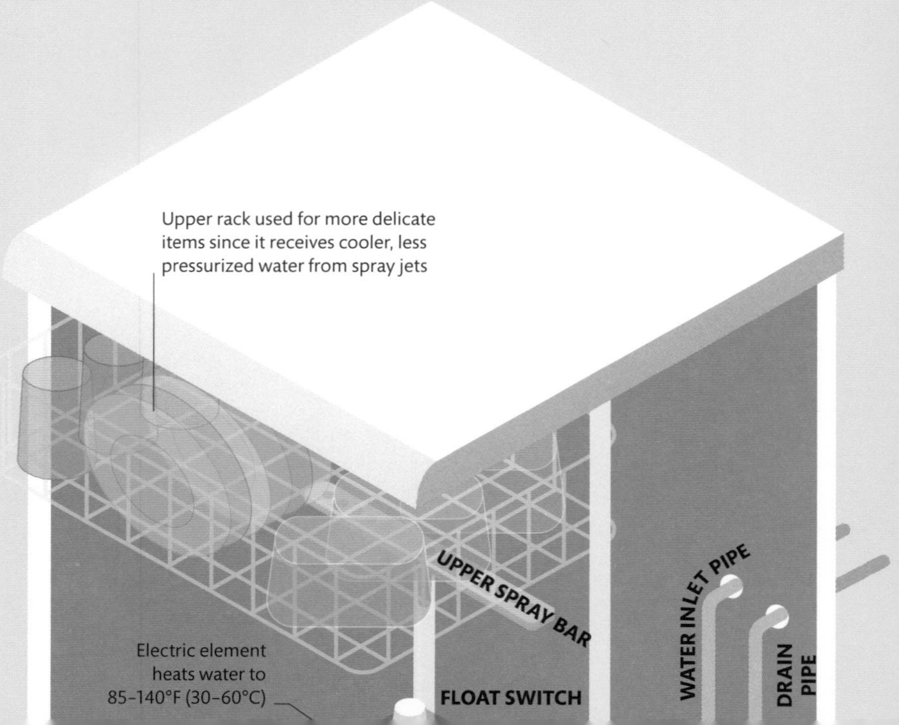
Upper rack used for more delicate items since it receives cooler, less pressurized water from spray jets
UPPER SPRAY BAR
WATER INLET PIPE
DRAIN PIPE
Electric element heats water to 85–140°F (30–60°C)
FLOAT SWITCH

High heat
A dishwasher uses high heat to clean household dishes and cutlery. Due to the heat, some items may melt or warp when washed in a dishwasher.

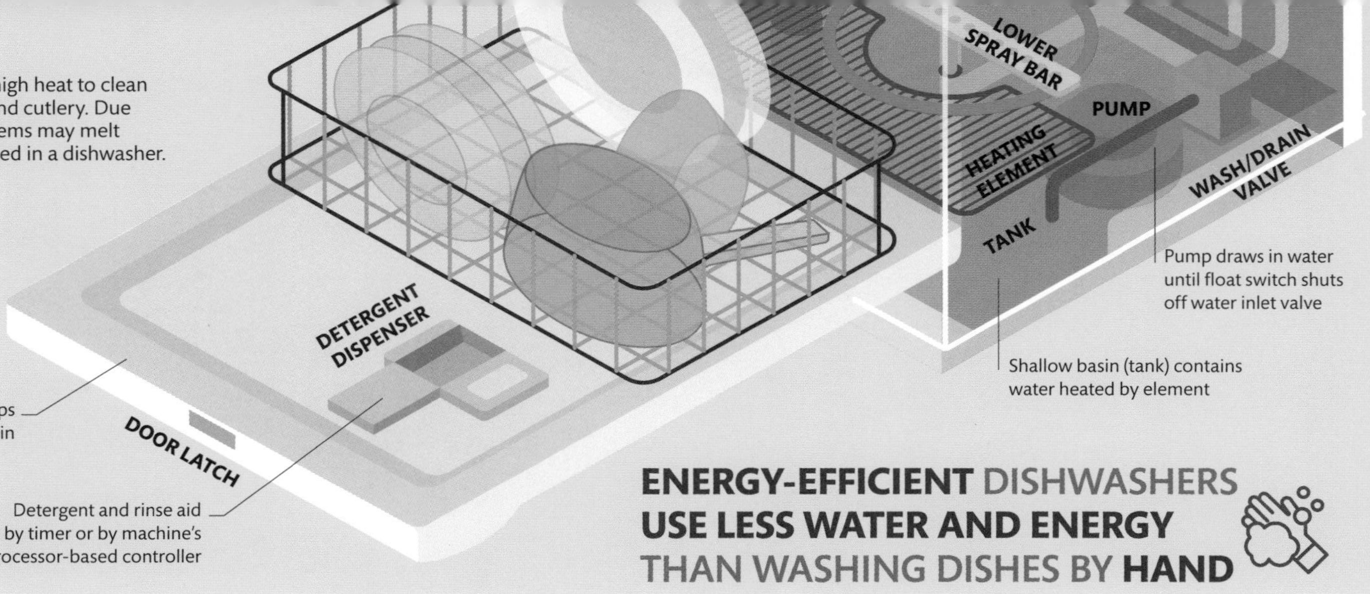
LOWER SPRAY BAR
PUMP
HEATING ELEMENT
WASH/DRAIN VALVE
TANK
Pump draws in water until float switch shuts off water inlet valve
Shallow basin (tank) contains water heated by element
DETERGENT DISPENSER
DOOR LATCH
Watertight seal keeps all moisture in
Detergent and rinse aid released by timer or by machine's microprocessor-based controller

ENERGY-EFFICIENT DISHWASHERS USE LESS WATER AND ENERGY THAN WASHING DISHES BY HAND

Water softeners

Some areas have hard water that inhibits detergents, leaves streaks, and damages heating elements. Hard water contains a higher concentration of minerals such as calcium and magnesium compounds. An ion exchanger passes hard water through resin beads loaded with sodium ions. Unwanted ions are attracted to the beads and exit the water as they displace the beads' sodium ions, leaving the water softened and lower in minerals.

KEY ▲ Calcium ⬟ Magnesium ◆ Sodium ■ Iron ⬢ Manganese

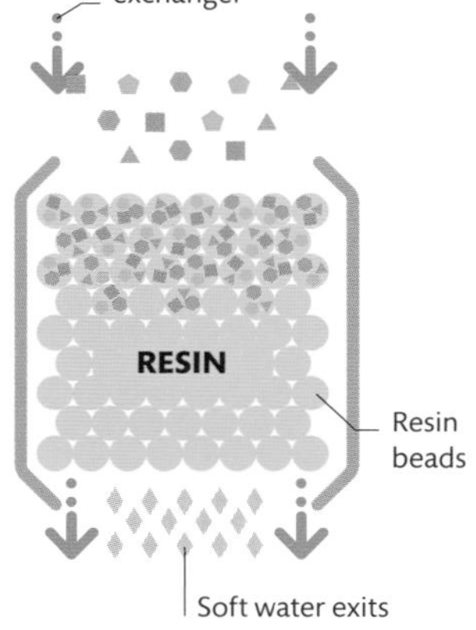
Hard water enters exchanger
RESIN
Resin beads
Soft water exits

Softening cycle
As hard water flows through the tank of resin beads, its ions are attracted to the beads, displacing sodium ions.

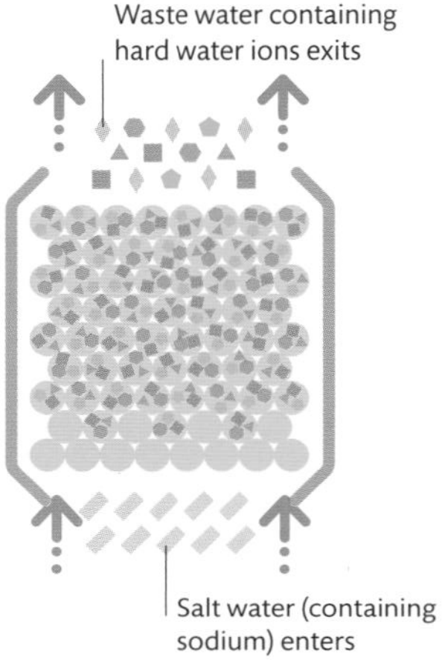
Waste water containing hard water ions exits
Salt water (containing sodium) enters

Regeneration cycle
Salt water is flowed through the resin beads, replenishing the beads with sodium ions and displacing magnesium, calcium, and other unwanted ions.

DISHWASHER TABLETS

Detergent tablets contain a cocktail of chemicals with different roles. These include chlorine and oxygen bleaches to dissolve food stains, as well as enzymes that attack the bonds between atoms of protein and starch molecules in foods, making them easier to wash away.

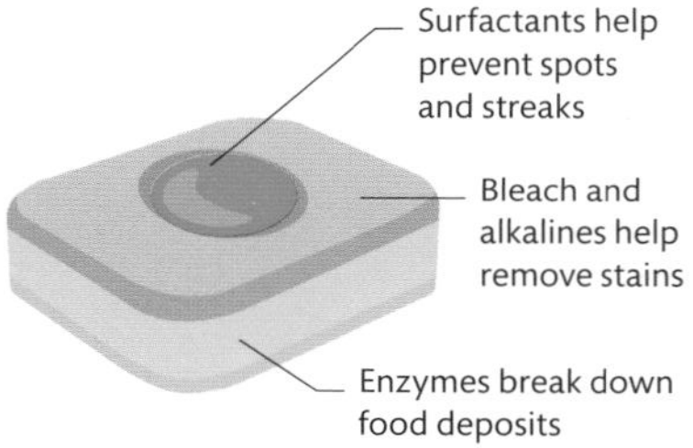
Surfactants help prevent spots and streaks
Bleach and alkalines help remove stains
Enzymes break down food deposits

Refrigeration

Refrigerators and air-conditioning units cool interior spaces by transferring heat energy through the movement of special chemicals around a coiled circuit of pipes.

3 Refrigerant expands The liquid flows through an expansion valve that lowers the refrigerant's pressure, making it expand and cool. It passes into the pipes of the evaporator running inside the fridge.

2 Cooling the refrigerant The gas travels through thin, coiled pipes in the condenser, where metal vanes transfer heat from the refrigerant to the surrounding air. The refrigerant becomes a liquid.

Refrigerators

Refrigerators are effectively heat pumps that move heat energy from cold areas to warm areas—the opposite direction to usual heat flow. A closed system of pipes circulates refrigerant (see panel, opposite), which changes state through compression and expansion and draws heat out of the refrigerator's interior. Freezers work in the same way, just at lower temperatures.

WHAT TEMPERATURE SHOULD A FRIDGE BE SET TO?

A fridge should be kept at about 39°F (4°C). Temperatures above this may not inhibit the growth of bacteria on food.

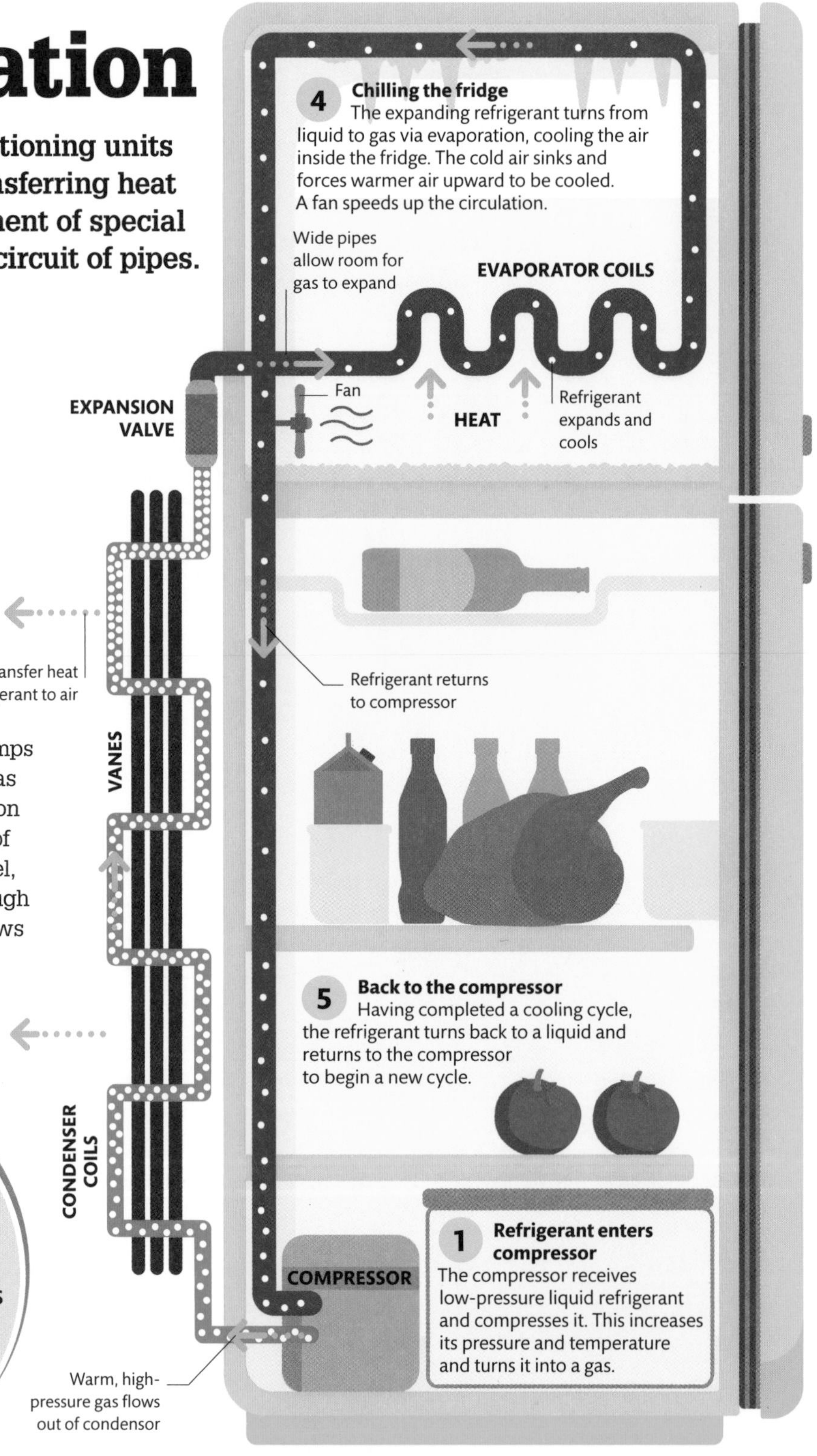

EXPANSION VALVE
Vanes transfer heat from refrigerant to air
VANES
CONDENSER COILS
Warm, high-pressure gas flows out of condensor
Wide pipes allow room for gas to expand
EVAPORATOR COILS
Fan
HEAT
Refrigerant expands and cools
Refrigerant returns to compressor
COMPRESSOR

4 Chilling the fridge The expanding refrigerant turns from liquid to gas via evaporation, cooling the air inside the fridge. The cold air sinks and forces warmer air upward to be cooled. A fan speeds up the circulation.

5 Back to the compressor Having completed a cooling cycle, the refrigerant turns back to a liquid and returns to the compressor to begin a new cycle.

1 Refrigerant enters compressor The compressor receives low-pressure liquid refrigerant and compresses it. This increases its pressure and temperature and turns it into a gas.

Air-conditioning

Domestic air-conditioning (AC or A/C) units are designed to draw in warm air from a living space and then cool it by evaporation, in a similar process to that used in refrigerators. A closed circuit of refrigerant, moved around by a pump, cools warm air drawn in by a fan. The refrigerant then transports the transferred heat out of the building to an external condenser, where the heat dissipates into the air outside. The refrigerant starts another cycle by flowing through an expansion valve, which lowers its pressure and temperature, ready to cool more air. As the air indoors cools, droplets of water vapor condense into liquid, leaving the air less humid as well as cooler.

Home AC unit
An AC unit consists of indoor and outdoor components. The indoor part draws in warm air and cools it; the outdoor part expels the air's heat.

FANS AND AIR CONDITIONERS ACCOUNT FOR ABOUT **15 PERCENT** OF **US ELECTRICITY CONSUMPTION**

REFRIGERANTS

These substances change easily between gas and liquid as their temperature changes. As a liquid changes to a gas, the liquid that is left has less energy and becomes colder. Chlorofluorocarbons (CFCs) were widely used as refrigerants until they were found to damage the atmosphere's ozone layer. Hydrofluorocarbons (HFCs) are now used in household appliances.

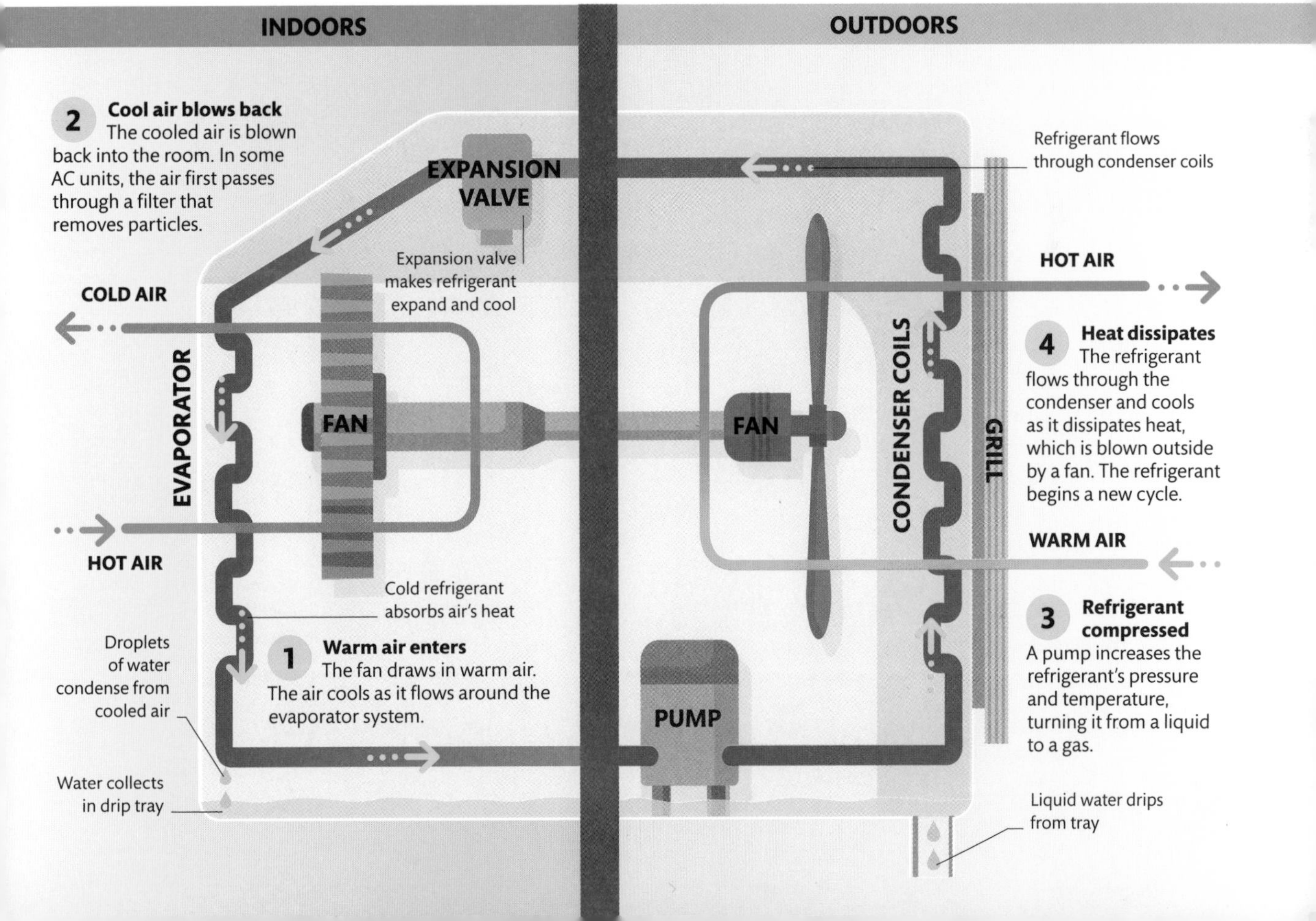
INDOORS
OUTDOORS
2 Cool air blows back
The cooled air is blown back into the room. In some AC units, the air first passes through a filter that removes particles.
EXPANSION VALVE
Expansion valve makes refrigerant expand and cool
COLD AIR
EVAPORATOR
FAN
HOT AIR
Cold refrigerant absorbs air's heat
Droplets of water condense from cooled air
1 Warm air enters
The fan draws in warm air. The air cools as it flows around the evaporator system.
Water collects in drip tray
PUMP
Refrigerant flows through condenser coils
HOT AIR
FAN
CONDENSER COILS
GRILL
4 Heat dissipates
The refrigerant flows through the condenser and cools as it dissipates heat, which is blown outside by a fan. The refrigerant begins a new cycle.
WARM AIR
3 Refrigerant compressed
A pump increases the refrigerant's pressure and temperature, turning it from a liquid to a gas.
Liquid water drips from tray

Vacuum cleaners

By creating a partial vacuum in its interior, a vacuum cleaner draws in a mixture of air and solid particles, including dirt. These are then separated from each other either by filtering or by centrifugal force.

Creating a vacuum

An electric motor spins a fan at high speed to drive air quickly out of the rear of the cleaner and lowers the air pressure inside. As the air pressure is lower inside the cleaner than the ambient air pressure outside, a partial vacuum occurs. In a conventional vacuum cleaner, the resulting suction force draws air containing dust, dirt, hair, and fibers through a porous bag that traps particles, leaving clean air to exit the machine.

WHAT IS A HEPA FILTER?

HEPA (high efficiency particulate air) filters are made of composite materials arranged to capture airborne particles as small as $^{3}/_{250,000}$ in (0.0003 mm) in diameter.

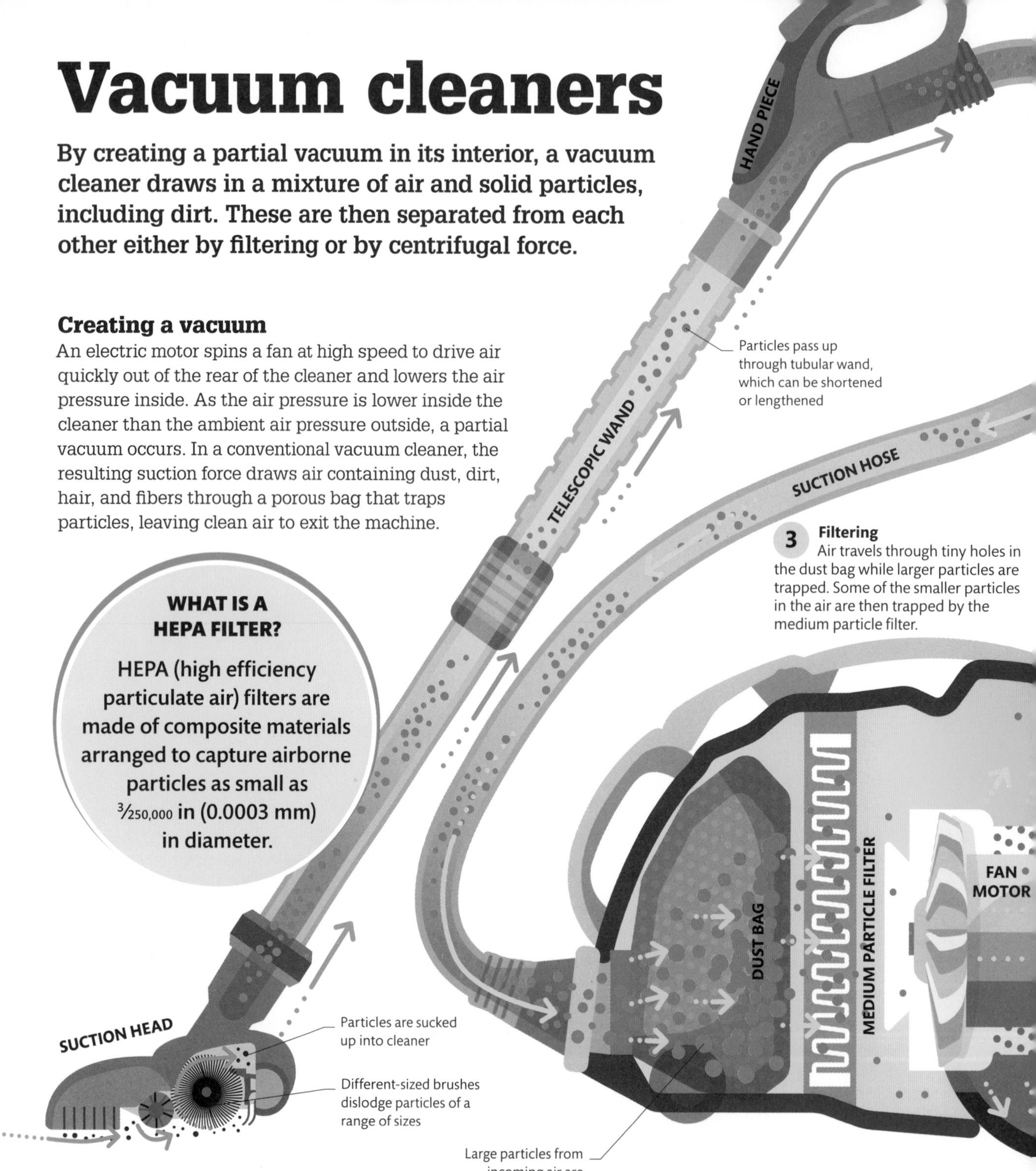

HAND PIECE
TELESCOPIC WAND
Particles pass up through tubular wand, which can be shortened or lengthened
SUCTION HOSE
DUST BAG
MEDIUM PARTICLE FILTER
FAN MOTOR
SUCTION HEAD
Particles are sucked up into cleaner
Different-sized brushes dislodge particles of a range of sizes
Large particles from incoming air are trapped in dust bag

3 Filtering
Air travels through tiny holes in the dust bag while larger particles are trapped. Some of the smaller particles in the air are then trapped by the medium particle filter.

2 Dirt drawn into wand
A series of rotating brushes in the suction head loosen dirt and dust, which is drawn up into the wand and then travels into the cleaner. Most cleaners have a variety of cleaning attachments.

1 Suction is created
The motor spins the fan rapidly to generate suction, which draws air in through the suction head and along the wand and suction hose into the vacuum cleaner body.

4 Air expelled
The air cools the motor as it passes by. It then travels through a HEPA filter to remove microscopic particles before leaving the cleaner.
HEPA FILTER
Motor spins fan at great speed, usually at hundreds or thousands of revolutions per minute

Cyclonic vacuum cleaners

This type of cleaner dispenses with a dust bag and does not suffer from filters getting clogged with large or medium particles during cleaning. It relies on vortices of air (known as cyclones) spinning the air to fling particles out of the airstream. HEPA filters remove tiny particles from the air. They should be cleaned or replaced every 6 months.

THE **MOTORS** OF SOME **CYCLONIC VACUUM CLEANERS** SPIN UP TO **120,000 REVOLUTIONS PER MINUTE**

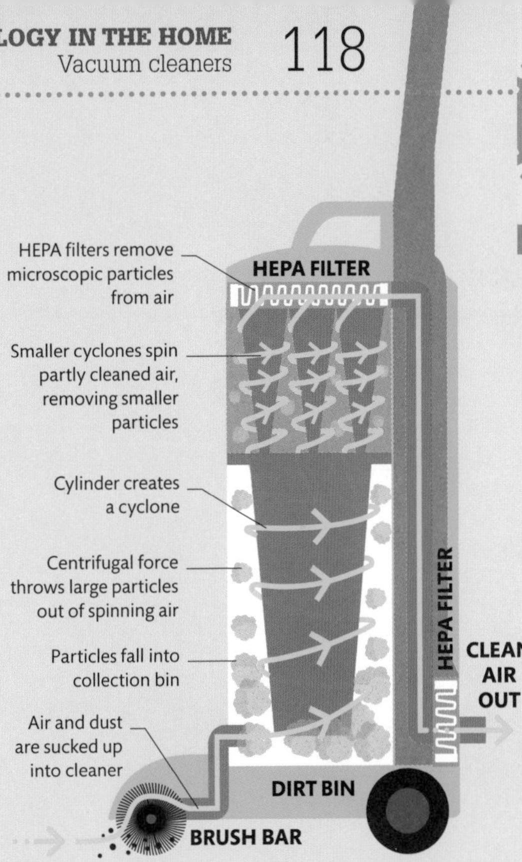
HEPA filters remove microscopic particles from air
HEPA FILTER
Smaller cyclones spin partly cleaned air, removing smaller particles
Cylinder creates a cyclone
Centrifugal force throws large particles out of spinning air
Particles fall into collection bin
Air and dust are sucked up into cleaner
HEPA FILTER
CLEAN AIR OUT
DIRT BIN
BRUSH BAR

Robotic cleaners

These mobile robots, driven by electric motors, navigate themselves around a living space as they clean floors. A package of sensors enables the robot to measure how far it has traveled and to detect obstacles. They also include cliff sensors, which spot sudden drops ahead such as stairs. After a cleaning session, the robot can guide itself to its charging station to recharge its batteries.

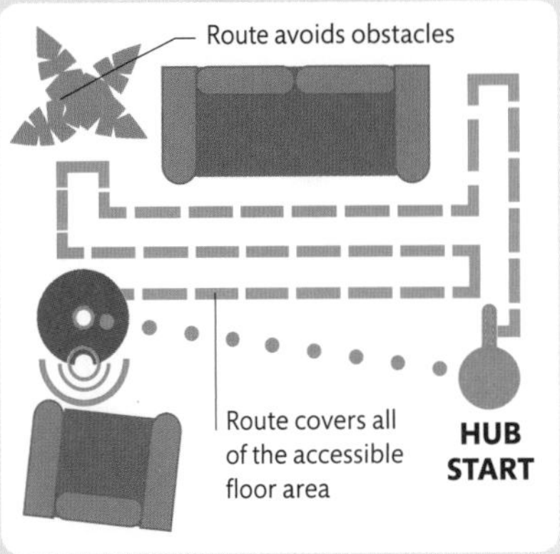
Route avoids obstacles
Route covers all of the accessible floor area
HUB START

Navigation
The robot's microprocessor-based controller runs software that plots a route around a room or rooms, ensuring total cleaning coverage. The robot keeps track of its location and can replot its route should an obstacle bar its way.

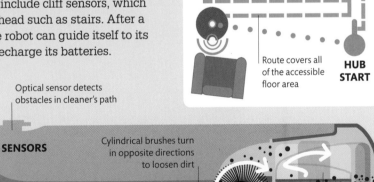

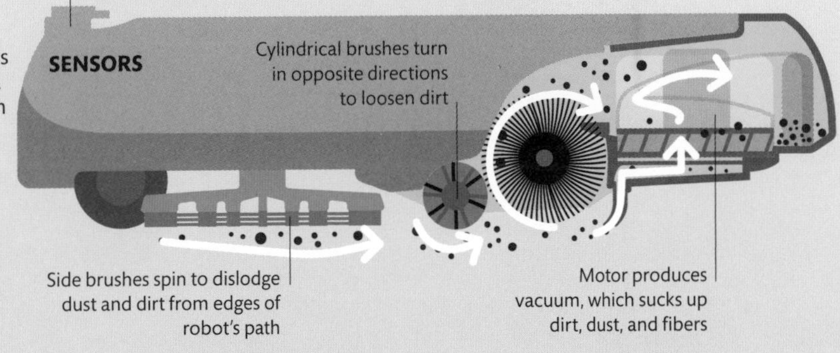
Optical sensor detects obstacles in cleaner's path
SENSORS
Cylindrical brushes turn in opposite directions to loosen dirt
Side brushes spin to dislodge dust and dirt from edges of robot's path
Motor produces vacuum, which sucks up dirt, dust, and fibers

Toilets

Toilets divert human waste for disposal or treatment at a sewage plant. More than 3 billion people have toilets in their homes that use water to flush and carry waste away.

Flushing toilets

A modern flushing toilet features a water storage tank, or cistern, and a water release mechanism that flushes waste away from a bowl and down piping leading toward a sewer system. Waste can be flushed using only the force of water falling under gravity to push waste down the pipe, or by using a siphon, which draws the water from the bowl (see below).

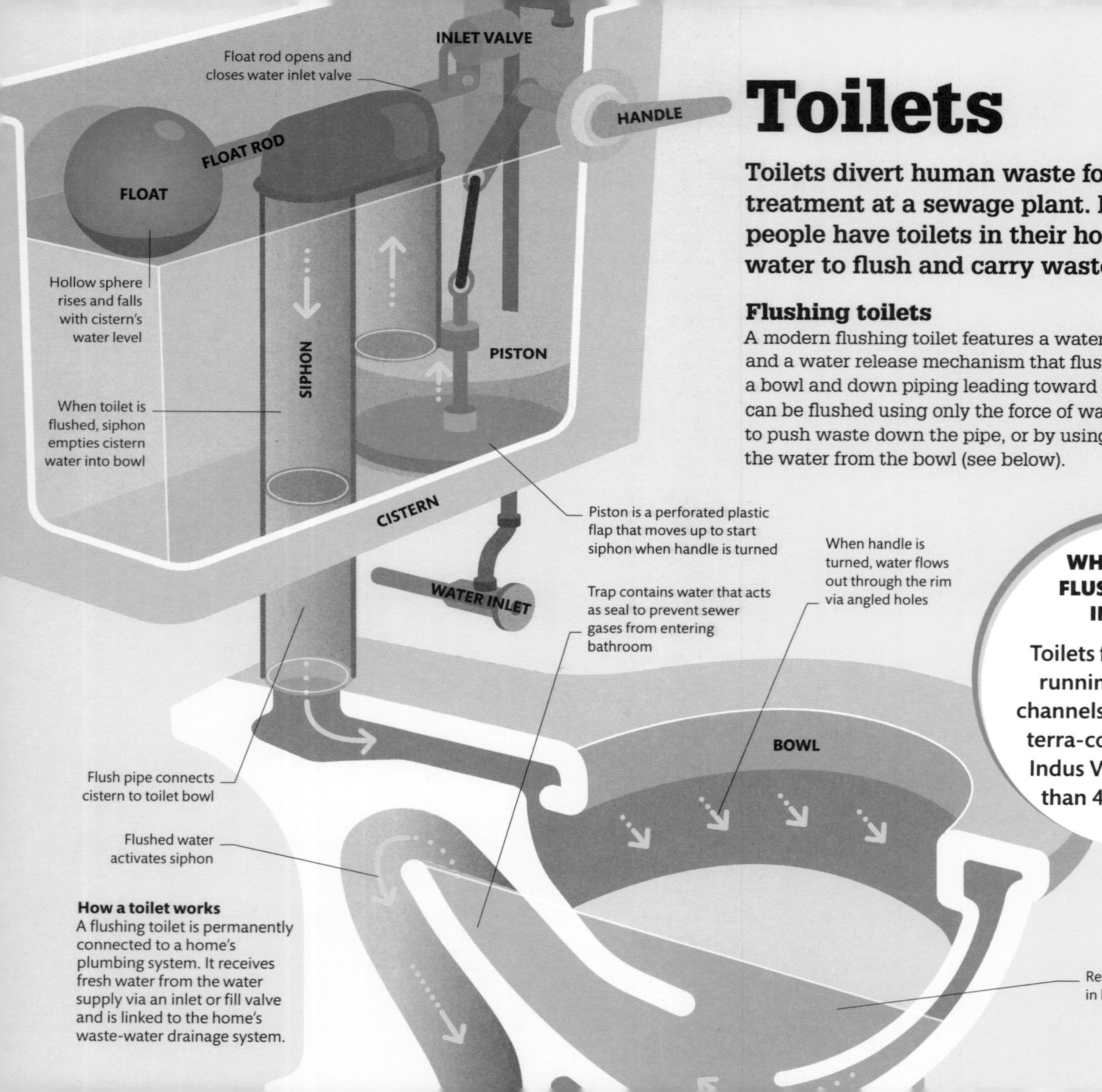
INLET VALVE
Float rod opens and closes water inlet valve
HANDLE
FLOAT ROD
FLOAT
Hollow sphere rises and falls with cistern's water level
SIPHON
PISTON
When toilet is flushed, siphon empties cistern water into bowl
CISTERN
Piston is a perforated plastic flap that moves up to start siphon when handle is turned
WATER INLET
Trap contains water that acts as seal to prevent sewer gases from entering bathroom
When handle is turned, water flows out through the rim via angled holes
BOWL
Flush pipe connects cistern to toilet bowl
Flushed water activates siphon
Reservoir of water held in bottom of bowl

How a toilet works
A flushing toilet is permanently connected to a home's plumbing system. It receives fresh water from the water supply via an inlet or fill valve and is linked to the home's waste-water drainage system.

WHEN WAS THE FLUSHING TOILET INVENTED?

Toilets flushed by water running into drainage channels or pipes made of terra-cotta were used in Indus Valley cities more than 4,000 years ago.

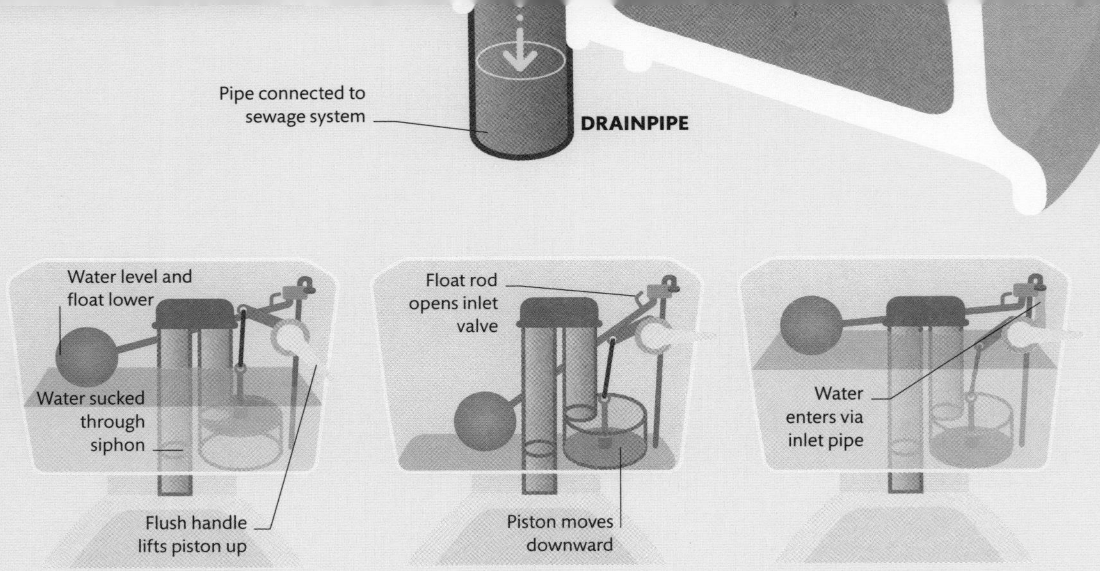

Pipe connected to sewage system
DRAINPIPE
Water level and float lower
Water sucked through siphon
Flush handle lifts piston up
Float rod opens inlet valve
Piston moves downward
Water enters via inlet pipe

1 Flushing
Turning the handle moves a lever to lift the piston upward. This forces water through the siphon tube, generating a suction force that pulls the rest of the cistern's water through the siphon and out into the bowl.

2 Emptying
The cistern empties rapidly, and water flows around the bowl and out through the drainpipe, carrying away the waste. The piston drops, and the float ball sinks, moving the float rod, which opens the inlet valve.

3 Refilling
With the inlet valve open, water enters the cistern. As the water level rises, so does the float ball. The float moves the float rod to close the inlet valve once the cistern has filled to the required level.

SIPHONS

Many toilets use siphoning to move water from the cistern to the toilet bowl or from the bowl to the drainpipe. Once some water is forced through the highest point of the upturned, U-shaped siphon, gravity and cohesive forces in the liquid help continue the siphoning action until there is no water left.

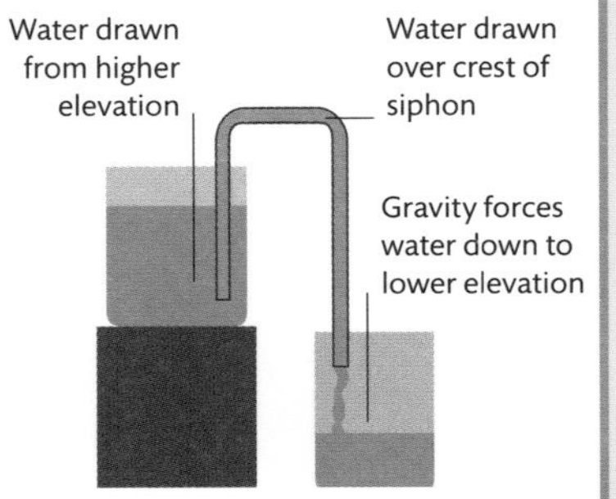

Water drawn from higher elevation
Water drawn over crest of siphon
Gravity forces water down to lower elevation

Composting toilets

At 1⅗–4¾ gallons (6–18 liters) per flush, a standard toilet's fresh water consumption can mount up over time, especially in a large household. In contrast, composting toilets use little or no water and do not place any demands on a municipal sewage system. Instead, these self-contained systems rely on the process of aerobic decomposition, in which bacteria, fungi, and, in some systems, earthworms break down the waste over a period of weeks or months into harmless, largely odor-free humus compost, which can be used as a natural fertilizer.

2.3 BILLION PEOPLE DO NOT HAVE BASIC SANITATION FACILITIES

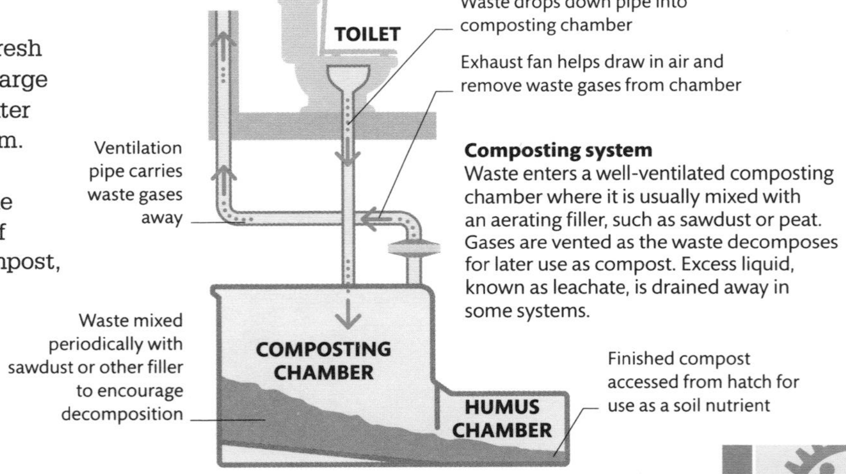

TOILET
Waste drops down pipe into composting chamber
Exhaust fan helps draw in air and remove waste gases from chamber
Ventilation pipe carries waste gases away
Waste mixed periodically with sawdust or other filler to encourage decomposition
COMPOSTING CHAMBER
HUMUS CHAMBER
Finished compost accessed from hatch for use as a soil nutrient

Composting system
Waste enters a well-ventilated composting chamber where it is usually mixed with an aerating filler, such as sawdust or peat. Gases are vented as the waste decomposes for later use as compost. Excess liquid, known as leachate, is drained away in some systems.

Locks

Locks are a form of secure bolt or clasp that require a specific key to open them. The key can be a physical object, a digital or numeric code, or a particular unique physical feature of a person. Among the most commonly used locks are cylindrical tumbler locks and combination locks.

CAN YOU PICK A LOCK USING HAIRPINS?

In some simple tumbler locks, it is possible to use a combination of hairpins or wires to push up all the pins and turn the cylinder.

Cylindrical tumbler locks

Commonly found in door fasteners and many padlocks, a tumbler lock consists of a barrel holding a cylinder—also known as a plug—that is able to turn. A series of chambers each contains a spring and pins of different lengths that prevent the cylinder from turning unless the correct key is inserted through an opening called the keyway.

3 ft (90 cm)

THE **LENGTH** OF **THE KEYS** USED TO OPEN AND CLOSE THE **BOMBPROOF DOOR** TO THE BANK OF ENGLAND'S GOLD VAULT

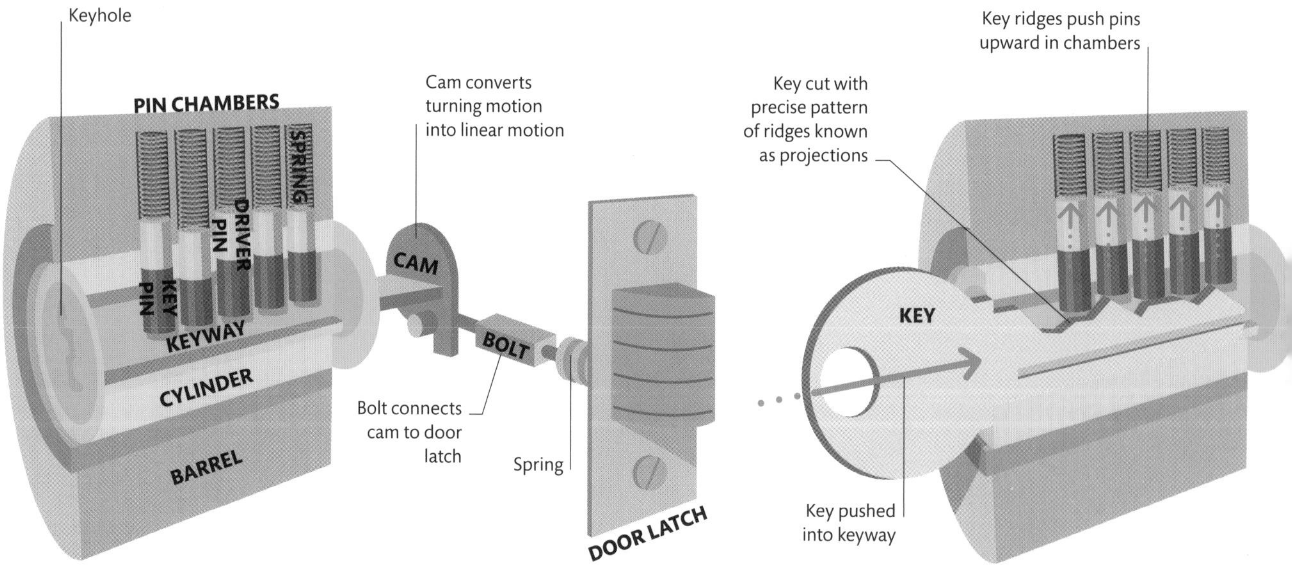

Keyhole
PIN CHAMBERS
SPRING
DRIVER PIN
KEY PIN
KEYWAY
CYLINDER
BARREL
Cam converts turning motion into linear motion
CAM
BOLT
Bolt connects cam to door latch
Spring
DOOR LATCH
Key ridges push pins upward in chambers
Key cut with precise pattern of ridges known as projections
KEY
Key pushed into keyway

1 Lock closed
In the locked position, the pins are pushed down their chambers by springs. This prevents the cylinder from turning, and the lock is closed.

2 Key inserted into lock
The key's ridges push the pins up precisely so that the tops of all the key pins align with the top edge of the cylinder.

Combination locks

A combination lock is a form of keyless lock that contains pins, similar to a tumbler lock, but mounted on a metal bar. Each pin is located behind a numbered wheel or dial that is turned manually. Only one unique combination of numbers will align all the holes in the wheels so that the pins can pass through and the lock can open.

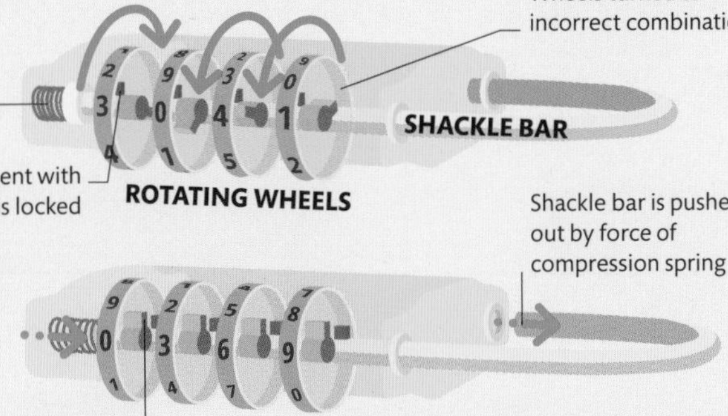
LOCKED
Wheels turned to incorrect combination
Compression spring
SHACKLE BAR
Pin out of alignment with hole, so wheel is locked
ROTATING WHEELS
UNLOCKED
Shackle bar is pushed out by force of compression spring
Pins line up with holes and can pass through wheels

Wheels and pins
Once the correct number sequence has been selected, and the holes in the dials are aligned, the compression spring forces the shackle bar outward into the unlocked position.

BIOMETRIC LOCKS

Some electronic locks use a person's physical features—such as fingerprints, the iris of the eye, or facial image—as the key to open the lock. A scanner identifies unique patterns in one of these features and stores them in a database of information associated with people allowed entry. When an approved person returns, recognition of these recorded patterns will open the lock.

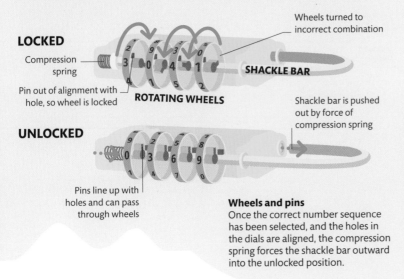

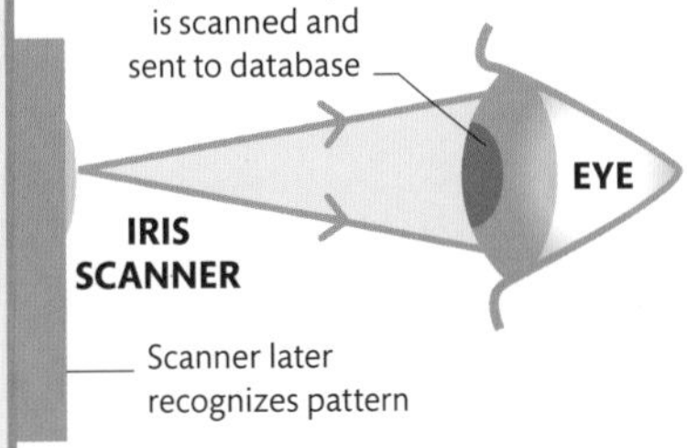
Iris pattern in eye is scanned and sent to database
EYE
IRIS SCANNER
Scanner later recognizes pattern

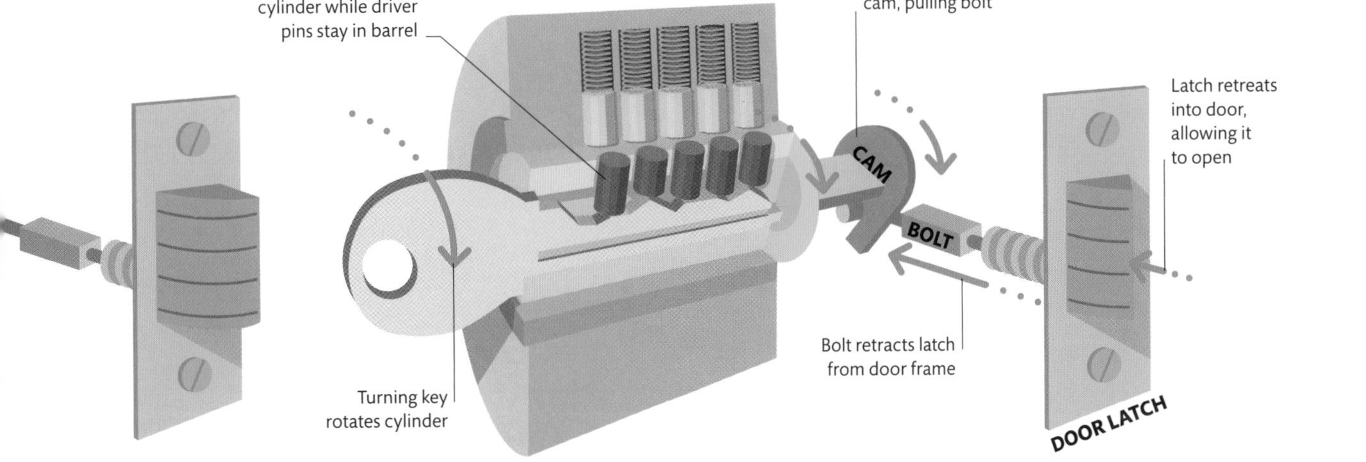
Key pins turn with cylinder while driver pins stay in barrel
Cylinder turns cam, pulling bolt
Latch retreats into door, allowing it to open
CAM
BOLT
Bolt retracts latch from door frame
Turning key rotates cylinder
DOOR LATCH

3 Latch opens
As the key rotates the cylinder, the cam changes the direction of force, retracting the bolt, which pulls the door latch back into an open position.

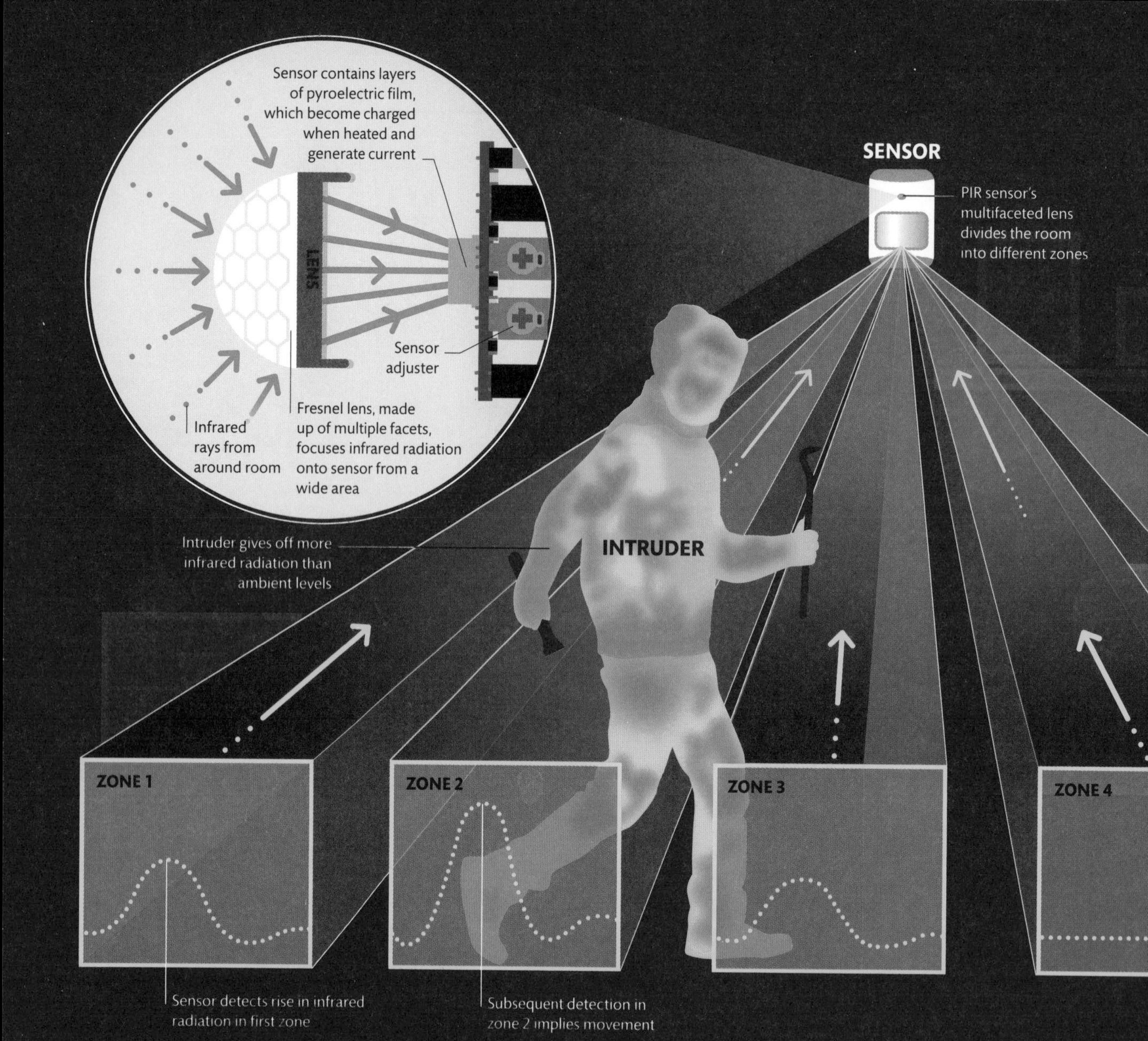
Sensor contains layers of pyroelectric film, which become charged when heated and generate current
LENS
Sensor adjuster
Infrared rays from around room
Fresnel lens, made up of multiple facets, focuses infrared radiation onto sensor from a wide area
SENSOR
PIR sensor's multifaceted lens divides the room into different zones
INTRUDER
Intruder gives off more infrared radiation than ambient levels
ZONE 1
ZONE 2
ZONE 3
ZONE 4
Sensor detects rise in infrared radiation in first zone
Subsequent detection in zone 2 implies movement

Security alarms

Technology has long played a key role in protecting homes and other buildings from intruders and burglaries. Modern alarm systems use various sensors to detect intruders, for example, by picking up their body heat or pressure from their footsteps or by responding to changes in the positions of doors or windows.

Passive infrared sensors

Everyone gives off infrared radiation at differing levels to their surroundings. Passive infrared (PIR) sensors detect changes in infrared emissions using thin layers of pyroelectric film. This film absorbs infrared radiation, which causes it to heat up and generate small electrical signals. A change in infrared levels across multiple areas of a room can signal the presence and movement of an intruder.

Movement detection
As an intruder moves across a room, he or she crosses different zones. The sensor picks up changes in infrared levels across the zones to detect movement.

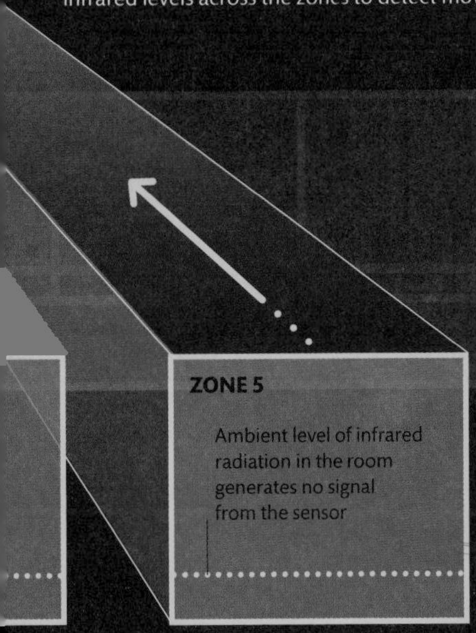
ZONE 5
Ambient level of infrared radiation in the room generates no signal from the sensor

WHERE IS THE BEST PLACE TO PUT A SECURITY SENSOR?

Choke points, such as hallways, where people have to pass through, are good locations, as are corners of rooms offering coverage of multiple entry points.

Contact sensors

The two parts of a magnetic contact sensor—one part fitted to a door or window, the other to its fixed frame—complete an electric circuit when closed. When the door or window is opened, and the contact between the two magnets is broken, the circuit is also broken. This sends a signal to the security alarm's controller, which interprets this as a possible unexpected entry.

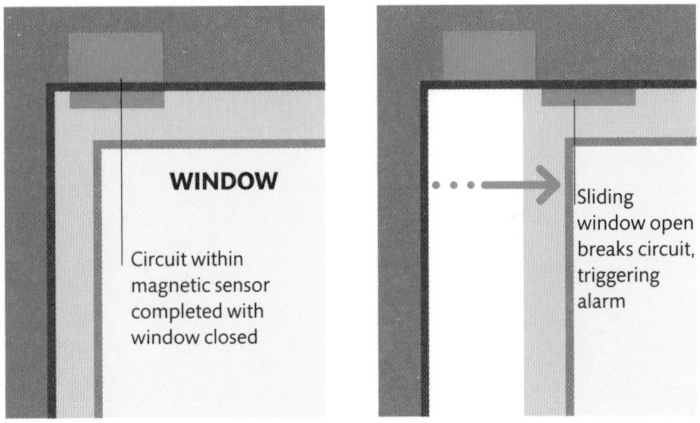
WINDOW
Circuit within magnetic sensor completed with window closed
Sliding window open breaks circuit, triggering alarm

34 percent
THE PROPORTION OF **BURGLARS** WHO ENTER THROUGH THE **FRONT DOOR**

CONTROL PANEL

An alarm system's controller enables the user to arm or disable the system by entering a specific numeric code. This central control point can also allow the user to enable only the security systems within certain zones or rooms. When armed, the controller monitors data sent by the sensors and, if triggered, sounds alarms, deploys any electronic locks, and may use wireless links to alert security guards or the police.

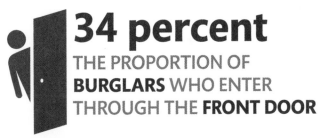

Fabrics

Fabrics are materials made from fibers obtained naturally or via chemical processing. A wide range of fabrics are manufactured, each with varying properties that may suit different needs, such as crease resistance, durability, water resistance, and elasticity.

Raw materials

Fibers for fabrics come from many natural sources, including plants grown as crops, such as cotton and flax, and animals such as sheep. Fossil fuel industries produce polymers (see p.78) that make a wide range of synthetic fabrics, including acrylic and polyester. Many of these are forced through a device called a spinneret to create long filaments that can be processed into yarn. The yarn is then knitted, woven, or bonded (see p.129).

WHAT IS THE MOST COMMON FABRIC IN THE WORLD?

Cotton makes up 30 percent of all the fibers produced for fabric. Cotton-growing accounts for 2.5 percent of global farmland use.

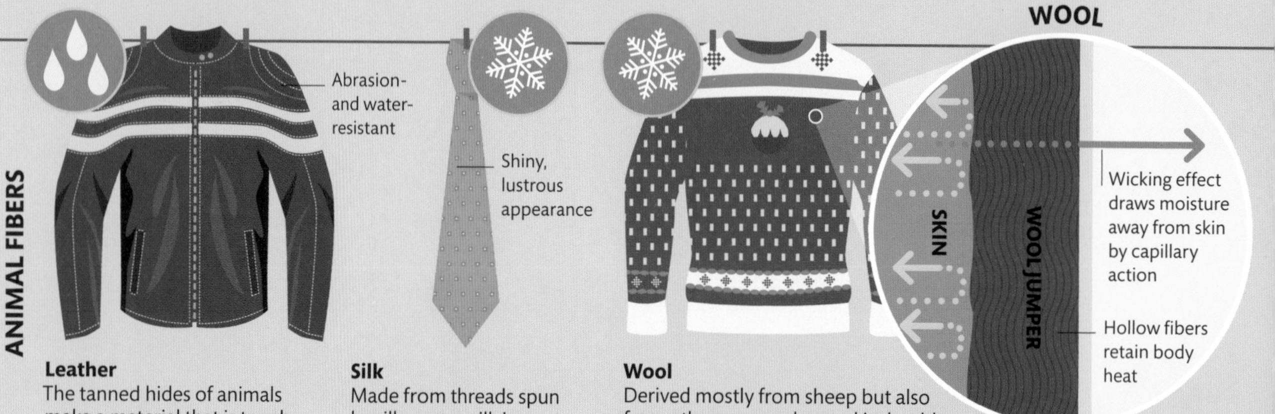

Leather
The tanned hides of animals make a material that is tough, durable, and does not tear easily. It is wind- and water-resistant but hard to sew.

Silk
Made from threads spun by silkworms, silk is lightweight, strong, offers good insulation, and holds its shape well.

Wool
Derived mostly from sheep but also from other mammals, wool is durable, moisture-repelling, and resistant to wrinkles and soiling. It also retains heat well and draws sweat away from the skin.

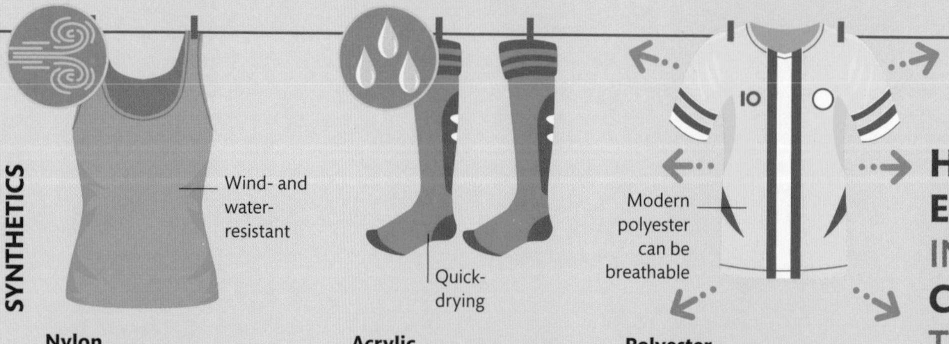

Nylon
A synthetic material derived from coal, nylon is made into smooth, lightweight, and highly elastic fabrics.

Acrylic
While lacking a natural feel, acrylic fabrics offer good insulation, wash easily, and retain their shape well.

Polyester
Produced from petroleum, this is resistant to stretching and shrinking and barely absorbs moisture.

HEATING ELEMENTS IN SOME **COATS** HELP THE WEARER TO KEEP WARM

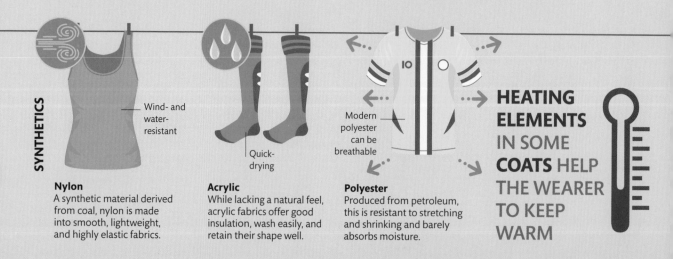

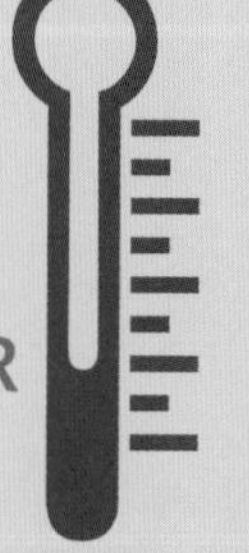

CARING FOR FABRICS

Fabrics all have different qualities and therefore have to be cared for in different ways. Most manufactured clothes contain labels giving instructions for care. These may suggest a fabric can be tumble dried, warn the owner about washing only at certain temperatures or avoiding ironing, or, in the case of a delicate fabric such as cashmere or viscose, state that it is to be dry cleaned only.

HAND WASH
MACHINE WASH
TUMBLE DRY
IRON
DRY CLEANING
DO NOT WASH

PLANT FIBERS

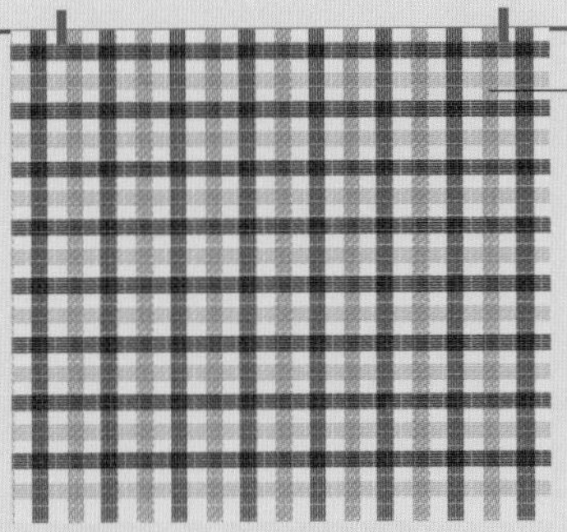
Colors stay bright because Rayon holds dyes well

Rayon
Developed as an alternative to silk and made from cellulose fibers derived mostly from wood pulp, this fabric is soft and comfortable. It dyes well but weakens when wet and is prone to abrasion.

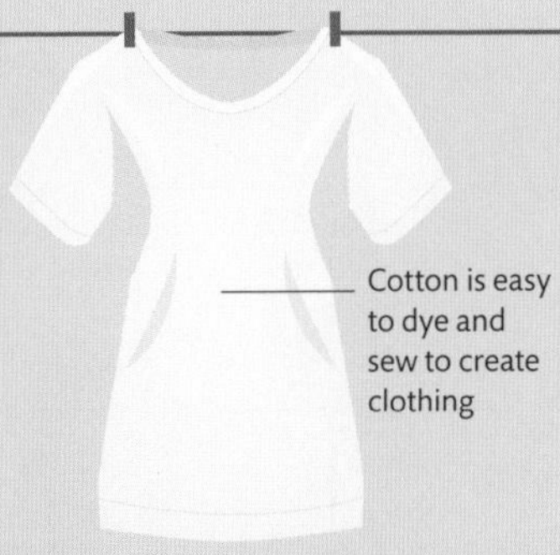
Cotton is easy to dye and sew to create clothing

Cotton
This versatile and common fiber can be knitted or woven into a range of fabrics that are durable, comfortable to wear, and breathable. It does wrinkle easily but is simple to wash and iron.

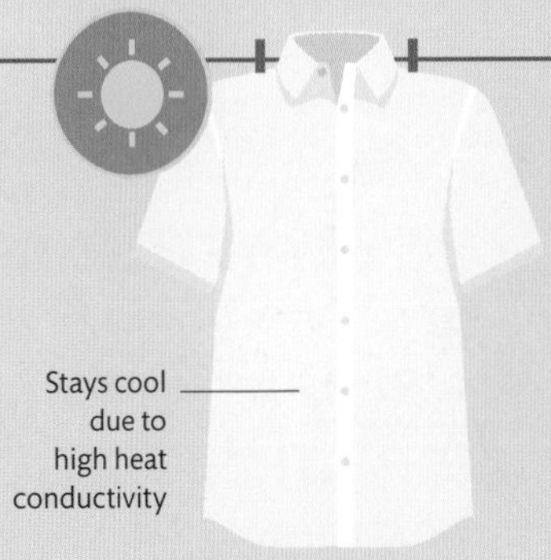
Stays cool due to high heat conductivity

Linen
The fibrous stem of the flax plant makes a fabric twice as strong as cotton. It is highly absorbent but dries quickly. Linen has low elasticity and creases rapidly but can be ironed easily.

New properties

New technologies can alter the properties of a synthetic or natural-fiber fabric. For example, polyester can be used to create swimwear that protects wearers from UV radiation in sunlight. Adding nanoparticles of certain substances can give fabric a new and useful attribute, such as using silver nanoparticles in sportswear and shoes to kill off the bacteria and fungi that cause the odors in sweat. Silica nanoparticles in a fabric repel stains and water by making the liquid bead and roll off more easily.

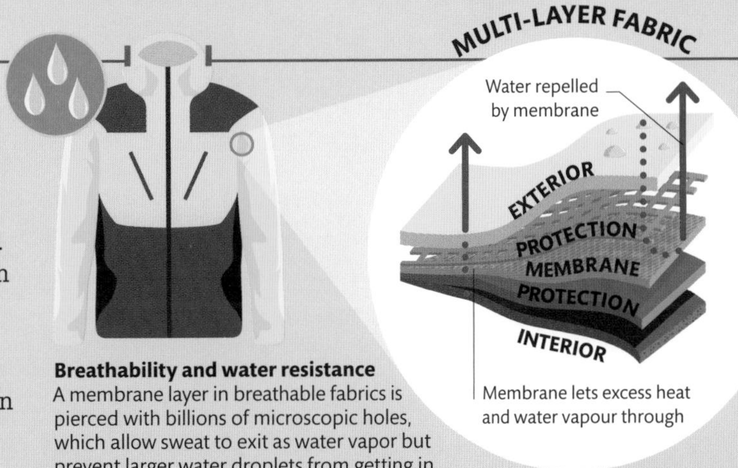
MULTI-LAYER FABRIC
Water repelled by membrane
EXTERIOR
PROTECTION
MEMBRANE
PROTECTION
INTERIOR
Membrane lets excess heat and water vapour through

Breathability and water resistance
A membrane layer in breathable fabrics is pierced with billions of microscopic holes, which allow sweat to exit as water vapor but prevent larger water droplets from getting in.

Clothing

For most of human history, clothing was handmade at home. Even today, when mass-produced clothing dominates most people's wardrobes, some people prefer to create their own garments or make their own alterations and repairs.

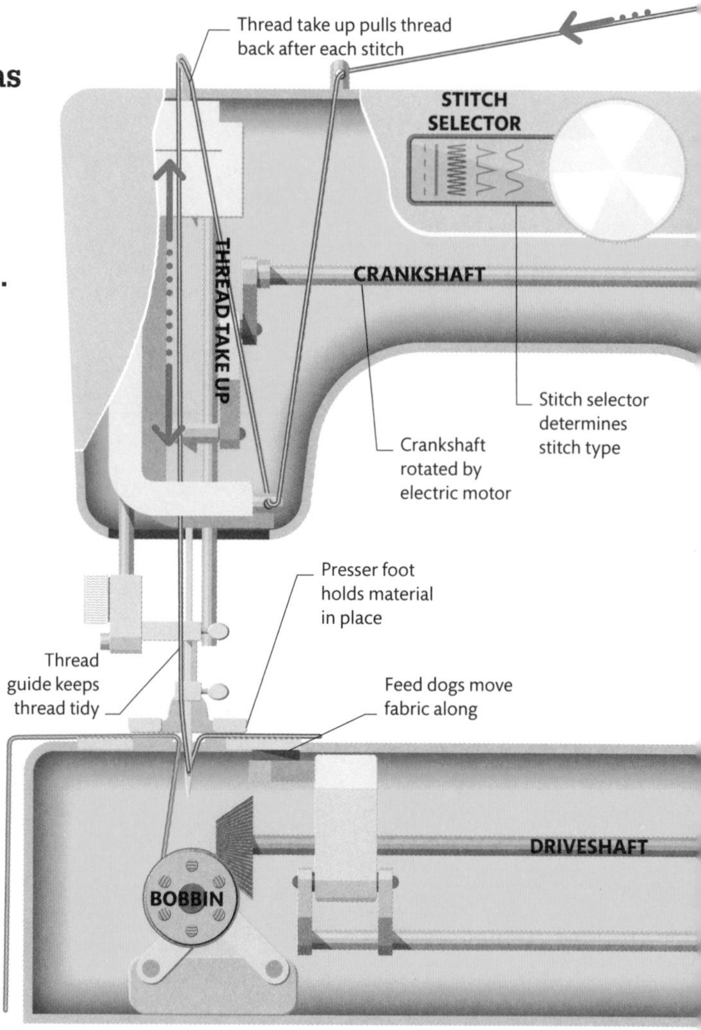

Thread take up pulls thread back after each stitch
STITCH SELECTOR
THREAD TAKE UP
CRANKSHAFT
Stitch selector determines stitch type
Crankshaft rotated by electric motor
Presser foot holds material in place
Thread guide keeps thread tidy
Feed dogs move fabric along
DRIVESHAFT
BOBBIN

Sewing machines

Sewing machines enable fast, accurate stitching to join fabrics together or produce a hem. Thread from a spool is guided through the needle, which moves up and down via a crank turned by the driveshaft. The driveshaft is powered by an electric motor. At the same time, feed dogs move the fabric in a synchronized fashion with the needle to produce a row of equal-sized stitches.

DOMESTIC SEWING MACHINES CAN HAVE A SEWING **SPEED** OF MORE THAN **1,000 STITCHES PER MINUTE**

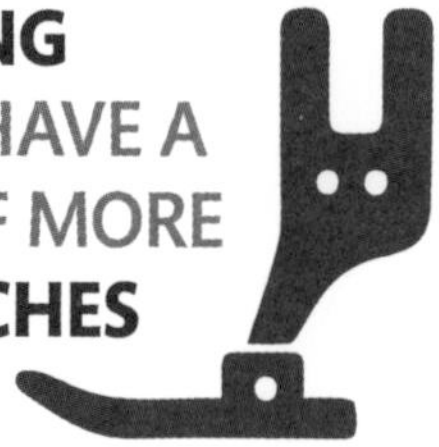

Making a stitch

A household electric sewing machine uses two threads to make a stitch. The machine's controls allow the user to change the size and the type of stitch used on fabric or clothing.

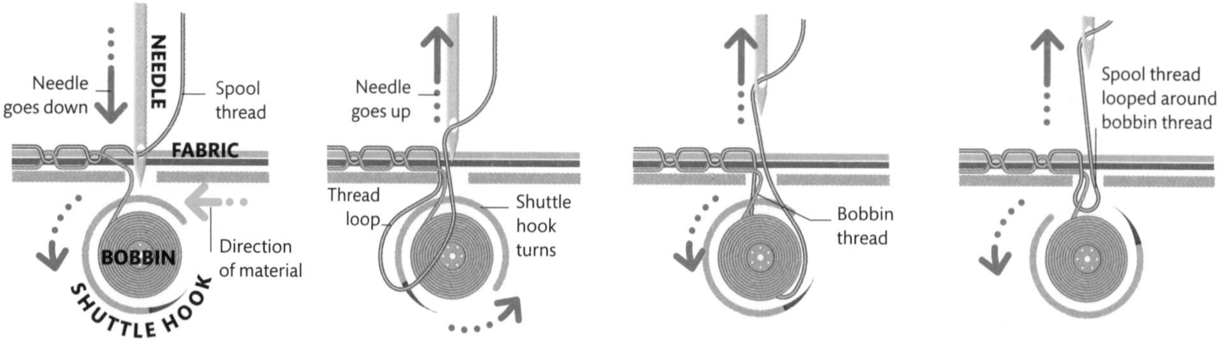

Needle goes down
NEEDLE
Spool thread
FABRIC
BOBBIN
Direction of material
SHUTTLE HOOK
Needle goes up
Thread loop
Shuttle hook turns
Bobbin thread
Spool thread looped around bobbin thread

1 Lowering the needle The needle lowers and penetrates the fabric, carrying the spool thread (blue) down to a spindle of thread, which is called a bobbin (red).

2 Hooking a loop As the needle travels upward, it leaves a loop of spool thread, which is caught by the shuttle hook as it rotates around the bobbin.

3 Carrying the thread The shuttle hook carries spool thread around the bobbin case, before the thread slips off the hook and around the bobbin thread.

4 Pulling the stitch Both threads are pulled up as the needle rises and the fabric advances. The threads are pulled into a stitch, which is tightened by the rising needle.

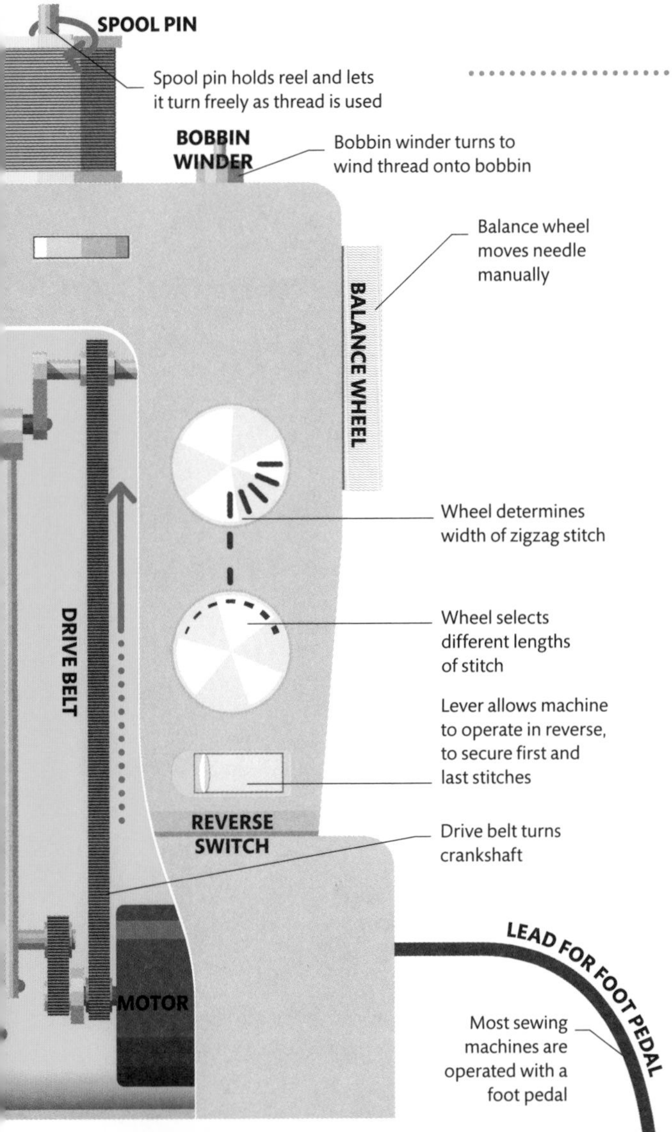
SPOOL PIN
Spool pin holds reel and lets it turn freely as thread is used
BOBBIN WINDER
Bobbin winder turns to wind thread onto bobbin
Balance wheel moves needle manually
BALANCE WHEEL
Wheel determines width of zigzag stitch
Wheel selects different lengths of stitch
Lever allows machine to operate in reverse, to secure first and last stitches
REVERSE SWITCH
Drive belt turns crankshaft
DRIVE BELT
MOTOR
LEAD FOR FOOT PEDAL
Most sewing machines are operated with a foot pedal

HOW FABRICS ARE MADE

Fabric is produced in a number of different ways. Woven fabrics are made from fibers or yarn interlaced at right angles. Knitted fabrics are made by looping long pieces of yarn together. Bonded fabrics are often made from webs of fibers melded together by heat, adhesives, or pressure.

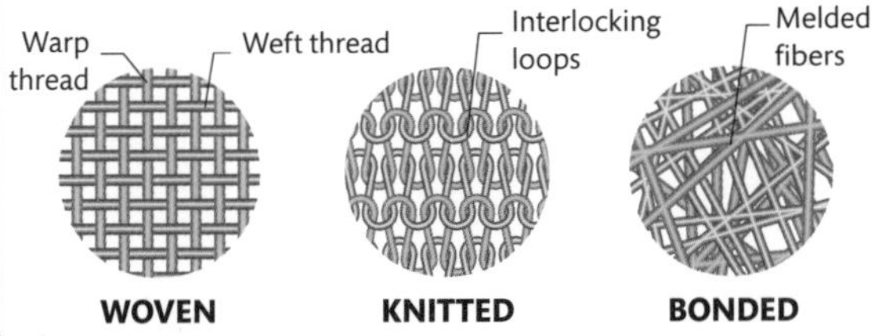
Warp thread
Weft thread
Interlocking loops
Melded fibers
WOVEN
KNITTED
BONDED

Fasteners

Clothing can be fastened in many ways from pop snappers to sewn-in magnets. Some fasteners, such as buttons, laces, and hooks and eyes, have been used for centuries. Others, such as the modern zipper and Velcro, are more recent inventions.

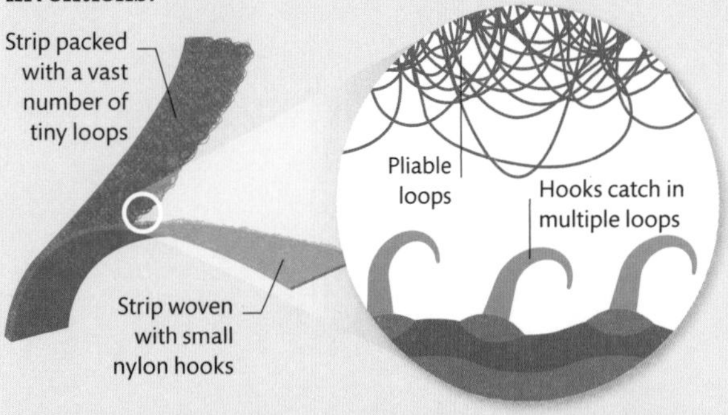
Strip packed with a vast number of tiny loops
Pliable loops
Hooks catch in multiple loops
Strip woven with small nylon hooks

Velcro

This fabric fastener is modeled on the tiny hooks of some seed burrs that stick stubbornly to fur and fabric. Velcro consists of two strips of nylon or polyester–one containing large numbers of tiny loops and the other with hooks that engage with the loops to provide a secure fastening.

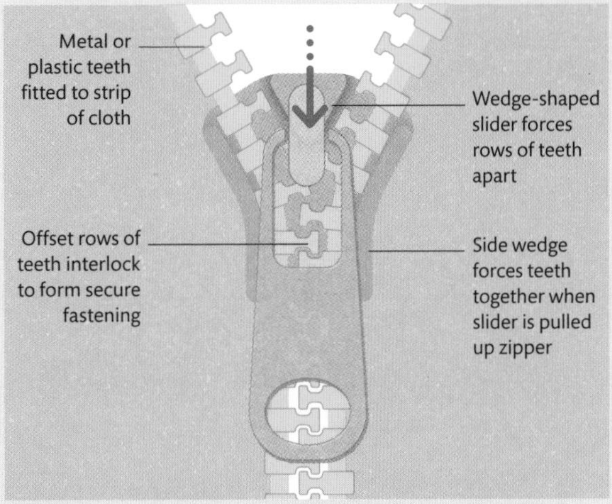
Metal or plastic teeth fitted to strip of cloth
Wedge-shaped slider forces rows of teeth apart
Offset rows of teeth interlock to form secure fastening
Side wedge forces teeth together when slider is pulled up zipper

Zippers

These ingenious fasteners feature two rows of staggered teeth. The Y-shaped channel inside the slider eases the teeth together when it is pulled up the zipper. Unzipping sees the central part of the slider act as a wedge, forcing itself between the two rows and prying the teeth apart.

THE WORLD'S LARGEST ZIPPER PRODUCER MAKES OVER 7 BILLION ZIPPERS EACH YEAR

Washing machines

Washing machines and tumble dryers both use powerful electric motors to automate and speed up manual tasks. There are two main types of washing machines: front loaders and top loaders.

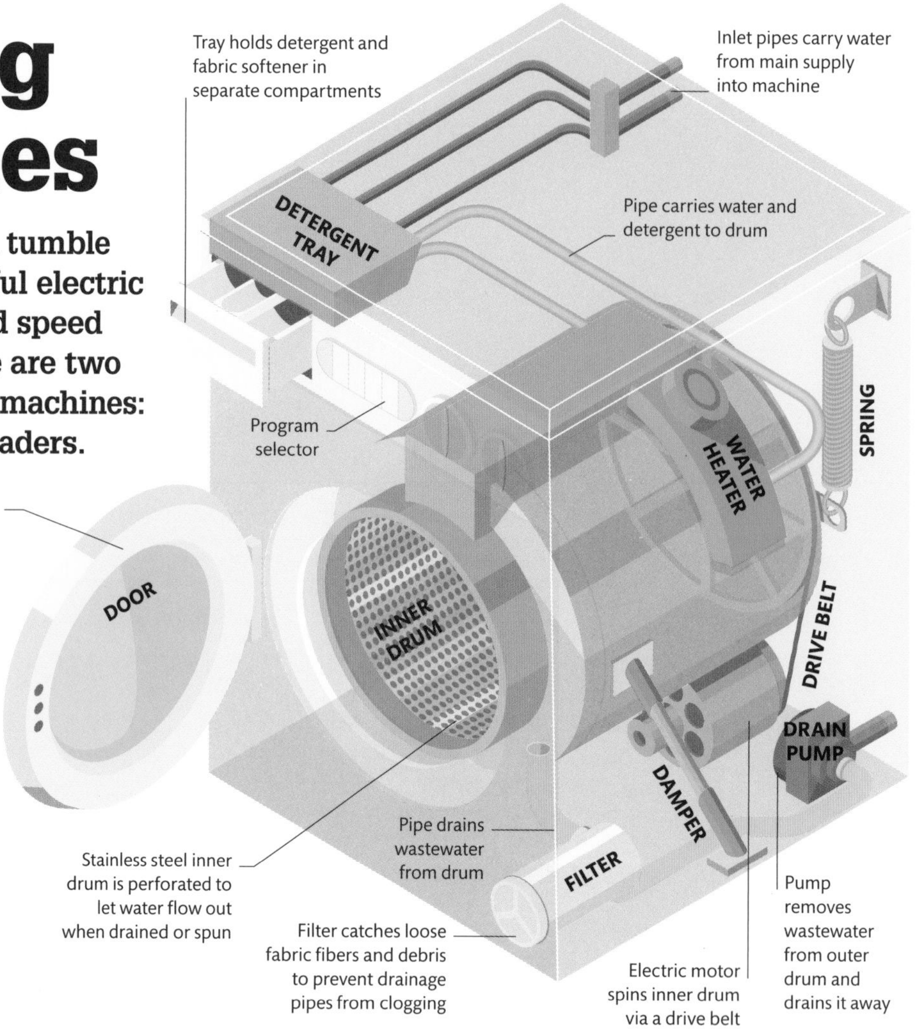

Tray holds detergent and fabric softener in separate compartments
Inlet pipes carry water from main supply into machine
DETERGENT TRAY
Pipe carries water and detergent to drum
Program selector
WATER HEATER
SPRING
Front-loading door has watertight seal and sensors that detect if it is shut fully
DOOR
INNER DRUM
DRIVE BELT
DRAIN PUMP
DAMPER
Pipe drains wastewater from drum
FILTER
Stainless steel inner drum is perforated to let water flow out when drained or spun
Filter catches loose fabric fibers and debris to prevent drainage pipes from clogging
Electric motor spins inner drum via a drive belt
Pump removes wastewater from outer drum and drains it away

Front loaders

An outer drum is held in place inside the washing machine by springs and shock-absorbing dampers. Within it, an inner drum is spun by a motor, either turning slowly to churn water, detergent, and clothes during a wash cycle, or spinning fast to remove water. A program governs the temperature of the water, duration of wash, and the rinsing and spin cycles.

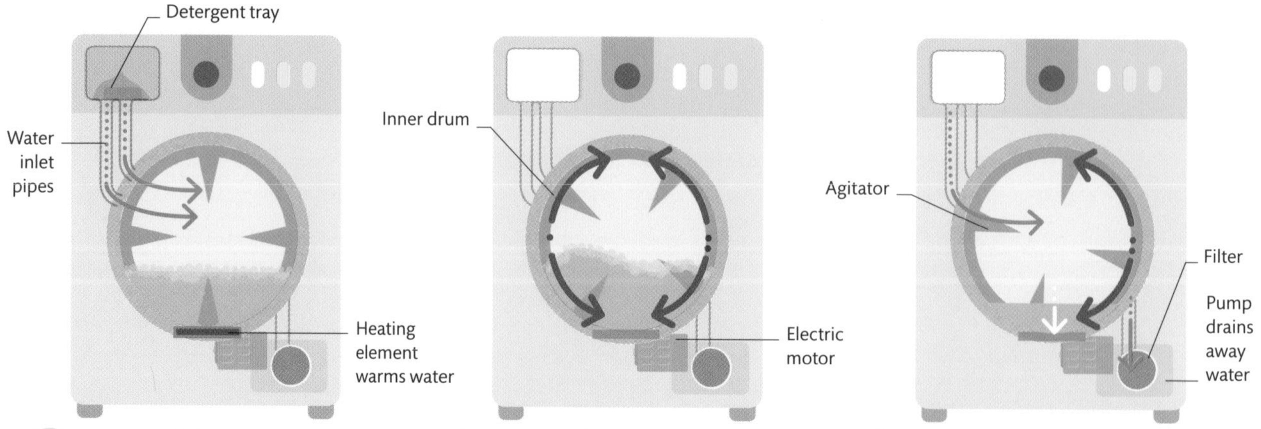

Detergent tray
Water inlet pipes
Heating element warms water
Inner drum
Electric motor
Agitator
Filter
Pump drains away water

1 Water and detergent fill drum Water enters the machine and passes through a detergent tray to wash detergent into the drum. Machines may be filled with hot and cold water or with cold water only.

2 Wash and drain The wash cycle starts once the desired water quantity and temperature are reached. A motor turns the inner drum back and forth through the water-detergent mix.

3 Rinse, agitate, and drain Wash water is drained away, and cold water fills the machine. Agitators in the inner drum help remove loosened dirt and any detergent remaining on clothes.

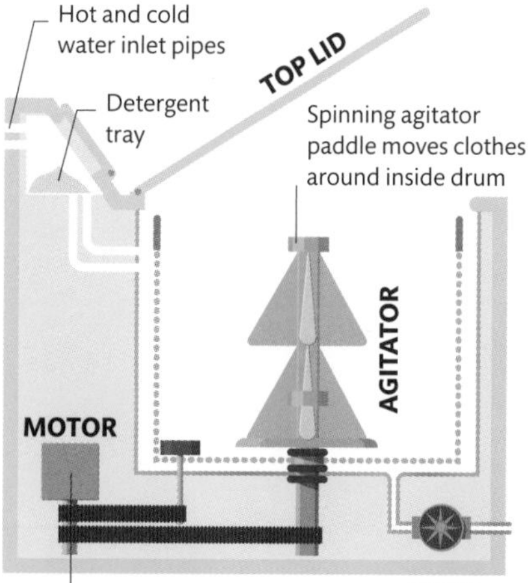
Hot and cold water inlet pipes
TOP LID
Detergent tray
Spinning agitator paddle moves clothes around inside drum
AGITATOR
MOTOR
Electric motor turns agitator
PUMP

Top loaders

These machines also possess an outer and inner drum, but neither drum moves during the washing cycle. Instead, the clothes and water-detergent mix are stirred thoroughly by a large rotating central agitator, which is powered by an electric motor. The same motor rotates the inner drum during the spin cycle to remove water from the load.

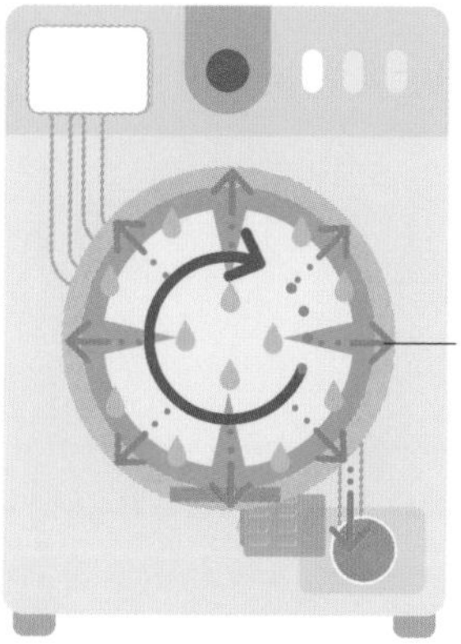
Water flung out of drum

4 Rapid spin and drain The motor rotates the inner drum at high speeds (300–1,800 rpm), flinging water out of the inner drum. Hot air may be blown into the drum to help the load dry.

Detergents

Most stains and dirt can be removed with hot water alone, but other particularly oily or greasy deposits require chemical help. Detergent molecules contain one acidic end, which is hydrophilic (attracted to water molecules), and a long hydrocarbon chain at the other, which is attracted to oil. Together, they attach to stains and help lift oil and grease away from fabrics.

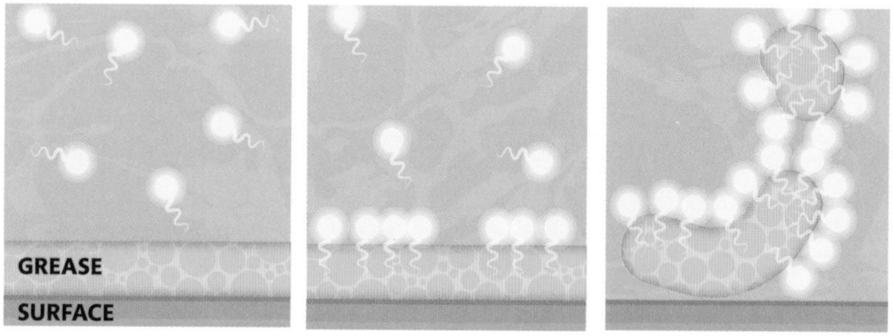
GREASE
SURFACE

1 Detergent released Detergent dissolves, and its molecules are mixed in water in the washing-machine drum, coming into contact with oil and grease stains in fabrics.

2 Attaching to dirt Repelled by water but attracted to oil, one end of the detergent molecules attaches to the stain. Multiple detergent molecules build up, engulfing the stain.

3 Dirt removed Agitation during the washing cycle and the pull of the hydrophilic end of the detergent molecules lifts the oil or grease out of the fabric, to be rinsed away.

SOME **WASHING MACHINES IN THE 1920S** WERE POWERED BY A **GAS-FUELED ENGINE** THAT BELCHED **EXHAUST FUMES**

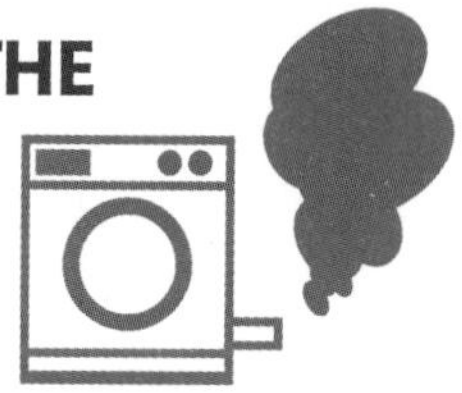

Tumble dryers

Wet clothes are placed in a tumble dryer's large drum, which turns slowly, powered by a belt-driven motor. On many models, the drum changes direction frequently to prevent clothes from bunching up. The laundry is tumbled from the drum's top to bottom, through dry, warm air blown into the drum by a fan, and heated by an electric element. The warm, moist air is then carried out through a vent—in some dryers, it first passes over a heat exchanger to extract heat energy.

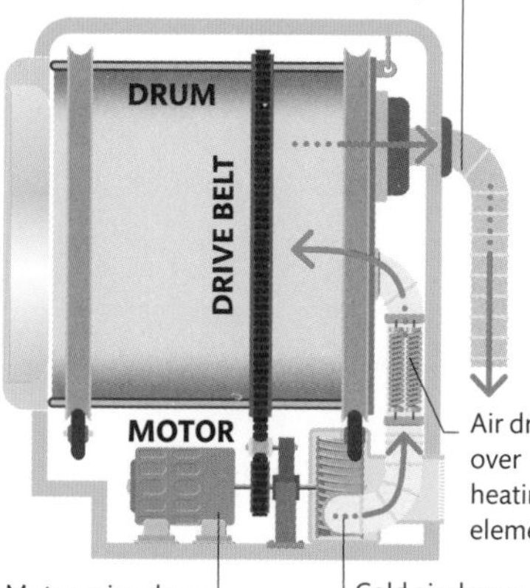
Vent hose carries moist air out of dryer
DRUM
DRIVE BELT
MOTOR
Air drawn over heating element
Motor spins drum via drive belt
Cold air drawn into dryer

Digital assistants

These versatile devices exist as apps on smartphones and as household hardware such as smart speakers. They use voice-recognition algorithms to understand a user's commands and questions. They then direct these requests over the Internet to, for example, run an entertainment application or access an information service.

TO MAKE THEM SOUND MORE HUMAN, SOME DIGITAL ASSISTANTS ARE PROGRAMMED TO INSERT PAUSES INTO THEIR SENTENCES

USER

1 Request sent A user speaks to send two requests to a smart speaker acting as a digital assistant. One is a command to alter the central heating of a home; the other is a question about what the weather is going to be like in Paris tomorrow.

2 Smart speaker Connected to the Internet, usually by a Wi-Fi link, this device recognizes and captures speech using its microphones. Analogue sound is processed into digital data and sent over the Internet to computer servers able to analyze and act on the requests.

How a smart speaker works

A smart speaker can broadcast speech or music streamed over the Internet and capture speech for voice-activated commands and questions. It transmits data via the Internet to and from servers in the cloud (see p.221) to provide responses to a user.

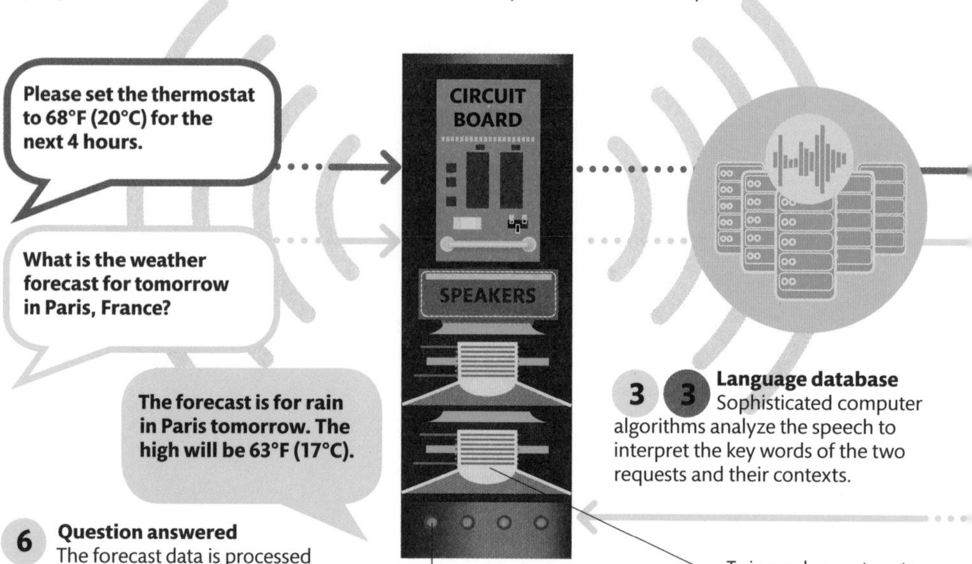
Please set the thermostat to 68°F (20°C) for the next 4 hours.
What is the weather forecast for tomorrow in Paris, France?
The forecast is for rain in Paris tomorrow. The high will be 63°F (17°C).
CIRCUIT BOARD
SPEAKERS

3 Language database Sophisticated computer algorithms analyze the speech to interpret the key words of the two requests and their contexts.

6 Question answered The forecast data is processed by the device service provider into speech files. These are broadcast through the digital assistant's amplifier and speakers for the user to hear.

Array of several microphones captures sound for processing by microprocessors in circuit board

Twin speakers—a tweeter for high pitches and a woofer for lower pitches—broadcast sound

WHAT WAS THE FIRST SMART HOME DEVICE?

In 1966, US engineer Jim Sutherland built the Echo IV smart home computer system, which was capable of controlling lighting, heating, and TVs.

The digital home

The rapid rise in computing power, the Internet, and embedded microprocessors in everyday devices allows millions of devices to be connected and controlled over computer networks. As more and more connected devices enter households, technology enables people to control many household tasks while they are away from home, such as adjusting the central heating thermostat via a smartphone app.

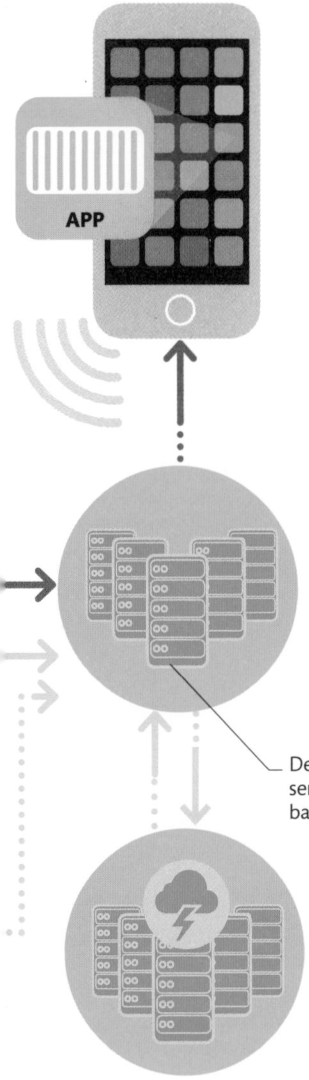
APP
Device service provider
sends weather information
back to smart speaker

5 Smartphone app
The heating request is sent to another digital device—in this case, the user's smartphone, which operates a smart heating app. The app controls the thermostat in the house and sends a signal back to the smart speaker to indicate that the user's request has been completed.

4 4 Device service provider
This software recognizes requests and directs them to the appropriate service, which may be another server in the cloud. The Paris request will be sent to a weather database. The heating request will be directed to an app on the user's smartphone.

5 Weather database
The device service provider accesses the weather database to find the forecast temperature and chance of rain in Paris. The data travels to the smart speaker via the device service provider.

THE INTERNET OF THINGS

Billions of devices embedded with microprocessors and communications technology can connect to the Internet, communicate with other machines or humans, and share data, such as through machine-readable QR (quick response) codes. This network of devices is known as the Internet of Things.

QR CODE

Biometric locks

Increasing numbers of digital devices, such as electronic door locks, replace physical keys with scanners. These capture a biometric feature of a person, such as the iris pattern or fingerprint. Software distills this image down to a unique pattern stored in a database. A match triggers a signal back to the lock, instructing it to open.

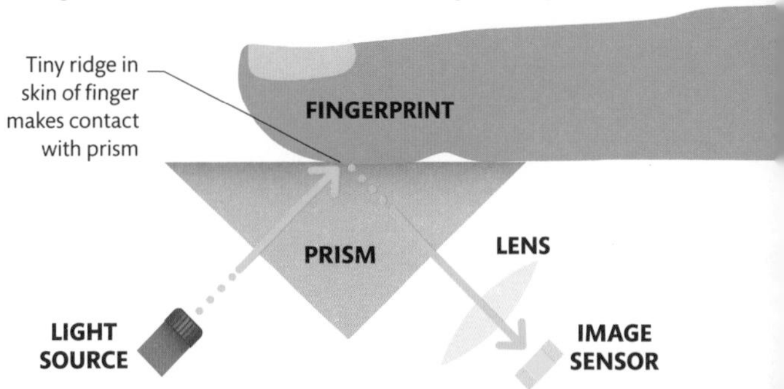
Tiny ridge in skin of finger makes contact with prism
FINGERPRINT
PRISM
LENS
LIGHT SOURCE
IMAGE SENSOR

1 Optical fingerprint scanner
LED light travels through a prism, bounces off the finger placed on the scanner, and is focused by a lens onto a digital image sensor such as a CCD chip. The sensor records the patterns of ridges and valleys that make up the fingerprint.

Distinguishing features of fingerprint are identified
FINGERPRINT
Digital template of fingerprint created

2 Analysis and algorithms
Software analyzes the fingerprint image, seeking out identifying features, such as joining lines (known as minutiae). The software uses an algorithm to create a digital template of the fingerprint.

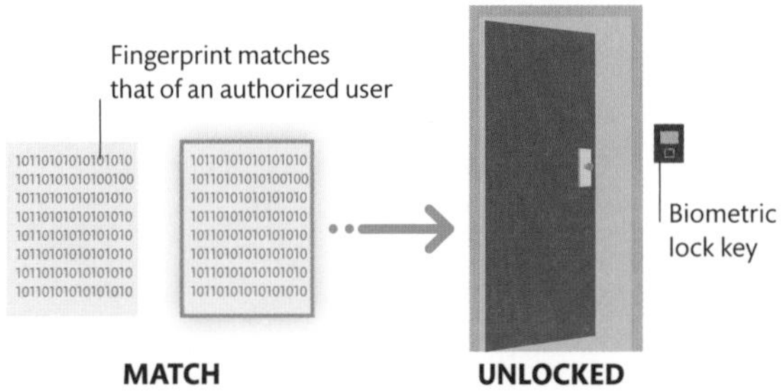
Fingerprint matches that of an authorized user
Biometric lock key
MATCH
UNLOCKED

3 Search and compare
The scanned pattern is sent to a database for comparison. If a match with an authorized user is found, an electronic signal will be sent back to the lock, instructing it to open to admit the person.

SOUND AND VISION TECHNOLOGY

Many technologies work with waves: microphones detect sound waves, while loudspeakers produce them; cameras detect light waves, while projectors produce them; and telecommunications use radio, light, and infrared waves to send and receive signals.

Longitudinal wave

Sound waves are longitudinal. This is because the variation of air pressure is backward and forward, going in the same direction as the wave is traveling.

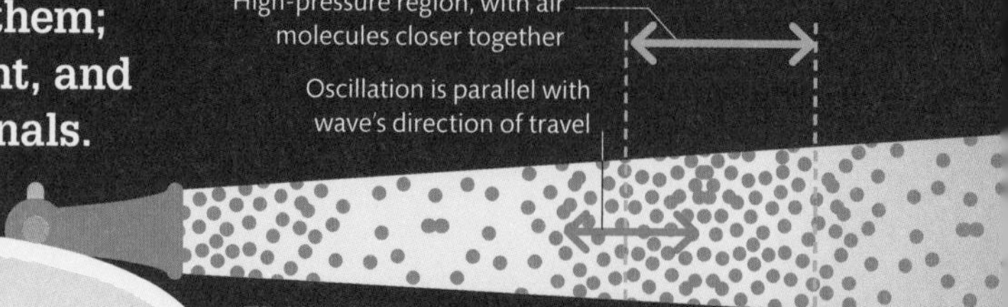

Sound and light waves

A wave is a disturbance that travels. The disturbance that makes sound waves is created by a vibrating object, such as a guitar string. The string produces variations in air pressure as it moves to and fro, and these pressure variations travel in all directions. Sound waves are longitudinal (see above). The disturbance that makes light waves, and other electromagnetic waves (see right and below), is created by particles that carry electric charge, such as electrons in atoms. This disturbance creates variations in the electric and magnetic fields. The variations are at right angles to the wave's direction—they are transverse waves.

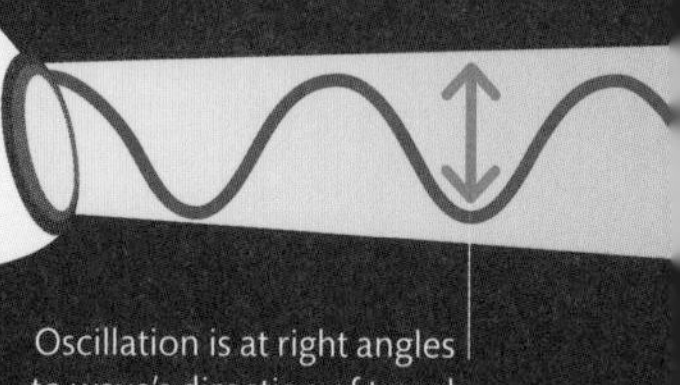

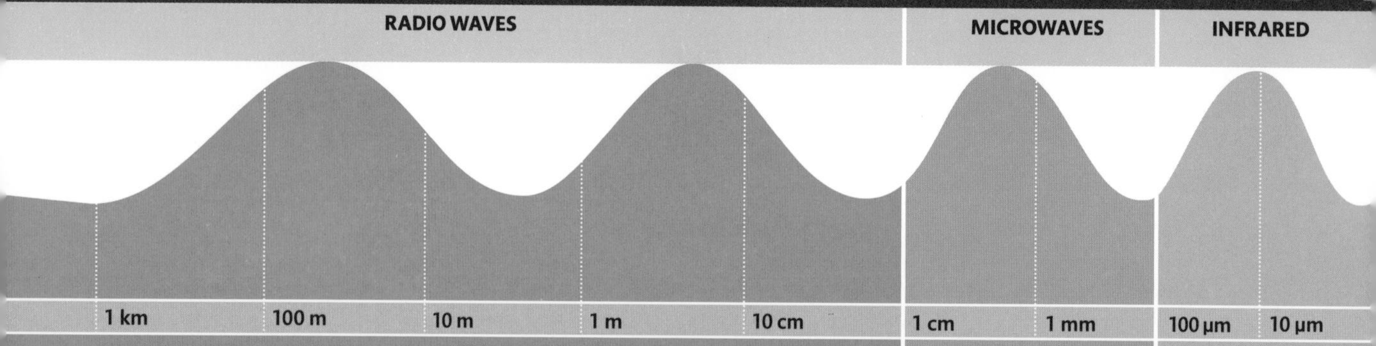

The electromagnetic spectrum

Light is electromagnetic radiation—waves created by disturbances in electric and magnetic fields. Our eyes are sensitive to a range of light from lower-frequency red light to higher-frequency blue light. But there are other kinds of electromagnetic radiation beyond the visible spectrum: radio waves, microwaves, and infrared radiation, with lower frequencies than visible light, and ultraviolet radiation, X-rays, and gamma rays, with higher frequencies.

Radio telescope
A dish antenna can be used to detect radio waves emitted by distant stars.

Microwave oven
Food heats up when high-energy microwaves excite the water molecules inside.

Remote control
A remote control uses pulses of infrared radiation to transmit digital control codes.

Transverse wave
Light waves are transverse: the variations in electric and magnetic fields are up and down and side to side, both at right angles to the wave's direction of travel.

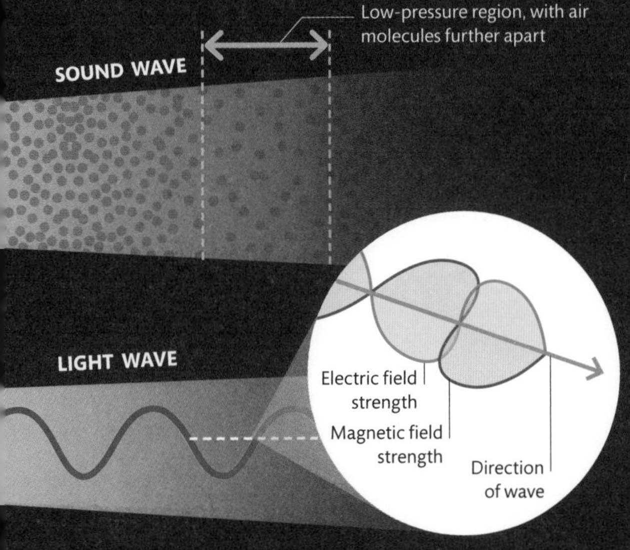

Measuring waves

All waves share measurable characteristics: speed at which they propagate (travel); amplitude (maximum intensity); frequency (how often the disturbance repeats); and wavelength (distance between waves).

Wave relationship
For a fixed wave speed, increasing the wavelength reduces the frequency and vice versa.

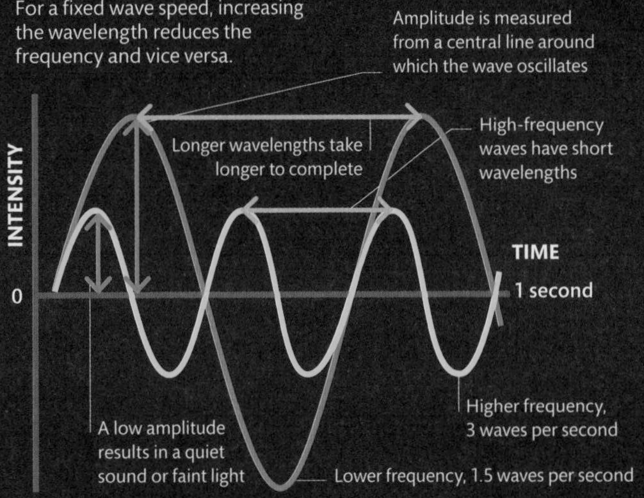

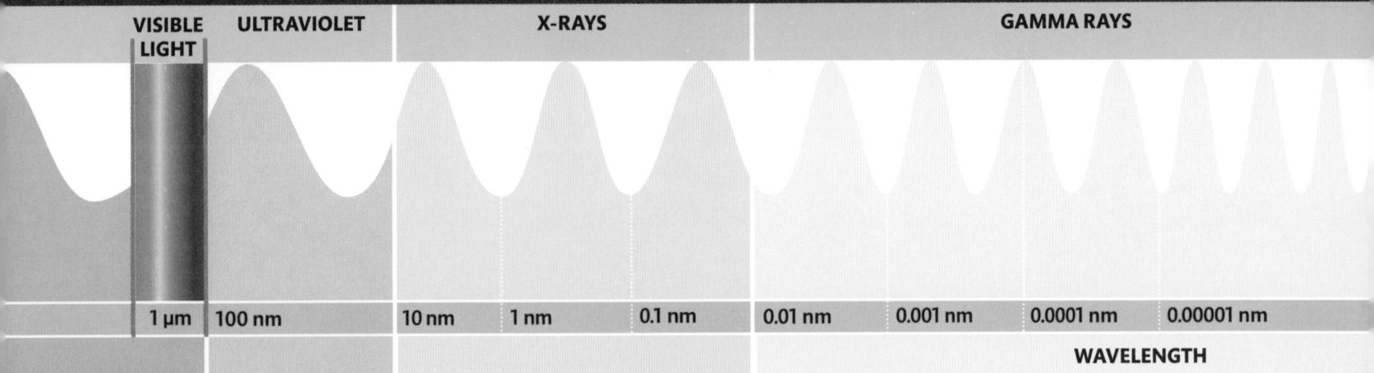

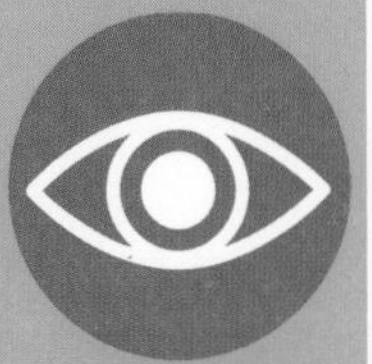

Human eye
Our eyes detect a narrow range of wavelengths as the color spectrum.

Disinfection
Certain wavelengths of UV light can be used to kill bacteria and sanitize objects.

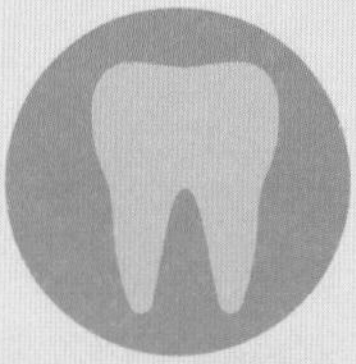

Dental X-ray
Short-wavelength X-rays pass through gum tissue to reveal the teeth underneath.

Vehicle inspection
High-energy gamma rays can penetrate vehicles to reveal images of dangerous items inside.

Using electromagnetic radiation
Humans put electromagnetic radiation to use in a range of technologies. The shortest wavelengths are measured in units such as micrometers (millionths of a meter, μm) and nanometers (billionths of a meter, nm).

Microphones and loudspeakers

A microphone creates an electrical wave called an audio signal. This electrical wave is a copy of the variations in air pressure of an incoming sound wave. When the audio signal is amplified, or strengthened, and fed through a loudspeaker, the original sound is reproduced and can be increased in volume.

SHOULD I WEAR EARPLUGS TO A CONCERT?

The loudspeakers at a pop concert can produce huge variations in air pressure that can damage your ear, so ear plugs are a good idea if you are close to them.

1 Diaphragm in
As a sound wave travels into the microphone, it passes through a layer of protective metal mesh before reaching the diaphragm, which is connected to a thin wire coil. High-pressure air pushes the diaphragm inward, moving the coil down.

2 Diaphragm out
Low-pressure air allows the diaphragm to move back again. As a result, the diaphragm moves to and fro with the rapid variations in pressure of any sound waves that hit it. As the diaphragm moves inward and outward, it takes the coil of thin wire with it.

3 Audio signal generated
The coil surrounds one pole of a permanent magnet, and the movement creates an electric current that flows first one way and then the other. This alternating current, the audio signal, is a copy of the variations of pressure in the sound wave.

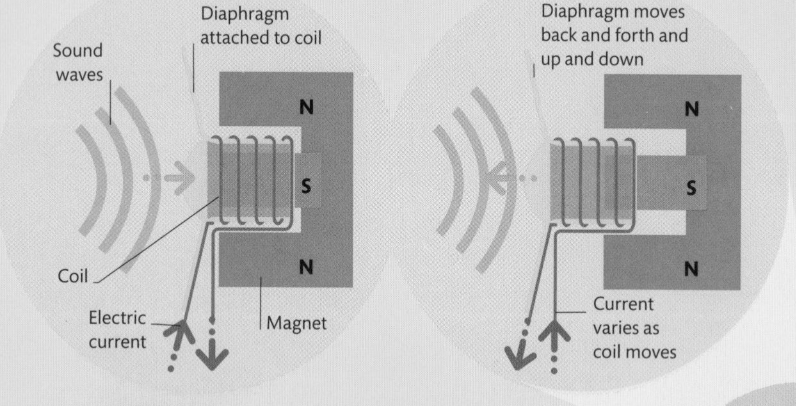

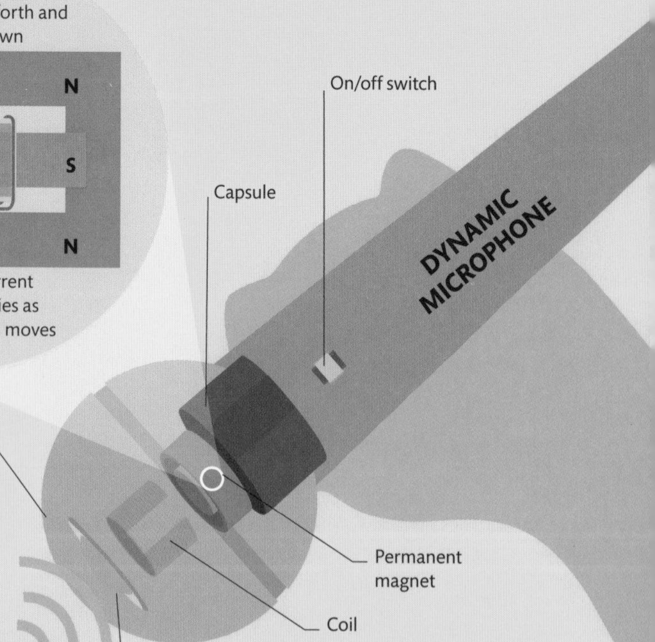

Catching sound waves

Sound is a disturbance of the air that travels out from its source in the form of waves of alternating high and low air pressure (see pp.136–137). The audio signal a microphone produces is a varying electric current: the variations in current match the variations of pressure in the sound wave. Inside a microphone is a thin membrane called a diaphragm. The sound waves make the diaphragm vibrate to and fro when they hit it—it is this movement of the diaphragm that creates the electrical signal.

Dynamic microphone
One commonly used type of microphone is the dynamic microphone. Inside it, the diaphragm vibrates a coil of wire positioned around a magnet, producing an alternating electrical current.

4 Signal amplification
The audio signal produced by a microphone is not powerful enough to produce sound in a loudspeaker. An electronic circuit called an amplifier boosts the signal.

Making sound

Loudspeakers use audio signals to reproduce sound. The audio signal might come straight from a microphone, or it might be output from the memory of a computer or smartphone where it has been stored. It might even arrive wirelessly, coded into radio waves. Wherever it comes from, it is too weak to produce a loud sound, and it must be strengthened (amplified) before reaching the loudspeaker.

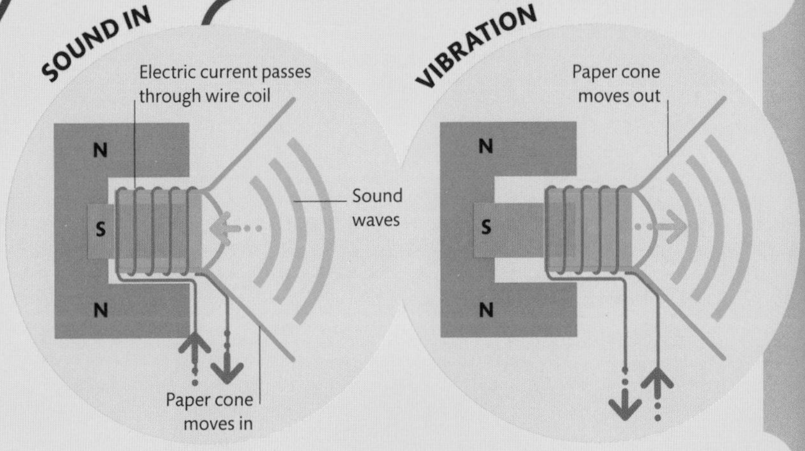

5 Sound out
The amplified audio signal is fed to a loudspeaker. The alternating current of the audio signal passes through a coil inside the loudspeaker, and that produces a varying magnetic field. The varying field causes the coil, and a paper cone attached to it, to vibrate in and out, reproducing the original sound waves.

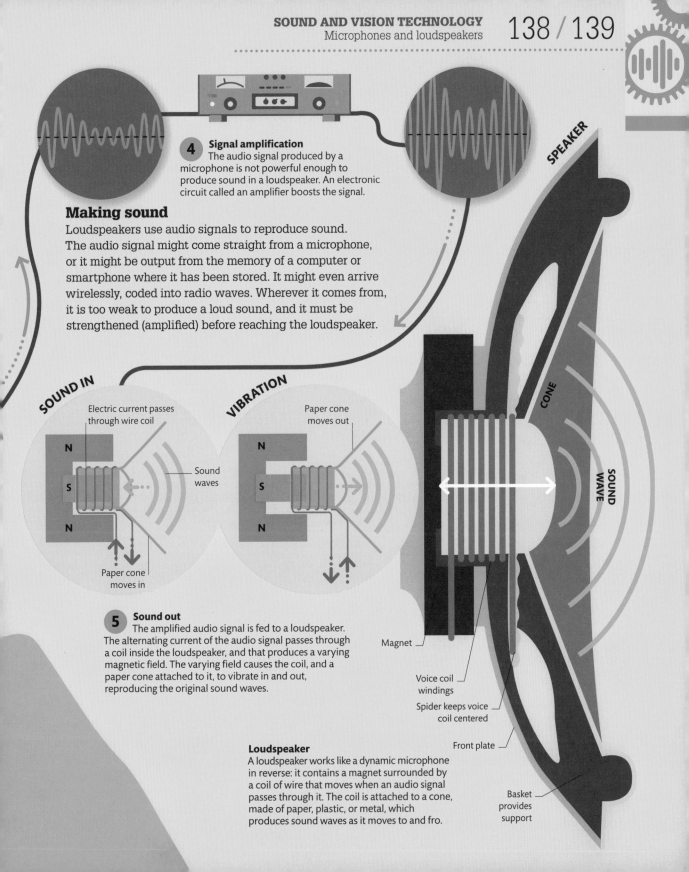

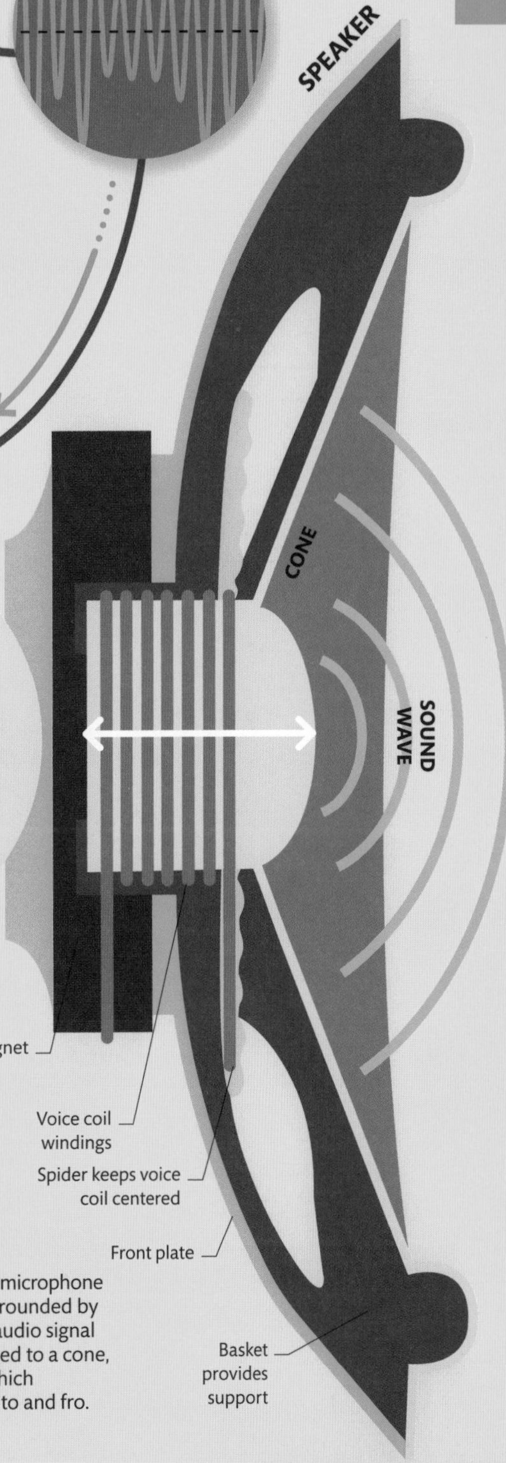

Loudspeaker
A loudspeaker works like a dynamic microphone in reverse: it contains a magnet surrounded by a coil of wire that moves when an audio signal passes through it. The coil is attached to a cone, made of paper, plastic, or metal, which produces sound waves as it moves to and fro.

Digital sound

Digital sound is stored as large collections of binary numbers. The numbers describe the oscillations of an audio signal—an electrical copy of the sound wave of the original sound. Playing back a sound involves using electronic circuits that can reconstruct the audio signal from the numbers and play it through a loudspeaker.

Analogue to digital to analogue

Digitization begins with an audio signal—an electrical copy, or analogue, of the sound wave. Typically, this comes from a microphone (see p.138). An analogue-to-digital converter measures the voltage of the audio signal thousands of times every second. It assigns each of the measurements, or samples, a number depending on the strength of the voltage. The numbers are stored in binary form (see p.158). To play back the sound, a sound signal must be produced and sent to a loudspeaker (see p.139) or headphones. This is done by a digital-to-analogue converter.

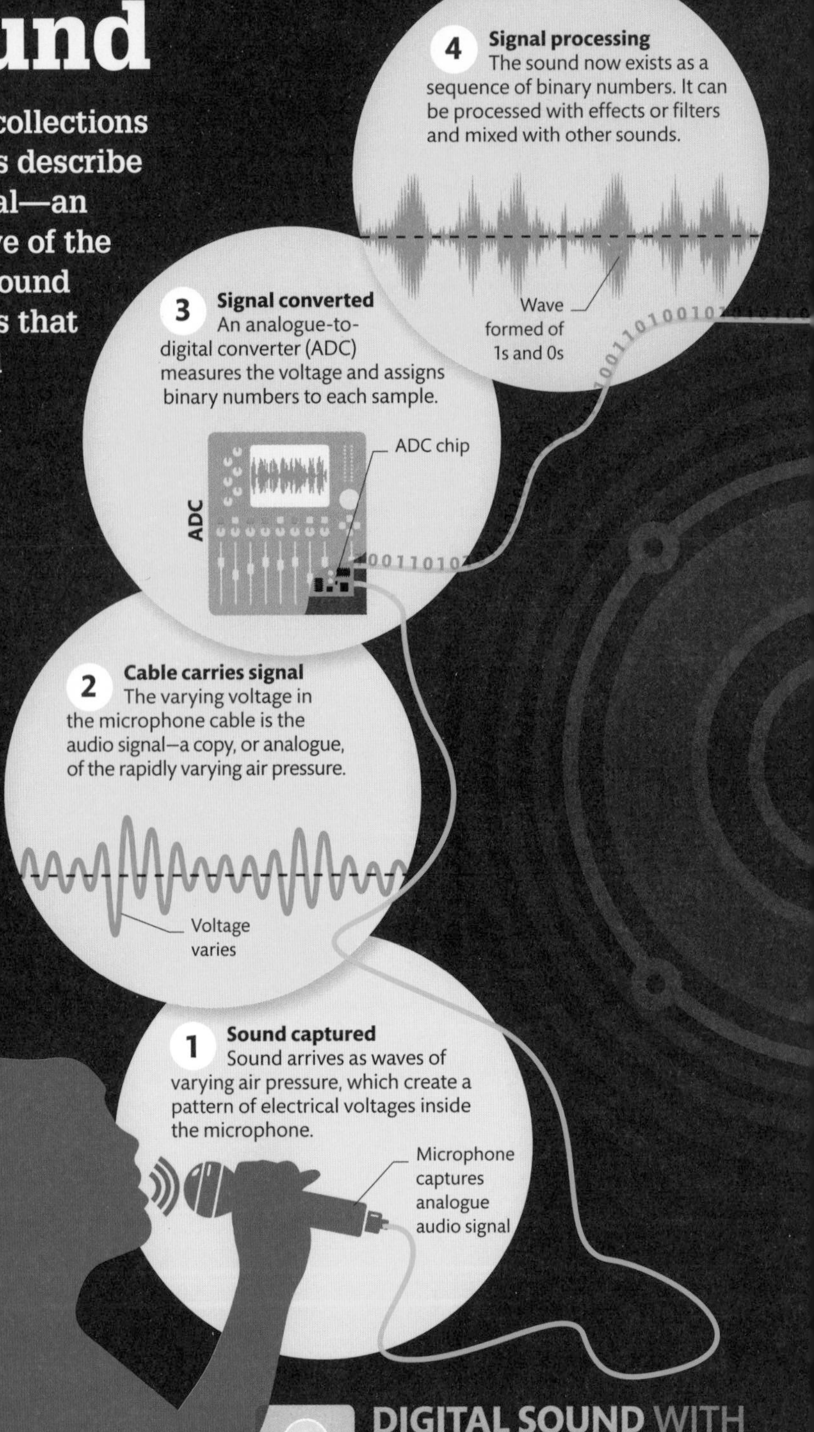
4 Signal processing
The sound now exists as a sequence of binary numbers. It can be processed with effects or filters and mixed with other sounds.
Wave formed of 1s and 0s
3 Signal converted
An analogue-to-digital converter (ADC) measures the voltage and assigns binary numbers to each sample.
ADC chip
ADC
2 Cable carries signal
The varying voltage in the microphone cable is the audio signal—a copy, or analogue, of the rapidly varying air pressure.
Voltage varies
1 Sound captured
Sound arrives as waves of varying air pressure, which create a pattern of electrical voltages inside the microphone.
Microphone captures analogue audio signal

WHAT IS COMPRESSED AUDIO?

Good-quality digital sound can take up huge amounts of storage. Compression reduces the amount of storage taken up, with little effect on sound quality.

DIGITAL SOUND WITH 16 BITS PER SAMPLE CAN MEASURE 65,536 LEVELS OF VOLTAGE

5 Storing sound
The sequence of binary numbers can be stored in the device's memory—on a hard disk or a USB flash drive, for example.

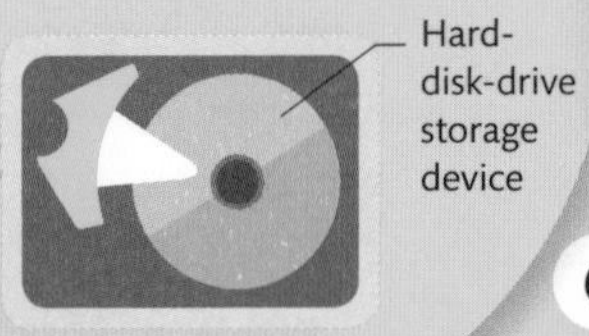

6 Re-creating sound
For playback, a processor retrieves the sequences from storage, ready to re-create the audio signal.

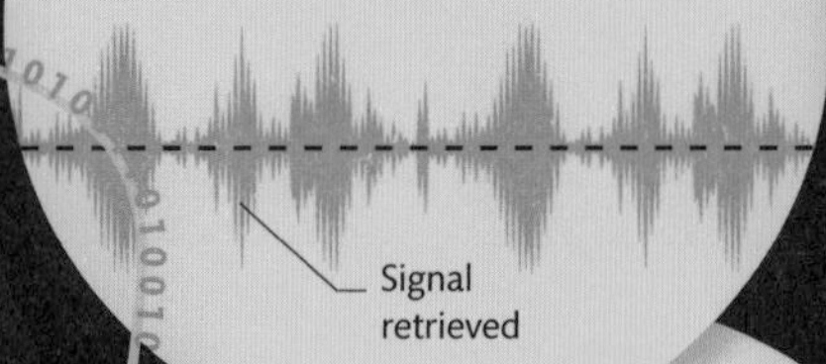

7 Back to analogue
A digital-to-analogue converter (DAC) uses the sequence of binary numbers retrieved from storage to re-create the audio signal.

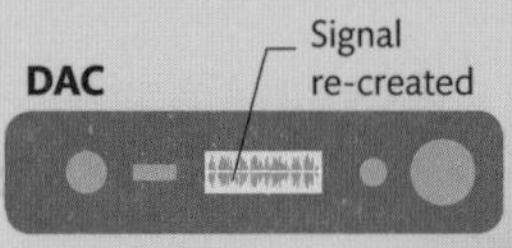

8 Amplifying the signal
The signal is now restored to analogue form so that it can drive an amplifier.

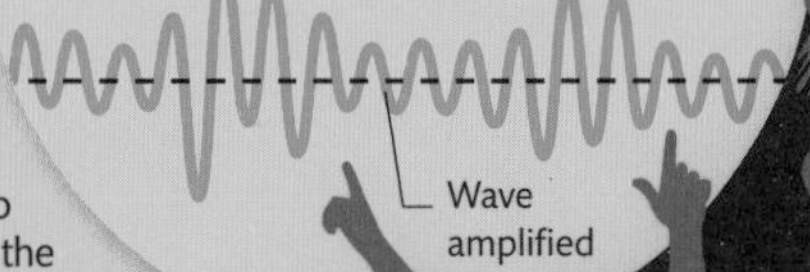

9 Playback
The amplified audio signal pushes a cone inside the speaker back and forth, creating sound waves of varying pressure.

Sound quality

The quality of digital sound depends on how many samples are used per second and how many bits (binary digits) are used to represent the number for each sample. The quality of sound on compact discs is standardized. It uses 44,100 samples per second and 16 bits per sample.

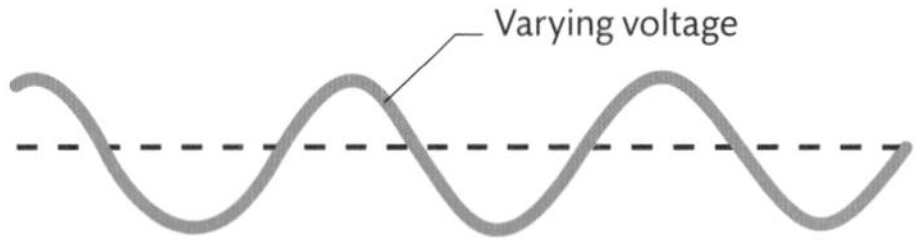

Original analogue audio signal
The audio signal the microphone produces is a smooth wave of varying voltage. It goes up and down hundreds or thousands of times each second.

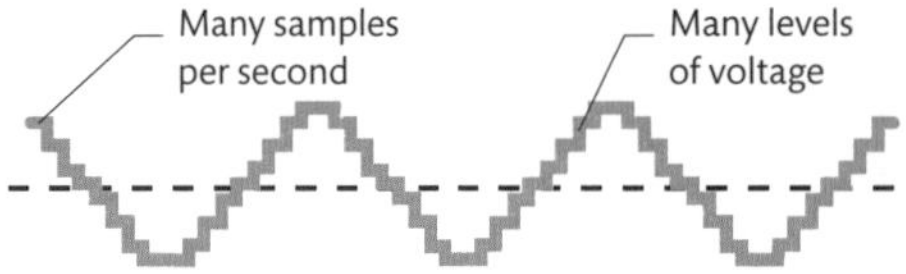

Good quality
Digital sound cannot re-create a perfect audio signal, but the more levels of voltage, and the more samples per second, the better.

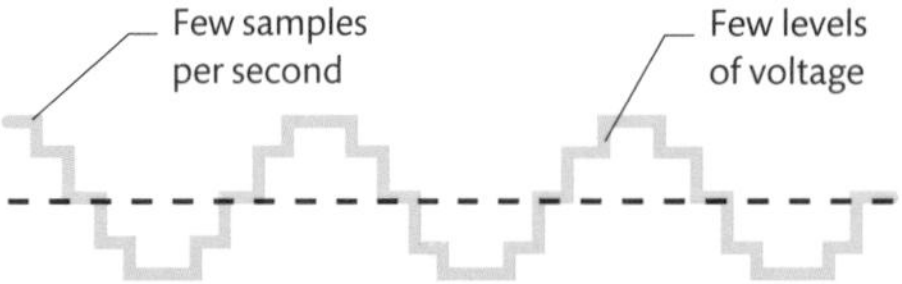

Poor quality
Poor-quality audio is choppy and distorted, because it has fewer bits per sample—which means fewer levels of voltage—and fewer samples per second.

ON THE PHONE

When you speak on the phone, the sound of your voice travels across the telephone network in digital form. A smartphone has an ADC and a DAC built in. For landline phones, the ADC and DAC are outside the home.

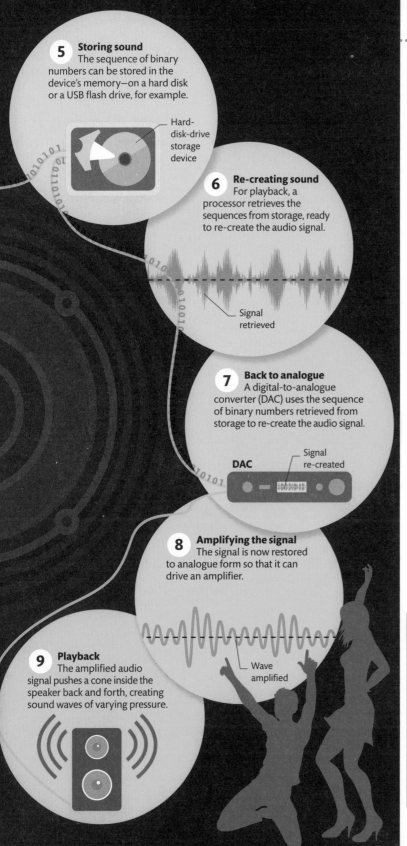

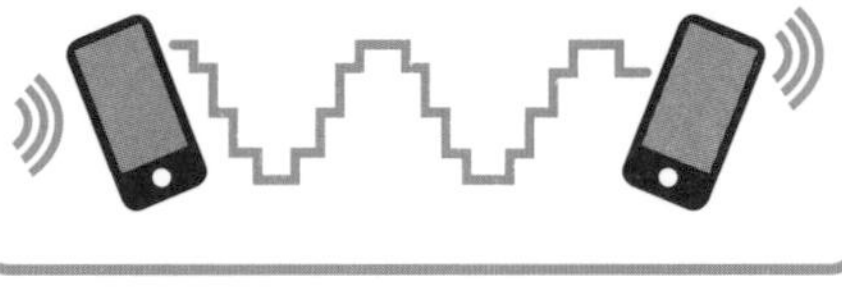

Telescopes and binoculars

We see things because light from them produces an image on the retina at the back of each eye. Faraway things produce only a small image on the retina. A telescope or a pair of binoculars produces a magnified image, which takes up more space on the retina.

WHAT DO THE TWO NUMBERS ON BINOCULARS MEAN?

On binoculars marked 10x50, 10 refers to the magnification and 50 to the diameter of the two objective lenses, in millimeters.

Telescopes

In a telescope, a lens or mirror, called the objective, focuses light from a distant object. This makes an image of the object inside the tube. The eyepiece lens magnifies the image. The longer the focal length (distance between the lens or mirror and the point where the rays meet) of the objective, the larger the image in the tube. The shorter the focal length of the eyepiece lens, the larger the image appears in your eye.

Reflecting telescope

In a reflecting telescope, the objective is a concave mirror. This reflects light back up the tube as it focuses it, and a flat mirror directs the light out into the eyepiece.

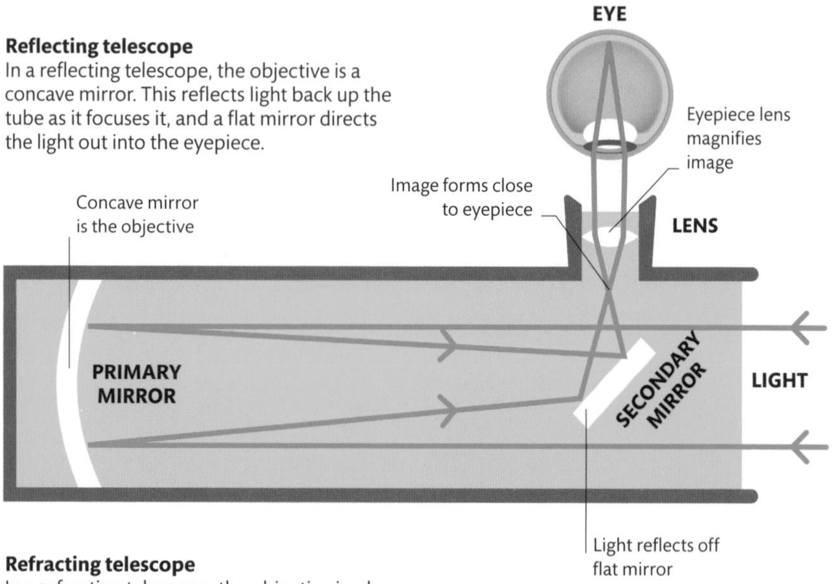
EYE
Eyepiece lens magnifies image
Image forms close to eyepiece
LENS
Concave mirror is the objective
PRIMARY MIRROR
SECONDARY MIRROR
LIGHT
Light reflects off flat mirror

Refracting telescope

In a refracting telescope, the objective is a lens. With only two lenses, the image produced is upside down, so some refracting telescopes contain more lenses to correct this.

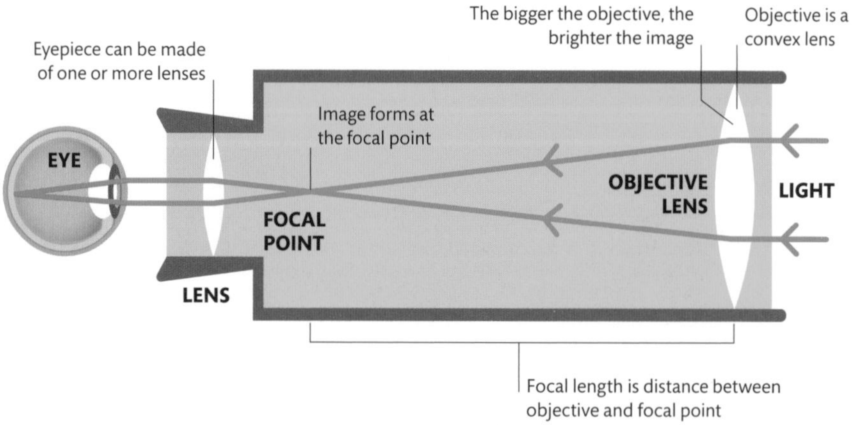
The bigger the objective, the brighter the image
Objective is a convex lens
Eyepiece can be made of one or more lenses
Image forms at the focal point
EYE
OBJECTIVE LENS
LIGHT
FOCAL POINT
LENS
Focal length is distance between objective and focal point

SPACE TELESCOPES

The atmosphere absorbs some of the light coming from distant planets, stars, and galaxies, and its turbulent motion reduces image quality. Space telescopes do not suffer these problems. Images are captured digitally and beamed back to Earth.

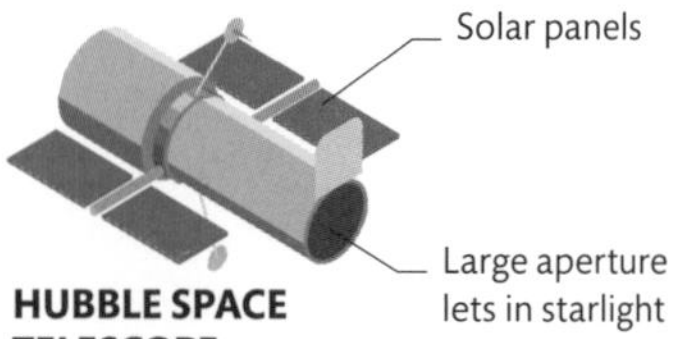
Solar panels
Large aperture lets in starlight

HUBBLE SPACE TELESCOPE

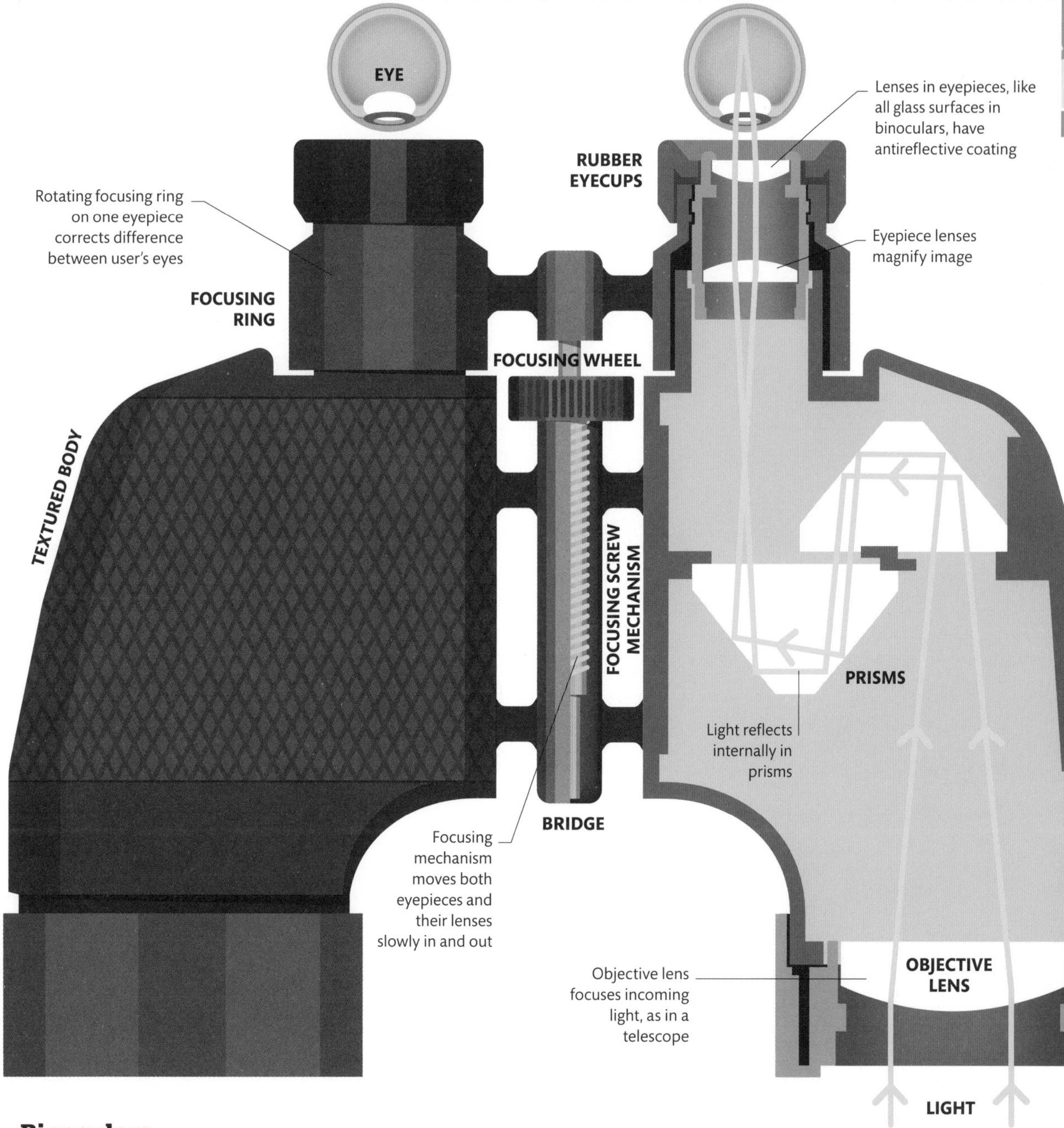
EYE
Lenses in eyepieces, like all glass surfaces in binoculars, have antireflective coating
RUBBER EYECUPS
Rotating focusing ring on one eyepiece corrects difference between user's eyes
Eyepiece lenses magnify image
FOCUSING RING
FOCUSING WHEEL
TEXTURED BODY
FOCUSING SCREW MECHANISM
PRISMS
Light reflects internally in prisms
BRIDGE
Focusing mechanism moves both eyepieces and their lenses slowly in and out
Objective lens focuses incoming light, as in a telescope
OBJECTIVE LENS
LIGHT

Binoculars

Binoculars consist of two refracting telescopes side by side—one for each eye. Two glass prisms in each tube turn the image the right way around and make it possible to have objective lenses with a long focal length in a short tube, by bending light back on itself twice. The smaller size makes binoculars easy to carry around, and the two eyepieces make for comfortable viewing.

THE **LARGEST OBJECTIVE LENS** IN A REFRACTING TELESCOPE **IS 40 IN (102 CM) IN DIAMETER.** IT IS IN **THE YERKES OBSERVATORY**

Electric lighting

Most electric lighting uses either fluorescent or LED lamps (bulbs). Far less energy-efficient incandescent light bulbs can still be found, although their use is declining.

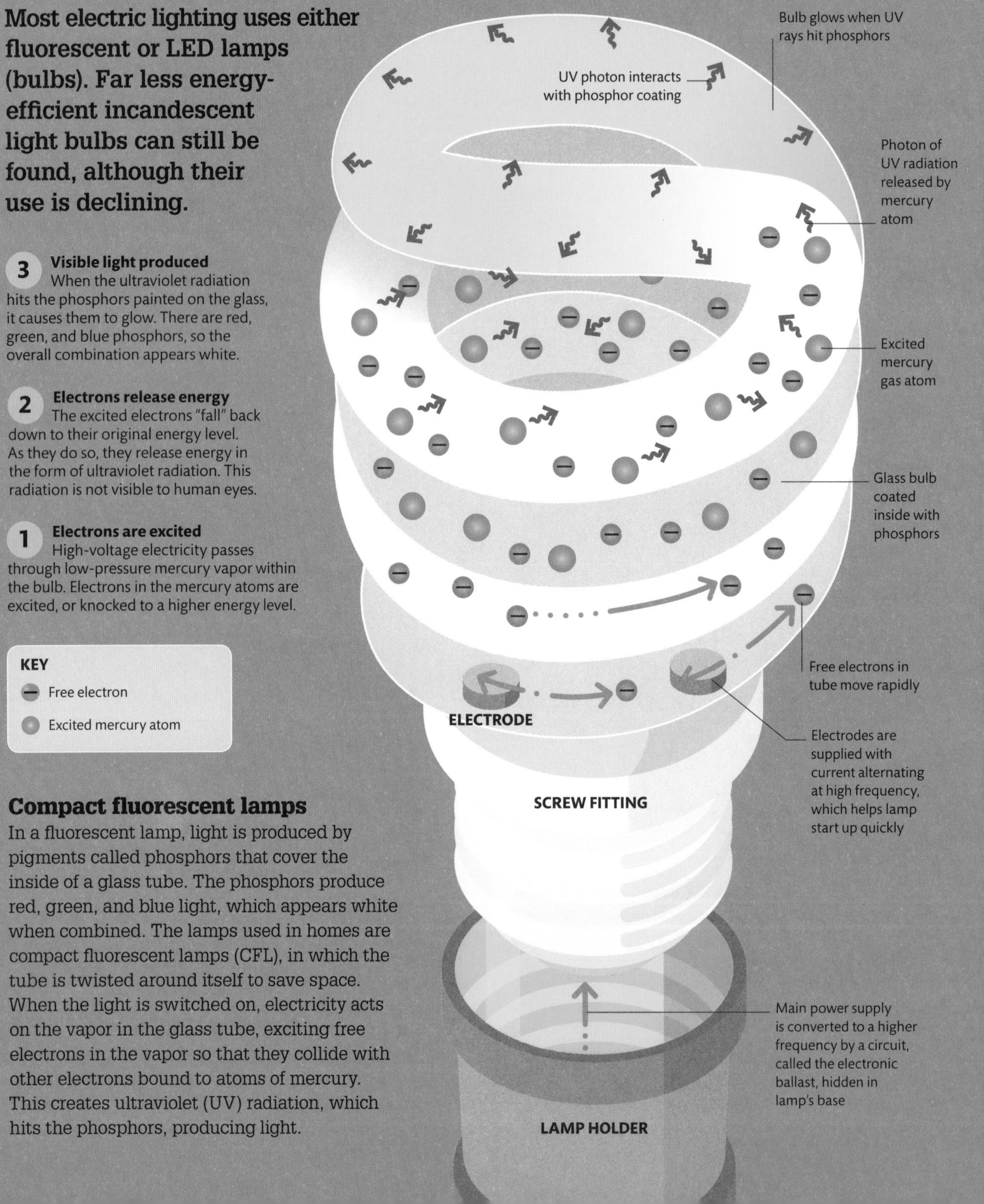

Bulb glows when UV rays hit phosphors
UV photon interacts with phosphor coating
Photon of UV radiation released by mercury atom
Excited mercury gas atom
Glass bulb coated inside with phosphors
Free electrons in tube move rapidly
ELECTRODE
Electrodes are supplied with current alternating at high frequency, which helps lamp start up quickly
SCREW FITTING
Main power supply is converted to a higher frequency by a circuit, called the electronic ballast, hidden in lamp's base
LAMP HOLDER
3 Visible light produced
When the ultraviolet radiation hits the phosphors painted on the glass, it causes them to glow. There are red, green, and blue phosphors, so the overall combination appears white.
2 Electrons release energy
The excited electrons "fall" back down to their original energy level. As they do so, they release energy in the form of ultraviolet radiation. This radiation is not visible to human eyes.
1 Electrons are excited
High-voltage electricity passes through low-pressure mercury vapor within the bulb. Electrons in the mercury atoms are excited, or knocked to a higher energy level.
KEY
Free electron
Excited mercury atom

Compact fluorescent lamps

In a fluorescent lamp, light is produced by pigments called phosphors that cover the inside of a glass tube. The phosphors produce red, green, and blue light, which appears white when combined. The lamps used in homes are compact fluorescent lamps (CFL), in which the tube is twisted around itself to save space. When the light is switched on, electricity acts on the vapor in the glass tube, exciting free electrons in the vapor so that they collide with other electrons bound to atoms of mercury. This creates ultraviolet (UV) radiation, which hits the phosphors, producing light.

LED lamps

In LED (light-emitting diode) lamps, light is produced by a sandwich of two kinds of semiconductor: n-type (negative) and p-type (positive). When connected to an electrical supply, electrons flow from the n-type to the p-type, releasing energy in the form of particles of light, called photons. In many household lamps, the LED produces blue light—some of which is absorbed by phosphors coating the LED. The phosphors produce yellow light, and the combination of blue and yellow appears white.

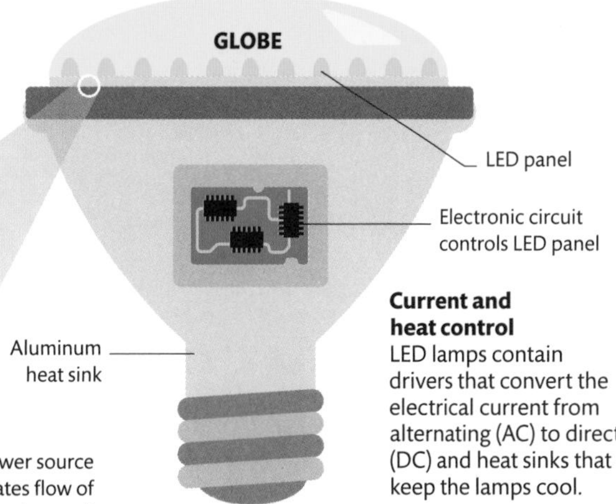
GLOBE
LED panel
Electronic circuit controls LED panel
Aluminum heat sink

Current and heat control
LED lamps contain drivers that convert the electrical current from alternating (AC) to direct (DC) and heat sinks that keep the lamps cool.

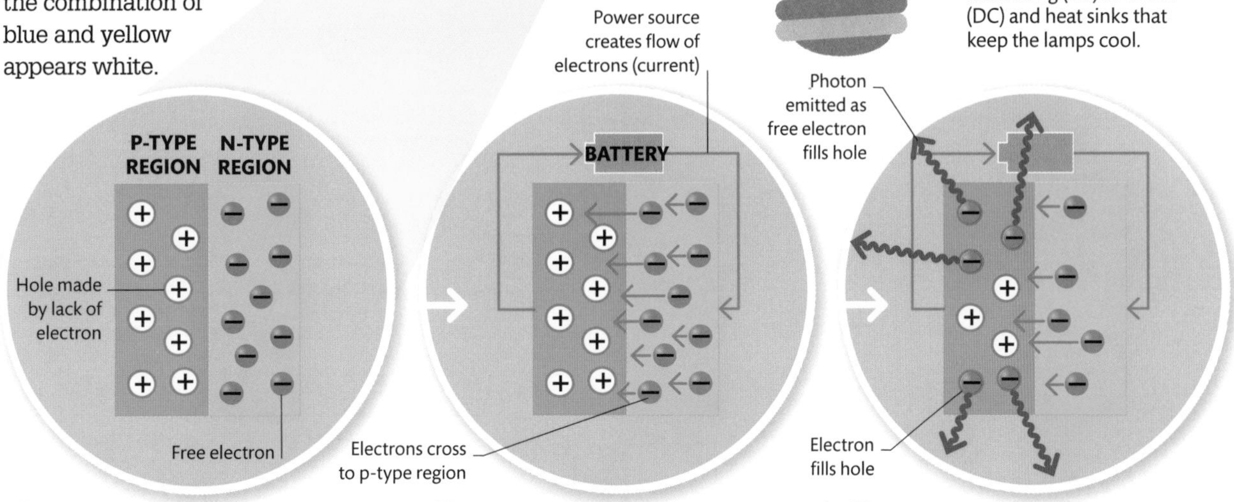
P-TYPE REGION
N-TYPE REGION
Hole made by lack of electron
Free electron
Power source creates flow of electrons (current)
BATTERY
Electrons cross to p-type region
Photon emitted as free electron fills hole
Electron fills hole

1 Semiconductors
The semiconductors in most LEDs are compounds of the element gallium. Adding traces of other elements creates n-type and p-type regions, in which there are too many or too few electrons.

2 Flow of electrons
Connecting an electrical supply across a junction between the regions pushes electrons from the n-type region into the p-type region, where they fill the holes that result from the lack of electrons.

3 Photons
When an electron fills a hole, it falls into a lower energy level in an atom of gallium, and its energy loss releases a photon. An LED produces billions or trillions of photons every second.

LIGHT SOURCES (EQUAL BRIGHTNESS)

CFL
Power consumption 18W
Average life span
8,000 hours

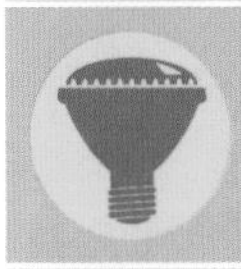

LED
Power consumption 9W
Average life span
25,000 hours

INCANDESCENT
Power consumption 60W
Average life span
1,200 hours

INCANDESCENT BULBS

Up to the end of the 20th century, the most common form of domestic electric lighting was the incandescent light bulb. Inside, a thin, coiled length of tungsten wire called a filament becomes white hot when an electric current flows through it. The wire does not burn, because the bulb is filled with inert gases rather than air. However, its incandescence produces light.

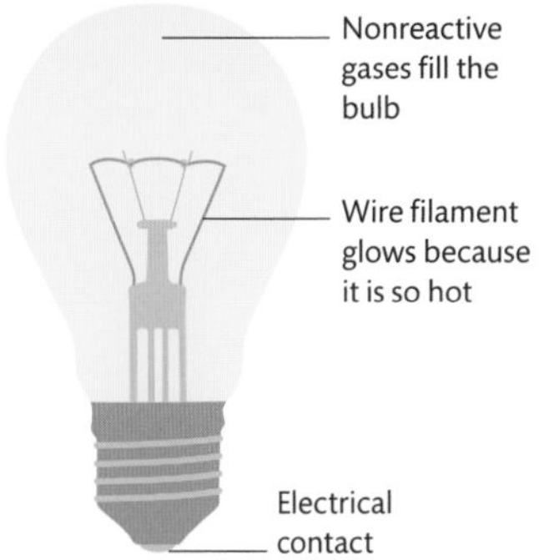
Nonreactive gases fill the bulb
Wire filament glows because it is so hot
Electrical contact

Lasers

A laser produces an intense beam of light that is collimated (all directed in a straight line, rather than spreading out) and coherent (all the waves are in step and all of the same frequency). The word "laser" stands for "light amplification by stimulated emission of radiation."

ARE LASERS USED AS WEAPONS?

Yes, there are a few systems already in use, in which high-power lasers are used to destroy targets. At present, though, most systems are still experimental.

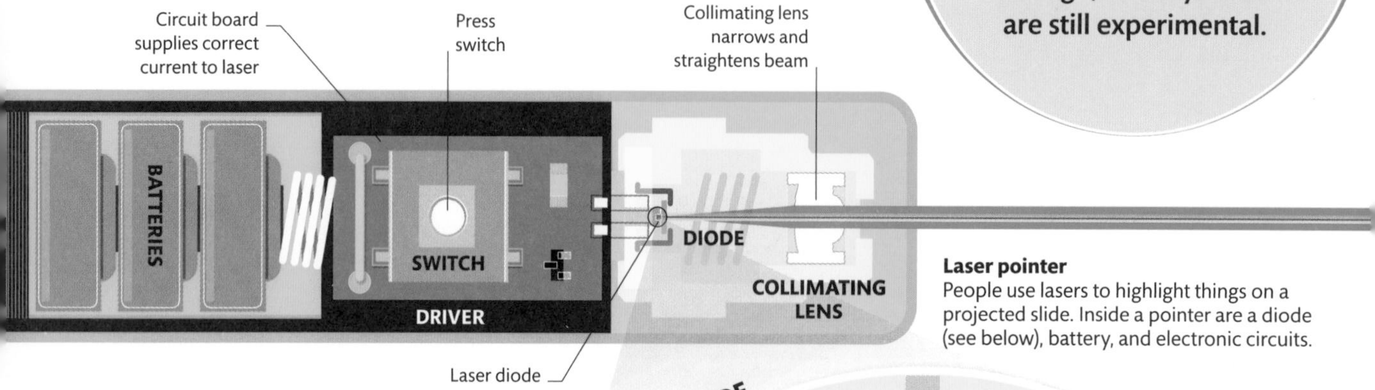

Circuit board supplies correct current to laser
Press switch
Collimating lens narrows and straightens beam
BATTERIES
SWITCH
DRIVER
DIODE
COLLIMATING LENS
Laser diode

Laser pointer
People use lasers to highlight things on a projected slide. Inside a pointer are a diode (see below), battery, and electronic circuits.

Solid-state lasers

The most common lasers are low-power "solid-state" laser diodes, in which the light is produced by a solid sandwich of semiconductor materials. The outer layers are made of silicon combined, or "doped," with other elements to conduct electricity, while the inner layers are undoped. When an electric current flows through the layers, it initiates a process that leads to the production of a burst of light, called a photon (see opposite). Laser diodes are used in devices such as fiber-optic cables, laser printers, and barcode readers.

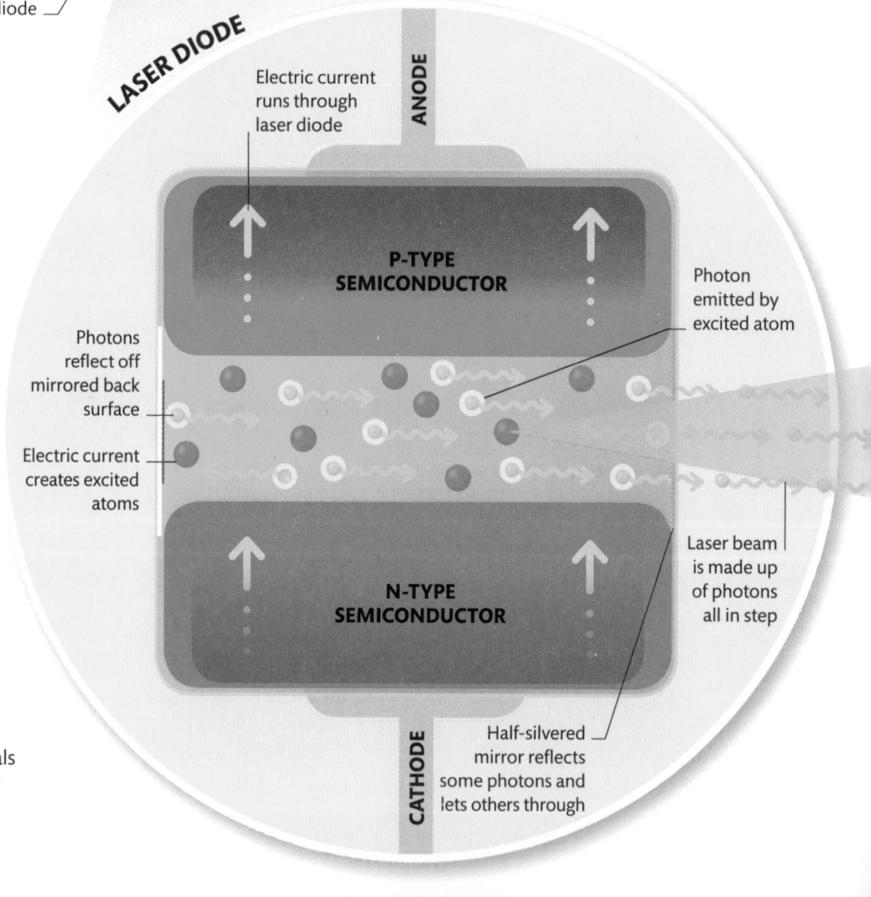

LASER DIODE
ANODE
Electric current runs through laser diode
P-TYPE SEMICONDUCTOR
Photon emitted by excited atom
Photons reflect off mirrored back surface
Electric current creates excited atoms
N-TYPE SEMICONDUCTOR
Laser beam is made up of photons all in step
Half-silvered mirror reflects some photons and lets others through
CATHODE

Laser diode
The two outer layers of semiconductor materials are "doped" n-type and p-type (see p.160). The center of the sandwich is undoped.

USES OF LASERS

Medical
Lasers are used for extremely precise cutting during surgery, cauterizing wounds, and for corrective eye surgery.

Surveying
Cheap, low-power lasers produce thin beams in straight lines, useful for builders and surveyors.

Welding
Some lasers can be automated for high-speed work, such as joining parts of car bodies, pots, and pans.

Manufacture
Lasers are used for precise cutting of fabrics for the clothing industry and etching out letters or numbers on keyboards.

Entertainment
Lasers provide light shows at concerts, often drawing intricate patterns in theatrical smoke. CD and DVD players also use lasers.

Telecommunications
Infrared laser diodes send digital information along optical fibers across worldwide networks.

GAS LASERS

Not all lasers are solid-state, semiconductor laser diodes. Many of the most powerful are gas lasers, in which the excited electrons are in atoms of a gas. Lasers with carbon dioxide gas as the lasing medium are used to cut and weld car parts, for example.

LASER BEAM

How photons are produced

The photons (particles of light) that make up a laser beam are created by a process called stimulated emission. They are produced by electrons in atoms in the lasing medium—in a solid-state laser diode, this is the undoped semiconductor in the semiconductor sandwich (see opposite). An electric current (or, in some lasers, a burst of light) excites (boosts) electrons to a higher energy level. When an electron falls back to the lower energy level, the extra energy is released as a photon. The photon travels through the lasing medium, stimulating more excited electrons to release photons. The color of the laser light depends on the difference in energy between the higher and lower levels.

LASERS CAN MEASURE THE DISTANCE FROM EARTH TO THE MOON TO WITHIN A FEW CENTIMETERS

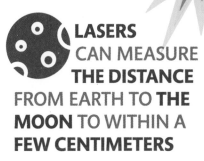

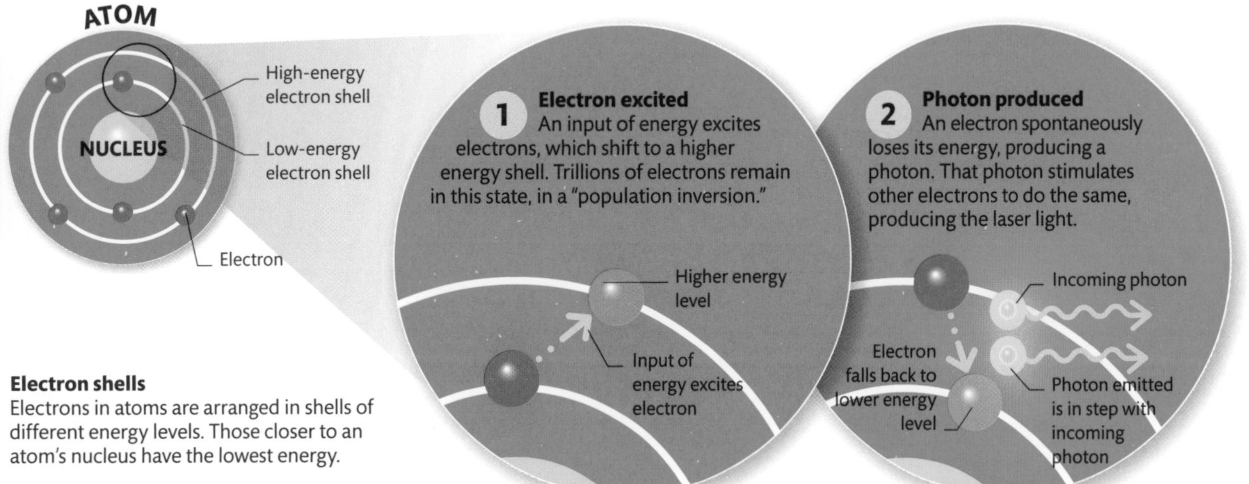
ATOM
High-energy electron shell
NUCLEUS
Low-energy electron shell
Electron
1 Electron excited
An input of energy excites electrons, which shift to a higher energy shell. Trillions of electrons remain in this state, in a "population inversion."
Higher energy level
Input of energy excites electron
2 Photon produced
An electron spontaneously loses its energy, producing a photon. That photon stimulates other electrons to do the same, producing the laser light.
Incoming photon
Electron falls back to lower energy level
Photon emitted is in step with incoming photon

Electron shells
Electrons in atoms are arranged in shells of different energy levels. Those closer to an atom's nucleus have the lowest energy.

Holograms

A hologram is a 3-D image made using laser beams. It is stored inside a photographic film as an interference pattern, which contains information about the object's surface. The image you see when you view a hologram has depth, and by moving your head, you are able to look at it from different angles.

ARE THE HOLOGRAMS OF MUSICAL PERFORMERS AT CONCERTS REAL HOLOGRAMS?

No, they are images created by mirrors—an illusion called "Pepper's Ghost."

Making a hologram

Holograms are made using laser light. Significantly, the light waves produced by a laser are all "in step" (see pp.146–147). To make the hologram, a laser beam passes through a splitter. Half of the light forms the reference beam, which is directed straight onto a photographic (light-sensitive) film. The other half forms the object beam, which bounces off the object to be pictured in the hologram. The reflected object beam falls onto the film, where its waves merge, or interfere, with the reference beam. The interference creates a pattern that holds information about the surface of the object—and this information can be extracted if light shines onto the film once it has been developed.

IF YOU **BREAK A HOLOGRAM** INTO **FRAGMENTS**, EACH FRAGMENT **CONTAINS THE WHOLE IMAGE**

Holographic image
A hologram is formed when two beams combine: the reference and object beams. The image is created when the pattern of interference between the beams is captured in a photographic plate.

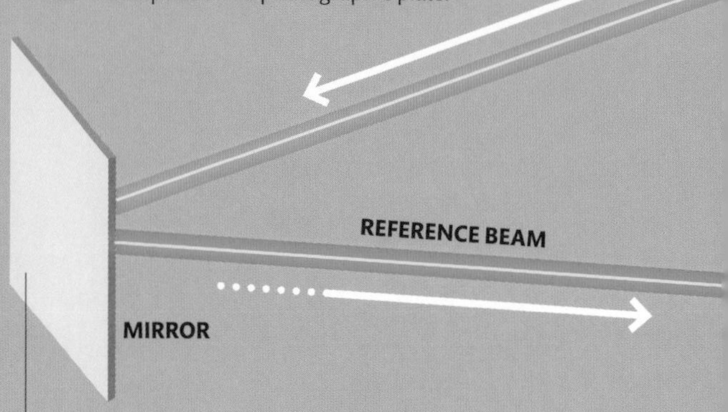

Mirror reflects reference beam onto diverging lens

3 Reference beam
Light that passes through the beam splitter bypasses the object. It reflects off another mirror that directs it toward the photographic film. First it passes through a diverging lens that broadens out the beam.

SECURITY HOLOGRAMS

The holograms on banknotes, credit cards, and concert tickets are designed to prevent those items from being forged. They are produced with laser beams but are visible in ordinary white light.

BANK CARD

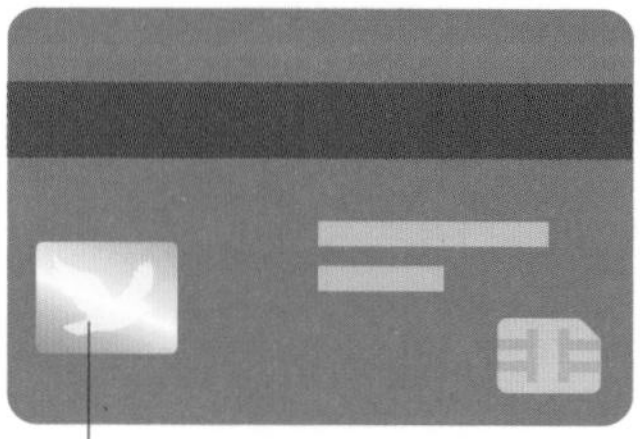

White light reflection hologram

Viewing a hologram

The hologram described above is called a transmission hologram. Another kind is a reflection hologram. It is similar, but there is no beam splitter: the reference beam passes through the film then reflects off the object, which is positioned behind the film, to become the object beam. When the photographic film is developed, it looks dark, with strange lines on it; there is no sign of an image. To view a reflection hologram, a laser beam passes through the film, reflecting off the interference pattern inside it and producing an image.

2 Object beam formed
Light that is reflected off the beam splitter forms the object beam. A mirror deflects it toward the object, but first it passes through a lens that broadens out the beam. This is called a diverging lens.

1 Laser fires beam
The light from the laser emerges as a thin beam of coherent light waves. This means that they are all the same wavelength and are all in step with each other.

Holographic plate
Light waves that have reflected off the object's surface will be out of step with the waves of the reference beam. When the two light waves meet inside the photographic plate, they combine, or interfere. In some places, where the waves are still in step, they will reinforce each other; in others, where they are out of step, they cancel out.

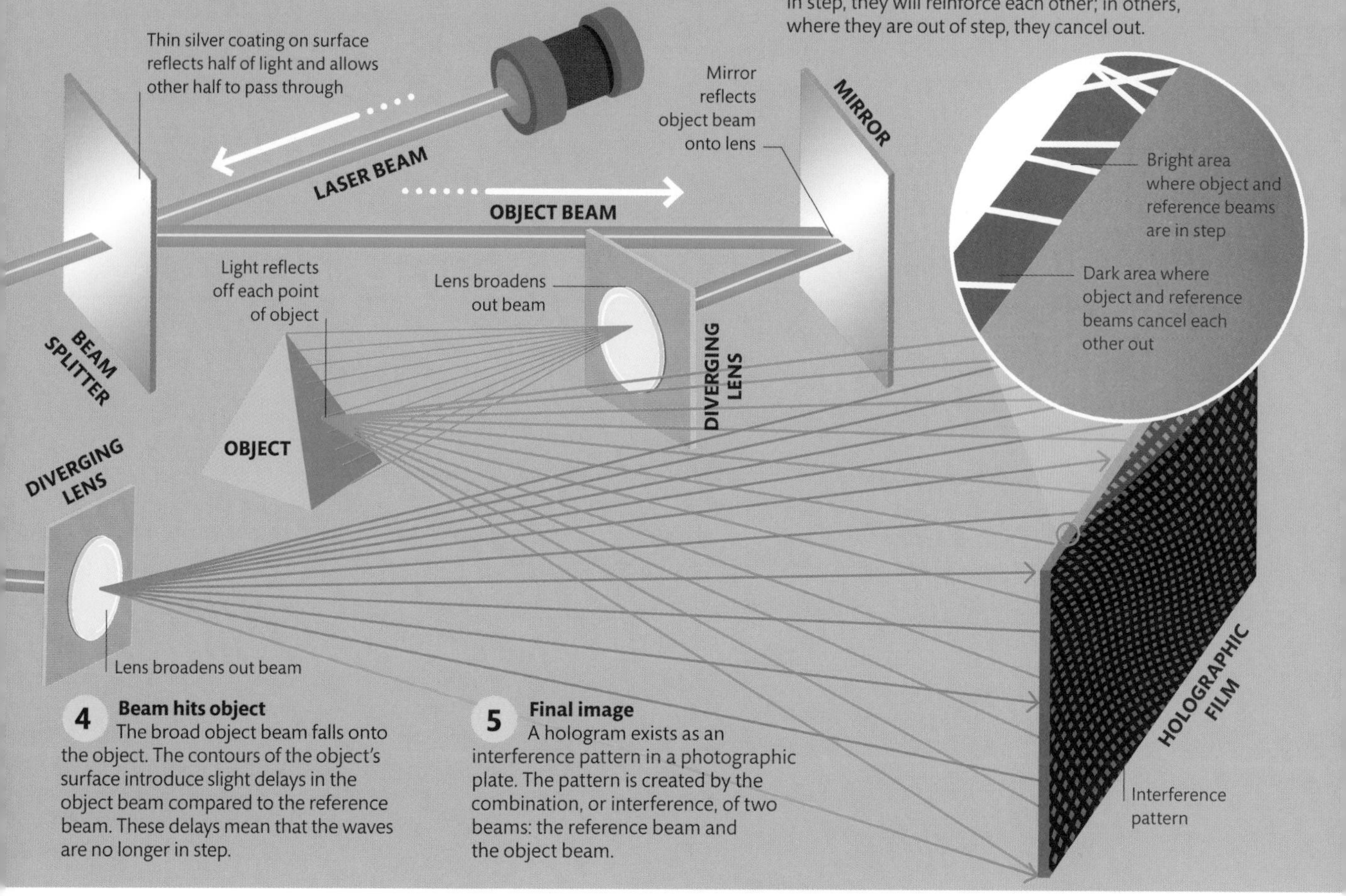
Thin silver coating on surface reflects half of light and allows other half to pass through
LASER BEAM
OBJECT BEAM
Mirror reflects object beam onto lens
MIRROR
Bright area where object and reference beams are in step
Dark area where object and reference beams cancel each other out
BEAM SPLITTER
Light reflects off each point of object
Lens broadens out beam
DIVERGING LENS
OBJECT
DIVERGING LENS
Lens broadens out beam
HOLOGRAPHIC FILM
Interference pattern

4 Beam hits object
The broad object beam falls onto the object. The contours of the object's surface introduce slight delays in the object beam compared to the reference beam. These delays mean that the waves are no longer in step.

5 Final image
A hologram exists as an interference pattern in a photographic plate. The pattern is created by the combination, or interference, of two beams: the reference beam and the object beam.

Viewing a transmission hologram
When light rays bounce off the interference pattern inside the photographic film of a transmission hologram, they recreate the pattern of light that bounced off the object. As a result, they form an image of the object behind the photographic plate. This virtual image has depth and can be viewed from any angle.

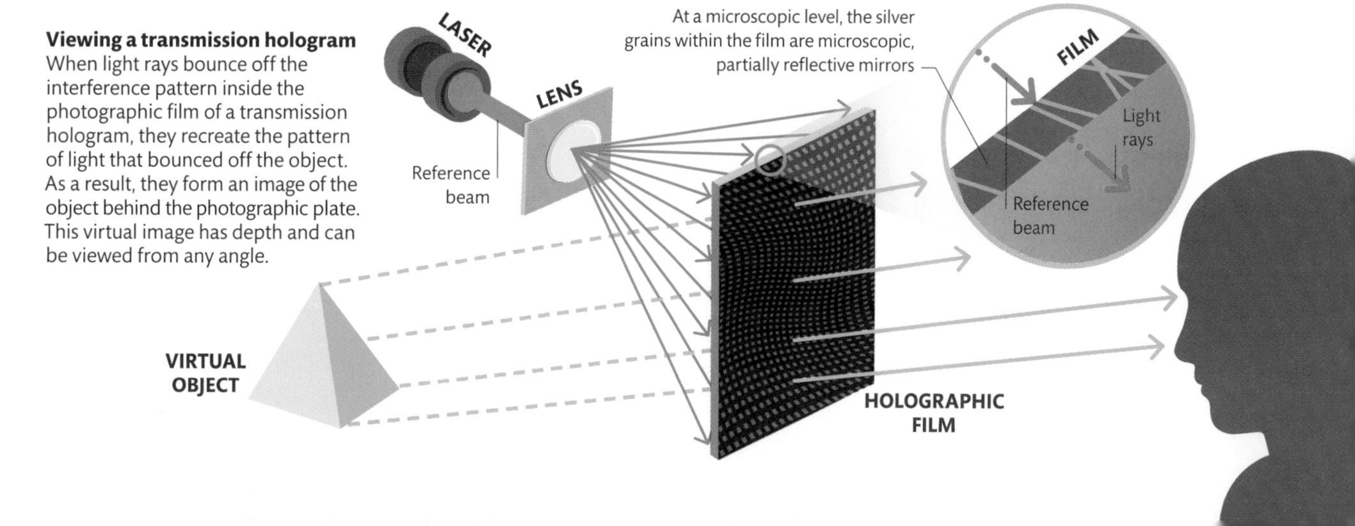
LASER
LENS
Reference beam
At a microscopic level, the silver grains within the film are microscopic, partially reflective mirrors
FILM
Light rays
Reference beam
VIRTUAL OBJECT
HOLOGRAPHIC FILM

Projectors

A projector produces 25, 30, or 60 bright images on a screen each second. Each image, or frame, is composed of thousands of pixels. There are a few different ways of producing the pixels, but the most common projector technology is DLP: digital light processing.

How a DLP projector works

Each pixel of the images a DLP projector produces is made of light that has bounced off one of thousands of tiny mirrors inside the projector. Each frame is made of red, green, and blue pixels shown one after the other. These three colors mixed together at different brightnesses can make any color. The instructions needed to mix them and produce the sequence of images on screen are encoded in digital form and streamed to the projector from a computer or stored on a memory card.

Inside a projector
A projector consists of a light source, filters to split the light into its component colors, and a series of mirrors and lenses to focus and enlarge the image.

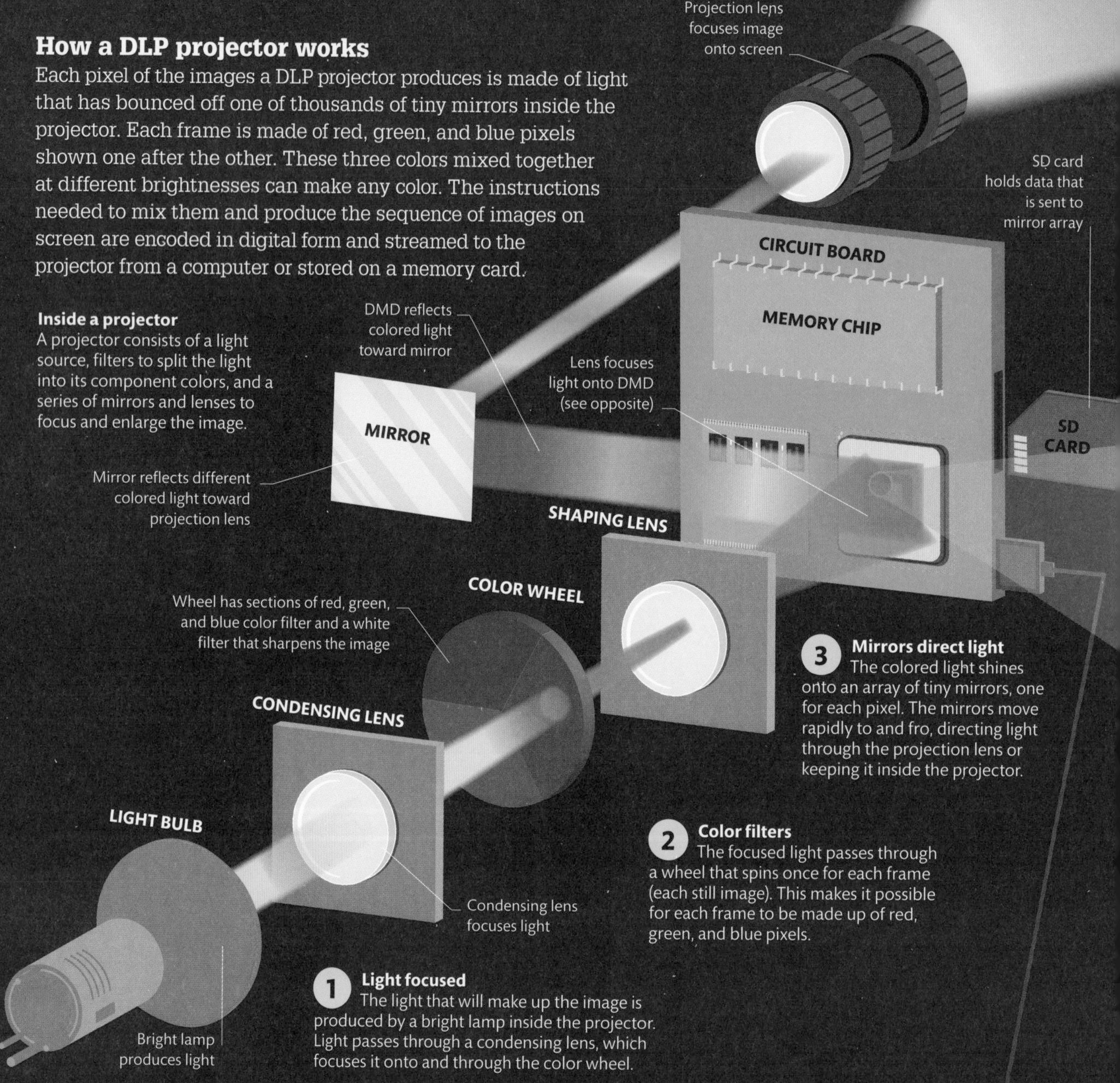

Projection lens focuses image onto screen
SD card holds data that is sent to mirror array
CIRCUIT BOARD
MEMORY CHIP
DMD reflects colored light toward mirror
Lens focuses light onto DMD (see opposite)
MIRROR
SD CARD
Mirror reflects different colored light toward projection lens
SHAPING LENS
COLOR WHEEL
Wheel has sections of red, green, and blue color filter and a white filter that sharpens the image
CONDENSING LENS
LIGHT BULB
Condensing lens focuses light
Bright lamp produces light

1 Light focused
The light that will make up the image is produced by a bright lamp inside the projector. Light passes through a condensing lens, which focuses it onto and through the color wheel.

2 Color filters
The focused light passes through a wheel that spins once for each frame (each still image). This makes it possible for each frame to be made up of red, green, and blue pixels.

3 Mirrors direct light
The colored light shines onto an array of tiny mirrors, one for each pixel. The mirrors move rapidly to and fro, directing light through the projection lens or keeping it inside the projector.

4 Image projected
Any light the mirrors have diverted through the lens is focused onto a screen. The light from all the mirrors makes up the projected image.

MOVIE PROJECTORS

Film carries moving images as a series of frames (still images). Inside a movie projector, the film is made to stop momentarily, while a rotating shutter allows light to pass through it—before the film moves on to the next frame.

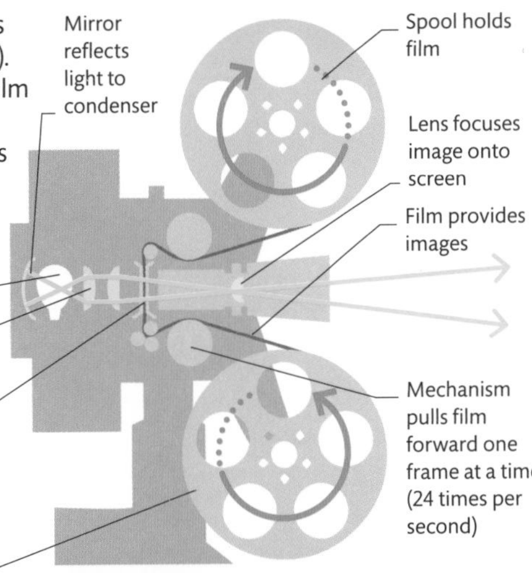
Mirror reflects light to condenser
Spool holds film
Lens focuses image onto screen
Film provides images
Light
Condenser focuses light on lens
Shutter flashes frames three times onto screen, to reduce flickering
Film winds onto second spool after passing through mechanism
Mechanism pulls film forward one frame at a time (24 times per second)

CAN I PROJECT IMAGES FROM MY SMARTPHONE?

Most projectors have a wireless connection that allows you to play content from smartphones and tablets. Some smartphones even have built-in projectors.

DMD mirrors
Each tiny mirror can swivel thousands of times each second—the more time it spends sending light through the lens, the brighter that pixel will be.

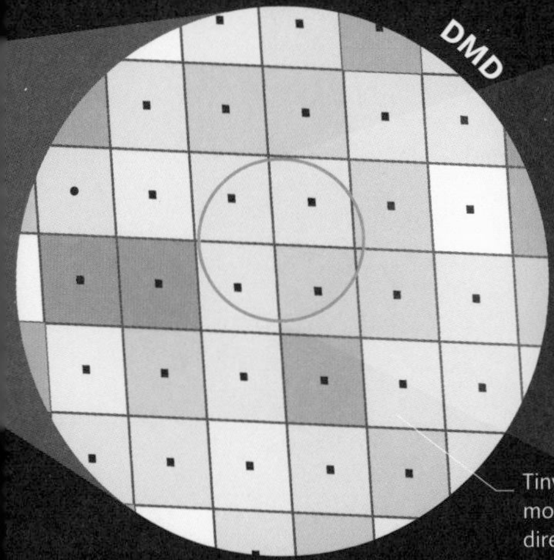
DMD
Tiny mirrors move to direct light

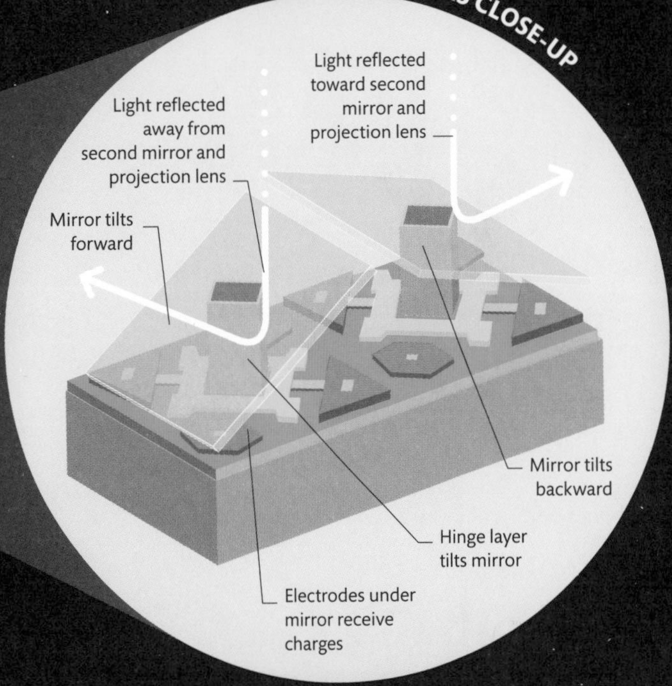
DMD MIRRORS CLOSE-UP
Light reflected toward second mirror and projection lens
Light reflected away from second mirror and projection lens
Mirror tilts forward
Mirror tilts backward
Hinge layer tilts mirror
Electrodes under mirror receive charges

Digital micromirror devices

At the heart of a DLP projector is a digital micromirror device (DMD). It houses thousands of tiny, moveable mirrors, which direct incoming light toward or away from the projecting lens. The projector's processor chip sends electric charges to tiny electrodes just underneath the corners of the mirrors—and the electric charges

TINY MIRRORS IN A DLP PROJECTOR MAY CHANGE THEIR TILT UP TO 5,000

Digital cameras

The digital cameras found in smartphones and tablets, and as stand-alone devices, all share three main features: lenses, which produce an image inside the camera; a light-sensitive chip, or sensor, that captures the image; and a processor that digitizes the image.

1 Focusing light
The lens focuses light to produce an image. It can be moved backward and forward, manually or automatically, to ensure that the subject of the photo is in focus.

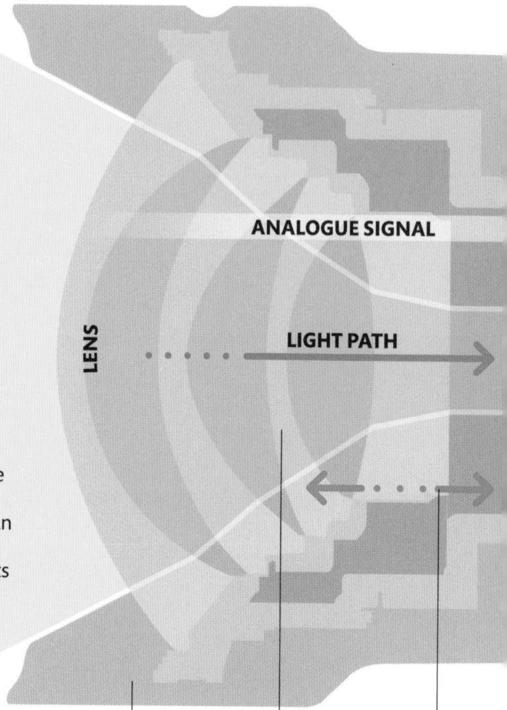

How a DSLR works

There are two main types of stand-alone digital cameras: compact and DSLR (digital single lens reflex). A compact camera has a main lens and, usually, a separate viewfinder. A DSLR has a mirror that directs light from the main lens up toward an eyepiece lens so that you can see directly through the camera's lens when lining up a shot. The mirror also acts as a shutter, flipping up when the shutter button is released to allow light to fall on the sensor.

THE BIGGEST **DIGITAL IMAGE** IN THE WORLD IS MADE UP OF **365 BILLION PIXELS**, STITCHED TOGETHER FROM **70,000** HIGH-RESOLUTION IMAGES

Capturing an image
A camera works a bit like the human eye—with a lens at the front that forms an image at the back. The image falls on an electronic sensor, which has millions of light-sensitive parts arranged in a grid.

PIXELS AND RESOLUTION

A digital image is composed of thousands or millions of dots called picture elements, or pixels. The more pixels, the higher the resolution and the sharper the image. Each pixel has binary numbers associated with it that determine how much red, green, and blue light should be displayed on a screen for that pixel.

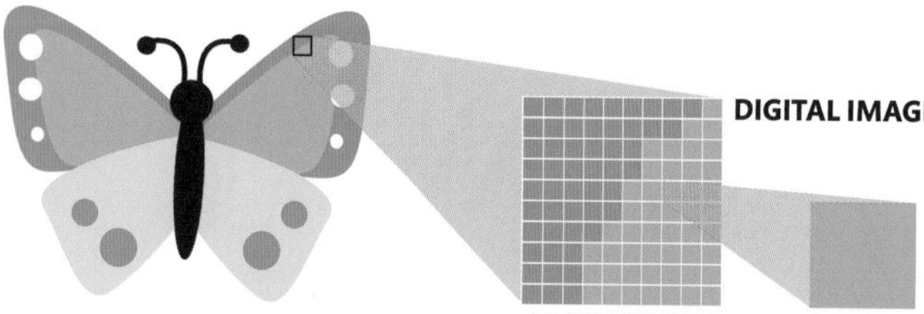

WHY ARE PHOTOS TAKEN AT NIGHT OFTEN BLURRY?

In low-light conditions, the shutter needs to stay open longer to collect enough light, so anything that moves during that time will appear blurred.

2 Light control
An adjustable ring called the iris diaphragm controls how much light makes it through to the sensor and also how much of the picture is in sharp focus.

4 Shutter opens
Behind the mirror is the shutter—in some cameras, the mirror acts as the shutter. When a picture is taken, the shutter flips up, allowing light onto the sensor. The longer it stays up, the more light gets through.

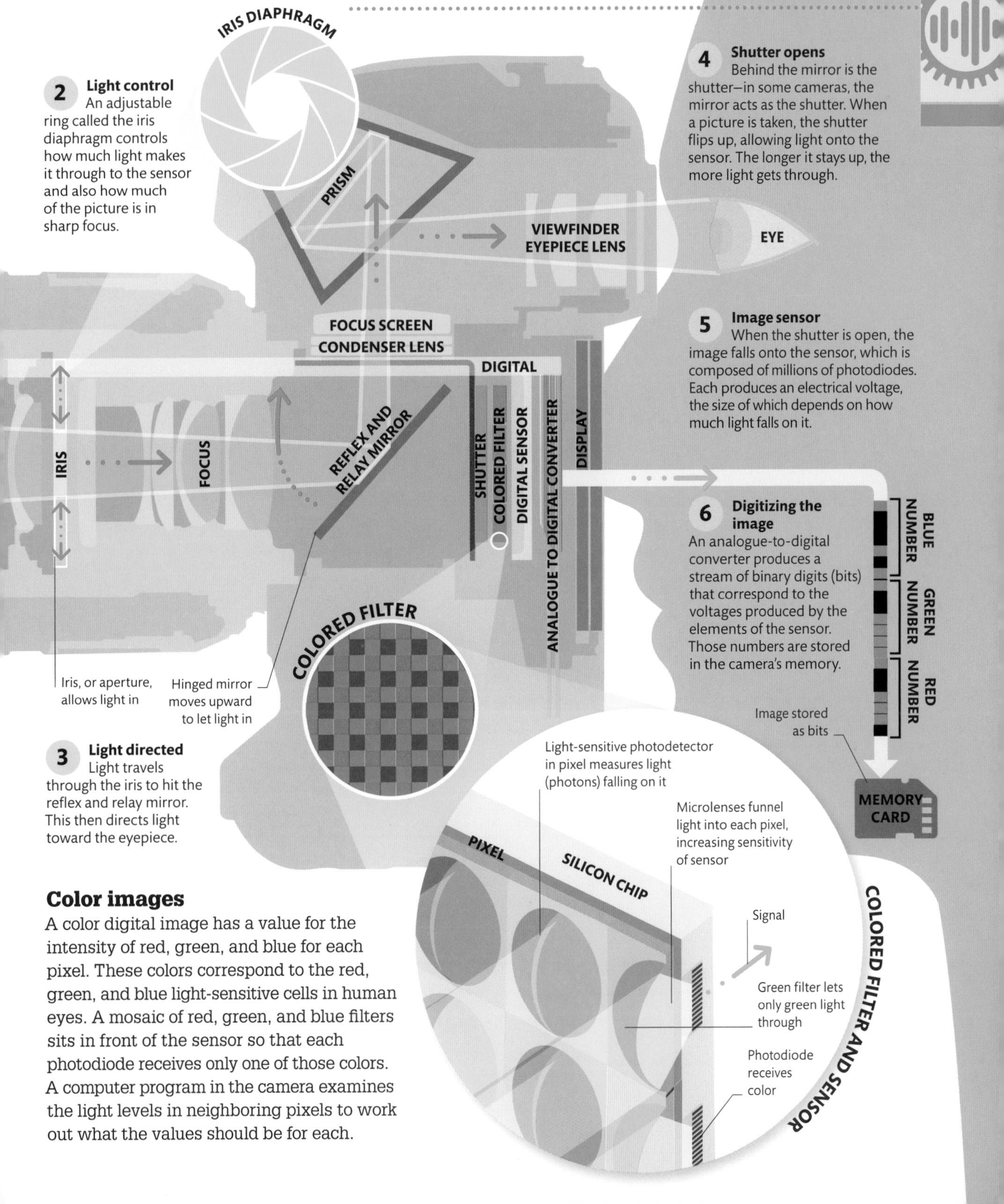

5 Image sensor
When the shutter is open, the image falls onto the sensor, which is composed of millions of photodiodes. Each produces an electrical voltage, the size of which depends on how much light falls on it.

6 Digitizing the image
An analogue-to-digital converter produces a stream of binary digits (bits) that correspond to the voltages produced by the elements of the sensor. Those numbers are stored in the camera's memory.

3 Light directed
Light travels through the iris to hit the reflex and relay mirror. This then directs light toward the eyepiece.

Color images

A color digital image has a value for the intensity of red, green, and blue for each pixel. These colors correspond to the red, green, and blue light-sensitive cells in human eyes. A mosaic of red, green, and blue filters sits in front of the sensor so that each photodiode receives only one of those colors. A computer program in the camera examines the light levels in neighboring pixels to work out what the values should be for each.

Printers and scanners

Computer printers enable us to output documents and photographs stored on a computer or other digital device, while scanners capture documents and photographs as digital images.

Inkjet printers

The most common type of printer uses a jet of ink droplets to form images and text on the printed page. In the printer, ink cartridges move back and forth, spraying ink onto the paper as it advances beneath them. A color image is made of millions of dots of four colored inks: yellow, magenta, cyan, and black. In many printers, the three nonblack inks are held in one cartridge. Each color is delivered individually, and they combine to give subtle variations in hue and tone. The cartridge heads have hundreds of holes through which the ink is forced.

Printer receives data from computer via Wi-Fi

2 Message received by printer Software inside the printer processes the document or image, taking account of the desired paper size. It also communicates back to the computer if ink levels are low or if there is no paper.

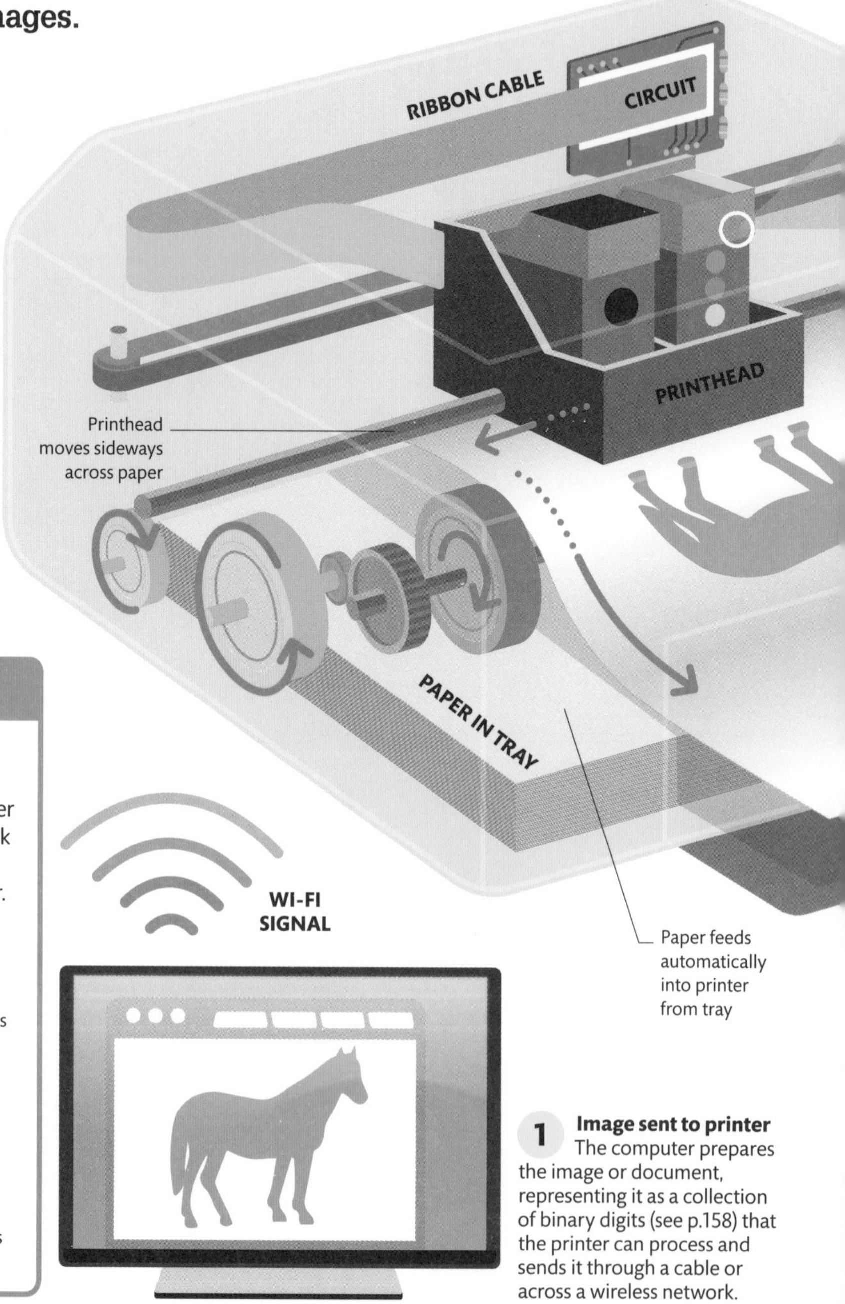

RIBBON CABLE
CIRCUIT
PRINTHEAD
Printhead moves sideways across paper
PAPER IN TRAY
WI-FI SIGNAL
Paper feeds automatically into printer from tray

1 Image sent to printer The computer prepares the image or document, representing it as a collection of binary digits (see p.158) that the printer can process and sends it through a cable or across a wireless network.

LASER PRINTERS

A laser scans across a rotating drum, which develops a negative charge where the beam hits. Positively charged toner sticks to the roller where the laser has hit. The roller transfers ink to the paper, which then passes through heated rollers, fusing the toner to the paper.

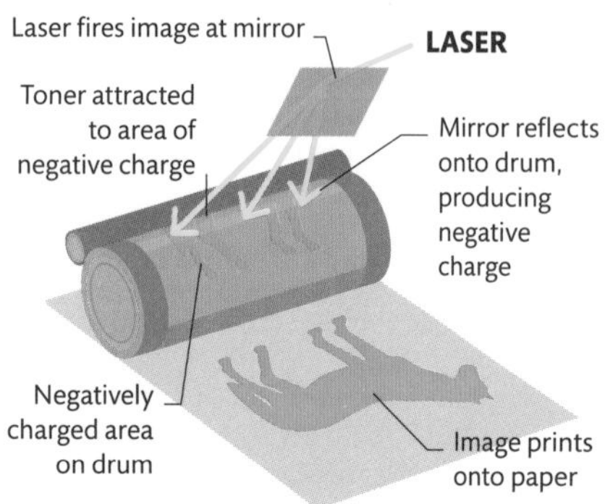

Laser fires image at mirror
LASER
Toner attracted to area of negative charge
Mirror reflects onto drum, producing negative charge
Negatively charged area on drum
Image prints onto paper

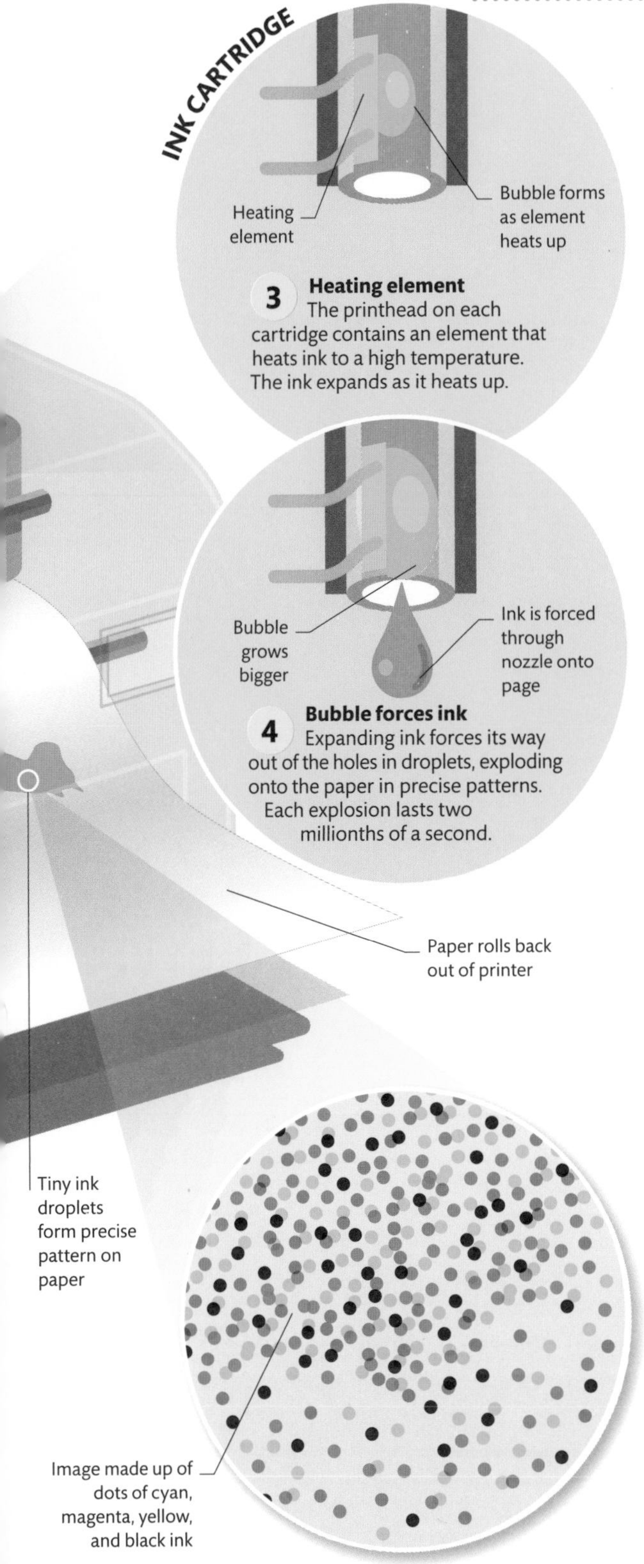
INK CARTRIDGE
Heating element
Bubble forms as element heats up
Bubble grows bigger
Ink is forced through nozzle onto page
Paper rolls back out of printer
Tiny ink droplets form precise pattern on paper
Image made up of dots of cyan, magenta, yellow, and black ink

3 Heating element The printhead on each cartridge contains an element that heats ink to a high temperature. The ink expands as it heats up.

4 Bubble forces ink Expanding ink forces its way out of the holes in droplets, exploding onto the paper in precise patterns. Each explosion lasts two millionths of a second.

How a scanner works

A scanner produces a digital image of any document laid face down on a glass scanner bed. The digital image is made of pixels (picture elements), just like the image produced by a digital camera (see pp.152–153). A bright strip lamp scans down the document. Light reflected off the document hits a CCD (charge-coupled device), which produces an electrical signal that varies depending on how much light it receives. The signals pass to an analogue-to-digital converter, which produces the binary numbers. The scanner then sends the digital image to the computer, through a cable or via Wi-Fi.

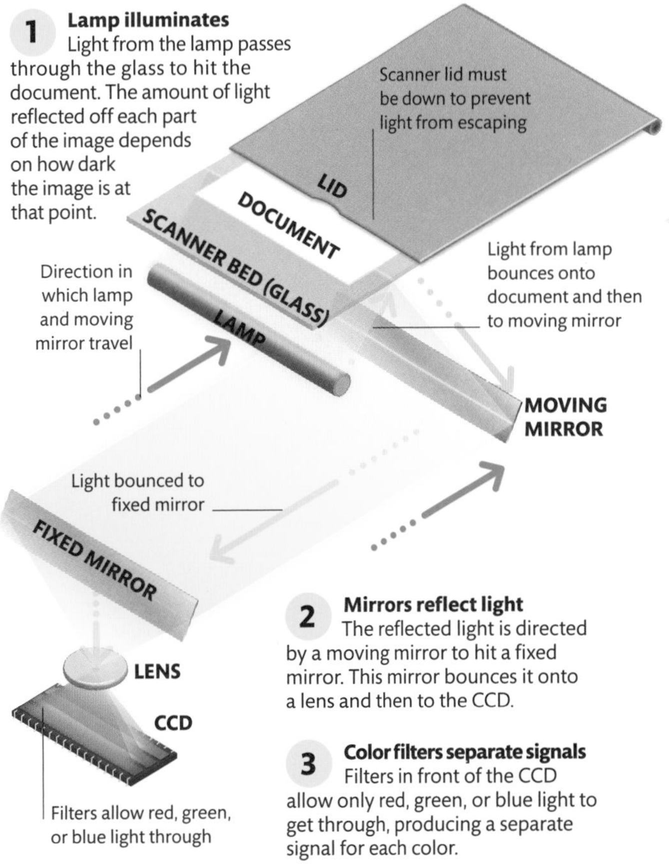
Scanner lid must be down to prevent light from escaping
LID
DOCUMENT
SCANNER BED (GLASS)
LAMP
Direction in which lamp and moving mirror travel
Light from lamp bounces onto document and then to moving mirror
MOVING MIRROR
Light bounced to fixed mirror
FIXED MIRROR
LENS
CCD
Filters allow red, green, or blue light through

1 Lamp illuminates Light from the lamp passes through the glass to hit the document. The amount of light reflected off each part of the image depends on how dark the image is at that point.

2 Mirrors reflect light The reflected light is directed by a moving mirror to hit a fixed mirror. This mirror bounces it onto a lens and then to the CCD.

3 Color filters separate signals Filters in front of the CCD allow only red, green, or blue light to get through, producing a separate signal for each color.

MOST PRINTERS LEAVE **MICRODOTS** CALLED **MACHINE IDENTIFICATION CODE** ON EVERY PAGE

COMPUTER TECHNOLOGY

The digital world

Most of the devices we use for communicating and for storing information are digital. These include computers, cameras, and radios. Inside a digital device, information is stored and processed as numbers.

Digitizing touch
The touch screen of a smartphone (see pp.204–205) or tablet produces two binary numbers that represent the coordinates of the point on the screen you touch.

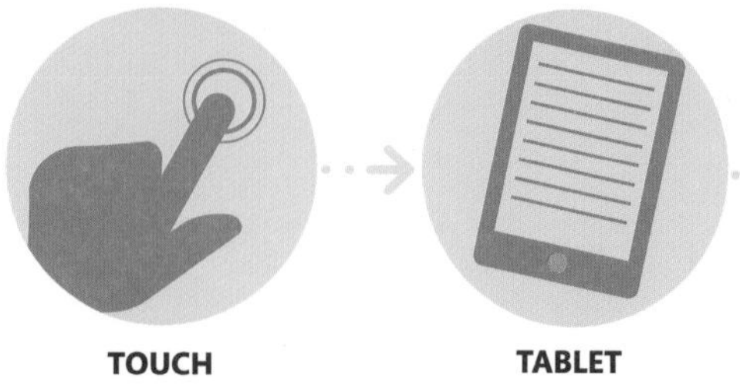

Digitizing sound
A circuit called an analogue-to-digital converter produces a stream of numbers that match the levels of voltage in an audio signal from a microphone (see pp.138–141) or musical instrument.

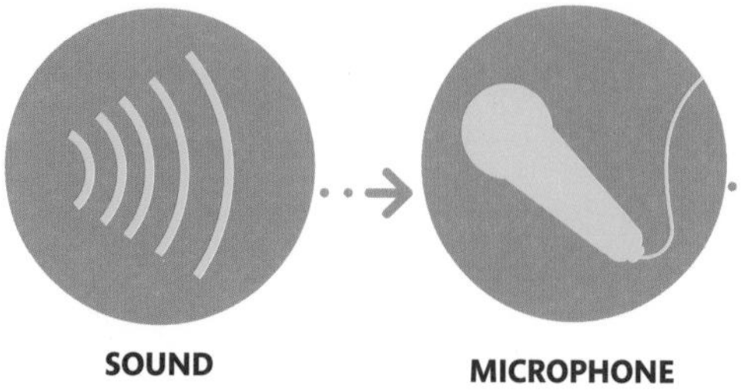

Digitizing images
A sensor inside a digital camera (see pp.152–153) produces numbers that correspond to the brightness of light at each picture element, or pixel, of an image.

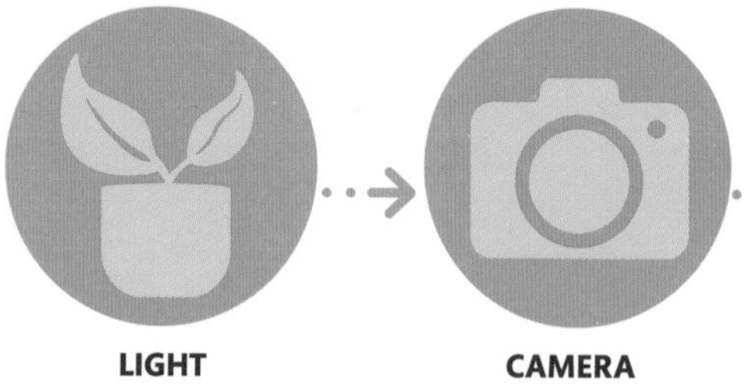

Digitizing information

The information digital devices store and process includes text, images, sound, and video—and the software that makes the devices work. This information is represented by binary numbers, which consist of two digits: 0 and 1. Any number can be represented by a set of binary digits, or bits. Creating a representation of information like this is called digitization.

Why binary?
Inside digital devices, the binary digits, 0 and 1, typically exist as electric currents (off and on) or electric charges (present or absent). Embedded in all digital devices are computers, which store and process these numbers.

Binary numbers

The binary number system is a place value system, just like the decimal system (0–9) that we use every day. But rather than 1s, 10s, 100s, 1,000s, and so on, the place values in the binary number system are 1s, 2s, 4s, 8s, and so on. Inside digital devices, electronic circuits produce electrical signals that represent binary digits, or bits. Most information is broken down into bytes: groups of eight bits.

THE **BINARY NUMBER SYSTEM** WAS DEVELOPED IN THE **17TH CENTURY**—LONG BEFORE IT WAS USED IN COMPUTING

Converting to binary
This example shows how the number we know in decimal as 23 can be expressed in binary.

Each column is worth double the column to the right

			32		16		8		4		2		1
DECIMAL	23	=	0 x 32	+	1 x 16	+	0 x 8	+	1 x 4	+	1 x 2	+	1 x 1
BINARY	010111		0		1		0		1		1		1

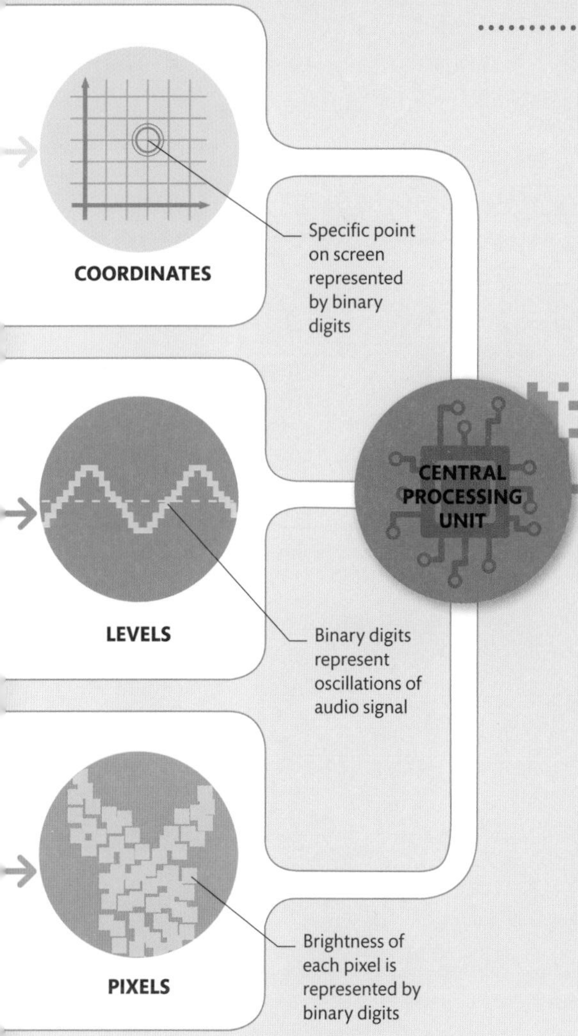

Digital signals
All the different ways in which information is digitized produce large collections of binary numbers that are processed by the central processing unit (CPU) of a computer embedded inside a digital device.

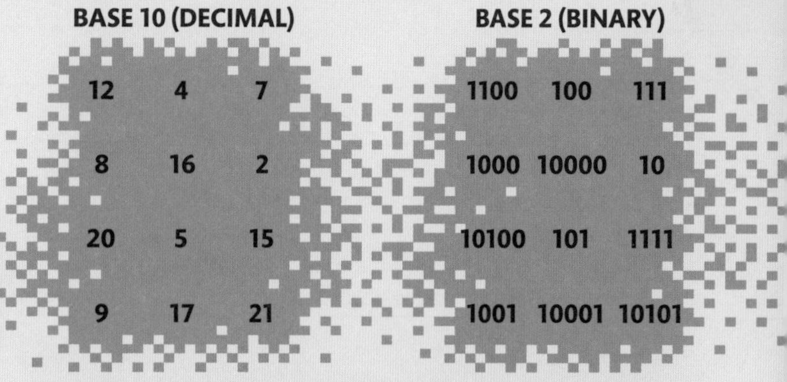

QUANTUM COMPUTING

All digital devices currently available use bits, which can take on only one value at a time, and contain computers that process instructions one at a time. Computer scientists and physicists are developing quantum computers that will use qubits, which can take on many values at the same time. By combining qubits, computers will be able to carry out a potentially infinite number of instructions, promising much faster devices in the future.

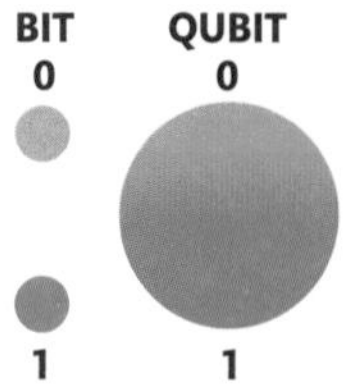

WHAT ARE DATA?

Data (singular datum) are pieces of information. In the digital world, data refers to any information stored and processed inside digital devices. It includes personal information about the users of digital devices.

UNITS OF DIGITAL INFORMATION		
Unit	**Size**	**Application**
Byte (B)	**8 bits**	A basic unit of information held by computers, a byte is equivalent to eight bits (binary digits).
Kilobyte (KB)	**1,000 bytes**	A short, simple text file on a computer would take up a few kilobytes.
Megabyte (MB)	**1 million bytes**	One million bytes (8 million bits) could represent one minute of digital sound.
Gigabyte (GM)	**1 billion bytes**	One billion bytes (8 billion bits) could represent 4,000 digital images.
Terabyte (TB)	**1 trillion bytes**	Computer hard disks are often this size, able to store large amounts of digital information.

Digital electronics

Inside digital devices, information is processed by transistors—electronic components etched onto small pieces of semiconductor material—in integrated circuits.

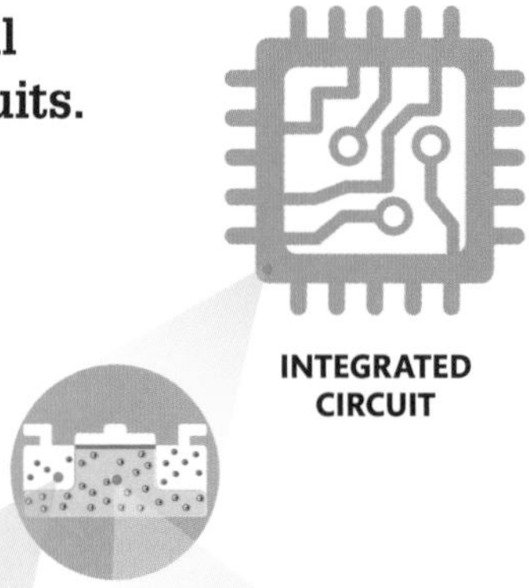

Semiconductors

Materials called semiconductors are at the heart of the digital world. The most common is the element silicon. Pure silicon is not a very good conductor of electricity, but adding impurities of other elements, or "doping," enables it to conduct an electric current, which is a flow of electric charge. By adding different elements to a semiconductor, its distribution of positive and negative charges can be precisely controlled to direct a current through it.

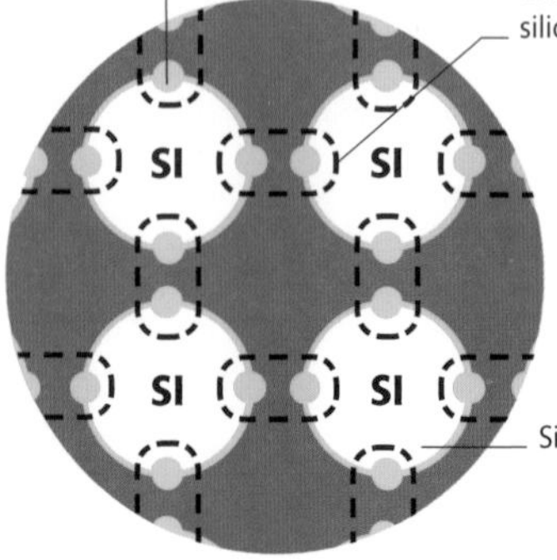

Silicon
Pure silicon can conduct electricity only when heat or light gives electrons enough energy to break free of their atoms.

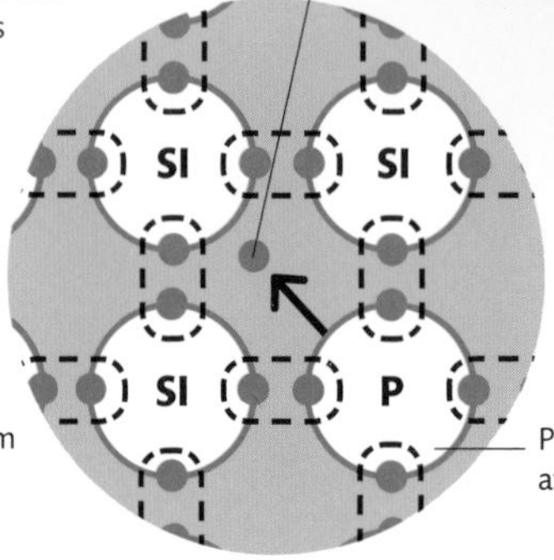

N-type (negative) silicon
Adding phosphorus atoms makes an n-type semiconductor with negatively charged electrons that are free to move.

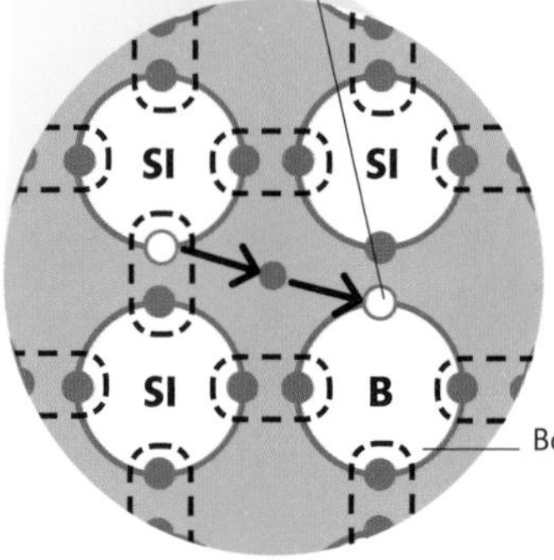

P-type (positive) silicon
Adding boron means there are not enough electrons. This leaves positively charged "holes" that can move through the silicon.

Transistors

The transistors found on integrated circuits are made from pure silicon doped precisely to produce n- and p-type regions. Current can flow through a transistor, from "source" to "drain," only if there is an electric field applied to a part called the "gate." Current flowing represents a binary "1"; no current, a "0."

Transistor is "off"
The source is connected to a negative voltage, pushing electrons toward the drain. But only "holes"—not electrons—can flow through an adjacent area of p-type silicon.

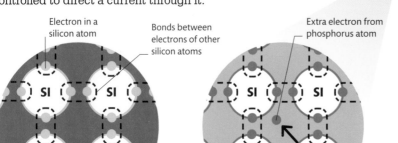

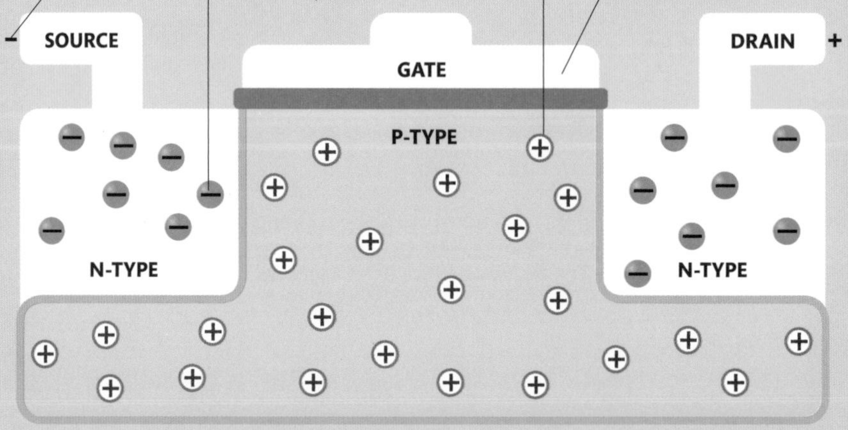

Digital integrated circuits

Integrated circuits (ICs), also known as "chips," typically contain billions of tiny transistors. Each one is either on or off (allowing current through or not), representing the binary numbers 1 and 0. Combinations of these numbers represent the letters, images, and sounds that make up the files on a computer, as well as the programs that make computers work.

CAN TRANSISTORS KEEP GETTING SMALLER?

Chip designers are close to the limit of how small silicon transistors can be, but with new materials, such as compound semiconductors, they will shrink them even further.

Types of digital integrated circuit
Digital integrated circuits are designed to do specific jobs. Electronics engineers fit them together with other components on a circuit board to make digital devices such as computers, tablets, smartphones, and digital cameras.

Analogue to digital
An analogue-to-digital chip takes information from the real world and codes it into collections of binary numbers.

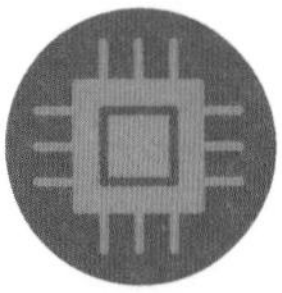

Microprocessor
Every digital device has an IC that processes programs—sets of instructions that make the device work.

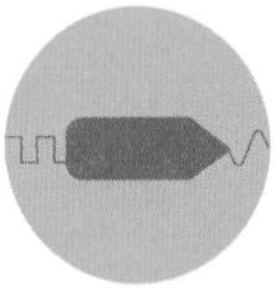

Digital to analogue
A digital-to-analogue chip processes digital sound (1s and 0s) to produce a signal that can be sent to loudspeakers.

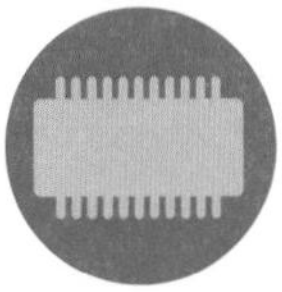

RAM chip
Random access memory (RAM) holds active programs and information to be processed.

Flash memory chip
Found in USB storage, digital cameras, and solid-state hard drives, flash chips can store huge amounts of information.

Graphics chip
Graphics chips send signals to the screen of a computer, smartphone, or tablet, rapidly refreshing the display.

System on a chip
An IC that contains all the circuits found on most other types of IC can be used as a stand-alone computer.

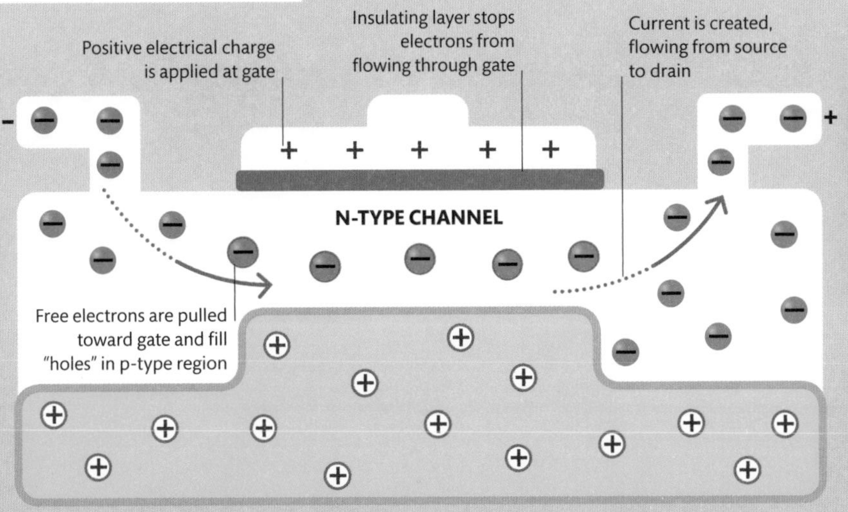
Positive electrical charge is applied at gate
Insulating layer stops electrons from flowing through gate
Current is created, flowing from source to drain
N-TYPE CHANNEL
Free electrons are pulled toward gate and fill "holes" in p-type region

EACH **TRANSISTOR** IN A MEMORY CHIP **STORES 1 SINGLE "BIT"**

Transistor is "on"
A positive charge at the gate attracts negatively charged electrons into the p-type region. These can become charge carriers that allow current to flow through the transistor.

Computers

Computers come in many shapes and sizes, including laptops, desktops, tablets, and smartphones. There are also computers embedded in all digital devices. Despite this variety, all computers work in the same way.

WHY DO COMPUTERS CRASH?

A computer can crash (freeze) for many reasons, but most common are mistakes in computer programs that mean instructions cannot be carried out.

Laptop computers

One of the most popular stand-alone computers is the laptop, or notebook computer. At the heart of a laptop—and any kind of computer—is the central processing unit (CPU), which carries out the instructions written into the programs the computer runs (see pp.164–165). The rest of the computer hardware is designed to enable information to be input to and output from the computer and includes circuits that connect wirelessly to networks of computers, including the Internet.

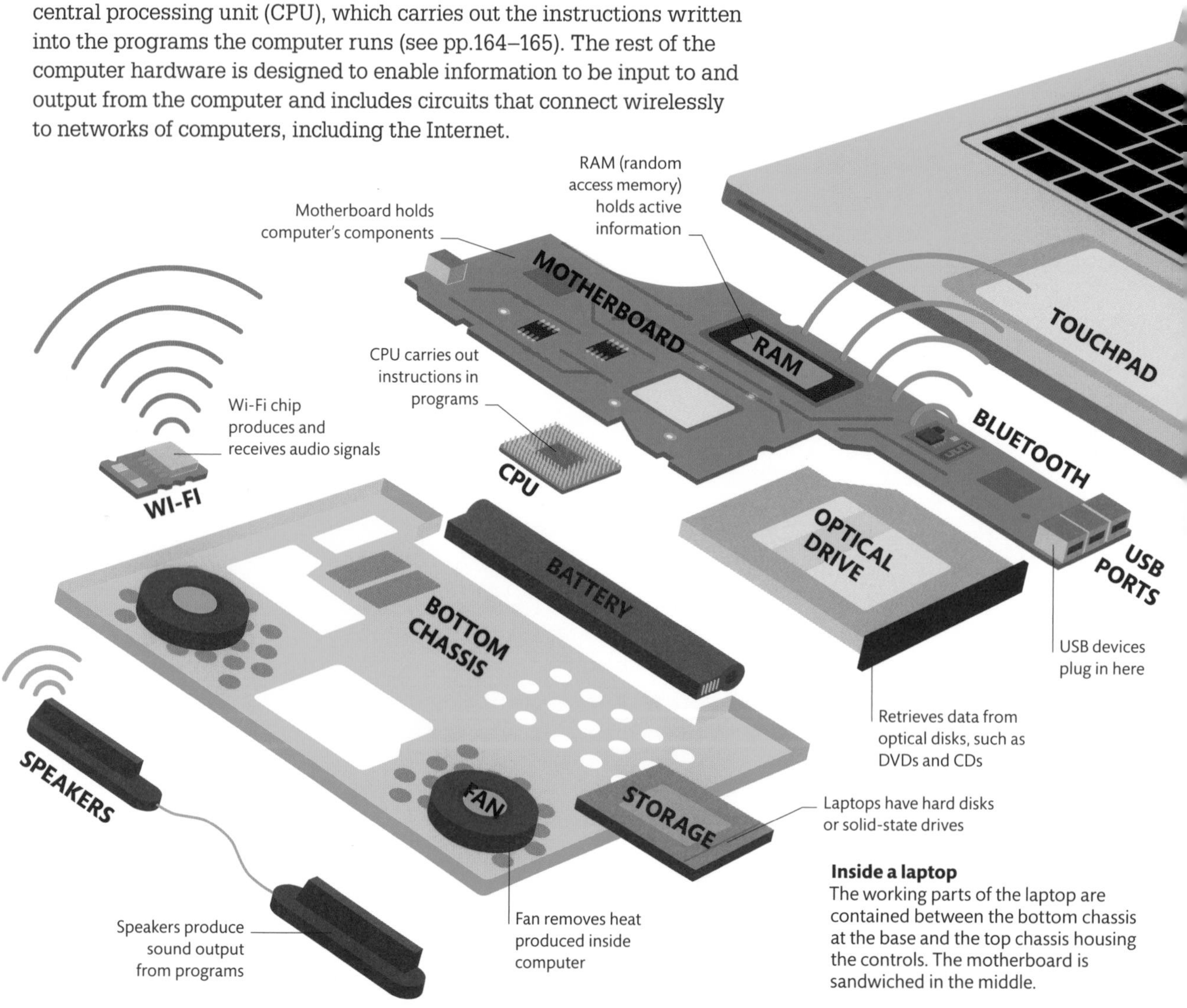

RAM (random access memory) holds active information
Motherboard holds computer's components
MOTHERBOARD
RAM
TOUCHPAD
CPU carries out instructions in programs
Wi-Fi chip produces and receives audio signals
WI-FI
CPU
BLUETOOTH
OPTICAL DRIVE
USB PORTS
BATTERY
BOTTOM CHASSIS
USB devices plug in here
Retrieves data from optical disks, such as DVDs and CDs
SPEAKERS
FAN
STORAGE
Laptops have hard disks or solid-state drives
Speakers produce sound output from programs
Fan removes heat produced inside computer

Inside a laptop
The working parts of the laptop are contained between the bottom chassis at the base and the top chassis housing the controls. The motherboard is sandwiched in the middle.

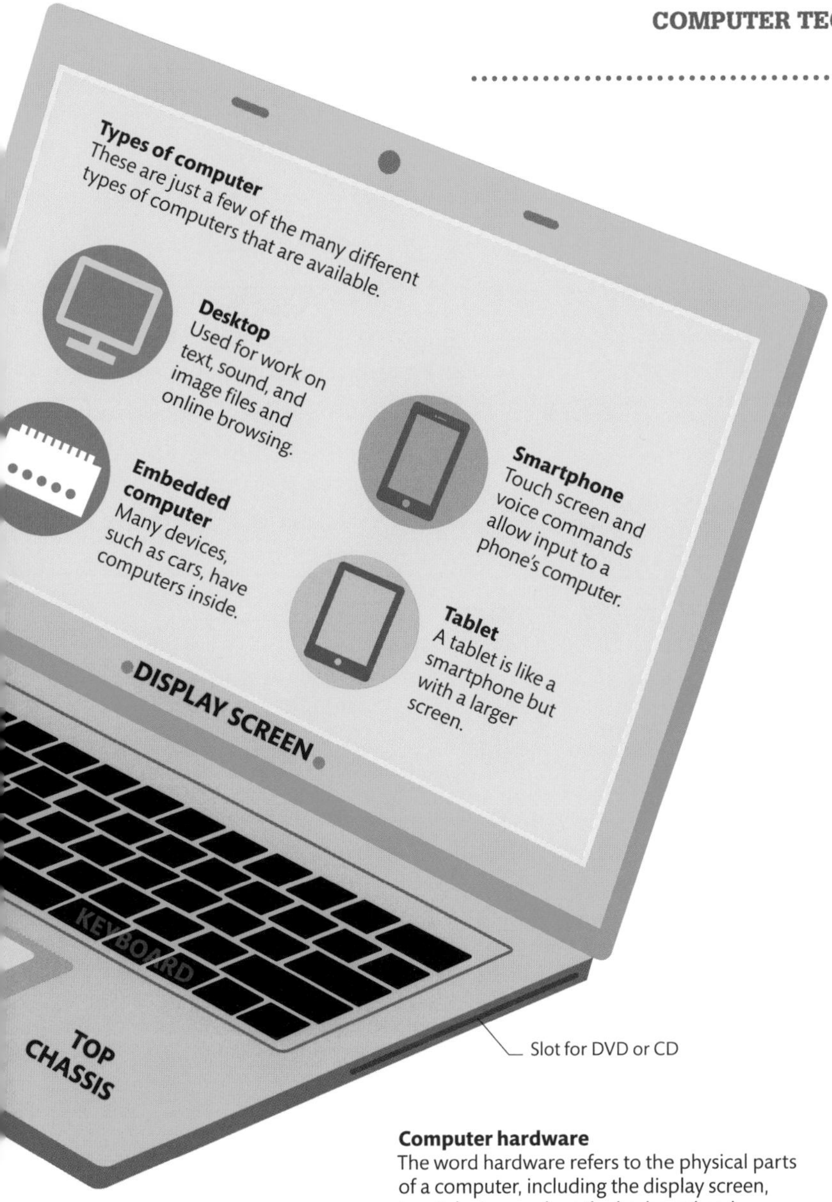
Types of computer
These are just a few of the many different types of computers that are available.
Desktop
Used for work on text, sound, and image files and online browsing.
Smartphone
Touch screen and voice commands allow input to a phone's computer.
Embedded computer
Many devices, such as cars, have computers inside.
Tablet
A tablet is like a smartphone but with a larger screen.
DISPLAY SCREEN
KEYBOARD
TOP CHASSIS
Slot for DVD or CD

Computer hardware
The word hardware refers to the physical parts of a computer, including the display screen, input devices such as the keyboard and touchpad, and all the electronic circuits that work together to make the computer function.

SUPERCOMPUTERS

A supercomputer is simply a very powerful computer—one that can process much more information, much more quickly, than a typical notebook or desktop computer. Supercomputers are used to predict the weather and to render the graphics for computer-generated scenes in films.

Storage

A computer's main memory is RAM (random access memory), but this stores only programs and information that are being processed at the time. Computer storage provides space to store programs and information that are not active, and it retains information even when the computer is turned off.

Storage media
Built-in storage on most computers is in the form of hard disks or flash storage (solid-state drives, SSDs), which typically hold between 250 GB (gigabytes) and 1 TB (terabyte). Removable storage, with less capacity, allows information to be transferred from one computer to another and includes USB drives.

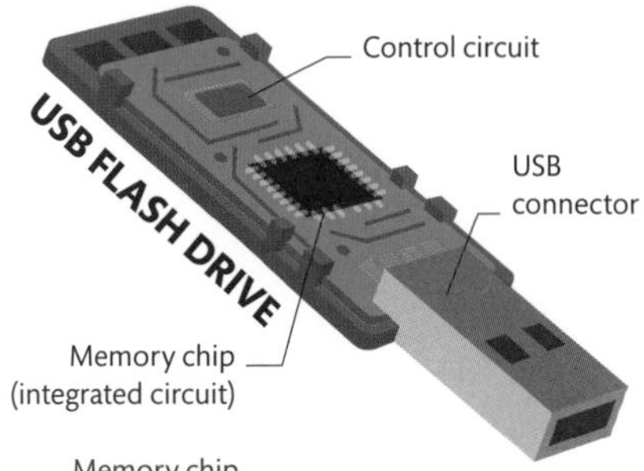
Control circuit
USB connector
USB FLASH DRIVE
Memory chip (integrated circuit)

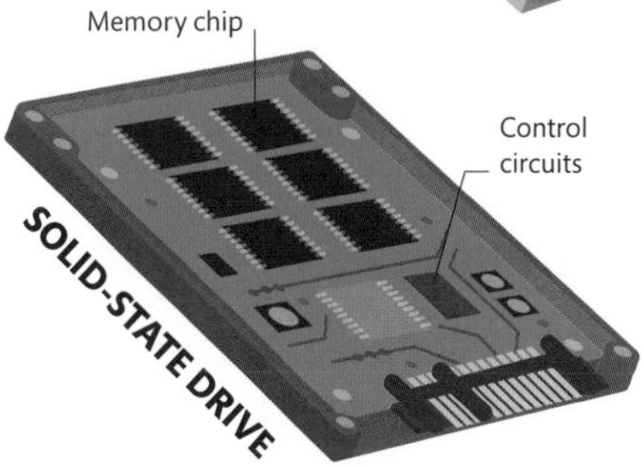
Memory chip
Control circuits
SOLID-STATE DRIVE

3 BILLION THE NUMBER OF **DESKTOP** AND **NOTEBOOK COMPUTERS** IN THE **WORLD**

How computers work

At the heart of every computer is an integrated circuit called the central processing unit (CPU). It communicates with the computer's main memory, input devices, and output devices.

WHAT IS A BUS?

In computing, buses are the connections between the various components that make up a computer. They include the cables that connect to input and output devices.

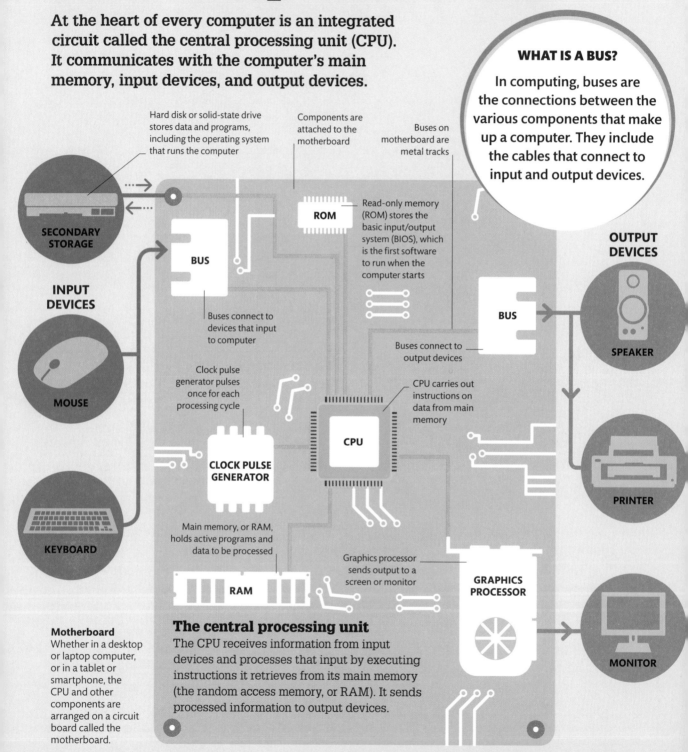

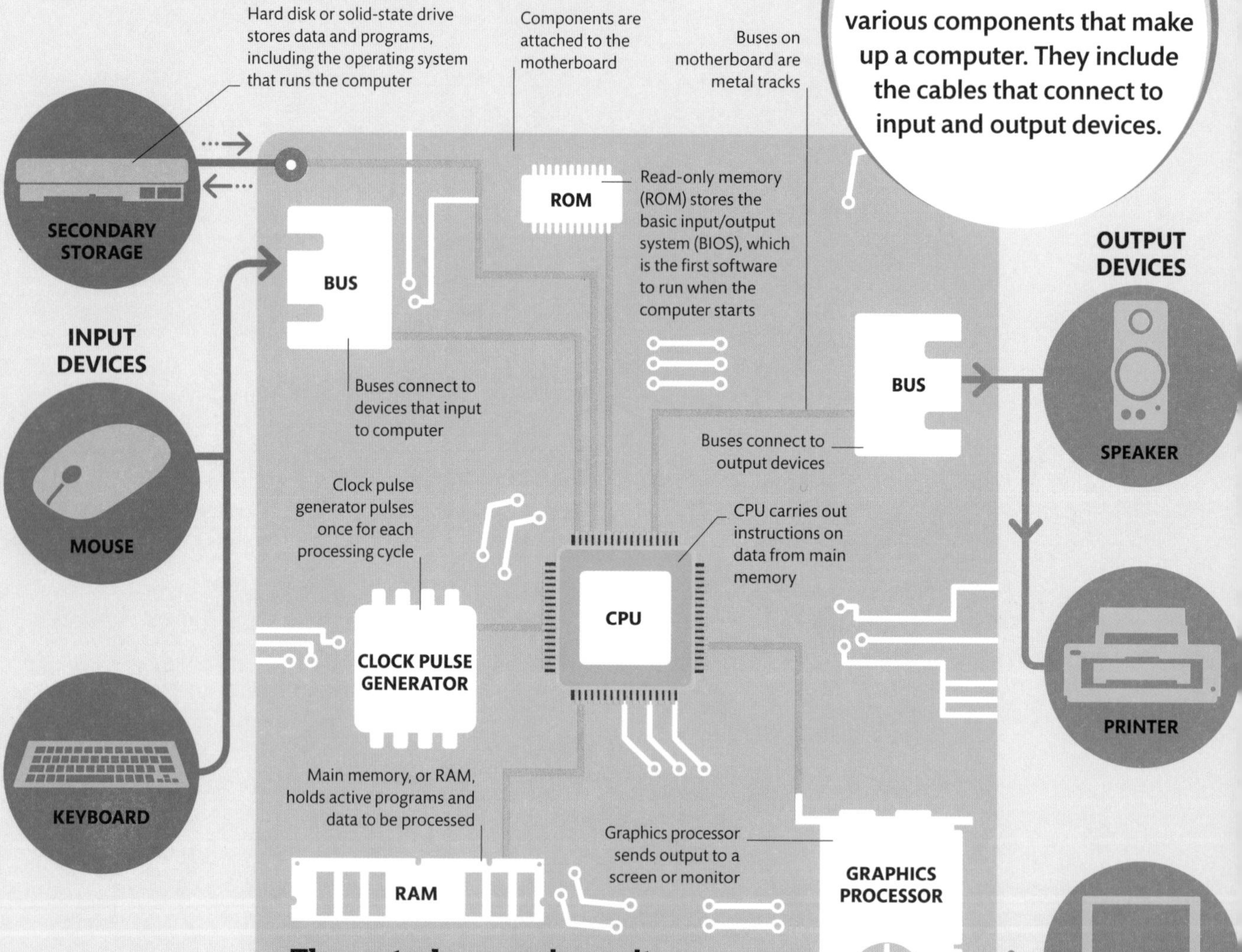
Hard disk or solid-state drive stores data and programs, including the operating system that runs the computer
Components are attached to the motherboard
Buses on motherboard are metal tracks
SECONDARY STORAGE
ROM
Read-only memory (ROM) stores the basic input/output system (BIOS), which is the first software to run when the computer starts
BUS
OUTPUT DEVICES
INPUT DEVICES
Buses connect to devices that input to computer
BUS
SPEAKER
Buses connect to output devices
MOUSE
Clock pulse generator pulses once for each processing cycle
CPU carries out instructions on data from main memory
CPU
CLOCK PULSE GENERATOR
PRINTER
KEYBOARD
Main memory, or RAM, holds active programs and data to be processed
Graphics processor sends output to a screen or monitor
RAM
GRAPHICS PROCESSOR
MONITOR

The central processing unit

The CPU receives information from input devices and processes that input by executing instructions it retrieves from its main memory (the random access memory, or RAM). It sends processed information to output devices.

Motherboard
Whether in a desktop or laptop computer, or in a tablet or smartphone, the CPU and other components are arranged on a circuit board called the motherboard.

How instructions are carried out

The CPU can carry out, or execute, only one instruction at a time. Retrieving and executing one instruction takes one cycle of processing time. In a typical CPU, there are billions of cycles per second, all coordinated by a clock—an electronic circuit that produces a stream of extremely rapid pulses.

Inside the CPU
An arithmetic logic unit (ALU) manipulates binary numbers, and a control unit directs the operations of the CPU. There are also registers—temporary storage locations—for the results of calculations.

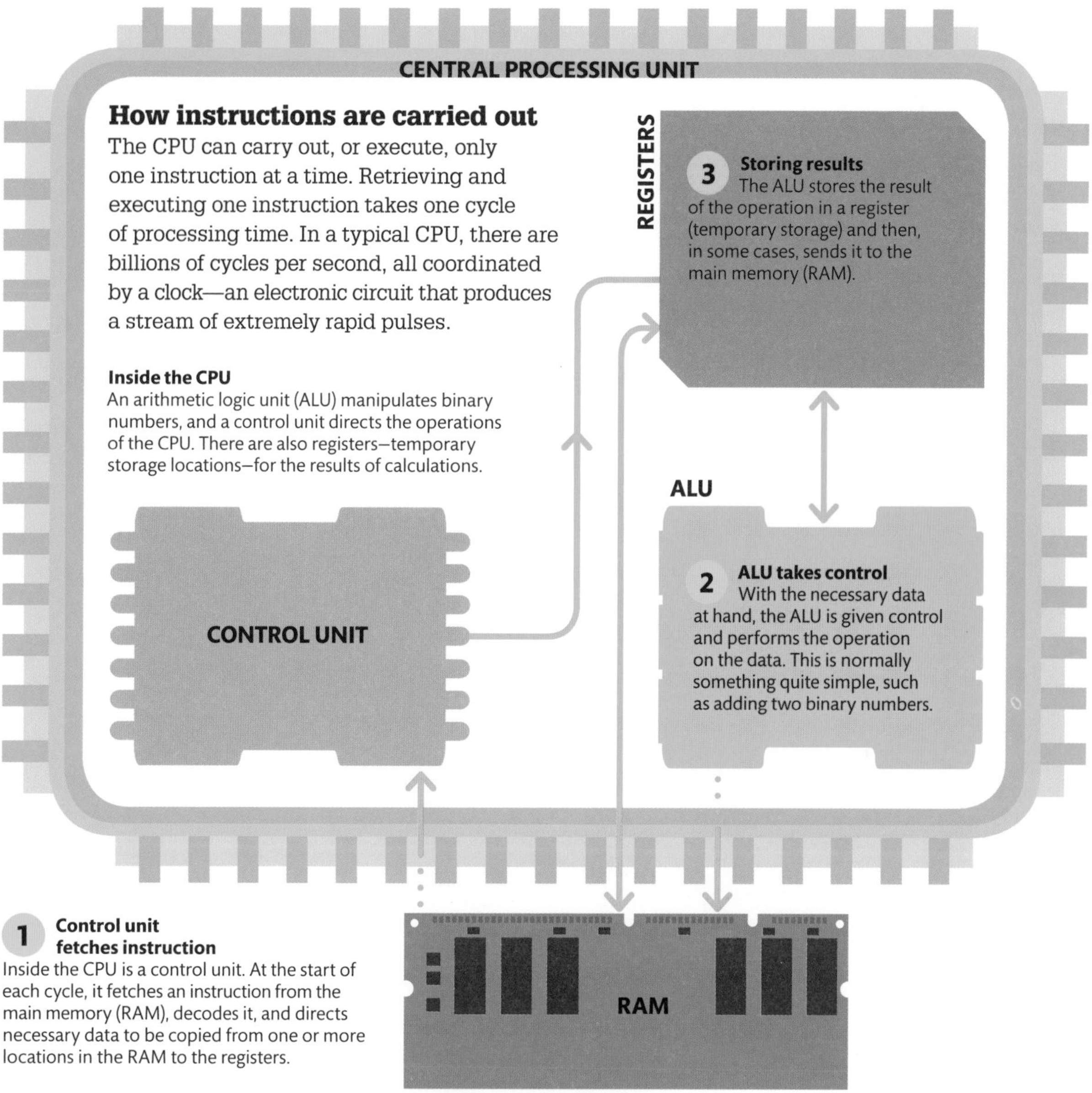

1 Control unit fetches instruction
Inside the CPU is a control unit. At the start of each cycle, it fetches an instruction from the main memory (RAM), decodes it, and directs necessary data to be copied from one or more locations in the RAM to the registers.

2 ALU takes control
With the necessary data at hand, the ALU is given control and performs the operation on the data. This is normally something quite simple, such as adding two binary numbers.

3 Storing results
The ALU stores the result of the operation in a register (temporary storage) and then, in some cases, sends it to the main memory (RAM).

MACHINE CODE

The data and instructions the CPU manipulates arrive as a stream of binary numbers—1s and 0s. This stream of numbers is called machine code and is broken into chunks, typically 32 or 64 binary digits (bits) long.

0 1 1 0 0 1 0 1
0 0 1 1 0 1 0 0
0 0 1 0 1 1 1 0
1 0 0 1 0 1 0 0

THE WORLD'S SMALLEST COMPUTER IS SMALLER THAN A GRAIN OF SALT

Keyboards and mice

A computer needs to be fed information before it can process it and then produce an output. Two important ways of inputting information—to interact with the computer directly—are the widely used keyboard and mouse.

DO I NEED A MOUSE PAD?

Optical mice do not work well on plain, shiny surfaces, because there is no detail for the mouse's camera to pick up. A mouse pad overcomes this problem.

Keyboard

While smartphones and tablets have touch-sensitive keyboards that appear on screen, desktop and laptop computers have keyboards with physical buttons. Inside the keyboard is a number of electric circuits—one for each key. The keys are simple switches, and pressing one completes that key's circuit. The electric current flows to an integrated circuit, which produces a set of binary digits (bits) unique to the key pressed.

Key layers

The most common type of keyboard today uses a technology called "rubber dome over membrane." A slider pushes two contacts together, and a rubber dome provides a force to return the key back to its normal position after it has been pressed.

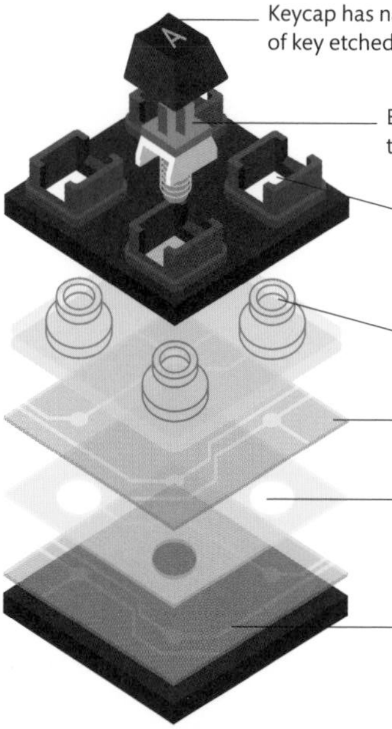

A
Keycap has name of key etched into it
Bottom of slider pushes two membranes together
Upper case has holes that guide movement of slider
Rubber domes push back up when key is released
Top membrane has electrical contacts
Holes in central membrane allow contacts to touch
Bottom membrane has electrical contacts

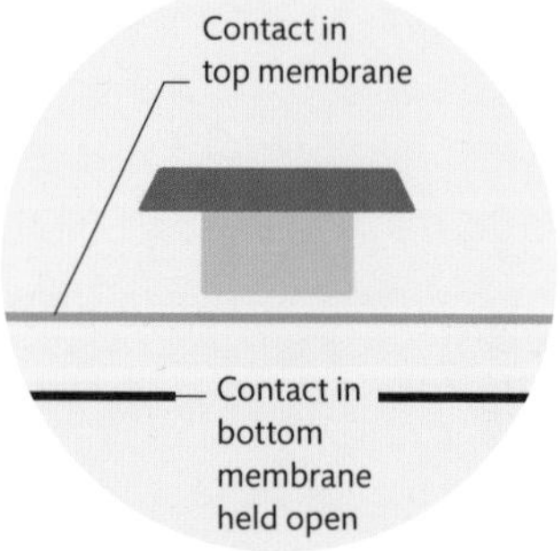

Contact in top membrane
Contact in bottom membrane held open

1 Key raised
Beneath each key on a computer keyboard are metal contacts. These contacts are normally held open until they are pressed down.

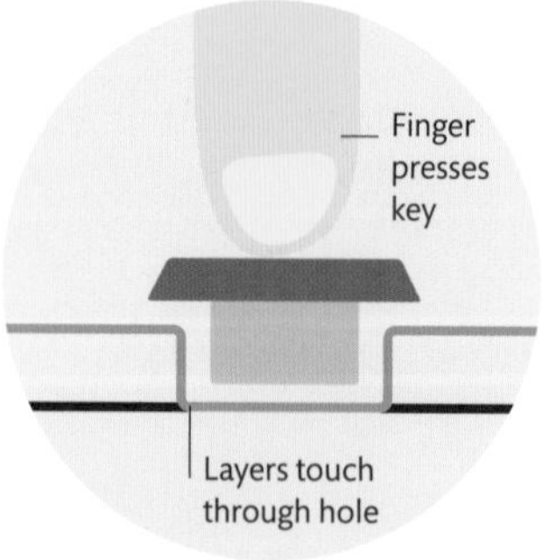

Finger presses key
Layers touch through hole

2 Key pressed down
Pressing the key closes the contacts, allowing electric current to flow through the circuit unique to that key. The current flows to an integrated circuit in the keyboard.

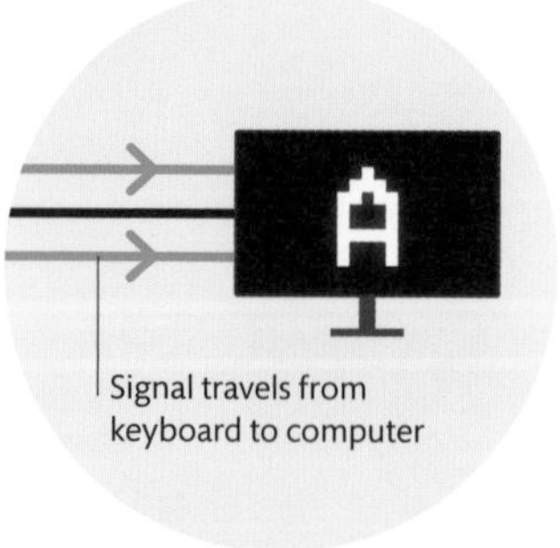

A
Signal travels from keyboard to computer

3 Signal sent to computer
The circuit recognizes which key has been pressed and sends a digital signal—a set of binary digits, or scan code—to the computer's main processor.

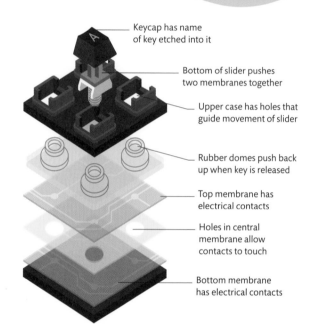

THE **FASTEST TYPING SPEED** EVER RECORDED WAS **216 WORDS PER MINUTE,** IN 1946

Optical mouse

A computer mouse allows you to move a pointer on the computer's monitor so that you can interact with documents and programs. Most computer mice are optical devices: they have a light inside that illuminates the surface below it and a tiny camera that creates an image of the surface. Circuits inside analyze the image and work out in which direction the mouse is moving, and how fast, and send that information to the computer.

COMMON CONNECTIONS

Mice and keyboards can be connected to a computer with cables or may be connected wirelessly, in which case information is coded into radio waves. The most common kind of wireless mouse uses Bluetooth technology.

Radio
Information travels via radio waves from an onboard transmitter to a receiver plugged into a USB port.

USB
Some mice and keyboards simply plug into the computer via a cable with a USB connector at the end.

Bluetooth
Information is sent from a cordless mouse or keyboard to a computer. This technology uses little power.

Built-in
Laptops have built-in keyboards and touch-sensitive touchpads, although external mice can be connected.

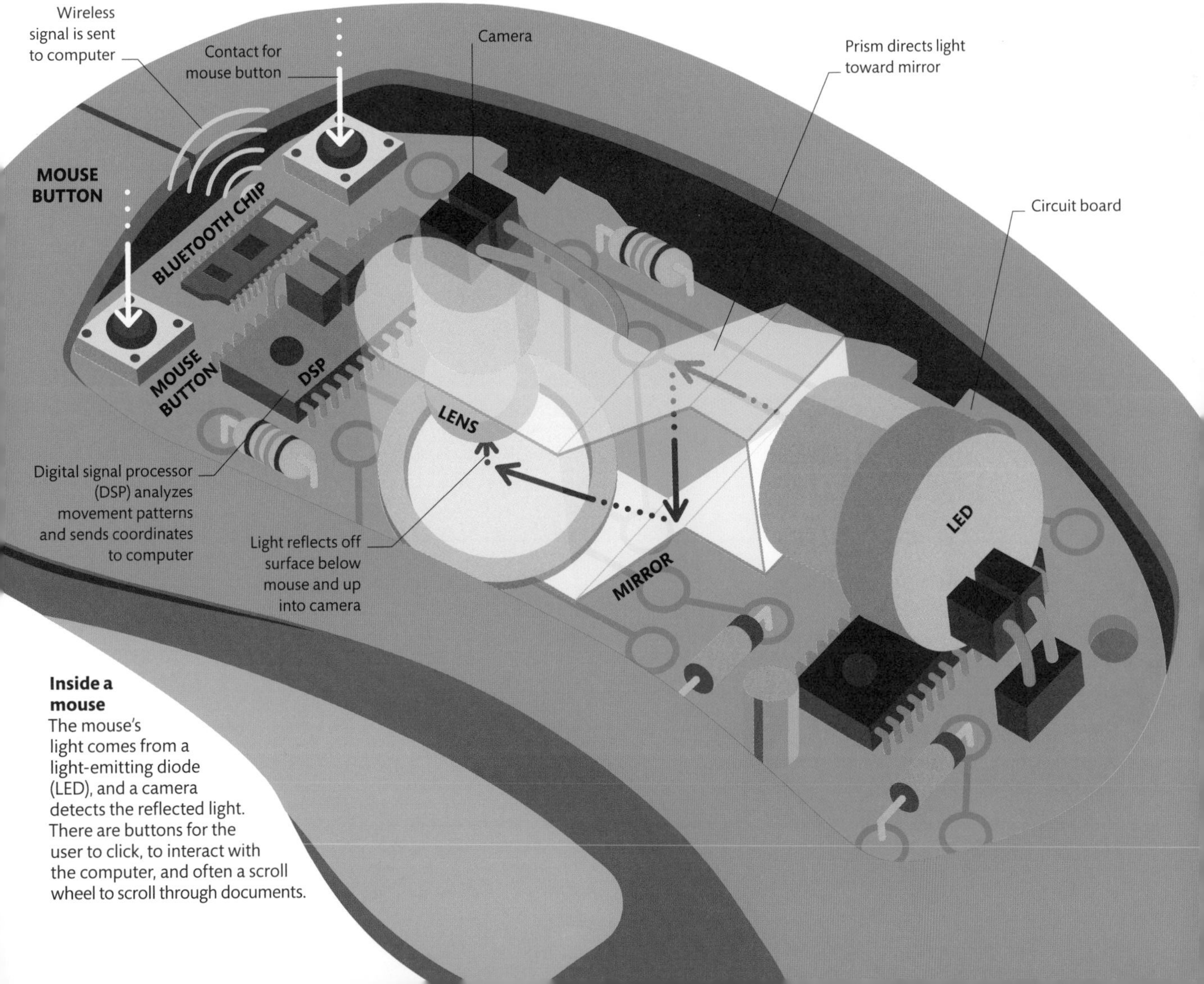
Wireless signal is sent to computer
Contact for mouse button
Camera
Prism directs light toward mirror
MOUSE BUTTON
BLUETOOTH CHIP
Circuit board
MOUSE BUTTON
DSP
LENS
LED
Digital signal processor (DSP) analyzes movement patterns and sends coordinates to computer
Light reflects off surface below mouse and up into camera
MIRROR

Inside a mouse
The mouse's light comes from a light-emitting diode (LED), and a camera detects the reflected light. There are buttons for the user to click, to interact with the computer, and often a scroll wheel to scroll through documents.

Computer software

The physical components of a computer are called hardware. Software is those parts you can't touch or hold—the programs, documents, sounds, and images. These exist as electric currents and charges that represent large collections of binary digits, 0s and 1s.

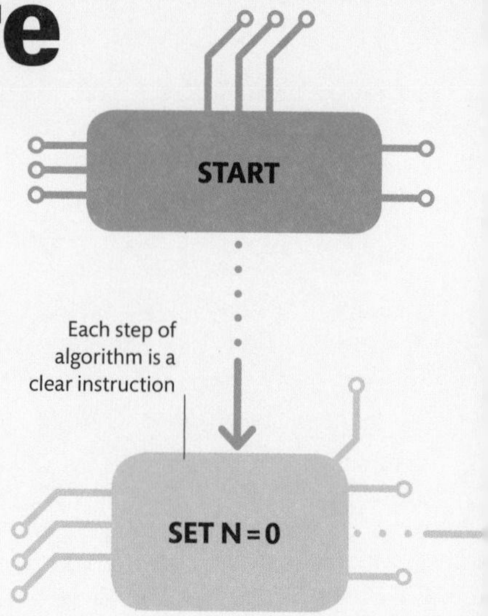

START
Each step of algorithm is a clear instruction
SET N = 0

Algorithms and programs

An algorithm is a set of carefully worked out steps that achieve a specific task. A computer program is a collection of simple algorithms. A computer runs a program in order, but it may need to halt or jump to a different part of the program, depending on the input or the results of a calculation. It may also run a particular part of a program over and over, until a particular condition is met.

APPLICATIONS

An application is a program that a user launches on purpose, such as a word processing or photo-editing program. Applications can be launched by clicking with a mouse or trackpad, touching a smartphone screen, or using a voice command. Other programs are launched automatically by the operating system.

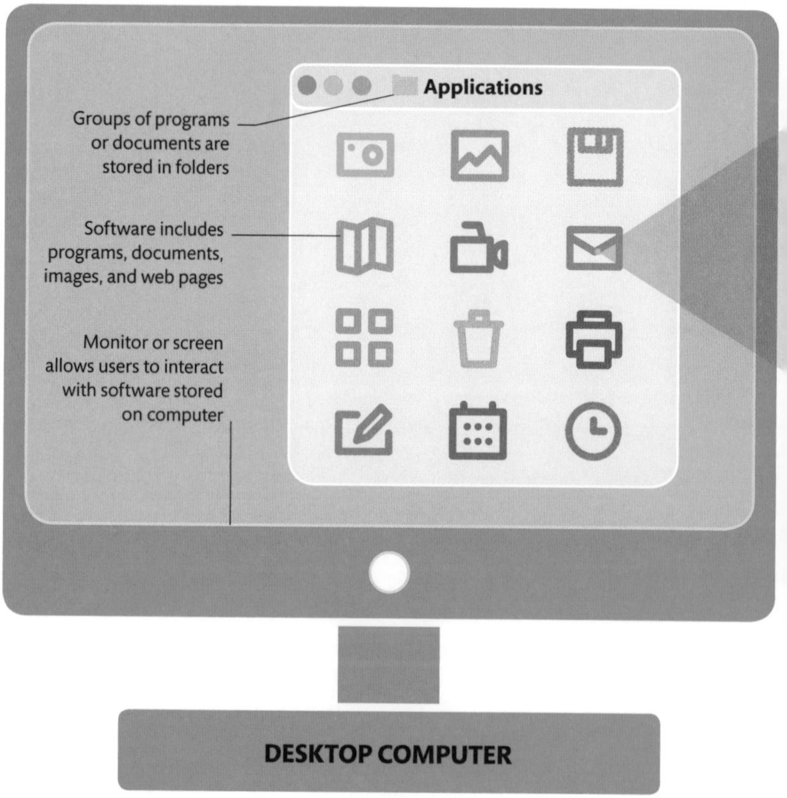

Applications
Groups of programs or documents are stored in folders
Software includes programs, documents, images, and web pages
Monitor or screen allows users to interact with software stored on computer
DESKTOP COMPUTER

HOW MANY TASKS CAN A COMPUTER DO AT ONCE?

Many programs run concurrently, but a computer can execute only one instruction at a time, carrying out a small part of each program in turn.

Operating systems

An operating system is always running whenever a computer is switched on. A kernel, the core program that interacts with programs that are open, directs the inputs and outputs to wherever they are needed.

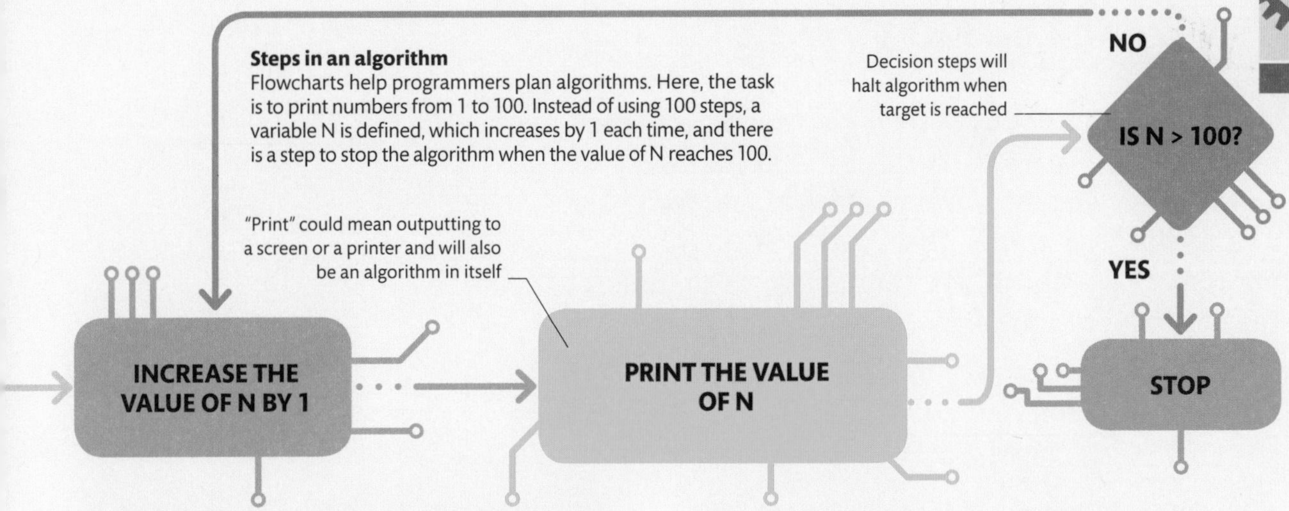
NO
IS N > 100?
YES
INCREASE THE VALUE OF N BY 1
PRINT THE VALUE OF N
STOP
Decision steps will halt algorithm when target is reached
"Print" could mean outputting to a screen or a printer and will also be an algorithm in itself

Steps in an algorithm
Flowcharts help programmers plan algorithms. Here, the task is to print numbers from 1 to 100. Instead of using 100 steps, a variable N is defined, which increases by 1 each time, and there is a step to stop the algorithm when the value of N reaches 100.

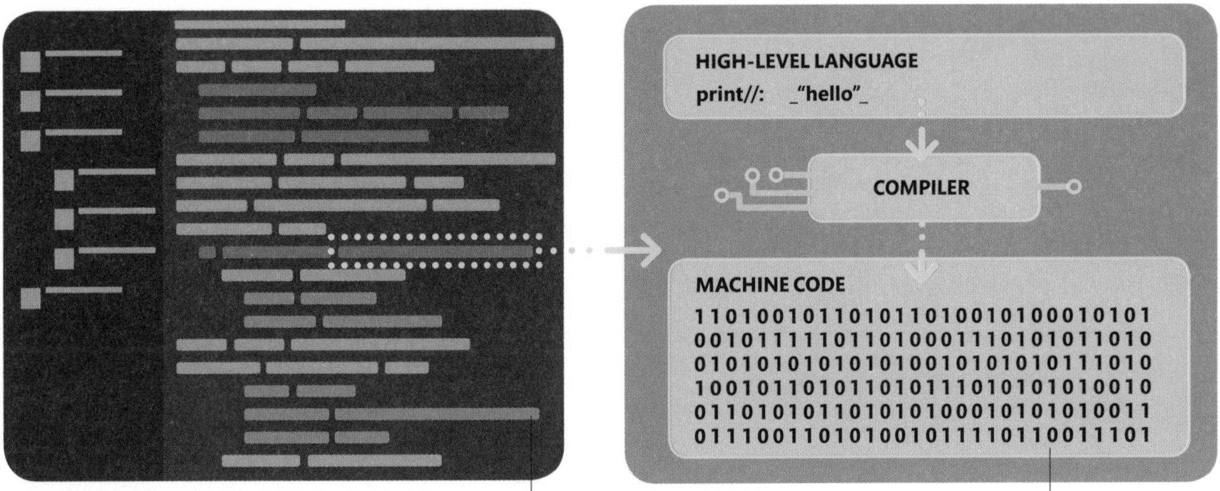
HIGH-LEVEL LANGUAGE
print//: _"hello"_
COMPILER
MACHINE CODE
1101001011010110100101000101 01
0010111110110100011101010110 10
0101010101010100101010101110 10
1001011010110101110101010100 10
0110101011010101000101010100 11
0111001101010010111101100111 01
Application written, or coded, in high-level language
High-level language translated into machine code

From high-level language to machine code
A compiler translates source code, written in high-level programming languages, into machine code. The result is an executable file consisting of binary numbers.

Programs and code

Programs are written, or coded, in words and symbols that humans can read and write. These words and symbols are known as high-level languages—such as Java and C++. The full set of instructions that make up a program is known as the source code. A computer processor cannot understand high-level languages, only binary numbers. Source code is translated by a program called a compiler into a set of on-and-off electric currents in the memory and processor that represent binary numbers. This is called machine code.

THE COMPUTER IN **NASA'S SPACE SHUTTLE** USED **LESS CODE** THAN MOST OF TODAY'S **CELL PHONES**

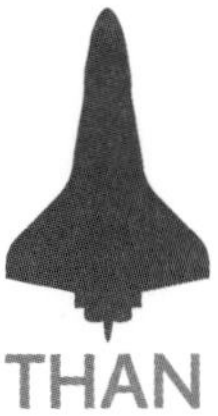

Artificial intelligence

Artificial intelligence (AI) is any technology that enables computers to do things that humans consider intelligent, such as recognizing patterns and solving problems. One goal of AI is for computers to "think" for themselves—to make their own decisions and respond to situations.

HOW DOES SPEECH RECOGNITION WORK?

A computer can recognize the building blocks of speech, called phonemes, and work out the words it has heard spoken.

Machine learning

For a computer to make intelligent decisions in complex situations, it needs to be able to learn, adapt, and recognize patterns. This machine learning is usually achieved by using artificial neural networks—programs that mimic the way brain cells (neurons) work. A network of artificial neurons, arranged in layers, can process huge amounts of information at once and learn to perform tasks such as recognizing faces, handwriting, voices, and trends in social media or commerce.

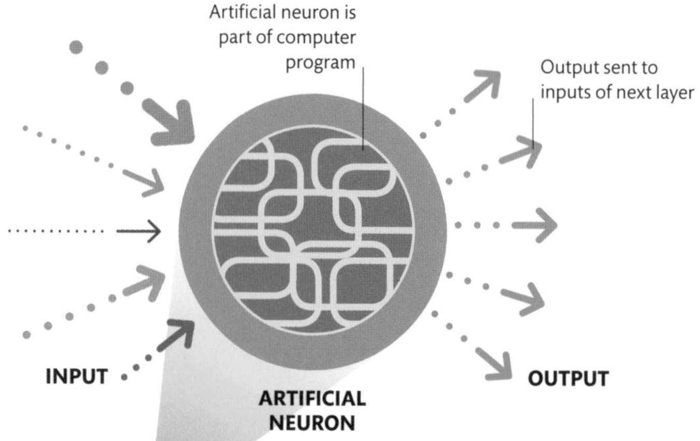

Artificial neuron is part of computer program
Output sent to inputs of next layer
INPUT
ARTIFICIAL NEURON
OUTPUT

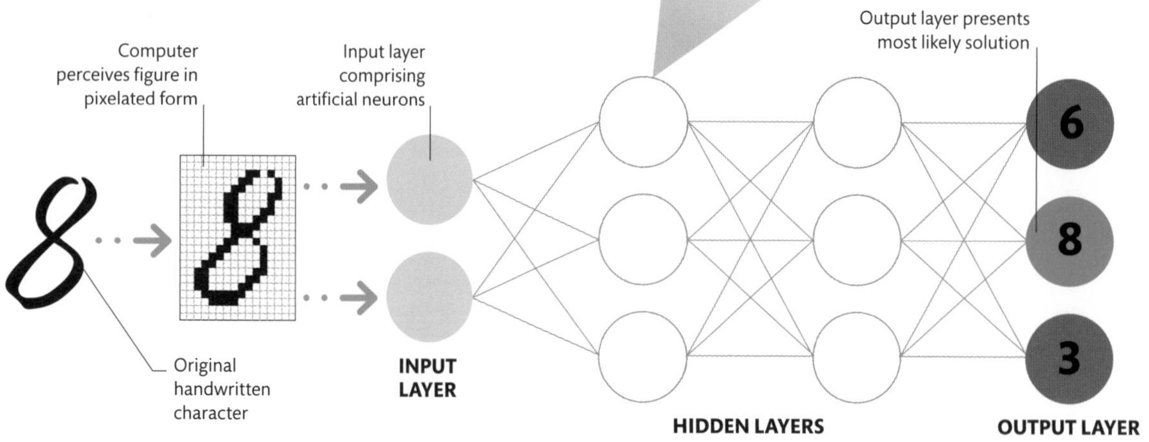

Computer perceives figure in pixelated form
Input layer comprising artificial neurons
Output layer presents most likely solution
6
8
3
Original handwritten character
INPUT LAYER
HIDDEN LAYERS
OUTPUT LAYER

Artificial neural network
Real neurons produce outputs based on inputs they receive from the senses and from other neurons—but they can change how they respond over time, depending on the inputs. Artificial neural networks work in the same way, and, like real ones, they are arranged in layers.

Input layer
The initial layer receives inputs. In this example, each neuron receives a number representing the brightness of an individual pixel from a digitized image of a handwritten character. Only two input neurons are shown here, but a real system would have many more.

Hidden layers
The output of each neuron in the input layer is also a number, the value of which depends on that of the input multiplied by a "weight." The weight changes as the network learns. The number passes on to neurons in several layers, each with its own weight.

Output layer
The outputs of the neurons in the hidden layers are passed to neurons in the output layer. In this network, there would be 10 output neurons—one for each of the digits 0 to 9. The network's "guess" for the character is the neuron that has the highest weighting.

COMPUTER CHESS PLAYER
Computer looks at every possible move

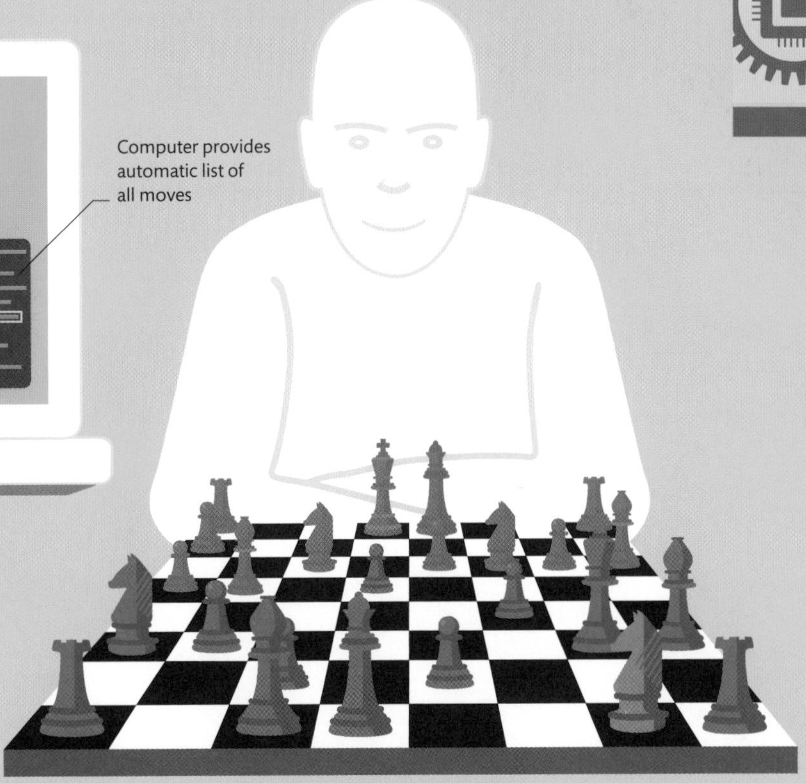

Computer provides automatic list of all moves
HUMAN CHESS PLAYER

Human versus computer
The human brain can look only a few moves ahead, while emotions and "gut feelings" may help, or sometimes hinder, the player. A computer looks at all the possible moves then picks the one that seems most promising. It can look many moves ahead for each scenario.

Playing games

Computers with AI can play games that require intelligence for humans to play, including complicated games such as chess. Powerful chess computers have even beaten the world's best human chess players. However, a game-playing computer can work only within the rules of the game; if anything happens that is outside the rules, the computer is unable to respond. Most game-playing computers follow programs that help them make the best move by analyzing all the possible moves and likely outcomes. In combination with machine learning (see opposite), AI systems can improve their skills at games.

IN 1997, THE COMPUTER **DEEP BLUE** FIRST BEAT THE CHESS WORLD CHAMPION **GARRY KASPAROV**

APPLICATIONS OF AI
Suggestions for music based on recent listening Machine learning finds songs that people with similar music tastes have chosen.
Best route for package deliveries In conjunction with digitized maps and traffic patterns, AI systems can save time and increase efficiency.
Helping doctors diagnose illnesses Fed a patient's symptoms, an AI system can search medical databases to suggest possible causes.
Self-driving cars Computers fed with images from onboard cameras, radar, and digital maps can drive cars safely.
Filtering out spam emails Instead of simply blocking certain senders' addresses, this system recognizes patterns and adapts to new trends.
Image recognition Artificial neural networks improve recognition of objects in digital images, even if the image is unclear.

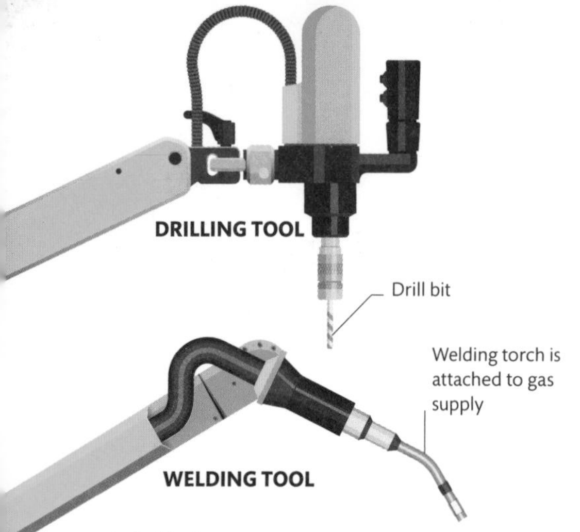
DRILLING TOOL
Drill bit
Welding torch is attached to gas supply
WELDING TOOL

End effectors
Many different kinds of tools can be fitted to the hand of a robotic arm, more correctly called the end effector. The most common is a gripper that can pick up, move, and drop small objects.

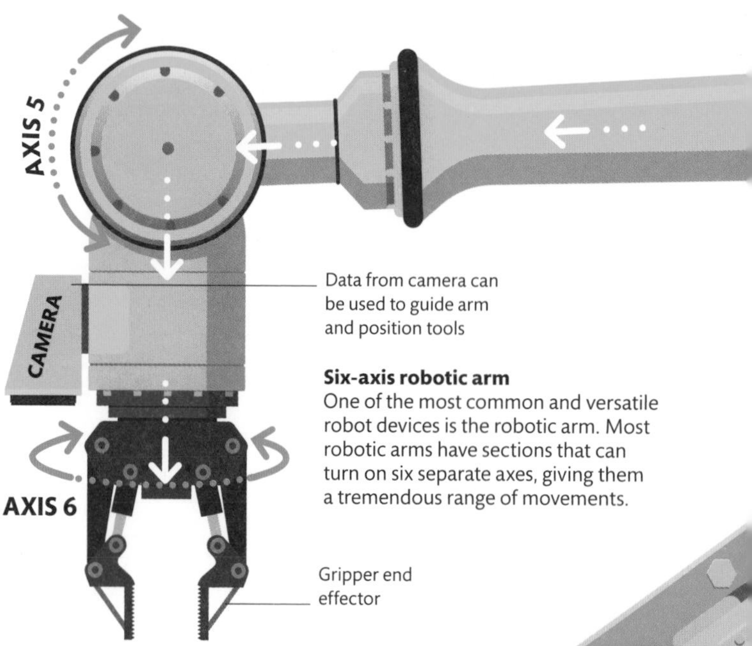
AXIS 5
CAMERA
AXIS 6
Data from camera can be used to guide arm and position tools
Gripper end effector

Six-axis robotic arm
One of the most common and versatile robot devices is the robotic arm. Most robotic arms have sections that can turn on six separate axes, giving them a tremendous range of movements.

How do robots work?

A robot is a computer-controlled machine that can do a range of tasks with little or no human intervention. Robots are used in factories and warehouses, in education, by the military, in the home, and just for fun.

How robots move

The parts of a robot that enable it to move and to manipulate objects are called actuators. The computer that controls a robot sends precise electric currents that make the actuators work. Most actuators are driven by a kind of electric motor called a stepper motor (see opposite). This kind of motor turns in small steps—making it possible for the parts of a robot to move precisely into the desired position. Some robots can also move around, using wheels, caterpillar tracks, or even legs.

CAN A ROBOT BE HACKED?

Yes, hackers can rewrite the programs in the computer that controls a robot. Making sure robots are safe and secure will become an important issue as robots become more common.

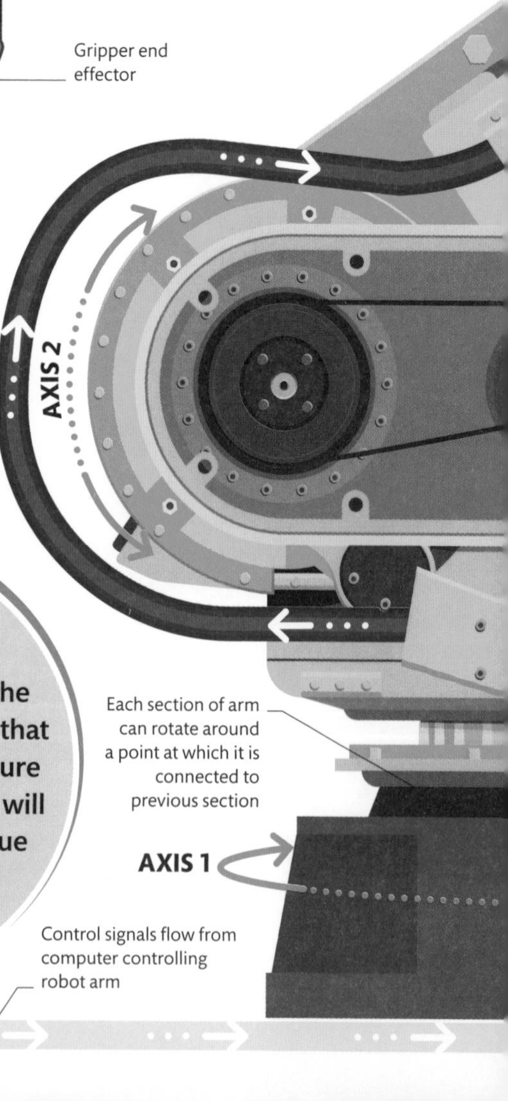
AXIS 2
Each section of arm can rotate around a point at which it is connected to previous section
AXIS 1
Control signals flow from computer controlling robot arm

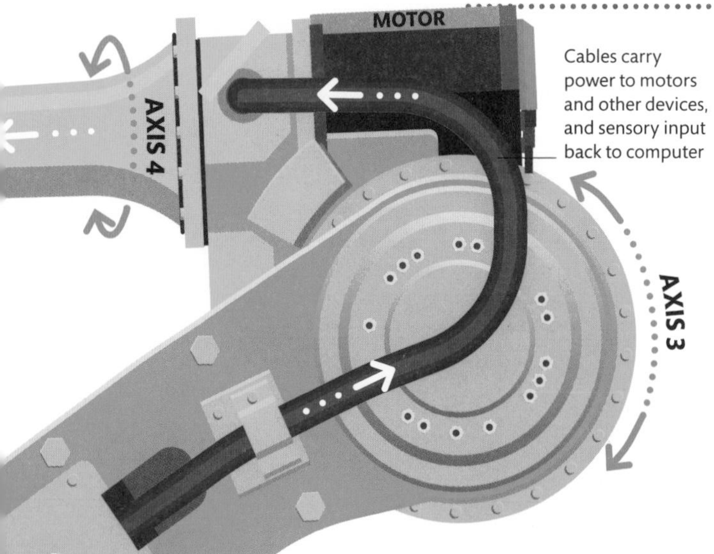
MOTOR
AXIS 4
Cables carry power to motors and other devices, and sensory input back to computer
AXIS 3

SENSING PRESSURE

The simplest type of pressure sensor used in robots is a foam pad that can conduct electricity, sandwiched between two metal plates. The plates are connected to a power supply. The more compressed the foam is, the more current flows through it.

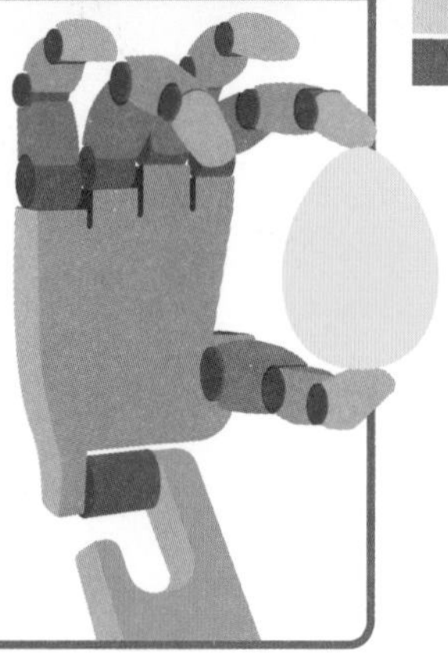

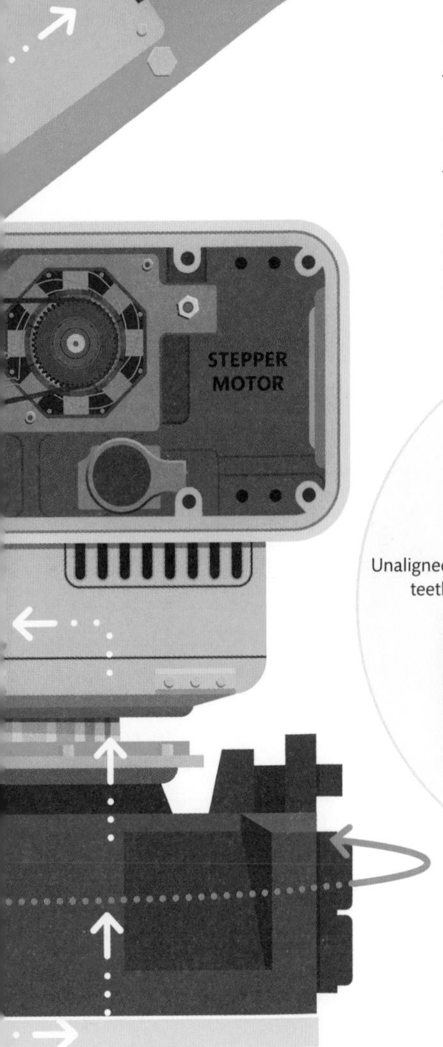
STEPPER MOTOR

Stepper motors

A stepper motor is made up of an inner rotating part (the rotor) and an outer static part (the stator). The rotor is a permanent magnet, and the stator is made of sets of electromagnets. The stator has fewer teeth than the rotor. Activating a set of electromagnets magnetizes the stator teeth with north and south poles. Magnetic attraction brings one set of teeth with opposing poles into alignment, while the matching poles are pushed out of alignment. By activating different sets of electromagnets, the rotor can be rotated in small increments at a time.

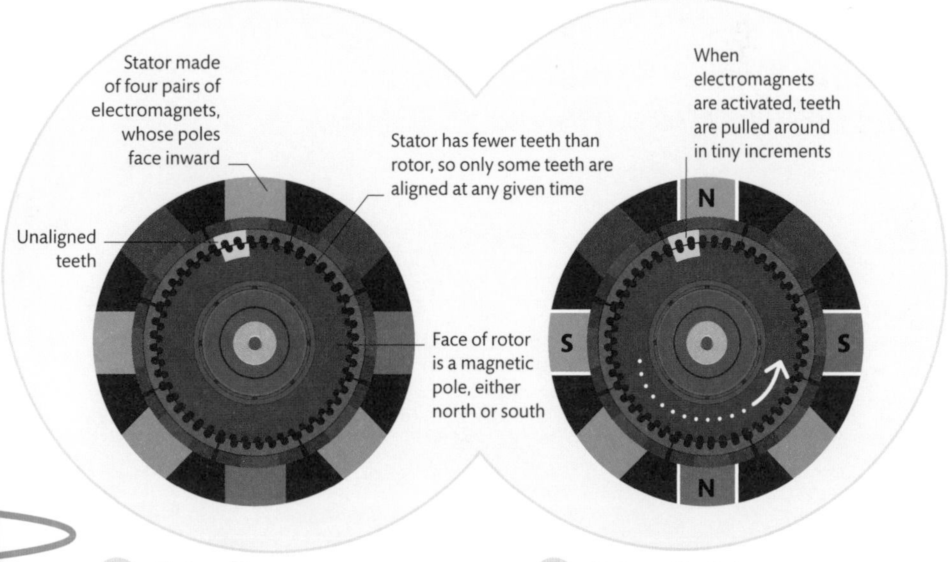
Stator made of four pairs of electromagnets, whose poles face inward
Unaligned teeth
Stator has fewer teeth than rotor, so only some teeth are aligned at any given time
Face of rotor is a magnetic pole, either north or south
When electromagnets are activated, teeth are pulled around in tiny increments
N
S
S
N

1 Motor off
A rotating magnetic rotor sits inside the stator, which is made of paired, stationary electromagnets. There are teeth on both the rotor and the stator.

2 Motor activated
When the electromagnets are activated, the magnetism drags the rotor around a tiny amount to make different teeth align. Each pair is activated in turn, moving the rotor.

What can robots do?

Some robots are completely autonomous, working without input from humans and making decisions based on the input they receive from their sensors. However, most robots are only semiautonomous.

Remote control
A robotic probe is controlled from Earth via radio signals—but can still do tasks unaided.

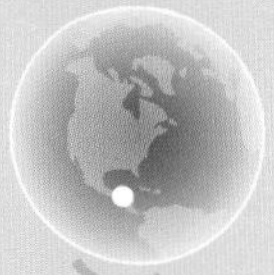

Semiautonomous robots

A semiautonomous robot has to be controlled, typically by a remote controller. However, the robot still needs an onboard computer to complete its tasks accurately—and many semiautonomous robots also make some of their own decisions based on their sensory inputs.

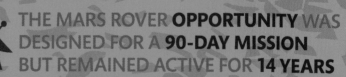

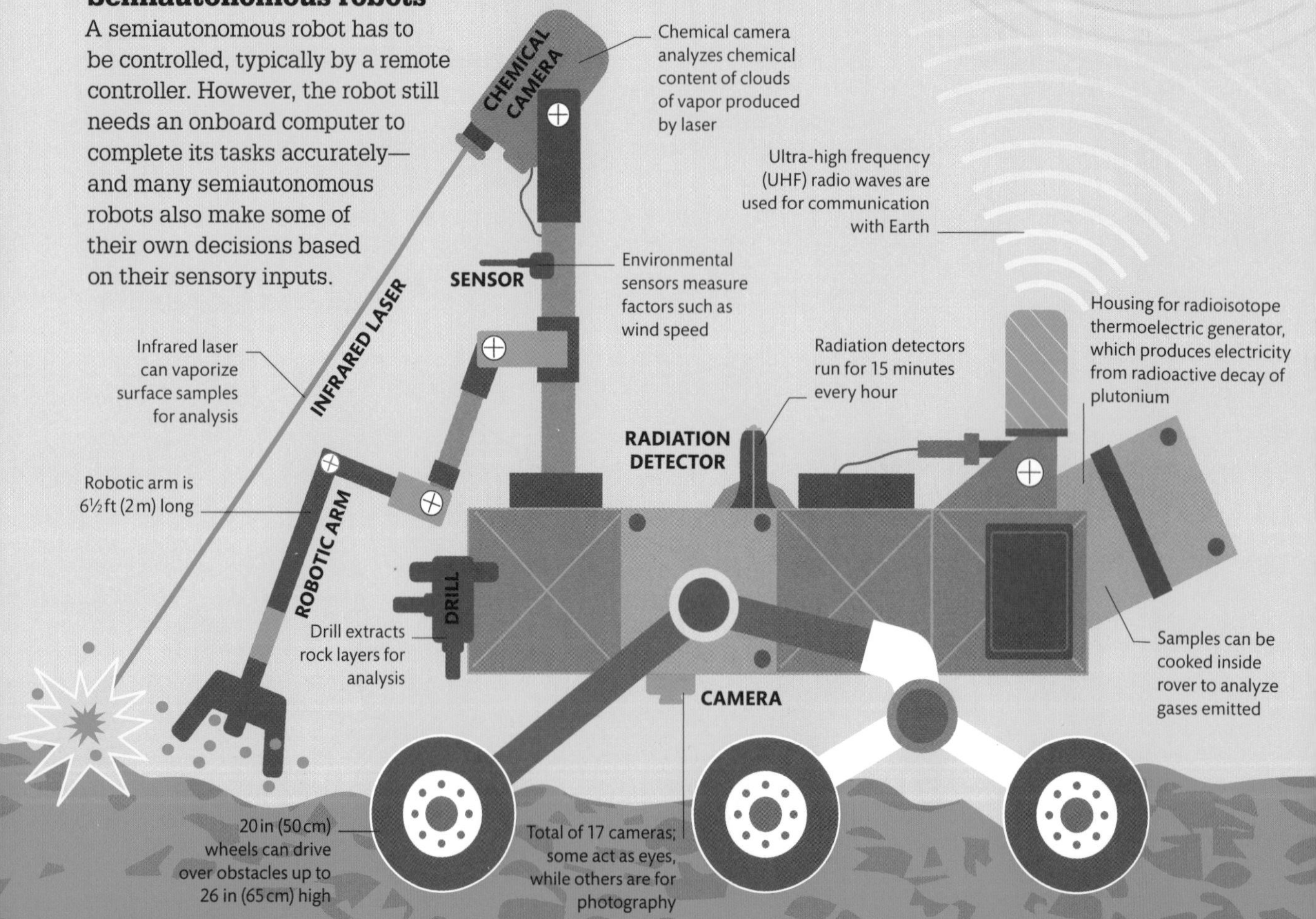
Signals may take between 4 and 24 minutes to reach Mars
CHEMICAL CAMERA
Chemical camera analyzes chemical content of clouds of vapor produced by laser
Ultra-high frequency (UHF) radio waves are used for communication with Earth
SENSOR
Environmental sensors measure factors such as wind speed
INFRARED LASER
Infrared laser can vaporize surface samples for analysis
Housing for radioisotope thermoelectric generator, which produces electricity from radioactive decay of plutonium
Radiation detectors run for 15 minutes every hour
RADIATION DETECTOR
Robotic arm is 6½ ft (2 m) long
ROBOTIC ARM
DRILL
Drill extracts rock layers for analysis
CAMERA
Samples can be cooked inside rover to analyze gases emitted
20 in (50 cm) wheels can drive over obstacles up to 26 in (65 cm) high
Total of 17 cameras; some act as eyes, while others are for photography

Mars Curiosity Rover
NASA's Mars Science Laboratory, known as Curiosity, is a six-wheeled robot designed to withstand the harsh Martian atmosphere. It uses a wide range of scientific instruments to collect data and send it back to Earth.

THE MARS ROVER **OPPORTUNITY** WAS DESIGNED FOR A **90-DAY MISSION** BUT REMAINED ACTIVE FOR **14 YEARS**

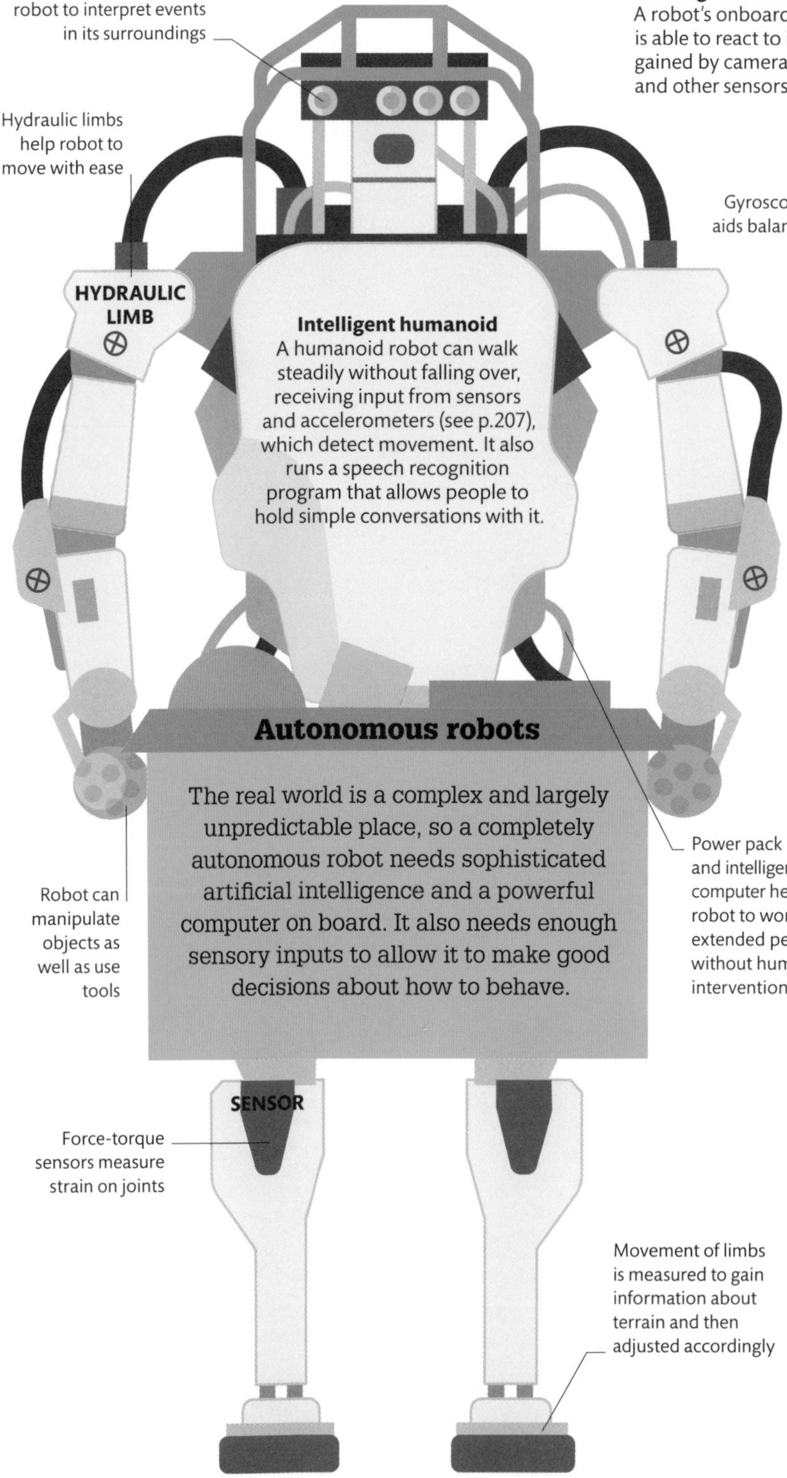
Variety of sensors allows robot to interpret events in its surroundings
Hydraulic limbs help robot to move with ease
HYDRAULIC LIMB
Robot can manipulate objects as well as use tools
Power pack and intelligent computer help robot to work for extended periods without human intervention
SENSOR
Force-torque sensors measure strain on joints
Movement of limbs is measured to gain information about terrain and then adjusted accordingly

Intelligent humanoid
A humanoid robot can walk steadily without falling over, receiving input from sensors and accelerometers (see p.207), which detect movement. It also runs a speech recognition program that allows people to hold simple conversations with it.

Autonomous robots

The real world is a complex and largely unpredictable place, so a completely autonomous robot needs sophisticated artificial intelligence and a powerful computer on board. It also needs enough sensory inputs to allow it to make good decisions about how to behave.

Sensing and seeing
A robot's onboard computer is able to react to information gained by cameras, lasers, and other sensors.

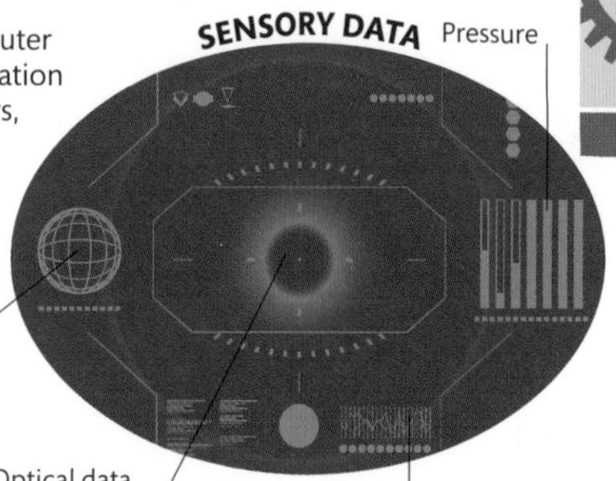
SENSORY DATA
Pressure
Gyroscope aids balance
Optical data from cameras
Infrared sensors detect nearby objects

TYPES OF ROBOTS		
Autonomous	Self-driving car	Uses cameras, other sensors, and satellite navigation
Autonomous	Vacuum cleaner	Cleans floor and returns to charging station
Autonomous	Factory robot	In predictable surroundings, a robot can work unaided
Semiautonomous	Rescue robot	Used during natural disasters but is remotely controlled
Semiautonomous	Missile	Can hit distant targets with little human control
Semiautonomous	Surgical robot	Controlled by a surgeon, making precise movements

EXOSKELETONS

People who do heavy lifting, such as factory workers, can use an exoskeleton for support. This is a powered suit with robotic actuators, such as motors and hydraulic rams, to enhance a person's arm and leg strength.

Virtual reality

Our brains perceive the world around us because they receive information from our senses—in particular, our eyes and our ears. By feeding our senses with sights and sounds generated inside a computer, via a virtual reality headset, our brains can perceive worlds that do not really exist—virtual worlds.

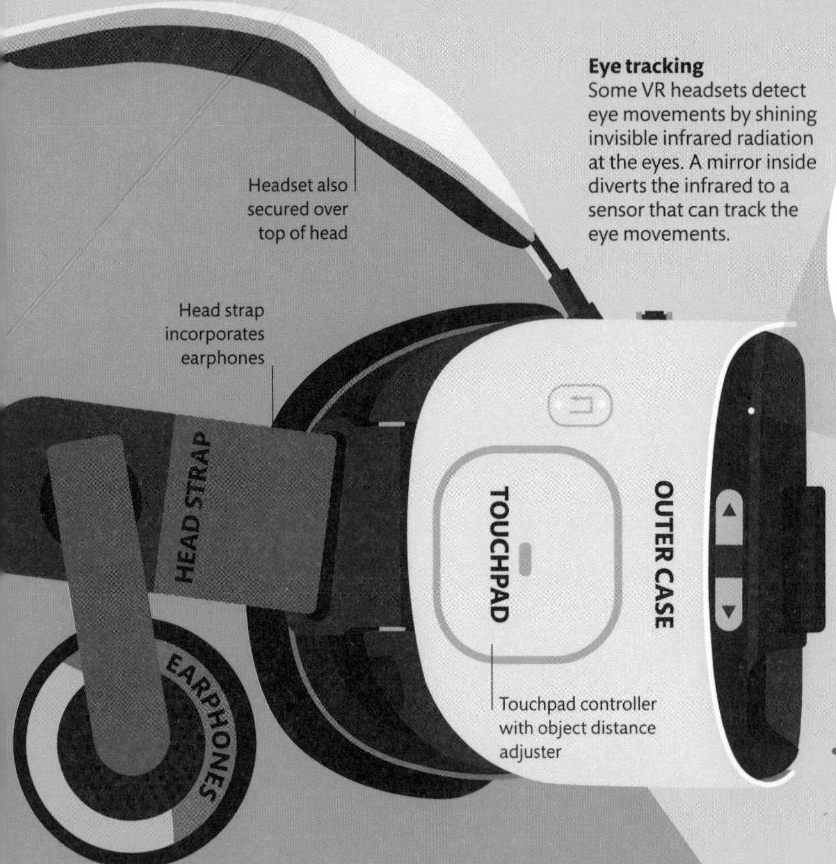
Headset also secured over top of head
Head strap incorporates earphones
HEAD STRAP
EARPHONES
TOUCHPAD
OUTER CASE
Touchpad controller with object distance adjuster

How a VR headset works

A virtual reality (VR) headset displays two views of the virtual world—one for each eye. This gives a sense of depth so that virtual objects appear at different distances, enhancing the feeling of presence. The headset detects the user's head position and movements—and, in some cases, eye movements—and feeds this information to the computer, which adjusts the view and allows the user to look around in the virtual world. Most headsets also contain stereo headphones so sounds from the virtual world can be heard.

Eye tracking
Some VR headsets detect eye movements by shining invisible infrared radiation at the eyes. A mirror inside diverts the infrared to a sensor that can track the eye movements.

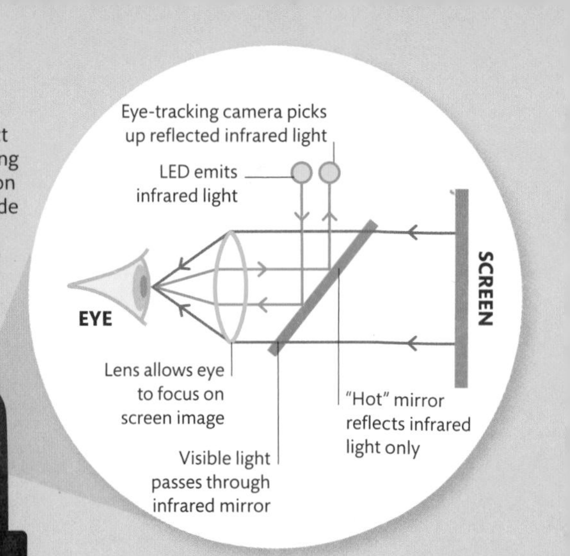
Eye-tracking camera picks up reflected infrared light
LED emits infrared light
SCREEN
EYE
Lens allows eye to focus on screen image
"Hot" mirror reflects infrared light only
Visible light passes through infrared mirror

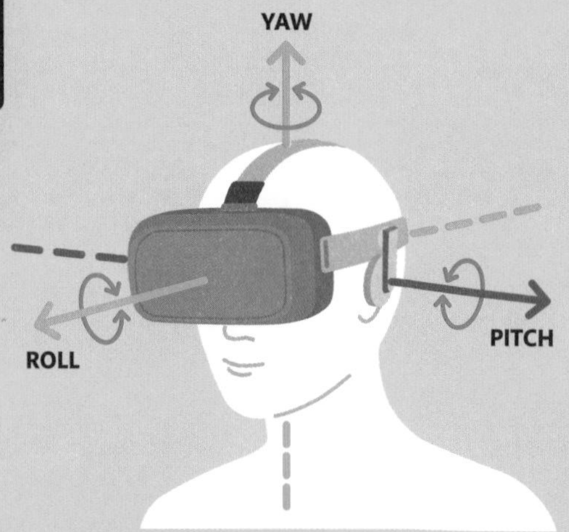
YAW
ROLL
PITCH

Head tracking
Inside a VR headset is a device called an accelerometer (see p.207), which detects head movements. The computer adjusts the view of the virtual world accordingly, so the user can look around the virtual world.

OMNIDIRECTIONAL TREADMILLS ARE BEING DEVELOPED SO VR USERS CAN **WALK FREELY IN VIRTUAL WORLDS**

AUGMENTED REALITY

A technology closely associated with VR is augmented reality. Typically used on a smartphone or tablet, an augmented reality app adds virtual objects into the live view from the device's camera. In this way, virtual objects appear in the real world. This can be useful in adventure games or in displaying information about buildings or vehicles in the real world.

Virtual worlds

The scenes that can be explored inside a VR headset are stored inside a computer. Most virtual worlds are created using computer-generated imagery (CGI) with three-dimensional modeling software that creates digital representations of the virtual objects and surfaces. The scene exists as a sphere, with the viewer at the center and objects all around. The VR headset displays only the part of the sphere at which the viewer is looking.

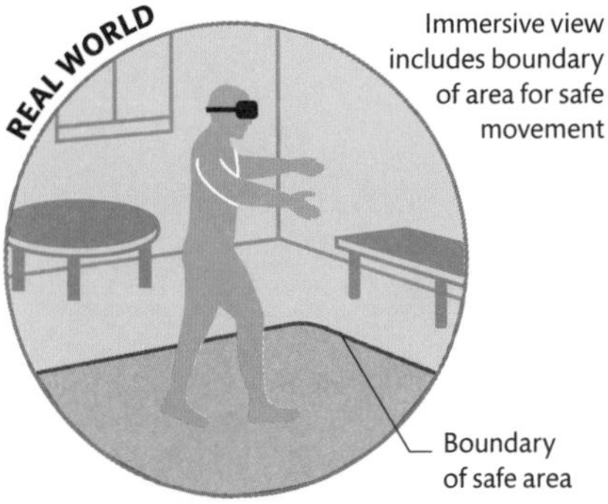

Real-world space
The real-world location can be anywhere—in a room, in a field, or on the beach. A VR headset blocks out the sights, and often the sounds, of the real world.

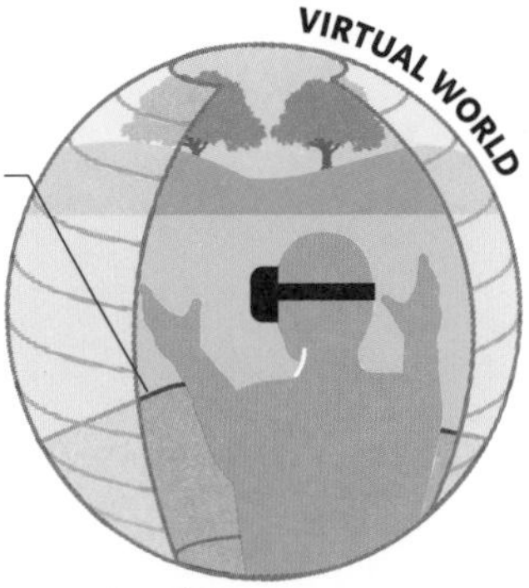

Immersive view
The screens inside the headset display a scene from a virtual world, and stereo headphones play virtual sounds, so that users feel as if they are really there.

Touch and feel

Some VR systems include gloves that allow interaction with some of the objects that appear in the virtual world. These gloves detect the movements of the real hands, and the computer displays virtual hands in the virtual world. At the fingertips are devices called actuators, which produce sensations that the user's brain perceives as pressure, so that he or she can "feel" and interact with the virtual objects.

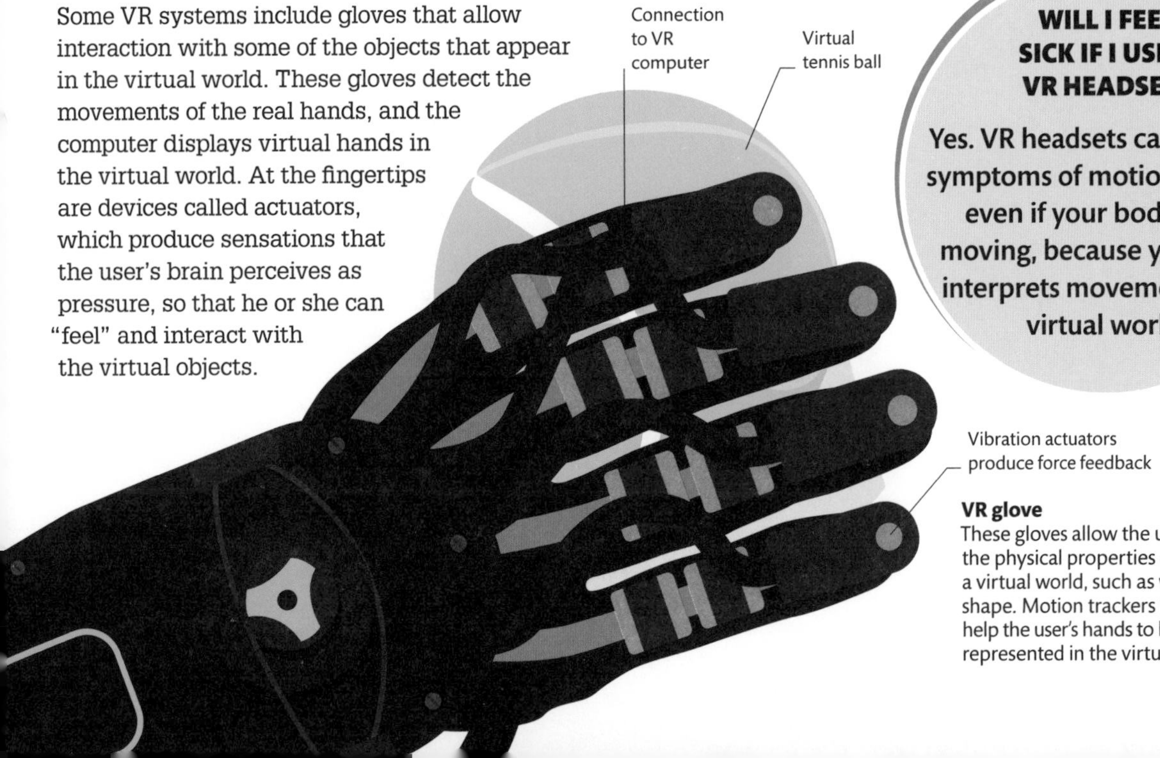

VR glove
These gloves allow the user to feel the physical properties of objects in a virtual world, such as weight and shape. Motion trackers in the fingers help the user's hands to be accurately represented in the virtual world.

WILL I FEEL SICK IF I USE A VR HEADSET?

Yes. VR headsets can produce symptoms of motion sickness, even if your body is not moving, because your brain interprets movement in the virtual world.

COMMUNICATIONS TECHNOLOGY

Radio signals

Radio waves are used to send and receive information across distances without wires or cables. We rely on radio signals for broadcasting, telecommunications, navigation, and computer networks.

Sending signals

Radio waves can contain information such as sound, text, images, and location data. This information is encoded into the wave by modifying different wave features, such as its frequency or amplitude (see opposite). To send information between locations, the radio signals are emitted by a transmitter using an antenna and travel through the air until they are picked up by receivers, also using antennas.

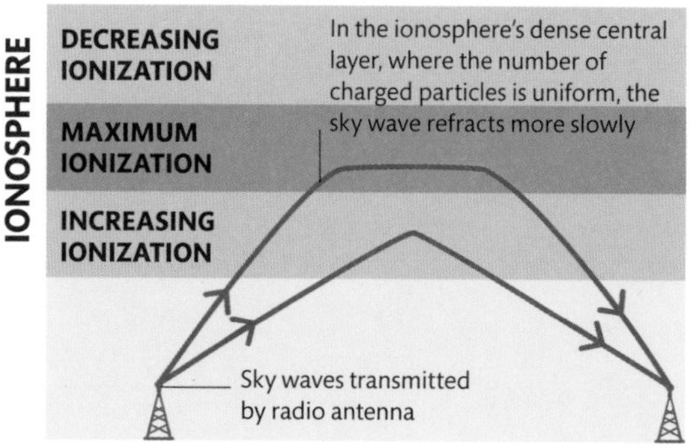
IONOSPHERE
DECREASING IONIZATION
MAXIMUM IONIZATION
INCREASING IONIZATION
In the ionosphere's dense central layer, where the number of charged particles is uniform, the sky wave refracts more slowly
Sky waves transmitted by radio antenna

Refraction in the ionosphere
When a sky wave is transmitted into the ionosphere, the electrically charged layer of Earth's atmosphere, it bends (refracts). The degree to which it is refracted is influenced by the angle of the wave, the wave's frequency, and the number of charged particles present in the layers of the ionosphere.

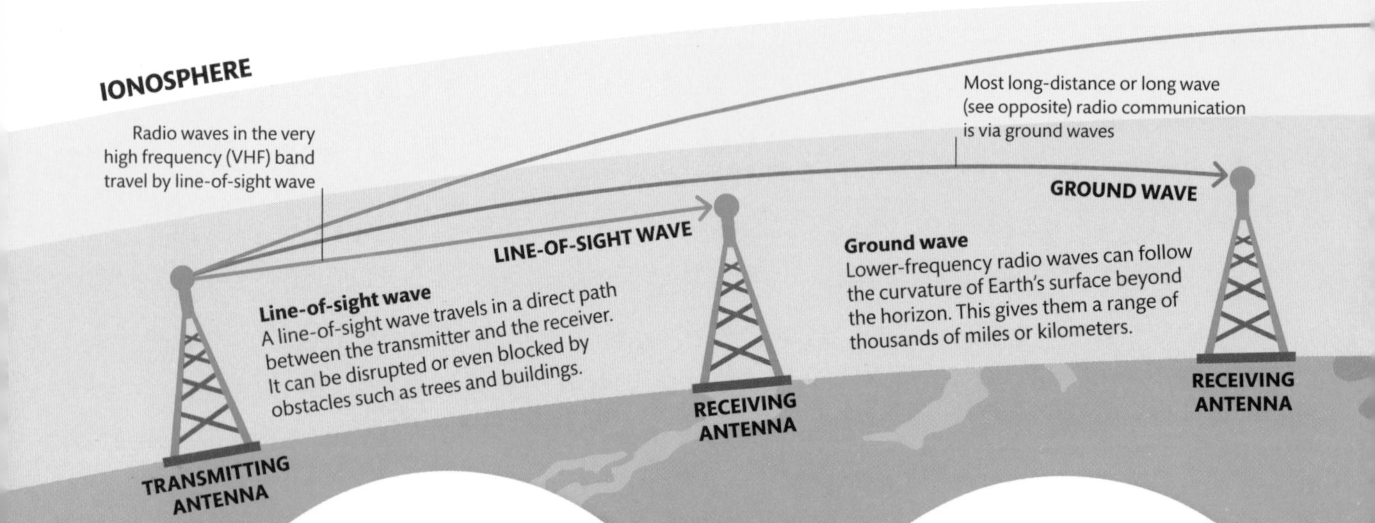
IONOSPHERE
Radio waves in the very high frequency (VHF) band travel by line-of-sight wave
Most long-distance or long wave (see opposite) radio communication is via ground waves
LINE-OF-SIGHT WAVE
GROUND WAVE
TRANSMITTING ANTENNA
RECEIVING ANTENNA
RECEIVING ANTENNA

Line-of-sight wave
A line-of-sight wave travels in a direct path between the transmitter and the receiver. It can be disrupted or even blocked by obstacles such as trees and buildings.

Ground wave
Lower-frequency radio waves can follow the curvature of Earth's surface beyond the horizon. This gives them a range of thousands of miles or kilometers.

1 Radio broadcast
A presenter speaks into a microphone, and the sound waves make a diaphragm inside the microphone vibrate. The microphone transforms these vibrations into an electric current.

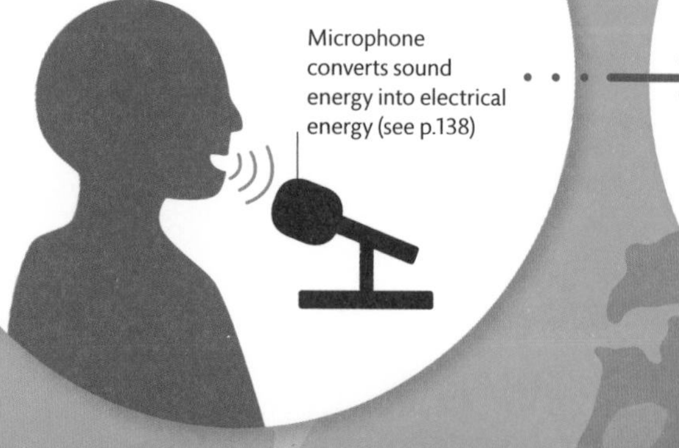
Microphone converts sound energy into electrical energy (see p.138)

2 Studio transmitter link
A studio transmitter link (STL) receives the audio signal from the broadcast studio and relays it to a transmission antenna. It does this via microwave links or fiber-optic cables.

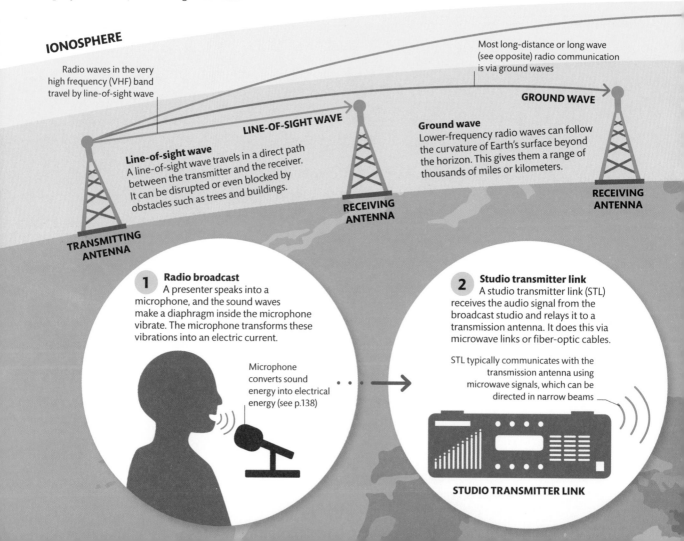

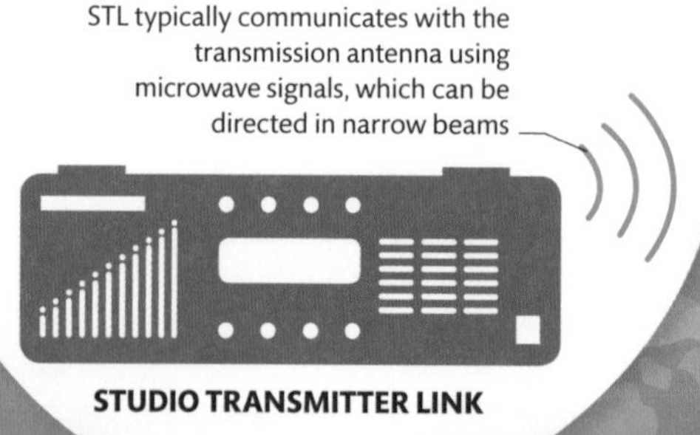
STL typically communicates with the transmission antenna using microwave signals, which can be directed in narrow beams
STUDIO TRANSMITTER LINK

Modulation

Information is encoded into a radio wave through modulation: combining an input wave with a wave of a single frequency called a carrier wave. In AM radio broadcasting, the amplitude of the wave is altered (amplitude modulation), and in FM radio, the frequency of the wave is changed (frequency modulation). For digital radio, there are many ways to combine the input and carrier wave (see p.182).

AM and FM
AM and FM waves both look and behave differently. FM range is less than AM, but the sound quality is better and less susceptible to interference or "noise."

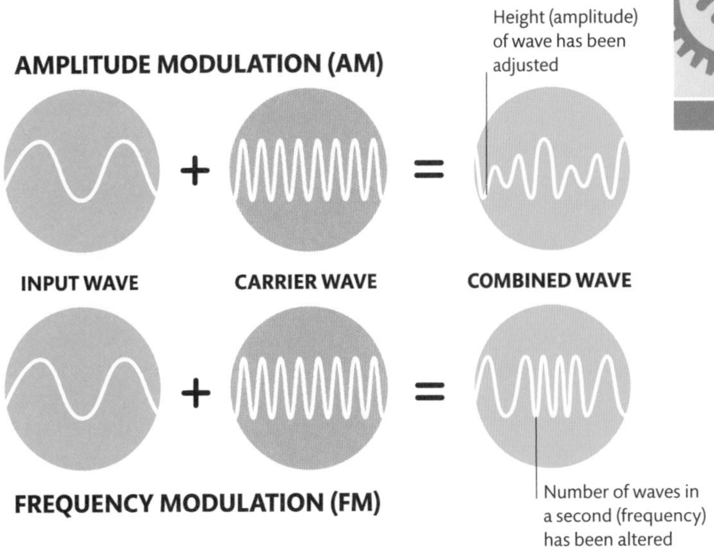
Height (amplitude) of wave has been adjusted
AMPLITUDE MODULATION (AM)
INPUT WAVE
CARRIER WAVE
COMBINED WAVE
FREQUENCY MODULATION (FM)
Number of waves in a second (frequency) has been altered

LIGHTNING PRODUCES VERY LOW-FREQUENCY RADIO WAVES CALLED WHISTLERS

Sky waves can cover a distance of up to 2,500 miles (4,000 km) in a single refraction from ionosphere

SKY WAVE

Sky wave
Some radio waves are refracted back to Earth's surface from the ionosphere, the charged layer of the upper atmosphere. These radio signals can travel vast distances.

EARTH'S SURFACE

RECEIVING ANTENNA

WHAT IS LONG WAVE?

Although not defined precisely, long wave refers to wavelengths greater than 3,300 ft (1,000 m), which are usually transmitted by ground waves.

3 Transmission signal
The current travels to the transmission antenna, making electrons vibrate rapidly back and forth. This generates varying electric and magnetic fields around the antenna, radiating electromagnetic waves.

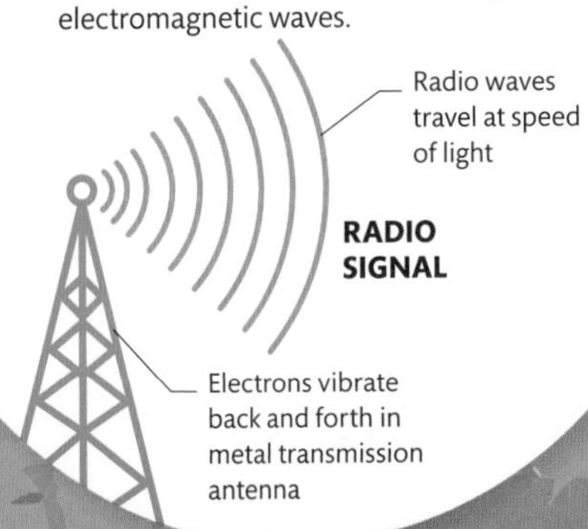
Radio waves travel at speed of light
RADIO SIGNAL
Electrons vibrate back and forth in metal transmission antenna

4 Radio broadcast received
The electric current passes through the radio speaker system, which causes a speaker cone to vibrate. The speaker emits sound waves, reconstructing the sound of the presenter's voice.

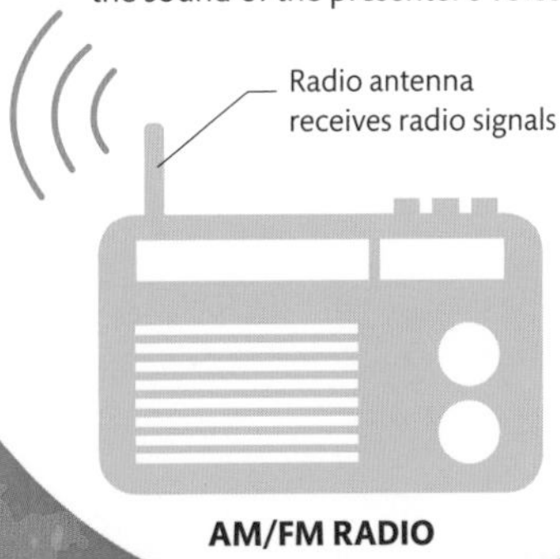
Radio antenna receives radio signals
AM/FM RADIO

Radios

A radio is a device that intercepts radio waves and converts them into a useful form. Broadcast radios receive audio programs transmitted by radio stations and play them through speakers.

How a radio works

A radio receives radio waves through an antenna, which converts them into small alternating currents. These currents are applied to a receiver, which filters out unwanted frequencies and amplifies the signals. The signals are then demodulated: the useful information-carrying signal is extracted from the carrier wave with which it was combined for transmission (see pp.180–181). Finally, the original audio program is played through speakers. Very simple radio receivers (tuned radio frequency receivers) carry out just these steps, but most radios perform additional processing.

STATIC IS CAUSED BY AMPLIFYING **RANDOM ELECTRICAL SIGNALS** BETWEEN BROADCAST FREQUENCIES

1 Antenna picks up radio signals
Radio waves emitted from a radio station's transmission antenna travel through the air and are intercepted by the radio's metal antenna. These waves apply a force to electrons in the metal, causing them to move rapidly back and forth, and generate an alternating current. This current is directed into the radio receiver.

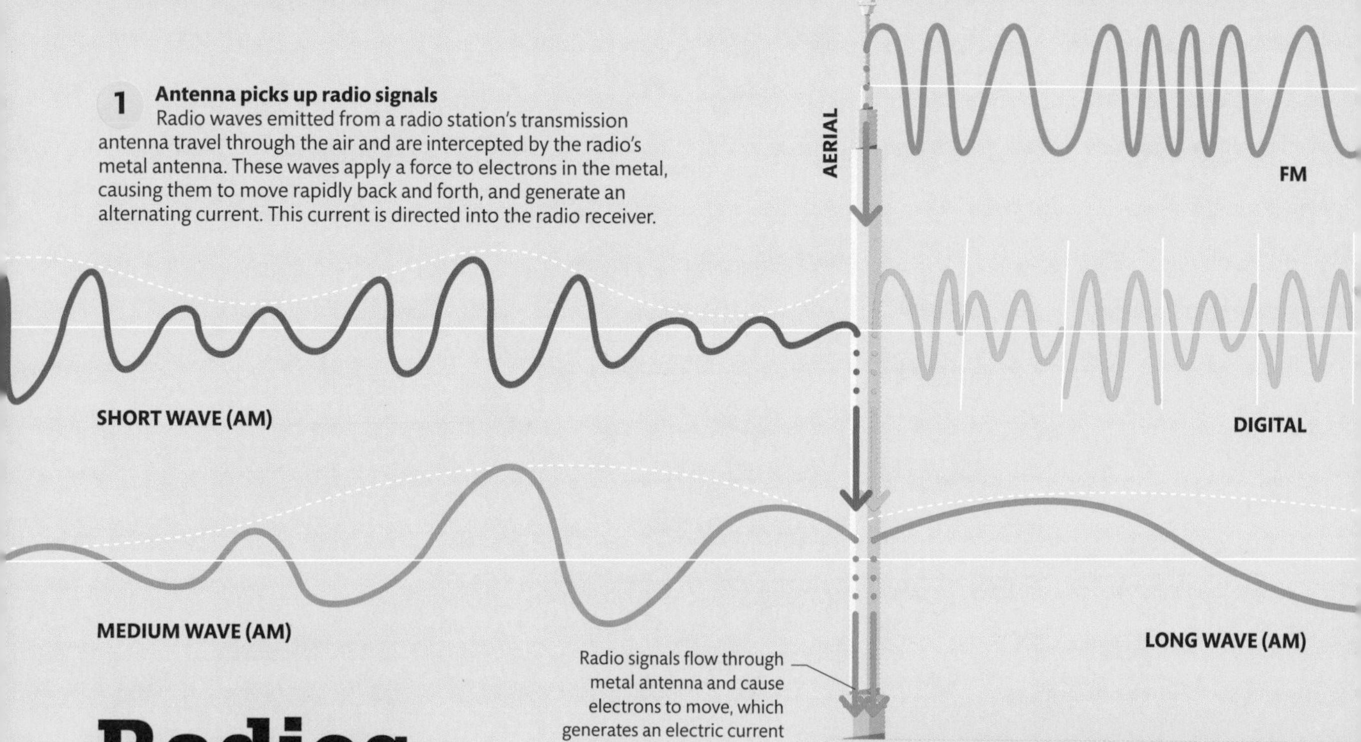
AERIAL
FM
SHORT WAVE (AM)
DIGITAL
MEDIUM WAVE (AM)
LONG WAVE (AM)
Radio signals flow through metal antenna and cause electrons to move, which generates an electric current

2 Tuning
Antennas can pick up radio waves of many frequencies, carrying programs from several radio stations. When a radio's tuner is adjusted, this changes the "passband"—the range of frequencies allowed through. Only signals of the desired frequency cause the circuit to resonate, passing them on to the rest of the receiver.

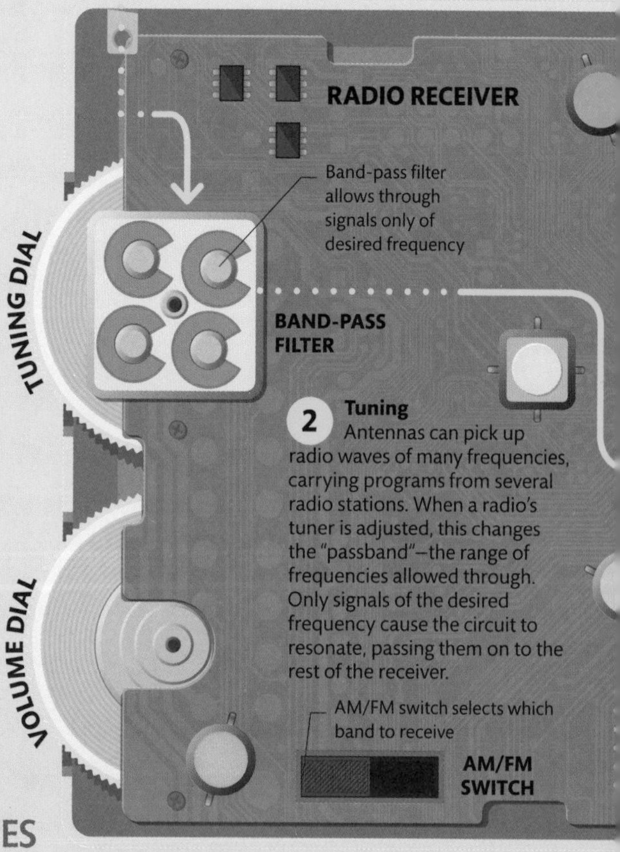
RADIO RECEIVER
Band-pass filter allows through signals only of desired frequency
TUNING DIAL
BAND-PASS FILTER
VOLUME DIAL
AM/FM switch selects which band to receive
AM/FM SWITCH

Digital radio

Digital audio broadcasting (DAB) is radio broadcasting using a digital signal. It is attractive to broadcasters because it allows them to make more efficient use of the radio spectrum compared with analogue radio. The original analogue signal is converted to a digital form before being compressed using formats such as MP2 and transmitted via digital modulation.

Digital modulation
After the analogue signal is converted to digital, the changes in frequency, amplitude, and phase are represented by binary digits. These signals are combined with analogue carrier waves (see p.181) to produce an analogue signal for transmission.

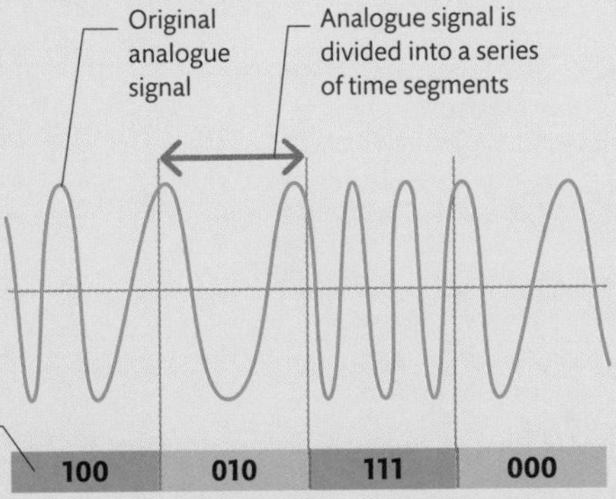
Original analogue signal
Analogue signal is divided into a series of time segments
Digital signal consists of a series of binary numbers, one for each time segment
100
010
111
000

RADIO TELESCOPES

A radio telescope is a type of radio receiver designed to intercept radio waves from astronomical sources, such as stars, nebulae, and galaxies. Radio telescopes require vast, sensitive antennas in order to collect signals emitted many light-years away.

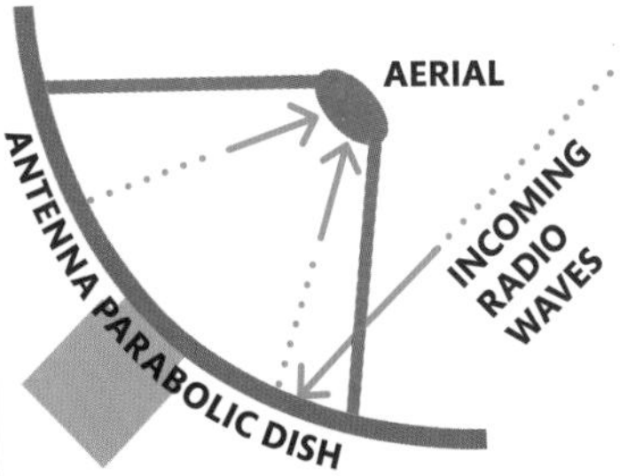
AERIAL
ANTENNA PARABOLIC DISH
INCOMING RADIO WAVES

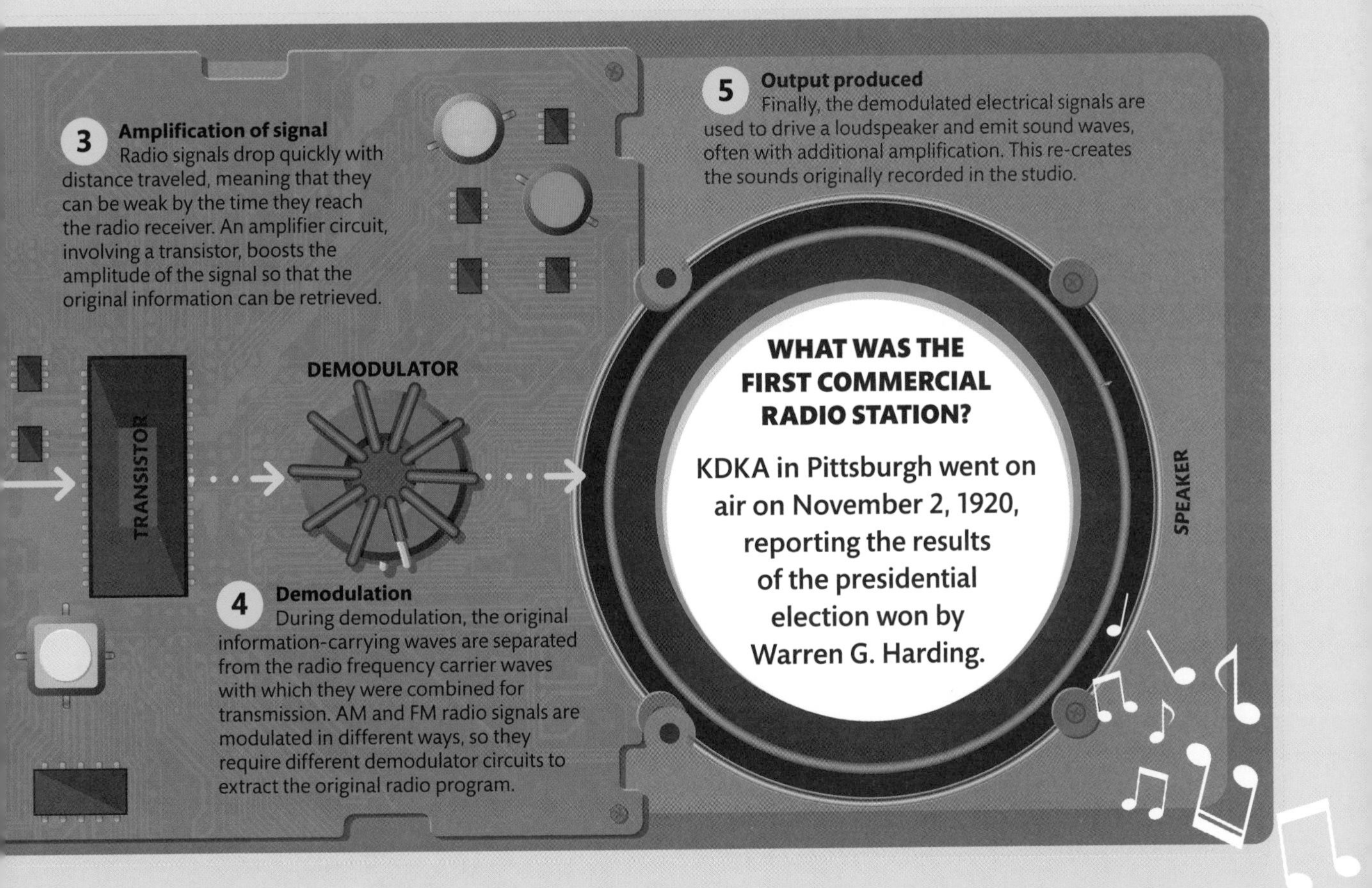
TRANSISTOR
DEMODULATOR
SPEAKER

3 Amplification of signal
Radio signals drop quickly with distance traveled, meaning that they can be weak by the time they reach the radio receiver. An amplifier circuit, involving a transistor, boosts the amplitude of the signal so that the original information can be retrieved.

4 Demodulation
During demodulation, the original information-carrying waves are separated from the radio frequency carrier waves with which they were combined for transmission. AM and FM radio signals are modulated in different ways, so they require different demodulator circuits to extract the original radio program.

5 Output produced
Finally, the demodulated electrical signals are used to drive a loudspeaker and emit sound waves, often with additional amplification. This re-creates the sounds originally recorded in the studio.

WHAT WAS THE FIRST COMMERCIAL RADIO STATION?

KDKA in Pittsburgh went on air on November 2, 1920, reporting the results of the presidential election won by Warren G. Harding.

Telephones

Telephones make conversation possible when people are too far apart to be heard directly. They convert sound waves into signals that can be rapidly transmitted to another telephone, where the speech is reproduced.

Telephone anatomy
Except for the development of keypads, the telephone's basic anatomy has not changed that much since its invention. It still features a speaker, microphone, and hook switch and a wall jack to connect it to the telephone network.

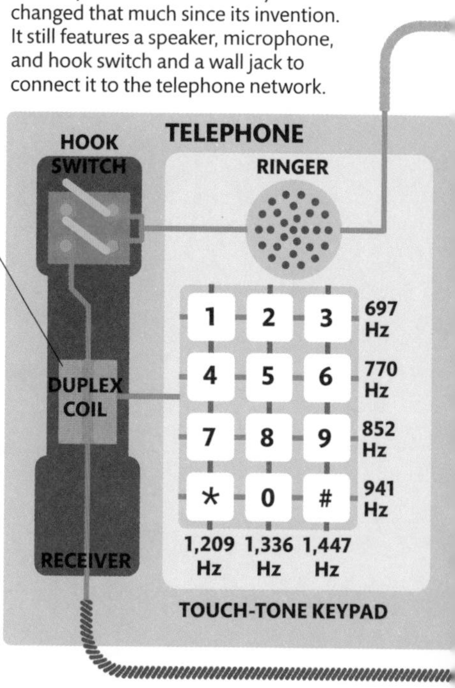

How telephones work

A person begins a call by picking up his or her handset and dialing a number to reach a recipient's telephone. Picking up the ringing telephone connects the speakers. The caller's speech travels through the telephone network in the form of electrical, optical, or radio signals, before being reproduced by the other telephone. Telephones contain both transmitters and receivers, which allow for two-way communication.

1 Connecting to an exchange
A switch called a hook switch connects and disconnects the telephone from the telephone network. Lifting up the telephone to make a call operates a lever, which forms an electrical connection between the handset and the local telephone exchange.

2 Dialing a number
Entering a digit on the keypad produces a distinct sound comprising two simultaneous frequencies, one high and one low. For example, key 7 produces a signal composed of tones with frequencies of 852 and 1,209 Hz. This unique sequence in a telephone number indicates to the exchange where the call should be directed to.

WHAT WERE THE FIRST WORDS SPOKEN ON THE TELEPHONE?

"Mr. Watson, come here; I want to see you," were the words uttered by telephone inventor Alexander Graham Bell on March 10, 1876, to his assistant.

Three transmission methods
Most information in the public switched telephone network is carried in the form of electrical, optical, or radio signals. These move much faster than the speed of sound.

1 Capturing a signal
A microphone in the mouthpiece converts sound waves into electrical signals of the same frequencies. These can travel through the telephone network in three different ways.

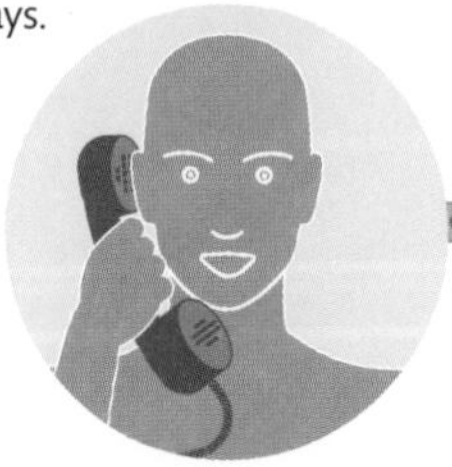

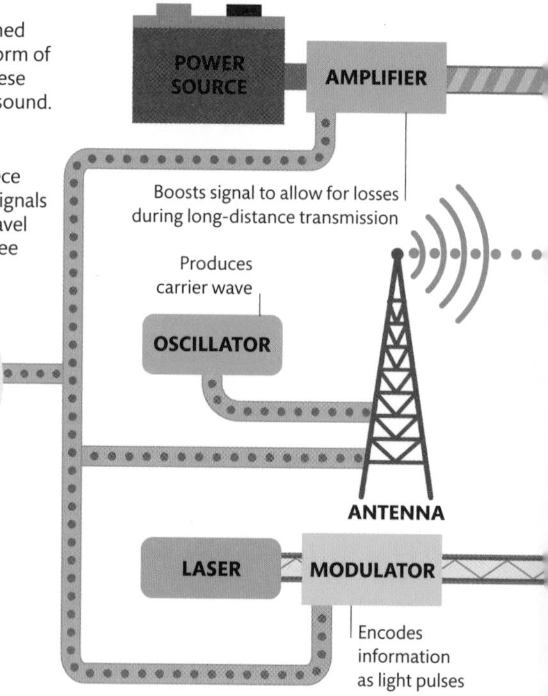

Delivery of sound

Telephone conversations sound natural when the signals travel quickly, with minimal delay. Sound waves are turned into electrical signals and travel through the telephone network as electrical or electromagnetic signals, before being transformed back into sound at their destination. This makes transmission so fast that it feels instantaneous, even on long-distance calls.

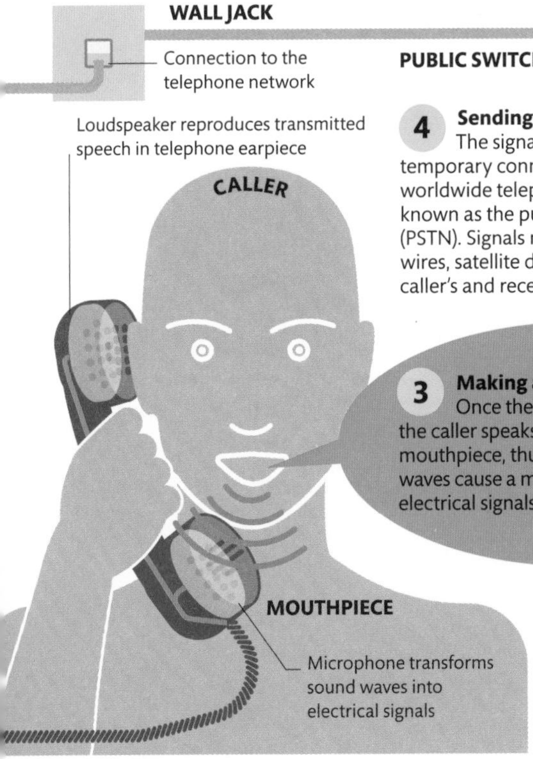
WALL JACK
Connection to the telephone network
PUBLIC SWITCHED TELEPHONE NETWORK
Loudspeaker reproduces transmitted speech in telephone earpiece
CALLER
MOUTHPIECE
Microphone transforms sound waves into electrical signals

4 Sending the sound signal
The signals travel rapidly through a temporary connection formed across a worldwide telephone communications network known as the public switched telephone network (PSTN). Signals may travel via fiber-optic cables, wires, satellite dishes, and cell towers between the caller's and receiver's telephones.

3 Making a sound signal
Once the telephones are connected, the caller speaks into a microphone in the mouthpiece, thus producing sound waves. The waves cause a membrane to vibrate and create electrical signals, which travel down the line.

RECEIVER
EARPIECE

5 Reproducing the sound
Inside the earpiece is a speaker. When it receives electrical signals, a membrane is driven back and forth at frequencies matching the current, causing the air to vibrate and produce sound waves.

"AHOY" WAS BELL'S SUGGESTED GREETING ON THE TELEPHONE BUT WAS REPLACED BY THOMAS EDISON'S SUGGESTION "HELLO"

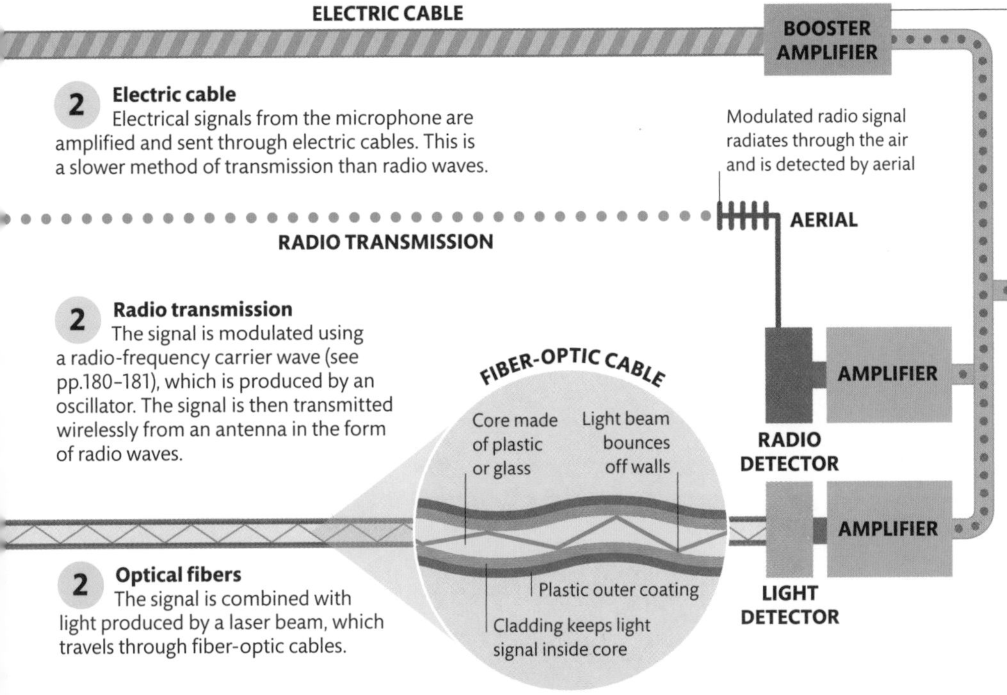
ELECTRIC CABLE
BOOSTER AMPLIFIER
Consists of transistors, which increase the power of the signal, extending its range
Modulated radio signal radiates through the air and is detected by aerial
AERIAL
RADIO TRANSMISSION
FIBER-OPTIC CABLE
Core made of plastic or glass
Light beam bounces off walls
Plastic outer coating
Cladding keeps light signal inside core
AMPLIFIER
RADIO DETECTOR
AMPLIFIER
LIGHT DETECTOR

2 Electric cable
Electrical signals from the microphone are amplified and sent through electric cables. This is a slower method of transmission than radio waves.

2 Radio transmission
The signal is modulated using a radio-frequency carrier wave (see pp.180–181), which is produced by an oscillator. The signal is then transmitted wirelessly from an antenna in the form of radio waves.

2 Optical fibers
The signal is combined with light produced by a laser beam, which travels through fiber-optic cables.

3 Sound signal arrives
The signals arrive at their destination and are passed on to a telephone receiver. The receiver demodulates the signal, extracting from it the useful information and reproducing the sound.

Telecommunications networks

Telecommunications networks are systems that enable the exchange of information, including Internet traffic, across great distances. These networks are made up of connected points that relay signals through a system of wires, cables, satellites, and other infrastructure to reach their destination.

WHAT WAS THE FIRST TELECOMMUNICATIONS NETWORK?

The telegraph network was the first to enable long-distance communication. The first transatlantic cable was completed in 1858.

The telephone network

In the early days of the telephone, phones had to be permanently wired together for the callers to speak to each other. Now, they are connected to the public switched telephone network (PSTN) instead. During a call, a temporary connection is established between the two telephones through PSTN infrastructure, allowing for the high-speed exchange of sound information. This huge network is made up of the world's local, national, and regional telephone networks, linked with exchanges to allow communication between most telephones.

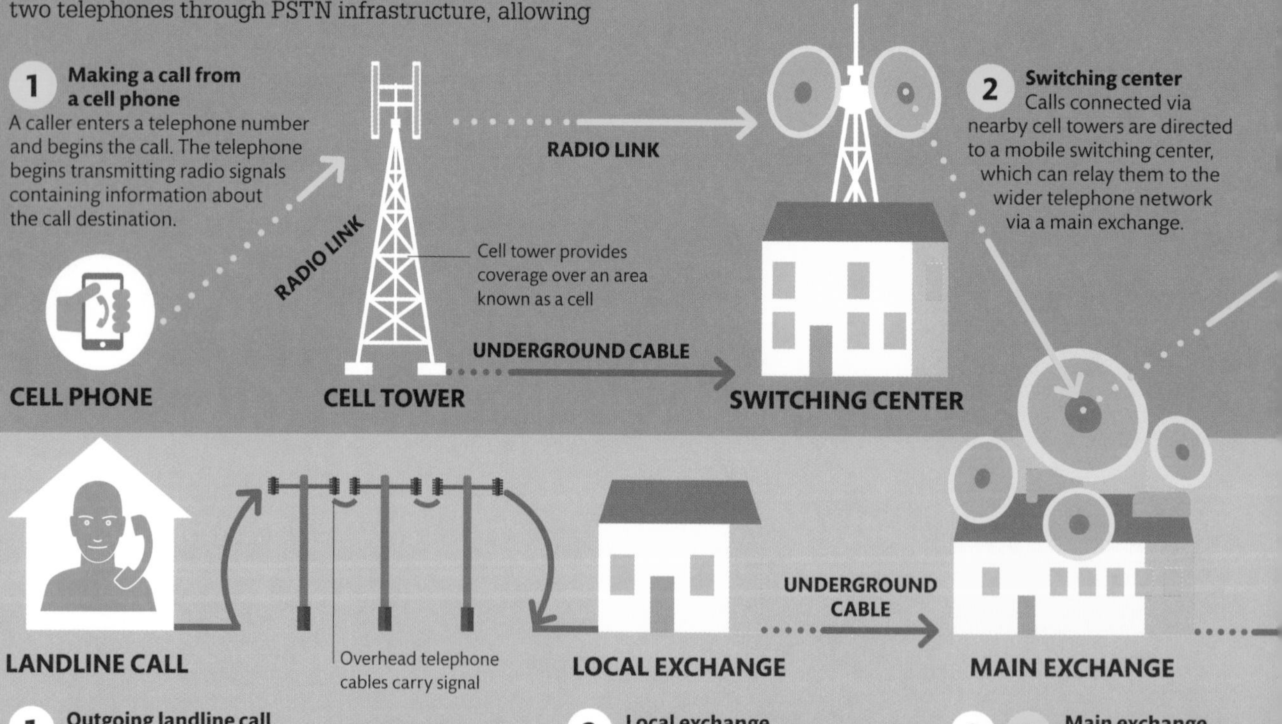

1 Making a call from a cell phone
A caller enters a telephone number and begins the call. The telephone begins transmitting radio signals containing information about the call destination.

2 Switching center
Calls connected via nearby cell towers are directed to a mobile switching center, which can relay them to the wider telephone network via a main exchange.

1 Outgoing landline call
A caller picks up the handset, making an electrical connection to the local exchange. As the caller enters a telephone number, signals indicating the call destination are sent down the line.

2 Local exchange
A local exchange connects telephones in a local area. If it detects a call destination that is further away, it relays the call to a main exchange.

3 3 Main exchange
Nonlocal cell phone and landline calls are passed to a main telephone exchange, which is capable of directing calls across much greater distances.

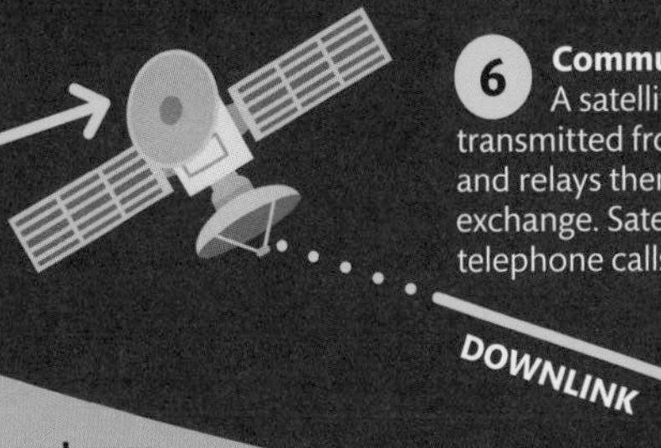

6 Communications satellite
A satellite receives radio signals transmitted from a ground-based station and relays them back to Earth to another exchange. Satellites are used infrequently for telephone calls due to the delay on the signals.

5 International exchange
An international telephone exchange links national telephone networks to the rest of the PSTN, allowing for communication across borders using international dialing codes.

UNDERWATER CABLE

Undersea cables carry telephone traffic between land-based stations, which could be separated by entire oceans

4 Relay tower
Tall relay towers receive and retransmit signals in order to establish wireless telecommunications channels between distant telephone exchanges.

Telephone infrastructure
An international cell phone call and a long-distance landline call share most of the same infrastructure, including a main exchange. However, to cover the significant distances required, making an international call may involve transmitting signals by underwater cable or, rarely, radio waves, while many landline calls are made using just electric and fiber-optic cables.

DIAL-UP INTERNET

Dial-up is a form of Internet access that uses the PSTN. The user's computer sends information down the telephone line to the Internet, via his or her Internet service provider (ISP). This process requires a modem (short for modulator-demodulator) to encode and decode audio signals from the telephone line. Millions of people living in remote areas still rely on dial-up Internet.

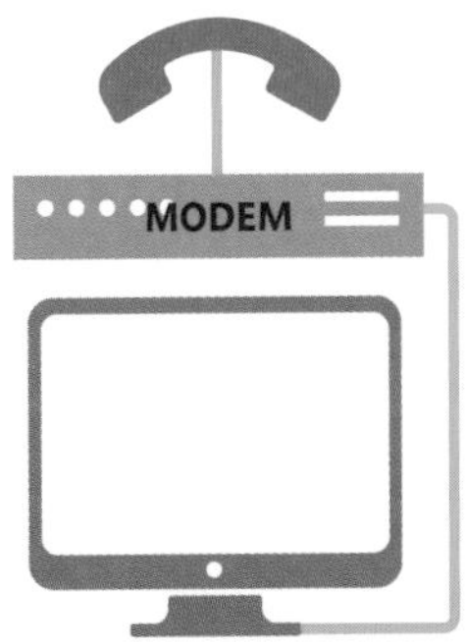

A local exchange connects to an electric utility box, which connects to each house with a landline telephone

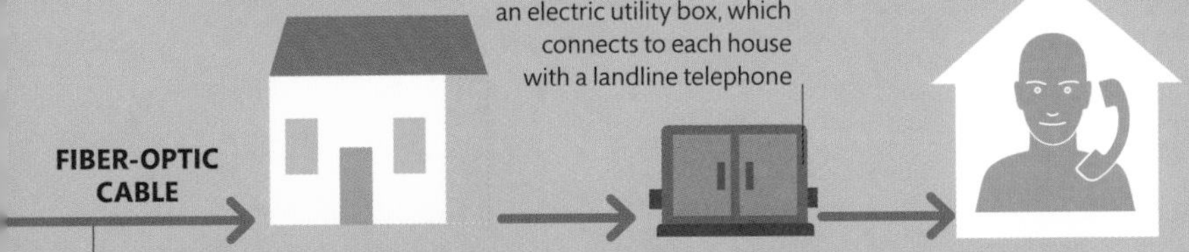

Often, underground fiber-optic cables (see pp.190–191) connect main exchange to local exchange

4 Incoming landline call
When the call reaches its destination, the recipient's phone rings. When it is picked up, a connection is established and the conversation can begin (see pp.184–185).

Television broadcasting

Television broadcasting enables video content to reach anyone with a television set. Before appearing on viewers' screens, television programs are transmitted using three kinds of broadcasting technology: terrestrial (using ground-based antennas), satellite, or cable.

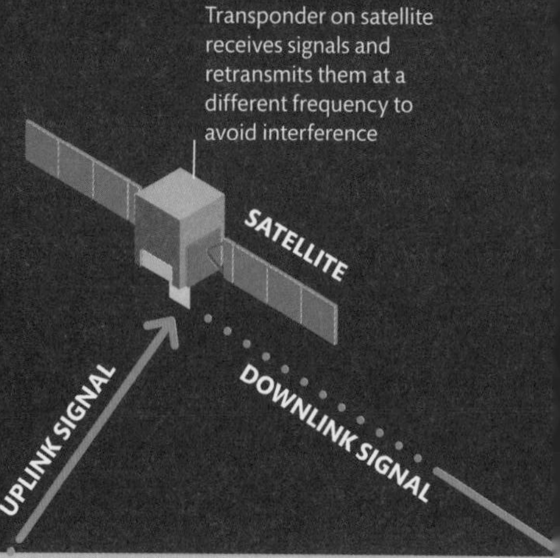
Transponder on satellite receives signals and retransmits them at a different frequency to avoid interference
SATELLITE
UPLINK SIGNAL
DOWNLINK SIGNAL

From studio to screen

Television scenes are captured with video cameras and microphones, which record visual and audio information as electrical signals. These signals, which contain instructions about exactly how television sets can reconstruct the scene, are modulated (see pp.182–183) and transmitted to viewers' homes via satellite, terrestrial, or cable broadcasting. Each television channel transmits its programs by using signals of different sets of frequencies.

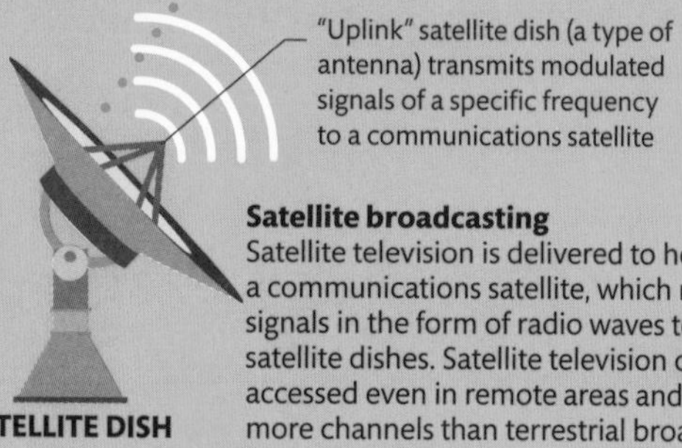
"Uplink" satellite dish (a type of antenna) transmits modulated signals of a specific frequency to a communications satellite
SATELLITE DISH

Satellite broadcasting
Satellite television is delivered to homes via a communications satellite, which relays signals in the form of radio waves to viewers' satellite dishes. Satellite television can be accessed even in remote areas and offers more channels than terrestrial broadcasting.

Converting scenes to signals
Modern cameras focus light onto a charge-coupled device, which measures and records the light across each point in a frame. This information—along with recorded sound—is converted into electrical signals ready for transmission.

TELEVISION BROADCAST

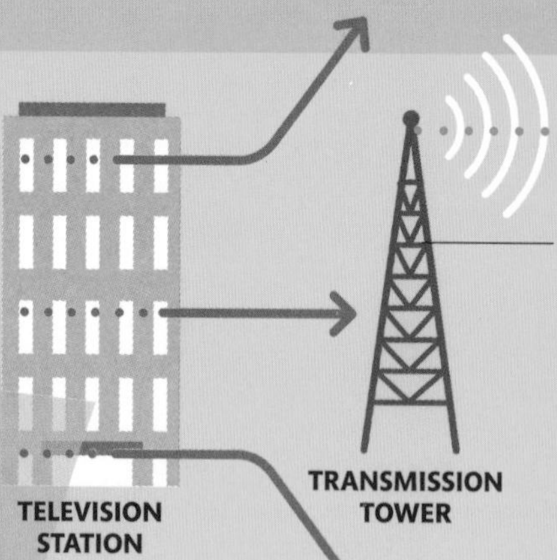
TELEVISION STATION
TRANSMISSION TOWER

Ground-based tower transmits analogue or digital signals in form of radio waves

Terrestrial broadcasting
Terrestrial broadcasting refers to signals transmitted directly from a television station to homes. Terrestrial television was the only available type of television broadcasting until the 1950s.

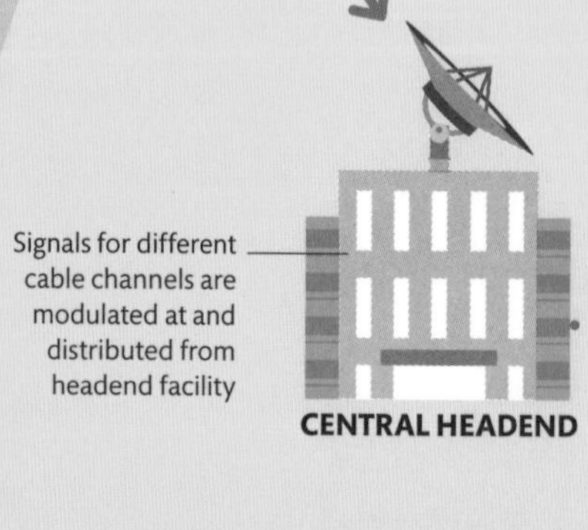
Signals for different cable channels are modulated at and distributed from headend facility
CENTRAL HEADEND

Cable broadcasting
Cable television is delivered to customers using optical signals transmitted through underground fiber-optic cables (see pp.184–185). The same cables may also be used for Internet access and telephone connections.

Analogue versus digital

Broadcasters are in the process of switching entirely from analogue to digital television, which converts data into binary code before it is reassembled back to its original form. Digital television allows for improved image quality, more efficient use of the radio spectrum, and therefore a greater choice of channels than analogue television.

Analogue signal	Digital signal
Analogue signals vary continuously in frequency, amplitude, or both	Digital signals represent a series of pulses comprising just two states: on (1) or off (0)
Video quality is degraded during copying	Video quality does not change in copying
Uncompressed video wastes bandwidth	Compression allows more channels
Aspect ratio (screen width: height) is 4:3	More cinematic aspect ratio of 16:9
Much redundant information is transmitted	Only useful information is transmitted
Interference or "noise" is seen by viewer	Interference is suppressed

Satellite dish on viewers' homes receives downlink signal
SATELLITE TELEVISION

WHEN THE SUN LINES UP BEHIND A SATELLITE, ITS **MICROWAVE RAYS** CAN DROWN OUT THE SIGNAL, CAUSING AN **OUTAGE**

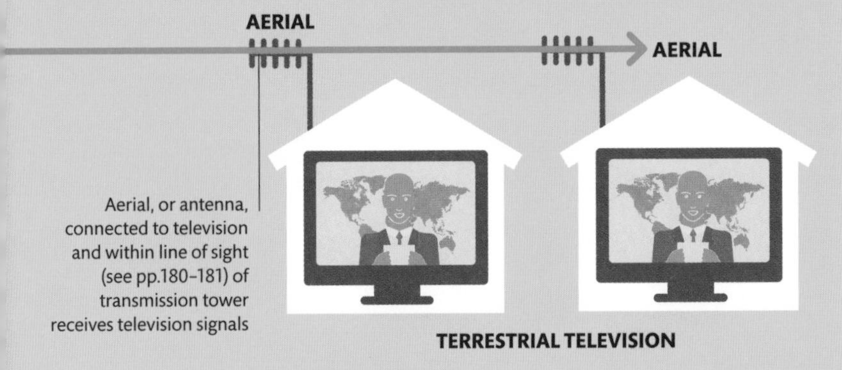
AERIAL
AERIAL
Aerial, or antenna, connected to television and within line of sight (see pp.180–181) of transmission tower receives television signals
TERRESTRIAL TELEVISION

RECORDED TELEVISION

Videocassette recorders, which became popular in the 1980s, allowed viewers to record television programs on reels of magnetic tape to be played back later. Video is now almost always stored digitally. Today, much television programming is available on demand after, during, or without scheduled broadcast, meaning that viewers can stream programs online at their convenience.

ON-DEMAND SMART TV BOX

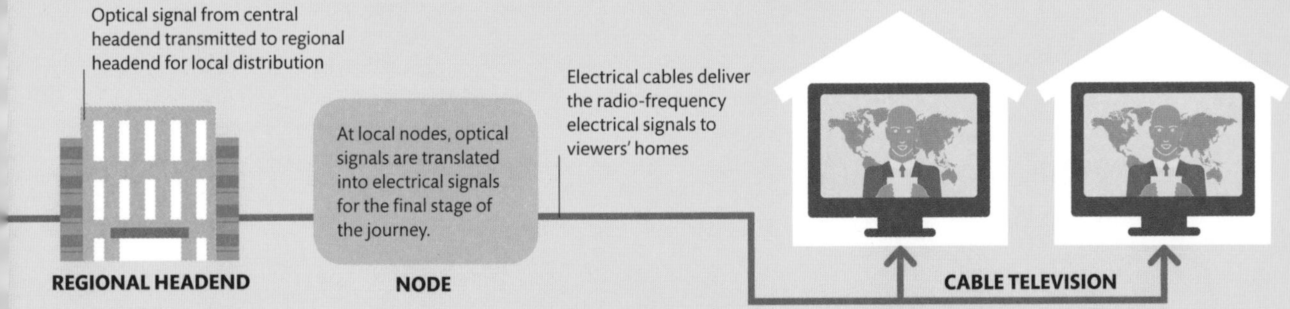
Optical signal from central headend transmitted to regional headend for local distribution
REGIONAL HEADEND
At local nodes, optical signals are translated into electrical signals for the final stage of the journey.
NODE
Electrical cables deliver the radio-frequency electrical signals to viewers' homes
CABLE TELEVISION

Televisions

A television set combines a receiver, display, and speakers to re-create video and audio transmitted by a broadcaster (see pp.188–189). Technological advances have led to slimmer televisions that produce higher-definition pictures and can connect to the Internet.

Flat screens

For decades, the only type of television was the cathode-ray tube (CRT) television, which created images using a vacuum tube to deflect beams of electrons onto a screen. These bulky devices have now been replaced by flat-screen televisions. Liquid crystal display (LCD) technology, which uses the optical properties of liquid crystal to create images, is incorporated into flat-screen televisions. In organic light-emitting diode (OLED) flat screens, a layer of organic matter produces light in response to an electric current. Each LED lights up individually so, unlike LCD screens, they do not need a backlight to supply light.

How OLED flat screens work
LEDs emit light when electrons move between a material with more electrons and a material with fewer electrons. OLEDs work similarly but are built using layers of organic (carbon-based) material.

1 Supplying electric charge
A thin-film transistor (TFT) array is placed beneath the OLED panel. Each pixel in the panel has at least three OLEDs, each of which is powered by its own transistor.

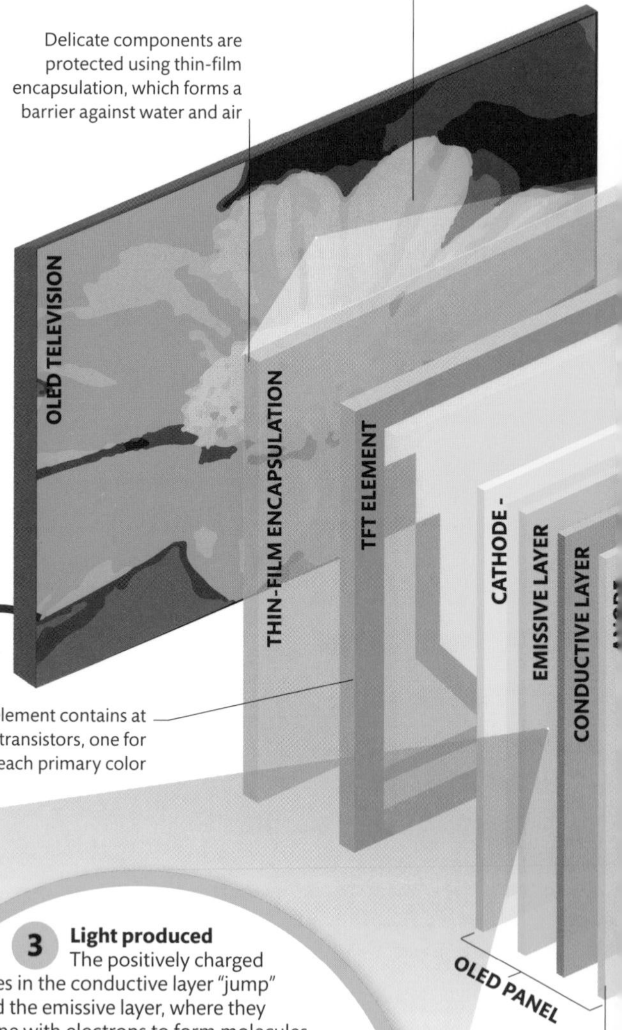
Accumulation of orange pixels in OLED screen
Delicate components are protected using thin-film encapsulation, which forms a barrier against water and air
OLED TELEVISION
THIN-FILM ENCAPSULATION
TFT ELEMENT
CATHODE -
EMISSIVE LAYER
CONDUCTIVE LAYER
Each TFT element contains at least three transistors, one for each primary color
OLED PANEL
OLED panel comprises a conductive layer and an emissive layer situated between two electrodes—the anode and cathode

2 Migrating electrons
The power source supplies electrons to the cathode and emissive layer, making the latter negatively charged. The anode and conductive layer lose electrons, which leaves behind "holes" and makes the conductive layer positively charged.

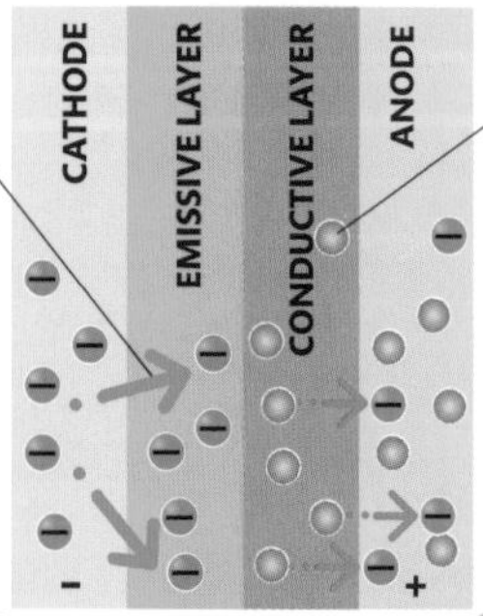
Electrons travel from cathode and enter emissive layer
CATHODE
EMISSIVE LAYER
CONDUCTIVE LAYER
ANODE
Holes created in conductive layer
-
+

3 Light produced
The positively charged holes in the conductive layer "jump" toward the emissive layer, where they recombine with electrons to form molecules. These molecules enter an "excited state," and when they relax, energy is released as light.

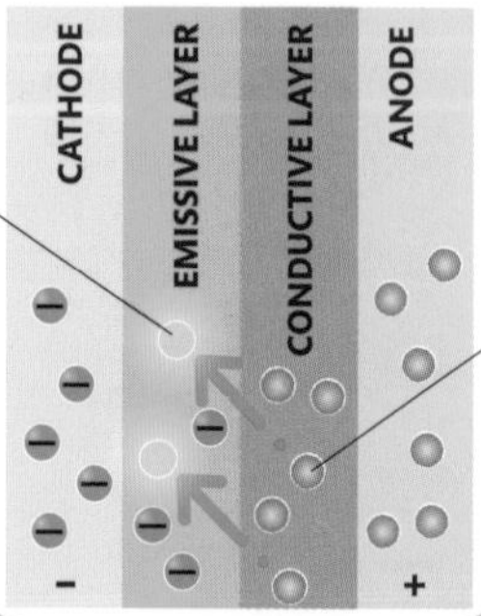
Burst of energy emitted as light
CATHODE
EMISSIVE LAYER
CONDUCTIVE LAYER
ANODE
Mobile "holes" can jump across the boundary between conductive and emissive layers
-
+

WHAT IS RESOLUTION?

Resolution describes how many pixels can be displayed on a screen. For example, high definition (HD) refers to 1,280 pixels in width and 720 in height.

SMART TV

A "smart" television functions primarily as a television but is able to connect to the Internet and other devices. In addition to playing broadcast television programs, it allows the user to watch Internet television, stream videos online, and download apps for other services. Apps can either be preloaded onto the smart TV or accessed through app stores.

Apps provide access to live TV and on-demand services
1
2
3
@
SMART TV

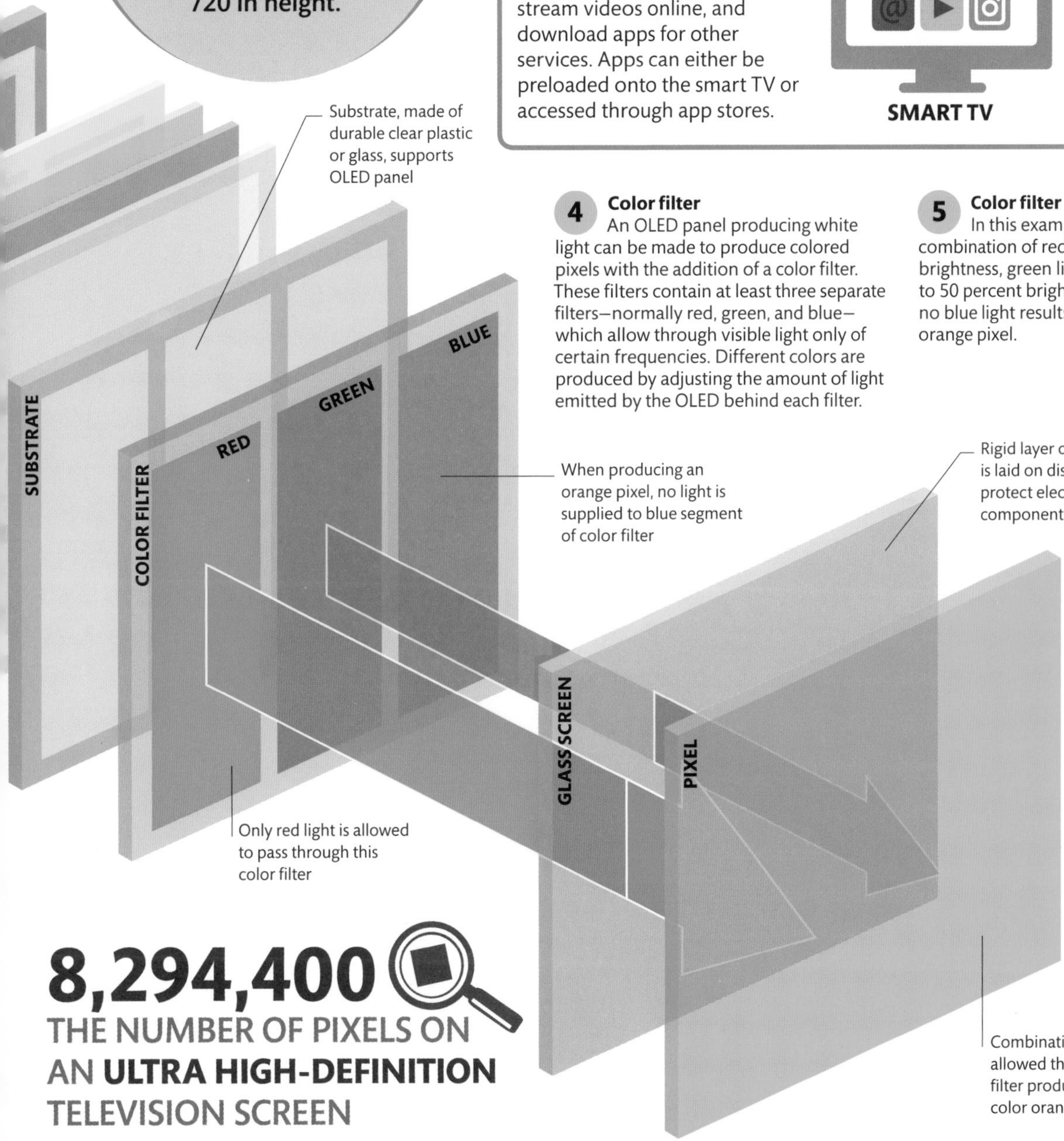
Substrate, made of durable clear plastic or glass, supports OLED panel
SUBSTRATE
COLOR FILTER
RED
GREEN
BLUE
When producing an orange pixel, no light is supplied to blue segment of color filter
Rigid layer of glass is laid on display to protect electronic components
GLASS SCREEN
PIXEL
Only red light is allowed to pass through this color filter
Combination of colors allowed through color filter produces the color orange

4 Color filter
An OLED panel producing white light can be made to produce colored pixels with the addition of a color filter. These filters contain at least three separate filters—normally red, green, and blue—which allow through visible light only of certain frequencies. Different colors are produced by adjusting the amount of light emitted by the OLED behind each filter.

5 Color filter
In this example, the combination of red light at full brightness, green light reduced to 50 percent brightness, and no blue light results in an orange pixel.

8,294,400
THE NUMBER OF PIXELS ON AN **ULTRA HIGH-DEFINITION** TELEVISION SCREEN

Uses of satellites

Although the first satellites were launched during the Cold War for the purposes of space exploration and defense, they are now engineered for a wide range of military and civilian uses. Most people make use of satellites every day without even realizing it.

Astronomical
Satellite-based telescopes are ideal for observing space because, unlike terrestrial telescopes, they are not obstructed by Earth's atmosphere.

Telephones
Satellite phones exchange signals with satellites instead of ground-based cell towers. They are often used in remote areas not covered by terrestrial signals.

Television
Many television broadcasters deliver programs via satellite. Viewers receive the signals using a satellite dish attached to the outside of their home.

Radio
Relaying radio programs via a satellite means that signals can be transmitted across entire countries.

Internet
Satellite Internet provides coverage in remote areas, but service can be slow due to the distance signals have to travel.

GPS navigation
Navigation devices can determine their position on Earth by exchanging information with satellites (see pp.194–195).

Weather
Some satellites are designed to monitor features of Earth's weather and climate. They transmit the data back to Earth for analysis.

Defense
Military satellites have a range of uses, including surveillance, navigation, and for sending encrypted communications.

Satellites

Artificial satellites are specialized man-made spacecraft launched into orbit around Earth and other planets in the solar system. They are vital in telecommunications because they can receive signals from the ground and amplify and retransmit them to distant parts of Earth.

Communications satellites

Communications satellites are designed to send and receive radio signals carrying audio, video, and other kinds of data. Relaying signals via satellite allows for rapid communication across huge distances. Signals of set frequencies are sent into space from ground stations and picked up by a satellite's antenna. A transponder processes the information and boosts the signal before relaying it down to other ground stations on Earth.

SPUTNIK 1 WAS THE FIRST SATELLITE IN SPACE, LAUNCHED BY THE SOVIET UNION ON OCTOBER 4, 1957

Anatomy of a communications satellite
Communications satellites feature extremely sophisticated equipment designed to cope for extended periods of time in the extreme conditions of space, where maintenance or repair is practically impossible.

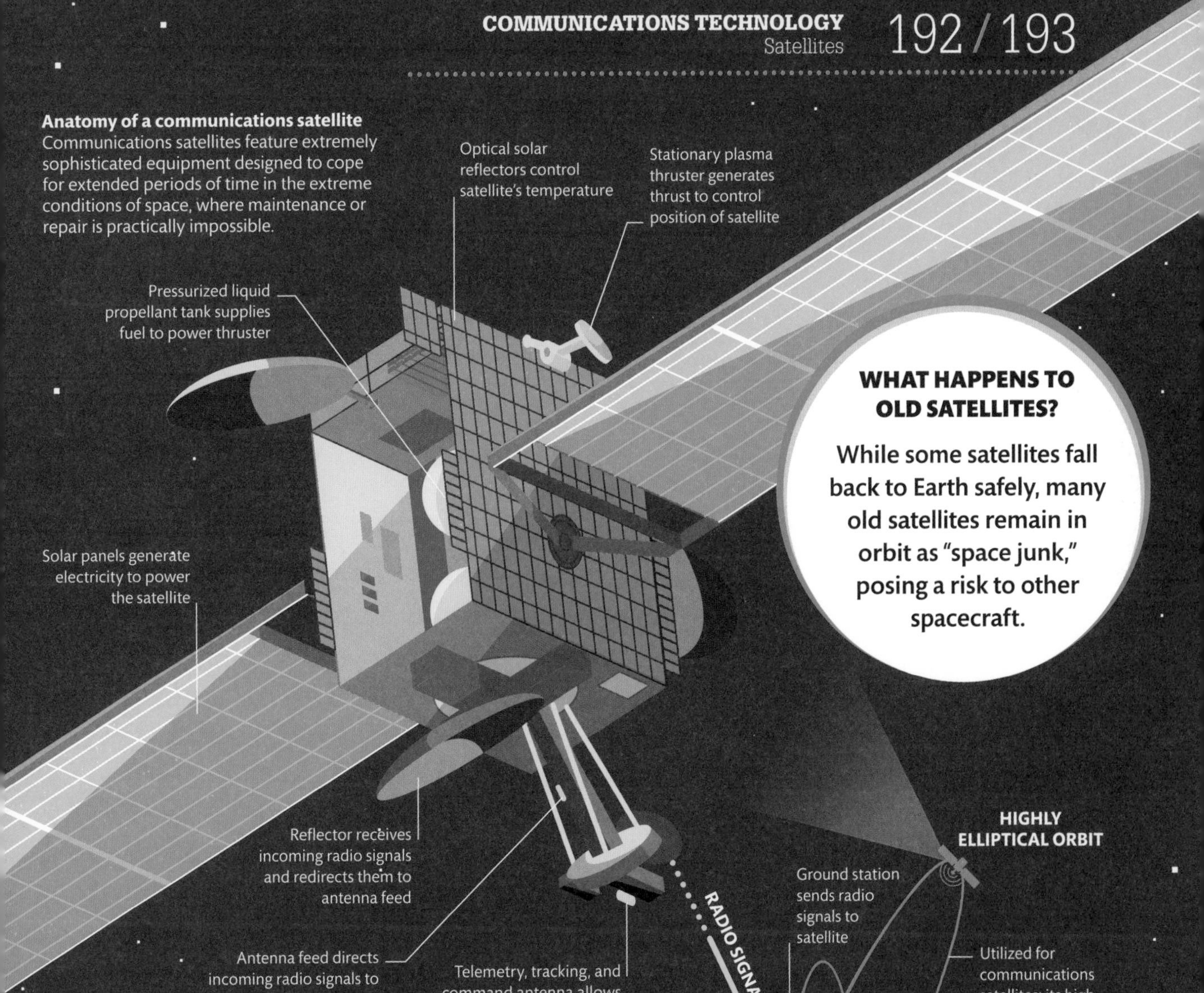
Optical solar reflectors control satellite's temperature
Stationary plasma thruster generates thrust to control position of satellite
Pressurized liquid propellant tank supplies fuel to power thruster
Solar panels generate electricity to power the satellite
Reflector receives incoming radio signals and redirects them to antenna feed
Antenna feed directs incoming radio signals to transponder for processing and sends outgoing signals back to Earth via reflector
Telemetry, tracking, and command antenna allows ground station to monitor and control satellite operations
RADIO SIGNAL

WHAT HAPPENS TO OLD SATELLITES?

While some satellites fall back to Earth safely, many old satellites remain in orbit as "space junk," posing a risk to other spacecraft.

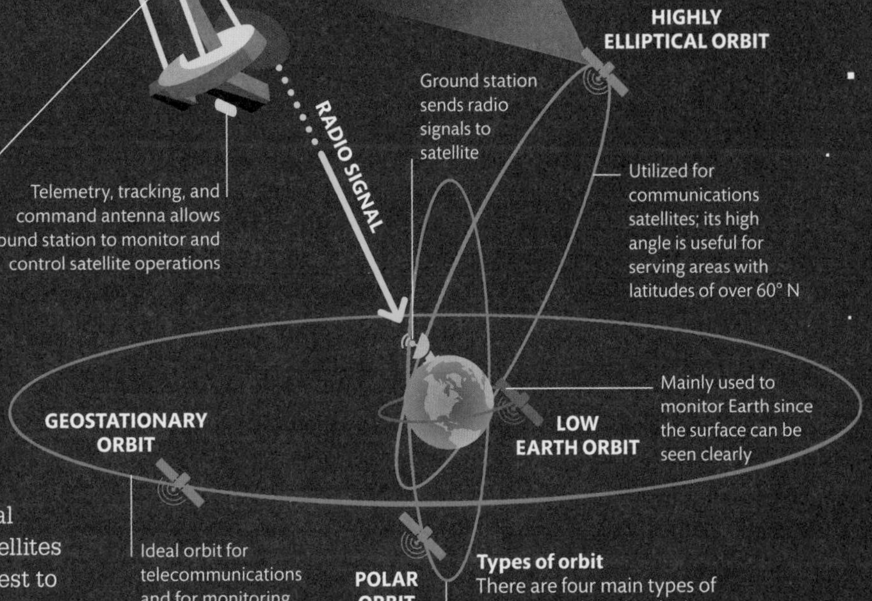
HIGHLY ELLIPTICAL ORBIT
Ground station sends radio signals to satellite
Utilized for communications satellites; its high angle is useful for serving areas with latitudes of over 60° N
GEOSTATIONARY ORBIT
Ideal orbit for telecommunications and for monitoring weather patterns
LOW EARTH ORBIT
Mainly used to monitor Earth since the surface can be seen clearly
POLAR ORBIT
Principally used to observe Earth

Satellite orbits

Satellites enter orbit if they launch at high enough velocity to overcome Earth's gravitational pull at the surface and then balance its weaker gravitational pull in space. Many communications satellites are in geostationary orbit. They travel west to east at the same rate as Earth spins, so they appear stationary above a point on the equator. Some satellites have polar orbits, crossing both poles on their journey around Earth.

Types of orbit
There are four main types of orbit around Earth, which are characterized by their shape, angle, and altitude. Most satellites are in low earth orbit, less than 1,250 miles (2,000 km) above the surface.

Satellite navigation

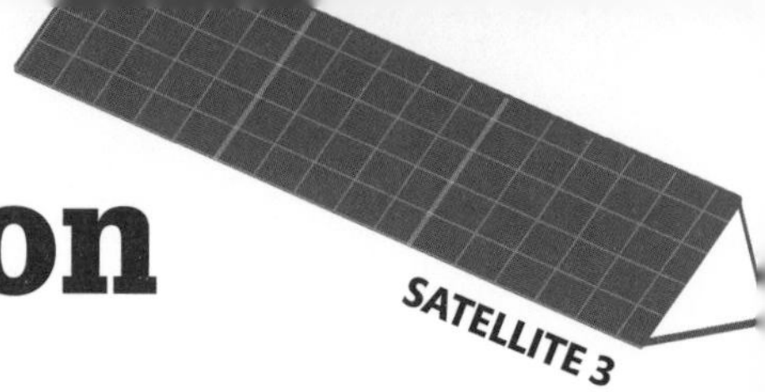

Satellite navigation systems, such as the Global Positioning System (GPS), can provide precise information about location. They rely on networks of satellites in orbit around Earth, which use radio signals to communicate with smartphones and other navigation devices.

Satellite navigation

Satellite navigation systems determine location using a number of small, orbiting satellites that are "visible" from anywhere in the world. Terrestrial radio stations, known as ground stations, track the paths of satellites. The satellites transmit radio signals containing time and position data back to Earth. A receiver picks up these signals and calculates the precise time taken for each signal to reach it. It can then work out its distance from the satellites and estimate its position.

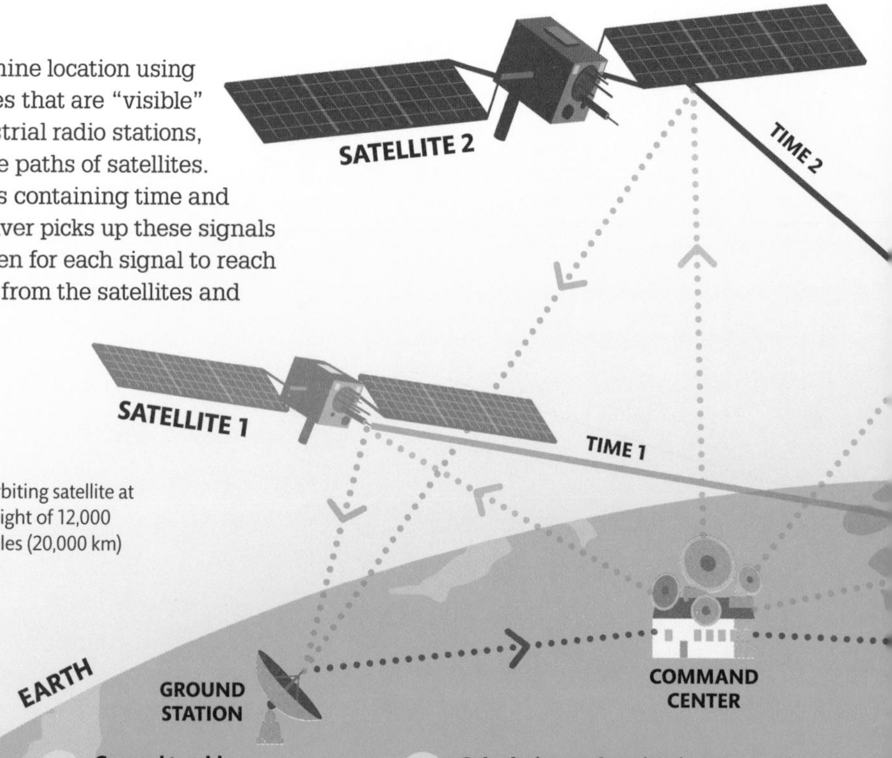

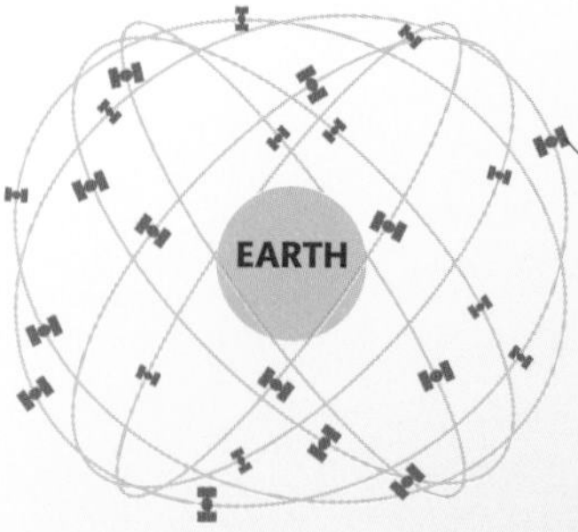

GPS constellation
GPS satellites circle Earth twice a day. To ensure at least four satellites are detectable anywhere on Earth, they are arranged in six orbital planes of equal size, each containing four satellites.

1 Ground tracking
Ground stations follow the satellites as they cross the sky above, collecting data and passing observations to the command center.

2 Calculating and navigating
The command center processes signals from the entire satellite network. It calculates the exact positions of all the satellites and sends them navigation instructions.

How trilateration works

Calculating the distance from one satellite places the receiver somewhere within a sphere around it. Finding the distance from other satellites narrows its possible position down to the area where the spheres intersect. This process is called trilateration.

Satellite 1
Calculating the distance from a single satellite places the receiver within the ground area intersected by a very large sphere.

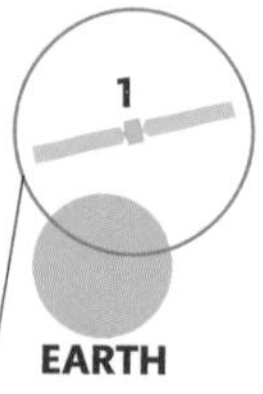

Satellite 2
Finding its distance from a second satellite reduces the possible area in which the receiver may be positioned to two points on an intersecting line.

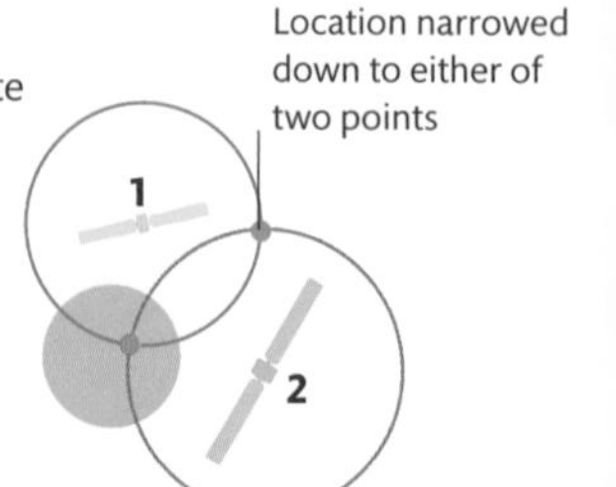

Time delays

According to Einstein's theory of special relativity, time is experienced differently by observers traveling at different velocities. Because time seems to pass more slowly for the fast-moving satellites, their clocks are programmed to run at a slightly different pace to clocks on the ground. Without this correction, satellite navigation would be inaccurate by several feet.

3 Location and time updates
The satellites repeatedly transmit radio signals to Earth with precise information about their location and the time of transmission. Signals from the satellites can be picked up by a GPS receiver.

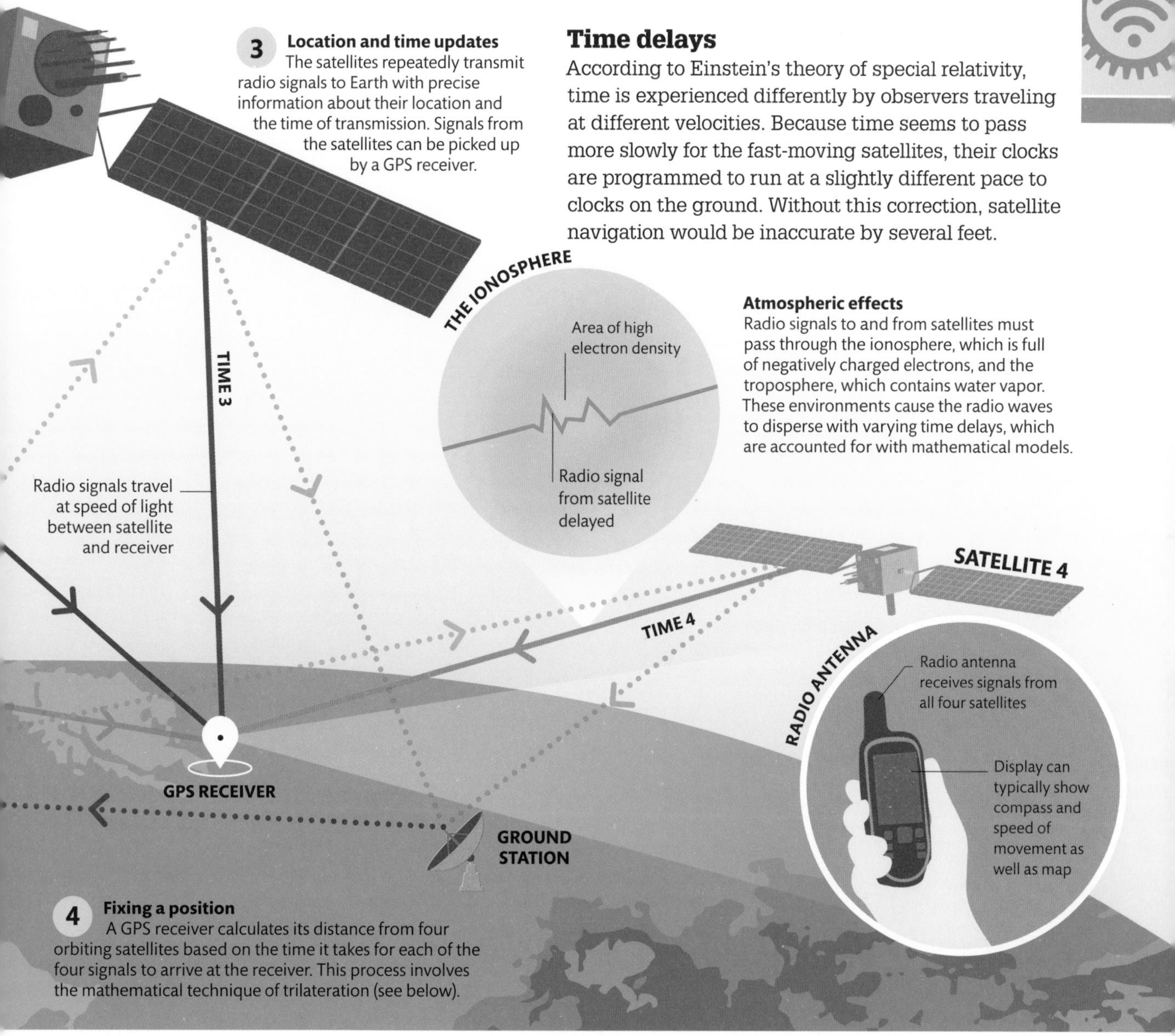
TIME 3
Radio signals travel at speed of light between satellite and receiver
THE IONOSPHERE
Area of high electron density
Radio signal from satellite delayed
SATELLITE 4
TIME 4
RADIO ANTENNA
Radio antenna receives signals from all four satellites
Display can typically show compass and speed of movement as well as map
GPS RECEIVER
GROUND STATION

Atmospheric effects
Radio signals to and from satellites must pass through the ionosphere, which is full of negatively charged electrons, and the troposphere, which contains water vapor. These environments cause the radio waves to disperse with varying time delays, which are accounted for with mathematical models.

4 Fixing a position
A GPS receiver calculates its distance from four orbiting satellites based on the time it takes for each of the four signals to arrive at the receiver. This process involves the mathematical technique of trilateration (see below).

Satellite 3
As it calculates its distance from a third visible satellite, the receiver narrows down its position to one possible location.

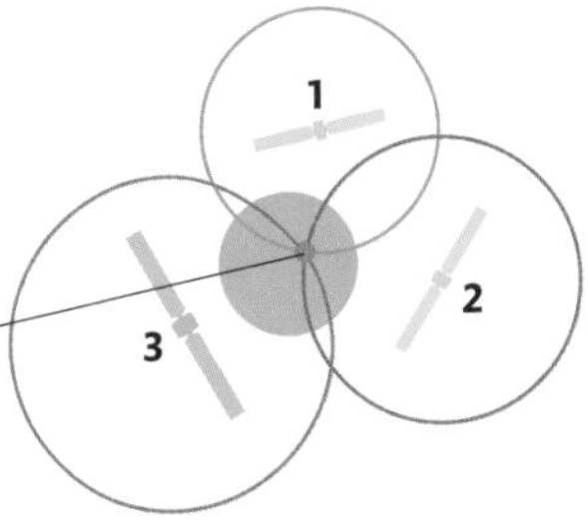
1
2
3
Receiver location can now be only a single point

Satellite 4
This satellite is used to correct the inaccurate position indicated by the receiver, because the clock built in to the receiver is not perfectly synchronized to the satellite clocks (see above).

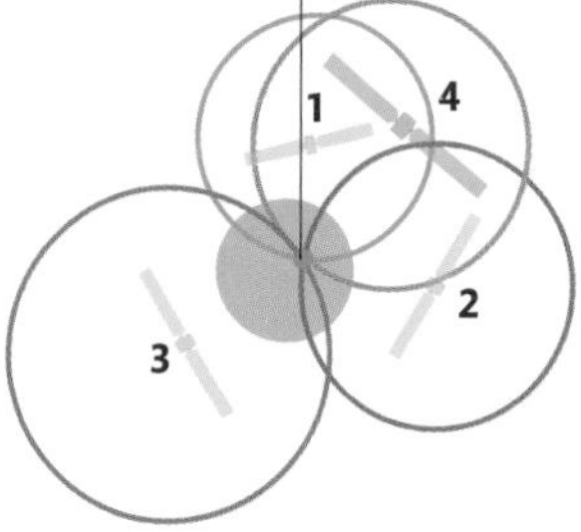
Position confirmed to within 3 ft (1 m)
1
4
2
3

The Internet

The Internet is a global network of connected computers that exchange data using a common set of rules. It supports important applications such as email and the World Wide Web.

A network of computers

Users can access the Internet from an end point such as a smartphone or a computer. These devices are normally connected to the Internet through an Internet service provider (ISP), which adds them to its network and also assigns a unique reference (called an IP address) to each device. These networks are in turn connected to other networks to form larger networks. The Internet is the collection of all these interlinked networks of computers, meaning that any computer on the Internet can connect to any other. When computers exchange data, layers of software manage the process that divides data into packets, which travel through routes of wires, fiber-optic cables, and wireless connections to reach their end destination.

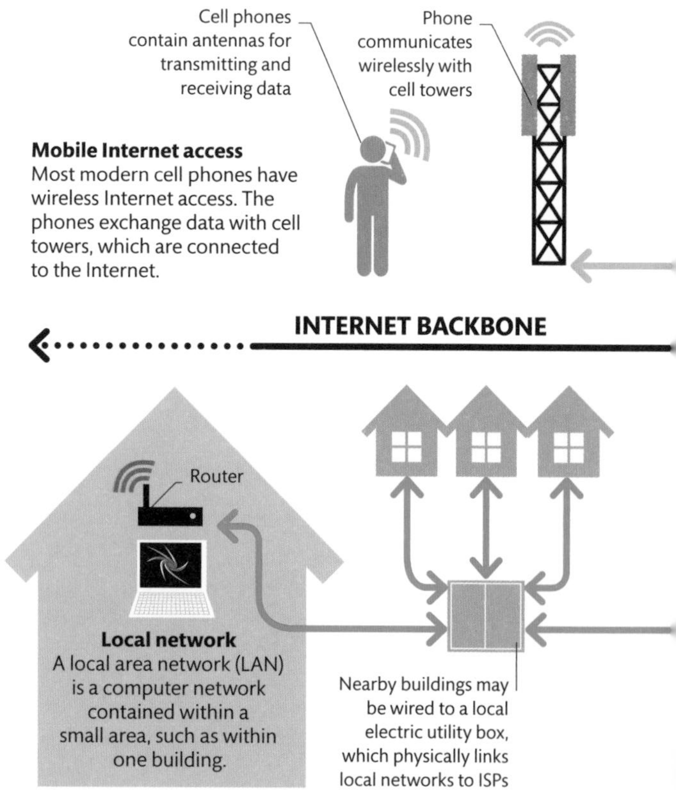
Cell phones contain antennas for transmitting and receiving data
Phone communicates wirelessly with cell towers
Mobile Internet access
Most modern cell phones have wireless Internet access. The phones exchange data with cell towers, which are connected to the Internet.
INTERNET BACKBONE
Router
Local network
A local area network (LAN) is a computer network contained within a small area, such as within one building.
Nearby buildings may be wired to a local electric utility box, which physically links local networks to ISPs

Routes for data

Older telecommunications networks rely on circuit switching to send and receive data, meaning that direct, wired connections are formed between end points during the exchange. Nowadays, packet switching is the primary method for exchanging data online. Software splits data up into sections called packets, which are labeled with their destination IP address and instructions for reassembly. These packets are directed to their end point via different routes and then reassembled at their destination. Packet switching allows for communication channels to be used far more efficiently, because different sets of data can travel through them at the same time.

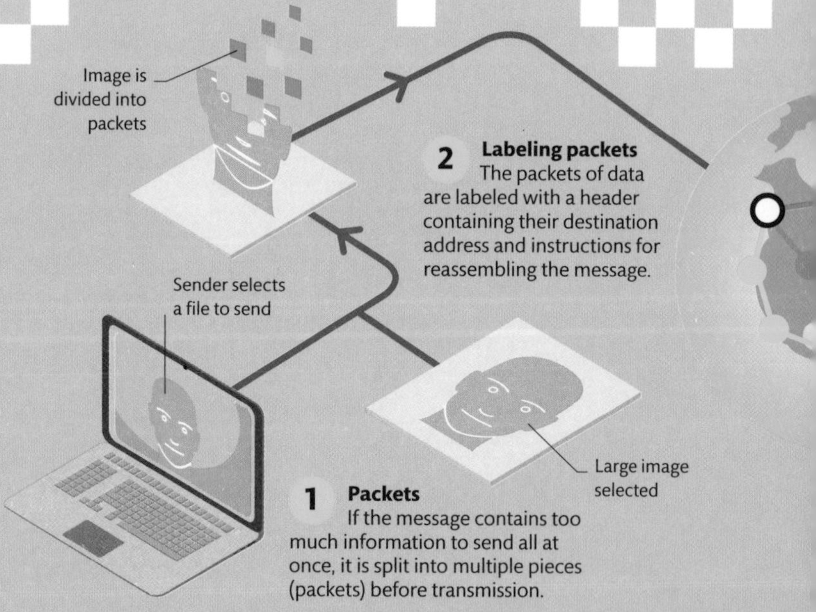
Image is divided into packets
2 Labeling packets
The packets of data are labeled with a header containing their destination address and instructions for reassembling the message.
Sender selects a file to send
Large image selected
1 Packets
If the message contains too much information to send all at once, it is split into multiple pieces (packets) before transmission.

56 PERCENT OF ALL **INTERNET TRAFFIC** IS FROM AN **AUTOMATED SOURCE**, SUCH AS HACKING TOOLS, SCRAPERS AND SPAMMERS, IMPERSONATORS, AND BOTS

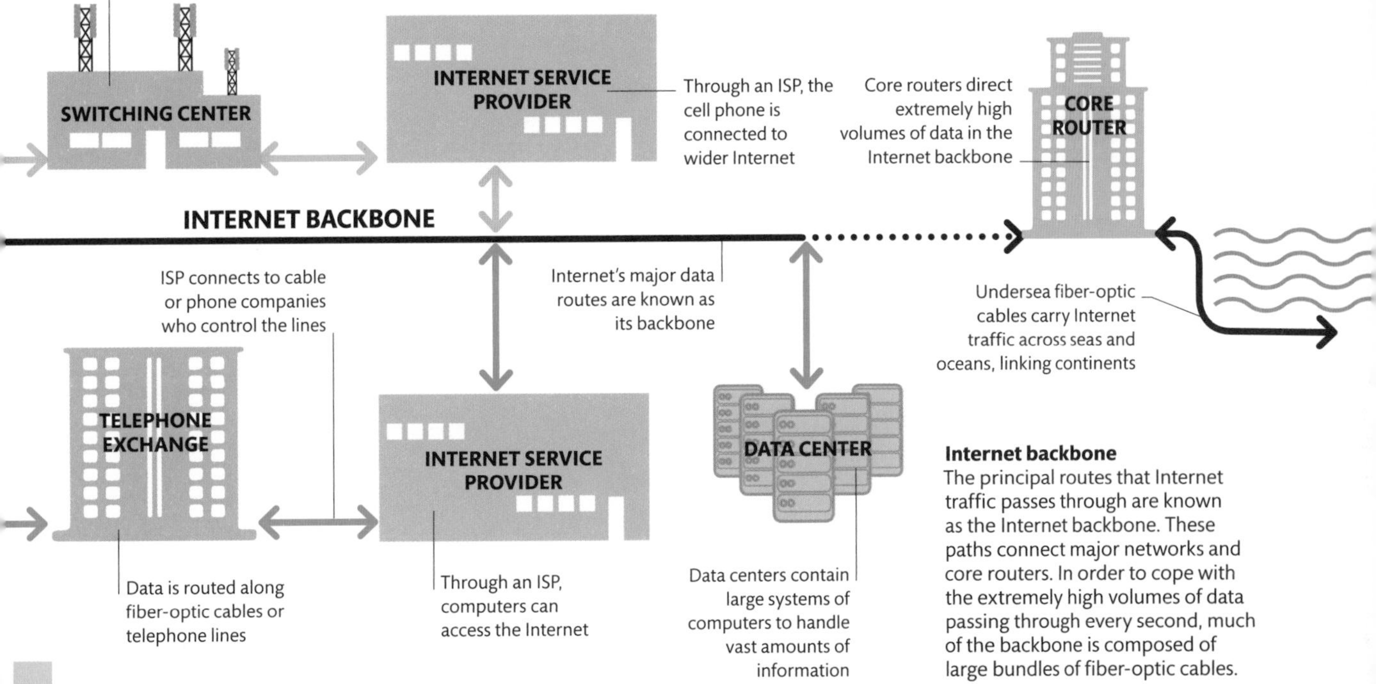
Data passes from towers to mobile switching center
SWITCHING CENTER
INTERNET SERVICE PROVIDER
Through an ISP, the cell phone is connected to wider Internet
Core routers direct extremely high volumes of data in the Internet backbone
CORE ROUTER
INTERNET BACKBONE
ISP connects to cable or phone companies who control the lines
Internet's major data routes are known as its backbone
Undersea fiber-optic cables carry Internet traffic across seas and oceans, linking continents
TELEPHONE EXCHANGE
INTERNET SERVICE PROVIDER
DATA CENTER
Data is routed along fiber-optic cables or telephone lines
Through an ISP, computers can access the Internet
Data centers contain large systems of computers to handle vast amounts of information

Internet backbone
The principal routes that Internet traffic passes through are known as the Internet backbone. These paths connect major networks and core routers. In order to cope with the extremely high volumes of data passing through every second, much of the backbone is composed of large bundles of fiber-optic cables.

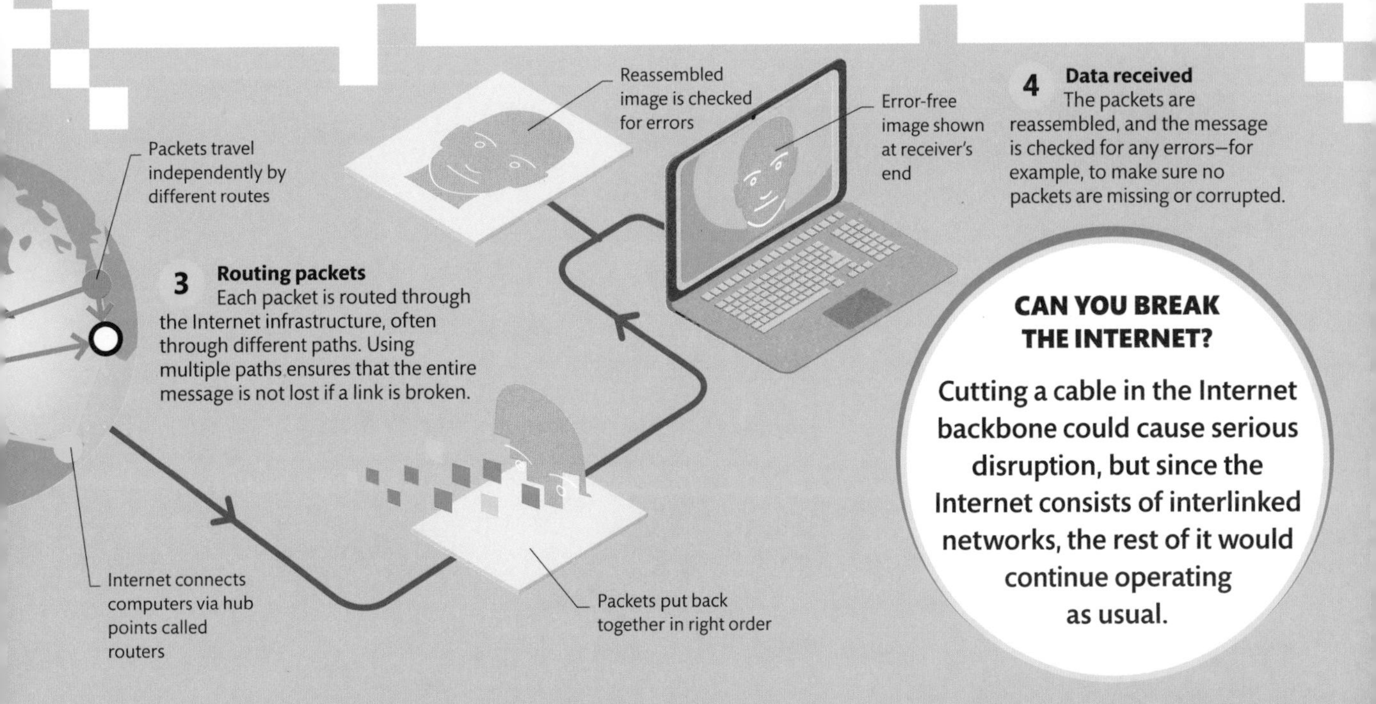
Reassembled image is checked for errors
Error-free image shown at receiver's end
Packets travel independently by different routes
Internet connects computers via hub points called routers
Packets put back together in right order

3 Routing packets
Each packet is routed through the Internet infrastructure, often through different paths. Using multiple paths ensures that the entire message is not lost if a link is broken.

4 Data received
The packets are reassembled, and the message is checked for any errors—for example, to make sure no packets are missing or corrupted.

CAN YOU BREAK THE INTERNET?

Cutting a cable in the Internet backbone could cause serious disruption, but since the Internet consists of interlinked networks, the rest of it would continue operating as usual.

The World Wide Web

The World Wide Web is a network of information that is accessed via the Internet (see pp.196–197). It consists of interlinked web pages formatted in a common language and identified by unique addresses.

How the World Wide Web works

The Web is a vast network of multimedia pages, navigated and downloaded using a program called a browser. Web pages are interlinked. A collection of linked and related web pages with a common domain name comprises a website. Each web page is identified by a unique uniform resource locator (URL), which specifies its location. Browsers retrieve these pages from servers as documents formatted using hypertext markup language (HTML) and render them as readable multimedia pages. The hypertext transfer protocol (HTTP) sets out the procedures for communication between browsers and servers on the World Wide Web.

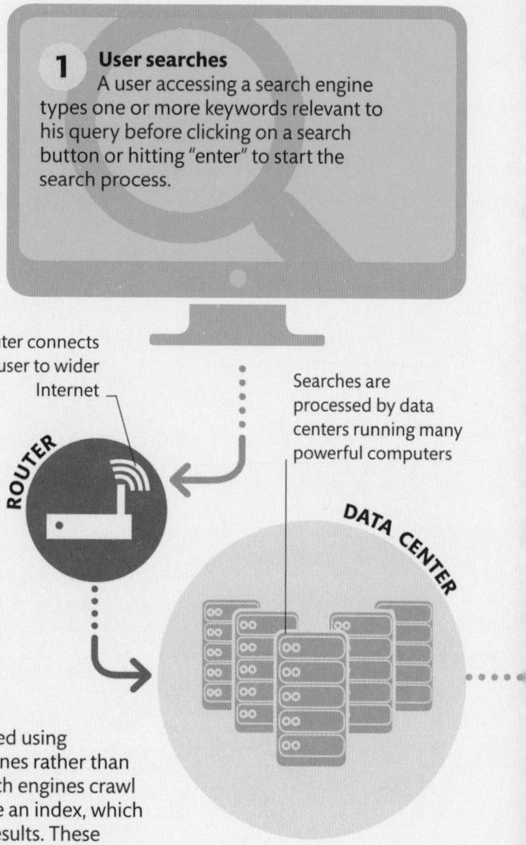
1 User searches
A user accessing a search engine types one or more keywords relevant to his query before clicking on a search button or hitting "enter" to start the search process.
Router connects user to wider Internet
ROUTER
Searches are processed by data centers running many powerful computers
DATA CENTER

2 Request
The search terms are sent to the wider Internet via a router. They are directed to the search engine's servers.

3 Search index
A computer scans the search engine's index for the most relevant and reputable pages containing those search terms.

Searching the Web
Web pages are often accessed using programs called search engines rather than entering URLs directly. Search engines crawl through web pages to create an index, which is used to generate search results. These results are presented as a list of relevant links.

HTML

HTML is a language used to design web pages. Browsers receive HTML documents from web servers and render them into readable web pages containing text and other media. Codes called HTML tags are used to add and structure content in the page; for example, <img> introduces an image, while <a> inserts hyperlinks to web pages, files, or email addresses.

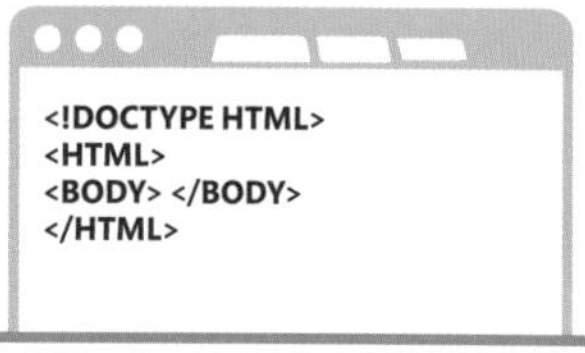
<!DOCTYPE HTML>
<HTML>
<BODY> </BODY>
</HTML>

Internet protocols

Hypertext transfer protocol (HTTP) is the universal set of rules underlying use of the World Wide Web. HTTP forms the basis for processing web documents and how servers, browsers, and other agents respond to commands. When a user enters a URL to access a web page, his browser looks for the web server's Internet address using the domain name system (DNS). It then sends a request to a web server, which issues a response with a status code containing information, such as whether the URL is valid so that the page can be loaded. A sequence of requests and responses is called an HTTP session.

HTTPS
Hypertext transfer protocol secure (HTTPS) uses the encryption protocol transport layer security (TLS). This gives the user privacy and security when browsing online.

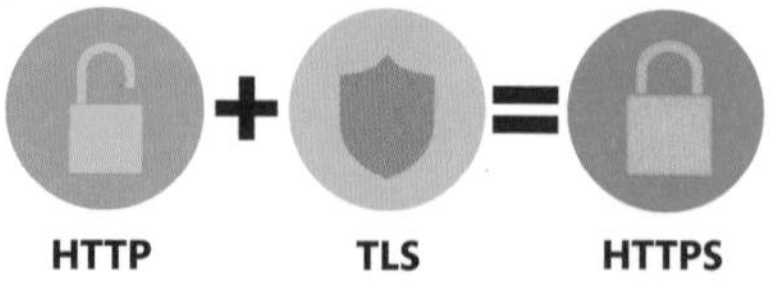
HTTP
TLS
HTTPS

4 Clicking on a link
The search engine compiles a web page listing the top results of the user's search. This list is returned to the user's computer and displayed by his browser. The user looks at snippets of text sampled from listed web pages to select a URL.

6 Viewing the page
The user's web browser receives the HTML document and uses this to render the web page to display text, images, and other media in a format that is useful to the user.

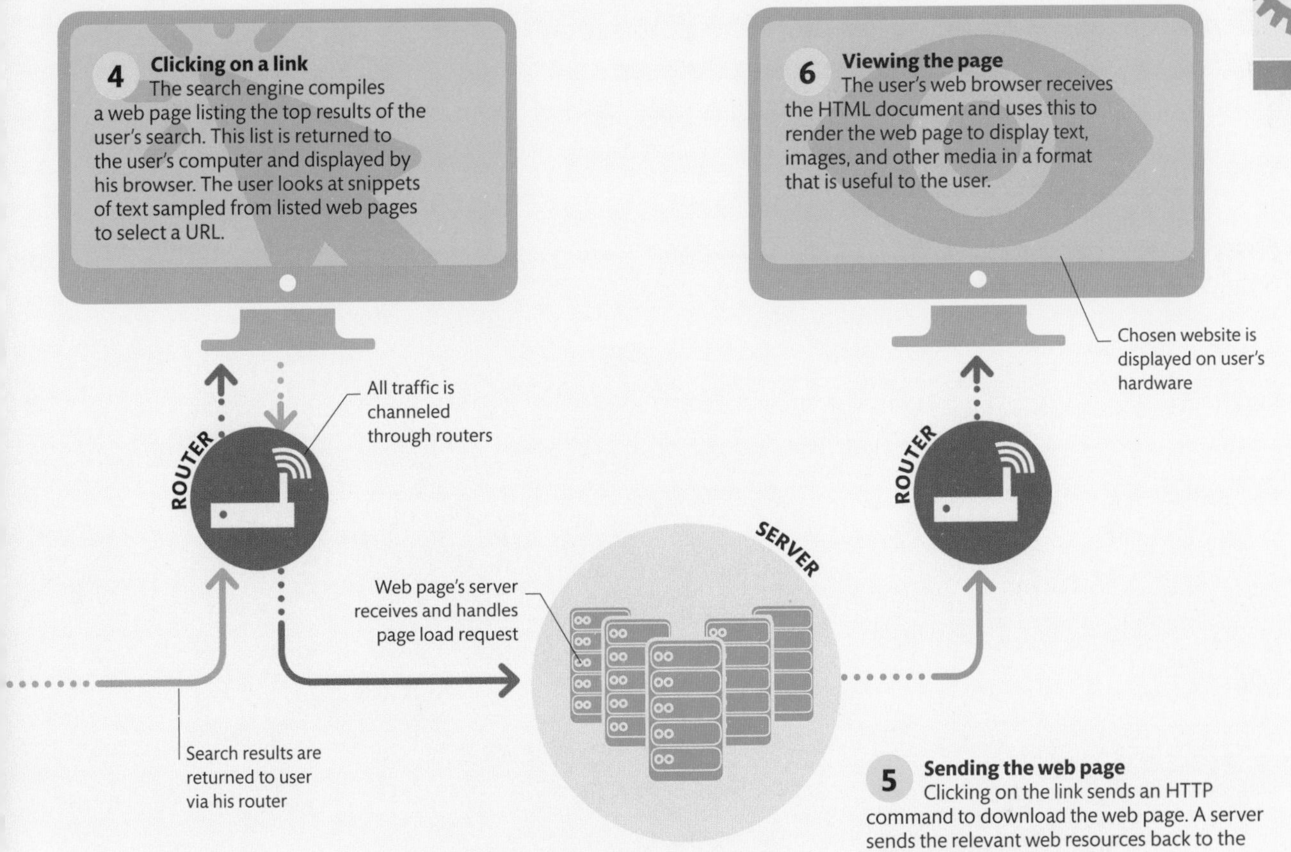
Chosen website is displayed on user's hardware
All traffic is channeled through routers
ROUTER
ROUTER
SERVER
Web page's server receives and handles page load request
Search results are returned to user via his router

5 Sending the web page
Clicking on the link sends an HTTP command to download the web page. A server sends the relevant web resources back to the user's computer via the Internet.

HTTP STATUS CODES		
Code	**Meaning**	**Description**
200	OK	Standard response for successful requests
201	Created	Request fulfilled and new resource created
301	Moved permanently	Resource permanently moved to a different URL
400	Bad request	Syntax of the request not understood by server
404	File not found	Document or file requested by client was not found
500	Internal server error	Request unsuccessful because of unexpected condition encountered by server
503	Service unavailable	Request unsuccessful due to server being down or overloaded
504	Gateway timeout	Upstream server failed to send a request in the time allowed

75 PERCENT OF PEOPLE NEVER SCROLL PAST THE **FIRST PAGE** OF SEARCH RESULTS

WHAT WAS THE FIRST WEBSITE?

The first website was created by Sir Tim Berners-Lee in 1991 for the European Organization for Nuclear Research (CERN).

Email

Electronic mail (email) is a method for exchanging messages using computers and other devices. Connecting to an email server allows a user to send and receive messages as well as other files in the form of attachments.

WHO SENT THE FIRST EMAIL?

Ray Tomlinson is credited with sending the first email, in 1971. While working for ARPANET, he developed a way for messages to be sent between computers.

How emails are sent

Emails are exchanged according to a set of rules, the Simple Mail Transfer Protocol (SMTP), which allows for communication across different devices and servers. When a user sends an email, the message is uploaded to an SMTP server, which communicates with the domain name system (DNS) to check the recipient's server address before delivery. An Internet domain is a group of addresses under the control of an individual or organization.

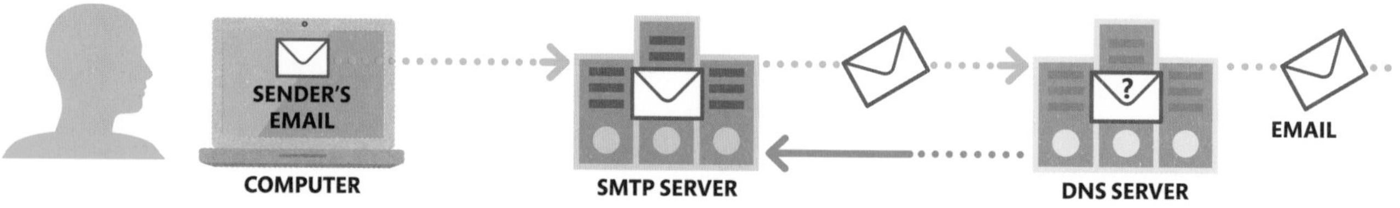

1 Email sent
The sender composes a message using a mail client: an application for composing, sending, and reading emails. He or she also enters the recipient's email address. Once the user presses send, the delivery process begins.

2 SMTP server
The message is sent to an SMTP server—the online equivalent of a post office. On this server, a mail transfer agent (MTA) checks the recipient's address and then looks up its domain.

3 DNS server
The MTA must communicate with a DNS, which translates domain names into IP addresses. The recipient's domain is checked to find his or her mail server. If it cannot be found, an error message is returned.

Spam and malware

Sending emails is cheap, so email is often used to send content to many users at once. Some unsolicited emails (called spam) are merely annoying, while others spread deliberately destructive software (malware). Once malware has been downloaded, it may disable, hijack, or alter computer functions, monitor activity, demand payments, encrypt or delete data, or spread to other computers. Email filters scan incoming emails for content indicating spam and malware.

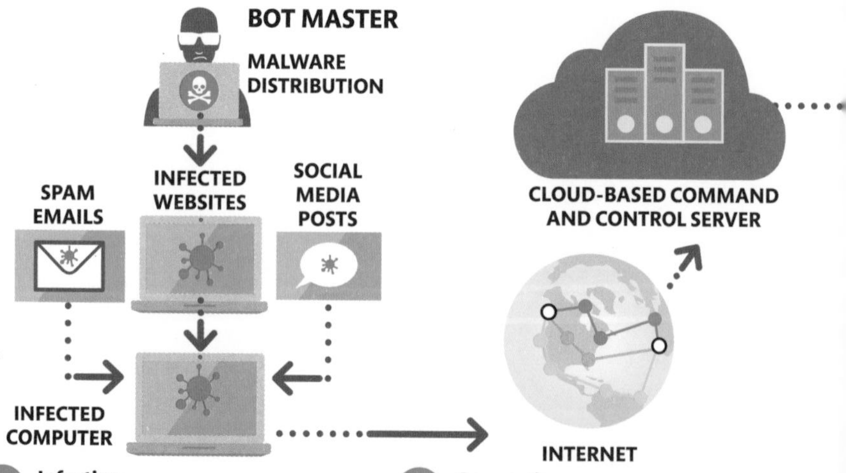

How a botnet works
Hackers who wants to anonymously carry out malicious activities online may breach the security of connected devices to create a network of devices they control: a botnet.

1 Infection
Hackers use malware that contains bots: applications that perform automated tasks. The malware is distributed, and if it is downloaded, it can infect the user's computer.

2 Connection
The bots discreetly instruct the infected computer to connect to a command and control (C&C) server. The hackers use this server to monitor and control the botnet.

EMAIL RETRIEVAL PROTOCOLS

Emails are sent between computers using SMTP, but to receive emails, the recipient uses an email client that follows either Post Office Protocol (POP) or Internet Message Access Protocol (IMAP). These two sets of rules handle received emails in different ways.

IMAP

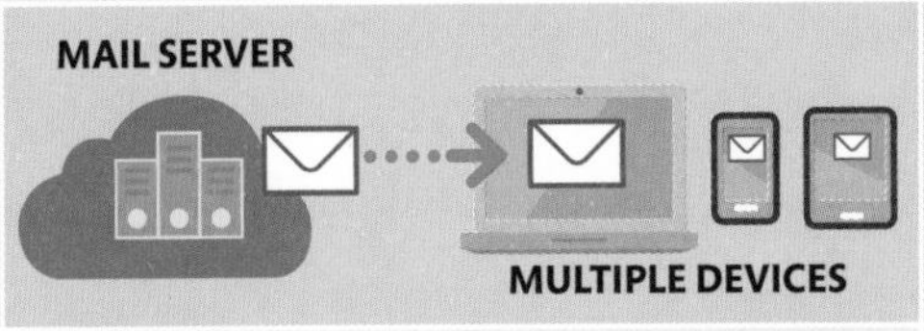

- The mail client syncs with server.
- Emails can be accessed and synced across many devices.
- Emails and attachments are not automatically downloaded to a device.
- Original sent and received messages are stored on a server.

POP3

- The mail client and server are not synced.
- Emails can be accessed only from a single device.
- Emails are automatically downloaded to the device then deleted from the server.
- Sent and received messages are stored on the device.

4 Email sent to delivery agent
If the recipient's mail server is found, the message is transferred to his or her mail delivery agent (MDA), using a transmission process described by the SMTP. The message may pass through several MTAs first.

5 Delivery agent passes on email
The MDA performs the final transfer in the process, taking the message from an MTA and sending it to the recipient's local device. It then files it in the user's correct email inbox.

6 Email received
The recipient opens his or her inbox and reads the new email. The way the email is accessed depends on the protocol (see above) adopted by the user's mail client.

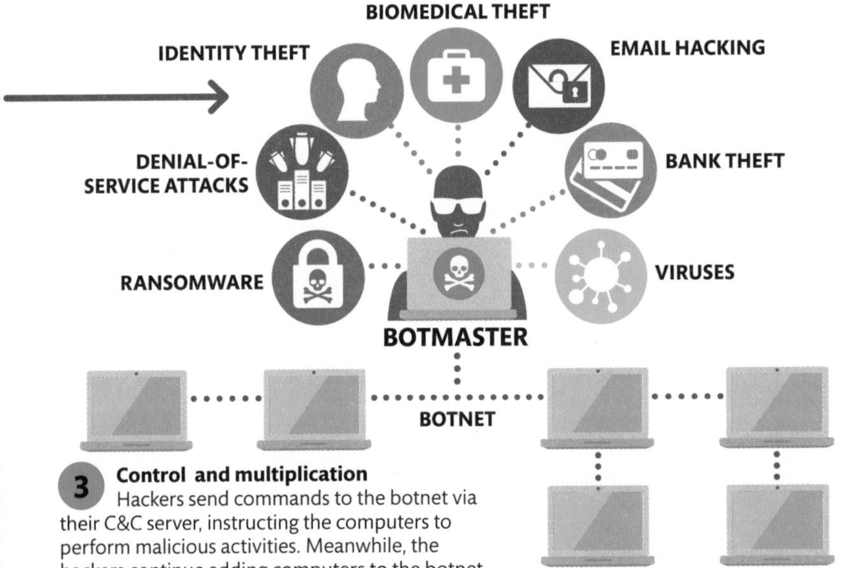

3 Control and multiplication
Hackers send commands to the botnet via their C&C server, instructing the computers to perform malicious activities. Meanwhile, the hackers continue adding computers to the botnet.

EMAIL ENCRYPTION

Email encryption prevents emails from being read by anyone other than the intended recipient by using public key encryption. The contents of an encrypted email can be decrypted only by using the correct mathematical key. At its simplest, the sender uses the recipient's public key to encrypt the message, which only the recipient can decrypt with his or her private (secret) key.

Wi-Fi

Wi-Fi uses radio waves to allow nearby devices, including cell phones, tablets, computers, printers, digital speakers, and smart TVs, to form connections and exchange data wirelessly. It is the most popular mode of mobile communication.

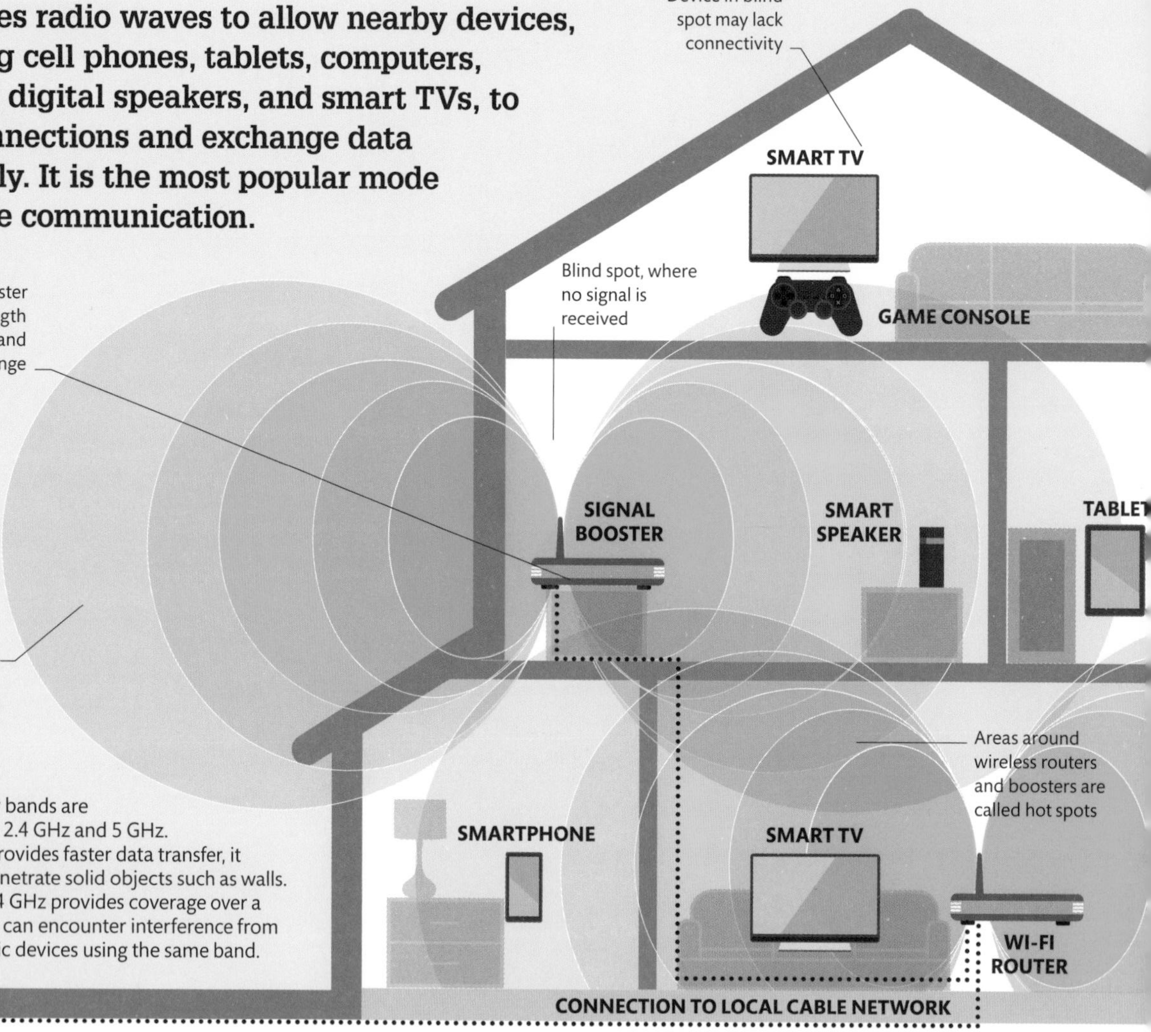

Device in blind spot may lack connectivity
SMART TV
GAME CONSOLE
Signal booster increases strength of signal and its range
Blind spot, where no signal is received
SIGNAL BOOSTER
SMART SPEAKER
TABLET
Signal strength reduces with distance from booster
Areas around wireless routers and boosters are called hot spots
SMARTPHONE
SMART TV
WI-FI ROUTER
CONNECTION TO LOCAL CABLE NETWORK

Wi-Fi bands
Two frequency bands are used for Wi-Fi: 2.4 GHz and 5 GHz. While 5 GHz provides faster data transfer, it struggles to penetrate solid objects such as walls. By contrast, 2.4 GHz provides coverage over a wider area but can encounter interference from other electronic devices using the same band.

How Wi-Fi works

Connecting a device to the Internet using Wi-Fi requires a built-in wireless adaptor—such as an antenna on a cell phone—that transforms digital data into radio signals. When a user sends some form of media, such as a text message or a photo, the adaptor encodes its digital form into radio signals and transmits it to a router. The router then translates the radio signals back into digital data, which it passes to the Internet through a wired connection. This process works the same way in reverse, allowing the wireless exchange of data between devices and the Internet.

Wi-Fi router
A router transfers data between connected devices and the Internet. It is wired to the Internet via its wide area network (WAN) port and connects to devices in the local area network (LAN) via its LAN ports, or wirelessly.

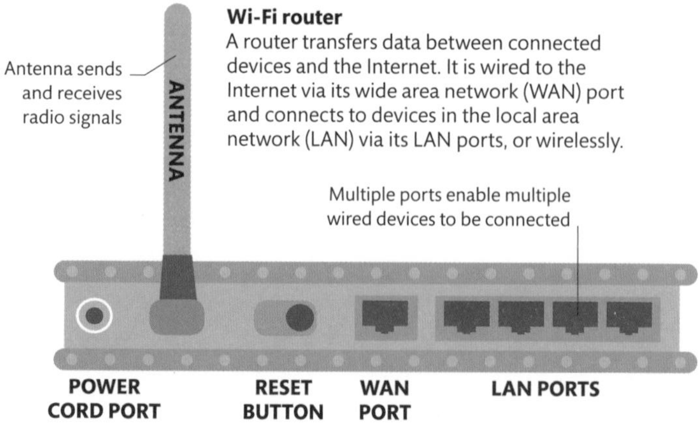

Antenna sends and receives radio signals
ANTENNA
Multiple ports enable multiple wired devices to be connected
POWER CORD PORT
RESET BUTTON
WAN PORT
LAN PORTS

WHAT IS BANDWIDTH?

Bandwidth refers to the amount of data that can be transmitted within a certain time. Higher bandwidth connections allow greater speeds of data transfer.

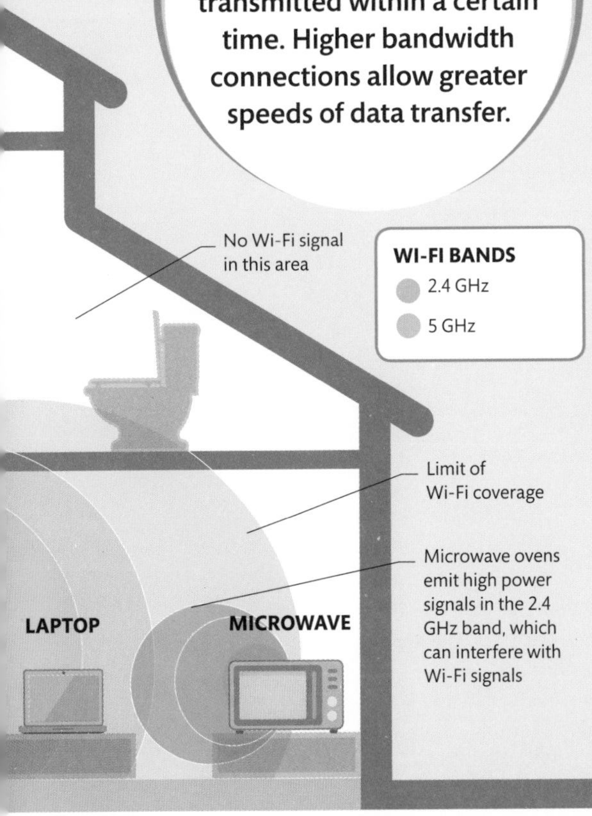
No Wi-Fi signal in this area
WI-FI BANDS
2.4 GHz
5 GHz
Limit of Wi-Fi coverage
Microwave ovens emit high power signals in the 2.4 GHz band, which can interfere with Wi-Fi signals
LAPTOP
MICROWAVE

Wi-Fi signals

The strength of a Wi-Fi signal falls rapidly with the distance between device and router. Wi-Fi range is usually up to 150 feet but can vary depending on frequency, transmission power, and antennas. Range tends to be smaller indoors due to the presence of obstacles, such as walls, although it can be extended using a signal booster.

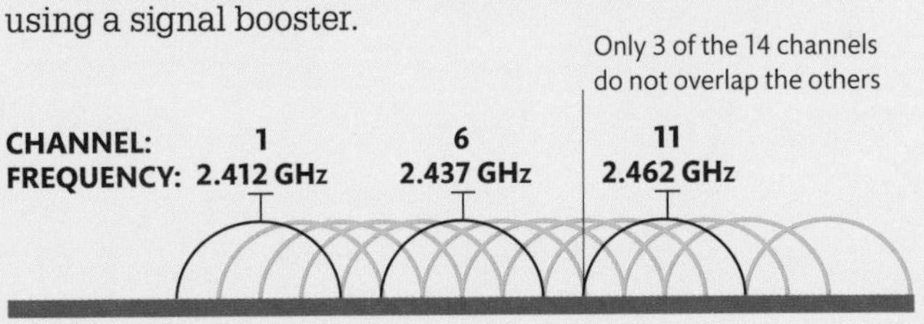
Only 3 of the 14 channels do not overlap the others
CHANNEL: 1 6 11
FREQUENCY: 2.412 GHz 2.437 GHz 2.462 GHz

2.4 GHz spectrum
Data is transmitted using specific frequencies (channels) that multiple devices can share. Using multiple channels enables more efficient communication, but in the 2.4 GHz band (shown here) many channels overlap, causing interference.

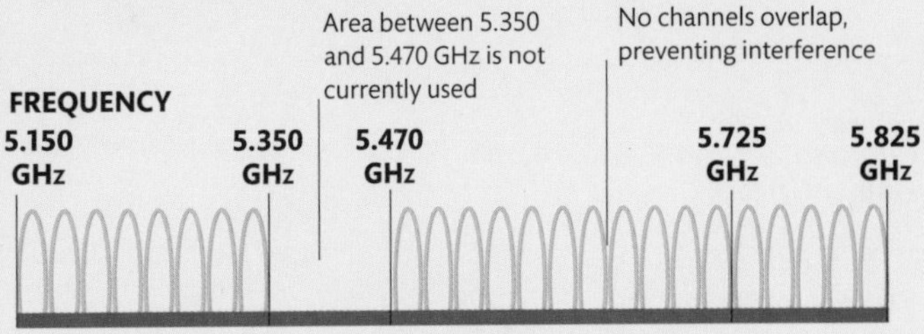
Area between 5.350 and 5.470 GHz is not currently used
No channels overlap, preventing interference
FREQUENCY
5.150 GHz
5.350 GHz
5.470 GHz
5.725 GHz
5.825 GHz

5 GHz spectrum
The 5 GHz spectrum has 24 nonoverlapping channels, making use of higher frequencies. This means that data can be transmitted through many channels simultaneously for greater efficiency and speed. Wi-Fi systems can use the area of the spectrum from 5.725–5.875 GHz but only for short-range, low-power devices.

HACKING WI-FI

Wireless Internet connections are vulnerable to hacking because a hacker can access a Wi-Fi network without having to be in the same building or needing to break through a firewall. Hackers can breach Wi-Fi security in various ways, including harvesting information transmitted and received by devices. A wireless network can be secured with Wi-Fi Protected Access. This relies on the user entering a verified password and works by producing new encryption keys for each data packet.

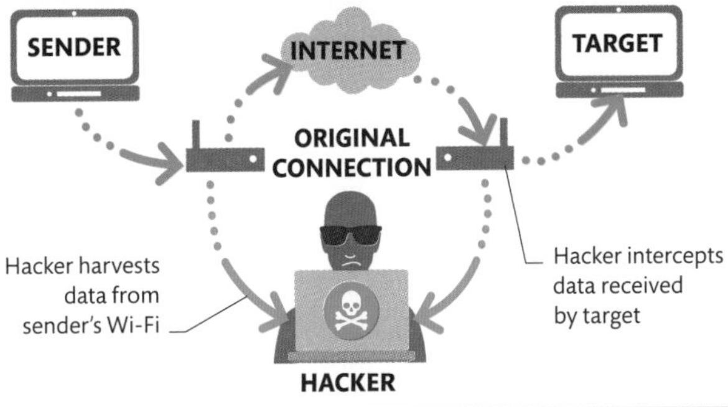
SENDER
INTERNET
TARGET
ORIGINAL CONNECTION
Hacker harvests data from sender's Wi-Fi
Hacker intercepts data received by target
HACKER

Mobile devices

A mobile device is a small, portable computing device. Most modern mobile devices can connect to the Internet (see pp.196–197) and other devices and are operated with a flat touch screen.

Mobile device components
A capacitive touch screen is made up of a layer of driving lines and a layer of sensing lines that form a grid on a glass substrate. This grid, which sits on top of the LCD display, is connected to a touch screen controller chip and the device's main processor.

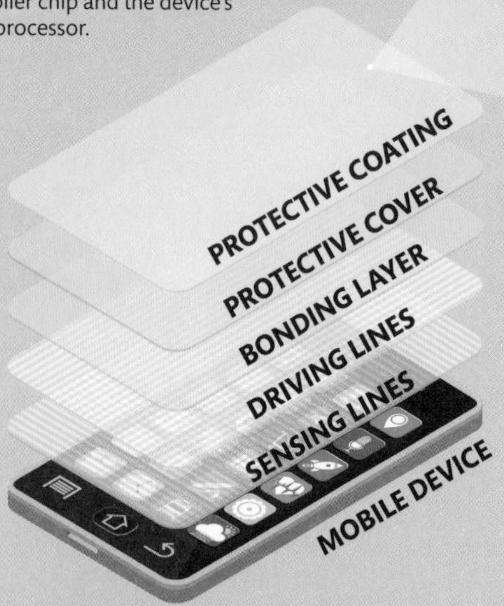
PROTECTIVE COATING
PROTECTIVE COVER
BONDING LAYER
DRIVING LINES
SENSING LINES
MOBILE DEVICE

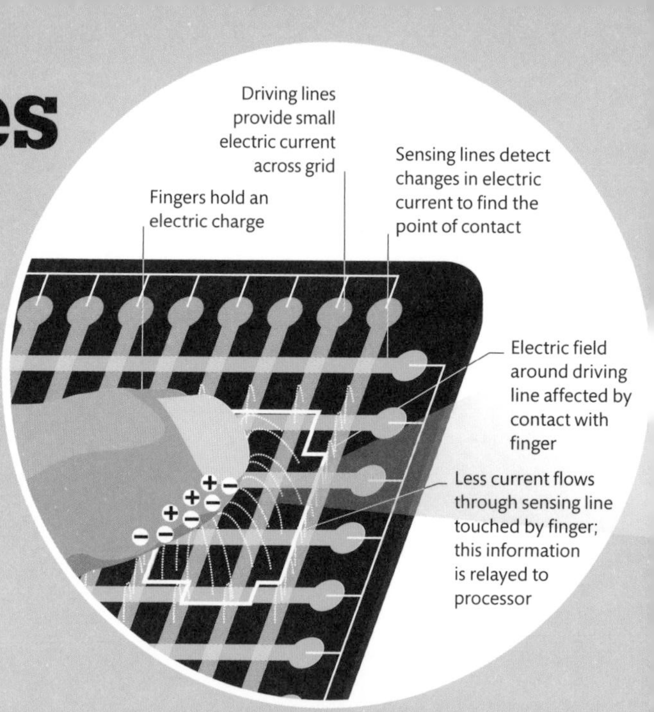
Driving lines provide small electric current across grid
Sensing lines detect changes in electric current to find the point of contact
Fingers hold an electric charge
Electric field around driving line affected by contact with finger
Less current flows through sensing line touched by finger; this information is relayed to processor

1 Screen touch
When a fingertip touches the screen, a small electric charge is pulled toward the electrically conductive finger. A fall in electric current is experienced across the grid, registering the touch.

Touch screens

There are two main types of touch screens: capacitive and resistive. Both allow users to interact directly with the elements displayed on their device with simple touches and gestures. The most common for mobile devices is a capacitive touch screen. It relies on the conductive properties of a fingertip or a stylus, making it more sensitive to gestures than other touch screens. Resistive screens rely on applying pressure to an outer layer of the screen, bringing two conductive layers made of transparent electrode film into contact.

BLUETOOTH IS NAMED AFTER A **KING** WHO UNITED VIKING TRIBES, AS IT WAS INTENDED TO **UNIFY COMMUNICATION** ACROSS DEVICES

Types of mobile devices

There are many types of mobile devices, which fulfill a range of applications. Some perform many functions, such as tablets, while others are designed for specific purposes, such as gaming or capturing video. Some mobile devices can be worn on the body for convenience and to collect data, for example, about the physical activity a person has performed each day.

Tablet
Tablets are flat mobile computers. They are larger than smartphones but share similarities.

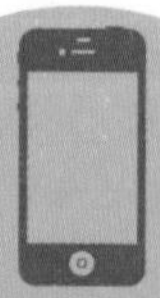

Smartphone
These devices have computing functions and Internet access as well as cellular connections.

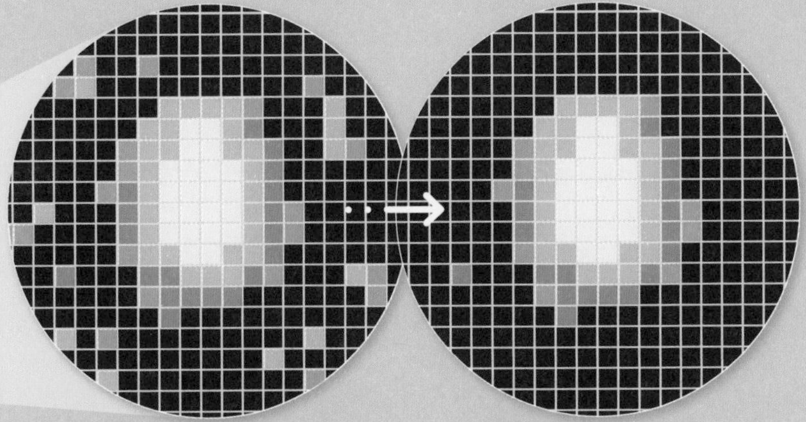

2 Raw data captured
Measurements of changes in the electric current are taken at every point in the grid. The points directly beneath the fingertip experience the largest drop.

3 Noise removed
Electromagnetic interference, or noise, must be filtered out to ensure a strong and stable touch response. This noise can come from external sources such as chargers.

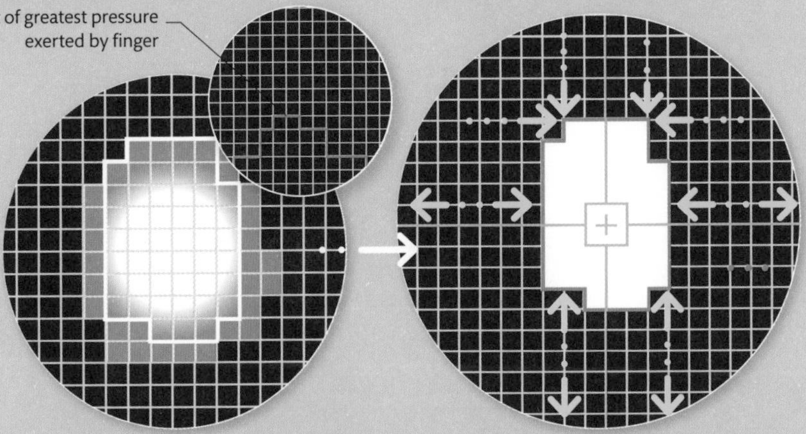

4 Pressure points measured
The size and shape of the areas of the grid in contact with the user's fingertips are identified to determine the points where the greatest pressure was applied.

5 Exact coordinates calculated
Electric signals from each point on the grid are sent to the device's main processor, which uses the data to calculate the precise position of the fingertip.

CONNECTIVITY

One of the most useful features of mobile devices is their ability to connect and communicate with other devices nearby. The devices could be physically connected, but it is usually more convenient to exchange data wirelessly using radio signals.

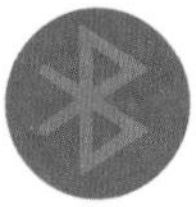

Bluetooth
Bluetooth uses radio waves to communicate over short distances. It allows wireless connectivity to other devices using radio signals, including Bluetooth headsets.

Wi-Fi
Wi-Fi (see pp.202–203) allows local networks of devices to communicate wirelessly via a router, which also connects to the Internet.

RFID
Radio-frequency ID tags—often attached to objects in stores or factories—emit unique radio waves, from which they can be identified by mobile devices.

NFC
Near-field communication allows two very close devices to communicate. This is used in contactless payment systems and keycards.

Smartwatch
These miniaturized computers feature many of the functions of a smartphone.

Gaming platform
Some gaming systems contain the screen, controls, speakers, and console within a single device.

e-Reader
e-Readers are designed for reading electronic books. Many use electronic paper (see pp.208–209).

PDA
Personal digital assistants (PDAs) are information managers. Most can access the Internet and work as a phone.

artphones

es are handheld computers with ve range of hardware and software functions. They are normally operated using a touch screen (see pp.204–205) that covers their front surface. Smartphones run mobile operating systems and can be customized by downloading and installing apps.

WHAT WAS THE WORLD'S FIRST SMARTPHONE?

IBM's Simon was the first smartphone and was released in 1994. It weighed 18 oz (510 g) and featured a modem for sending and receiving faxes.

What does a smartphone do?

Smartphones combine the features of a telephone and a small computer. They enable communication via a cellular network, Wi-Fi, Bluetooth, and GPS and are equipped with cameras, microphones, loudspeakers, and sensors, while millions of different services are available on app stores. The rise of these powerful, convenient devices has led many specialized devices to become defunct.

Loudspeaker
A miniaturized loudspeaker is built into the phone to provide sound for calls and media. It also enables a speakerphone function for hands-free calling.

Microphone
This allows a smartphone to function as a telephone. It also has a recording function and enables communication with digital assistants.

Camera
Virtually all smartphones have small, low-power cameras facing forward and backward. Most have a digital zoom feature and a flash produced by light-emitting diodes (LEDs).

Bluetooth
A Bluetooth chip allows the phone to connect wirelessly to other devices using radio signals. It also enables connection to Bluetooth headsets.

Satnav
A satellite navigation chip connects to a network of satellites in orbit, such as the US Global Positioning System (GPS). The satnav services are accessed through apps.

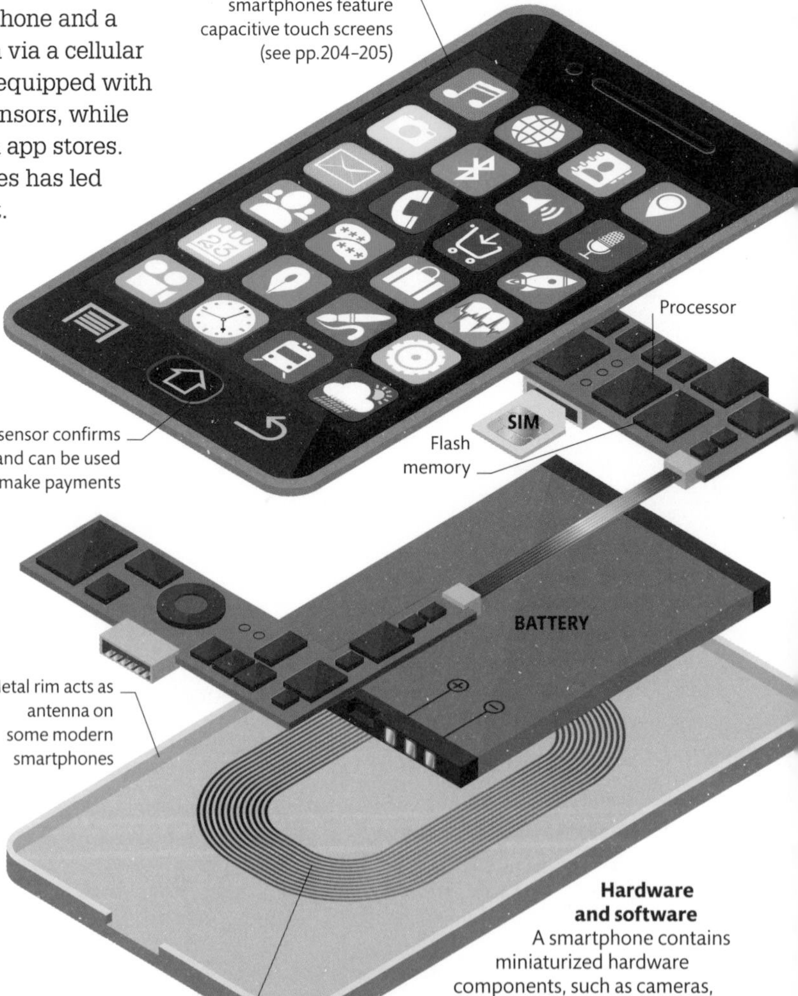
Most modern smartphones feature capacitive touch screens (see pp.204–205)
Processor
SIM
Flash memory
Fingerprint sensor confirms identity and can be used to make payments
BATTERY
Metal rim acts as antenna on some modern smartphones
Induction coils allow some smartphones to be charged wirelessly

Hardware and software
A smartphone contains miniaturized hardware components, such as cameras, which are run by a processor and powered by a rechargeable battery. The hardware features are very versatile thanks to apps: specialized programs that can be downloaded and installed onto the devices.

Messaging

Text messaging involves sending and receiving electronic messages via mobile networks. Most texts are exchanged using the Short Message Service (SMS), which permits the sending of short, text-only messages of up to 160 characters. However, the Multimedia Messaging Service (MMS) uses mobile networks to exchange messages containing photographs, videos, and audio clips.

How a text is sent

The sender's text is transmitted, via a cell tower, to a mobile switching center (MSC), which finds the address of the sender's short message service center (SMSC) and relays the text there. The SMSC checks whether the recipient is available. If so, it delivers the text via an MSC. If not, the text is stored until the recipient is available.

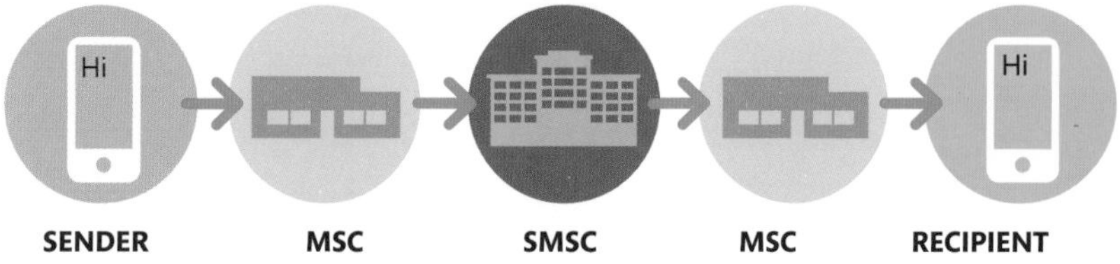

EVERY **SMARTPHONE** CONTAINS PRECIOUS METALS INCLUDING **GOLD, SILVER**, AND PLATINUM

Internet
Smartphones can connect to the Internet through Wi-Fi or cellular networks. Most phones now use 4G—the fourth generation of mobile technology, which allows much quicker loading speeds.

Games station
Smartphones can be used as portable game consoles. Unlike consoles, they do not include dedicated graphics cards; instead, they feature a powerful graphics processing unit, which renders images, animation, and video.

Address book
Most smartphones feature electronic address books where contact information is recorded. Some can pull in information via social media sites and email accounts and can be accessed via voice commands to digital assistants.

Payment systems
Smartphones can make contactless payments via various methods, including radio signals and magnetic signals imitating magnetic strips on bank cards. Payment usually requires an authentication process to confirm identity.

Music
Music can be downloaded from apps, streamed via Wi-Fi or a cellular connection, or imported from a user's collection. Smartphones support many music file formats, including MP3, AAC, WMA, and WAV.

Accelerometers

Many smartphones have miniaturized accelerometers, which measure acceleration. These sensors are used to detect the orientation of the device, so the display can change between landscape and portrait mode depending on how the device is held. They can also be used as an input for pedometers and mobile games.

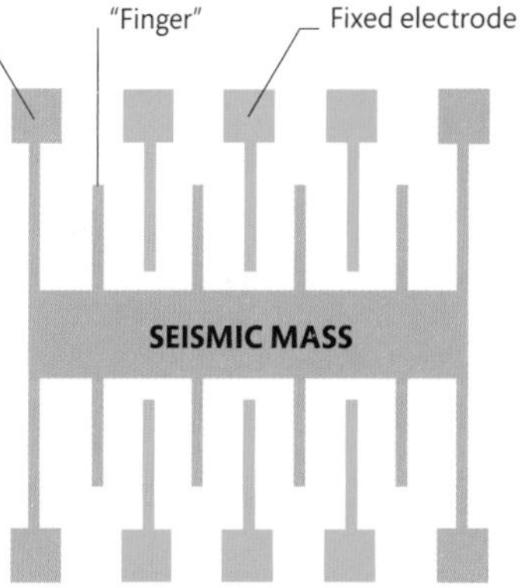

1 Phone not moving
Fixed electrodes sit between the "fingers" of a comblike seismic mass made of silicon. The electrodes and mass are connected to the battery, so the mass is electrically charged, creating an electric field between the "fingers." When the mass is still, no current flows.

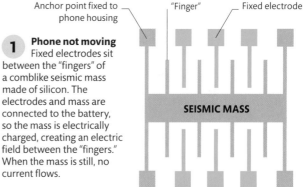

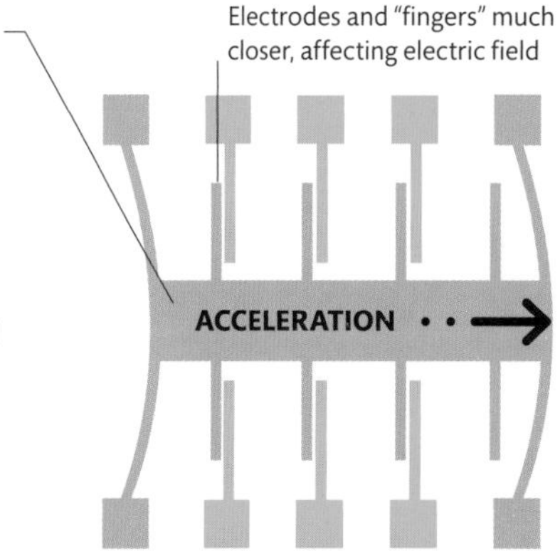

2 Movement detected
The mass deflects in response to motion, and its electric charge affects the electric field around the electrodes, creating a current. This information tells the processor how much the phone is moving and in which direction.

How electronic paper works

Inside electronic paper are thousands of tiny microcapsules, each containing black pigment particles and white pigment particles in a transparent, oily liquid. The black particles carry a negative electric charge, and the white particles have a positive charge. A small positive charge, provided by transistors under the display, attracts the black particles and repels the white ones. A negative charge does the opposite. The device's computer controls which kind of electric charge is presented where, building up black-and-white images and text on the display. If the electric charge is negative on one side and positive on the other, a single microcapsule will display half-white and half-black and appear gray.

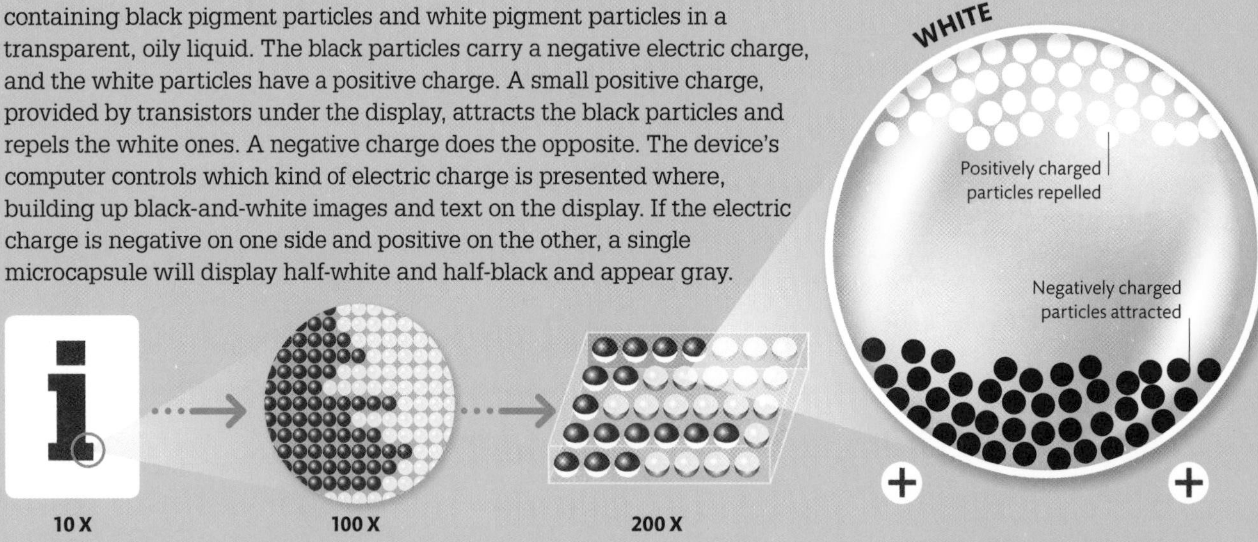

Microcapsules
The microcapsules that make up text and images on electronic paper are each about the same width as a human hair.

KEY
+ Positive charge — Negative charge

1 Black pigment particles are negatively charged, while the white particles are positive. A positive charge under the display attracts the black particles.

Electronic paper

Some e-readers display pages of text on screens made with electronic paper. Like real paper, electronic paper works with reflected light. This makes it better suited to reading text, because it causes less eye strain and works well in sunlight.

READING IN THE DARK

Electronic paper does not need its own light source, as computer screens do. For reading in the dark, however, many e-readers have LEDs along the side of the screen, to illuminate the display. The light travels across the inside of the transparent screen and is scattered downward onto the electronic paper.

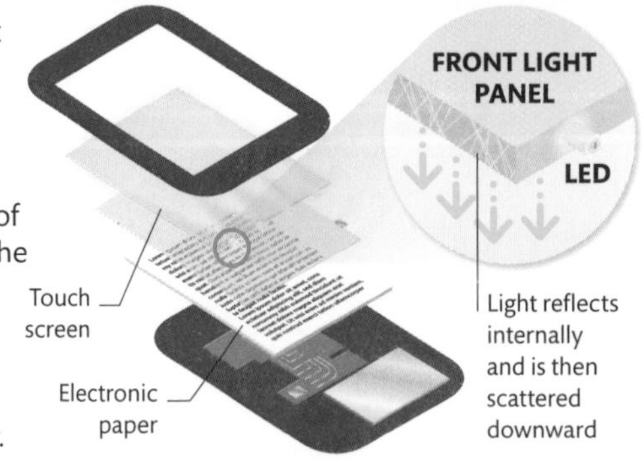

IS IT BETTER TO READ AN E-PAPER TABLET BEFORE BED INSTEAD OF AN LCD TABLET?

It may be. Using a tablet can make it harder to fall asleep, because the blue light it emits can disrupt the action of the sleep-regulating hormone melatonin.

ELECTRONIC INK TECHNOLOGY IS BEING USED TO DEVELOP **CLOTHES** WITH **CHANGING PATTERNS**

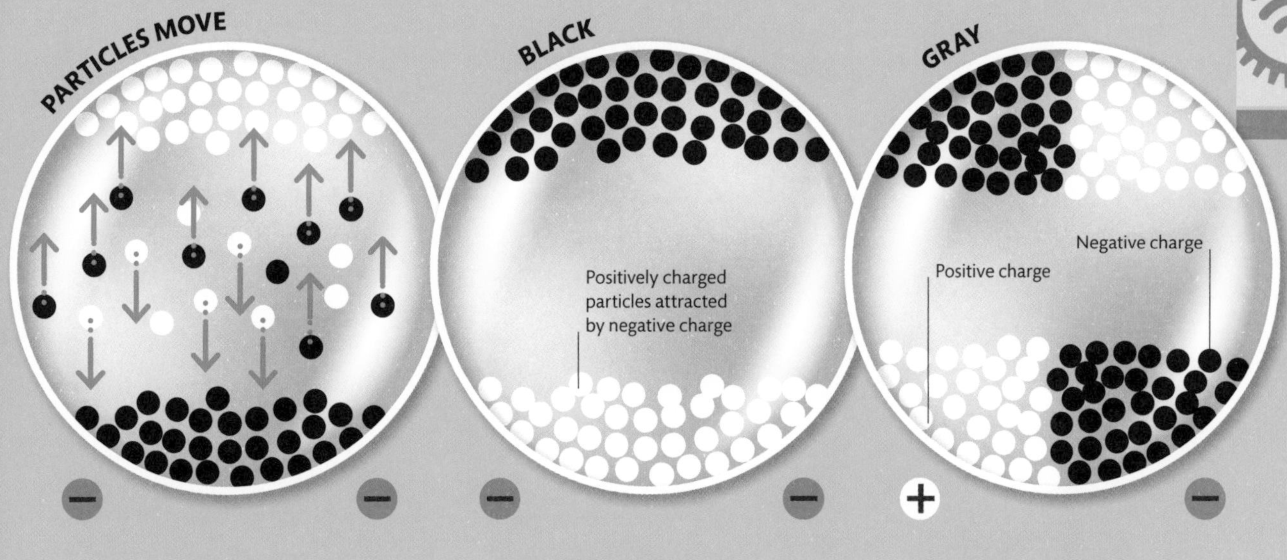
PARTICLES MOVE
BLACK
GRAY
Positively charged particles attracted by negative charge
Positive charge
Negative charge
−
−
−
−
+
−

2 When a negative charge is applied under the display, the positively charged white pigment particles switch places with the black particles.

3 The white pigment particles are attracted by the negative charge, while the negatively charged black particles are repelled and move away from it.

4 A computer inside the device controls which type of charge is applied where. A mixture of black and white particles will appear gray.

Electrowetting displays

Like electronic paper, electrowetting works by reflecting light. Electrowetting displays in color, and it can display video because it changes far more rapidly than electronic paper. Built onto a reflective white plastic sheet are thousands of tiny compartments, each containing a droplet of black liquid. Signals from a computer apply a voltage that causes the liquid to move back and forth like a curtain, absorbing light or allowing it to reflect.

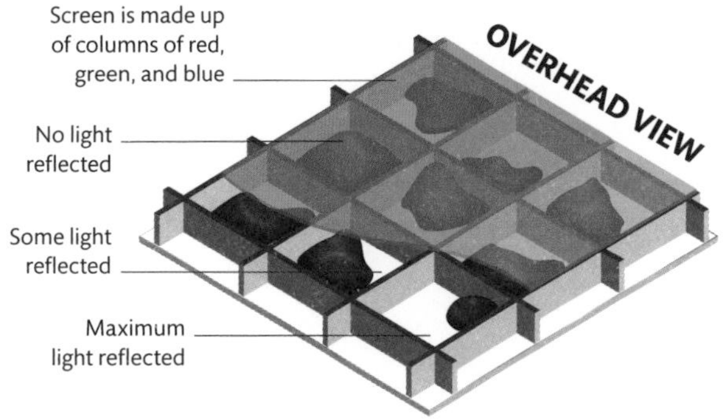
OVERHEAD VIEW
Screen is made up of columns of red, green, and blue
No light reflected
Some light reflected
Maximum light reflected

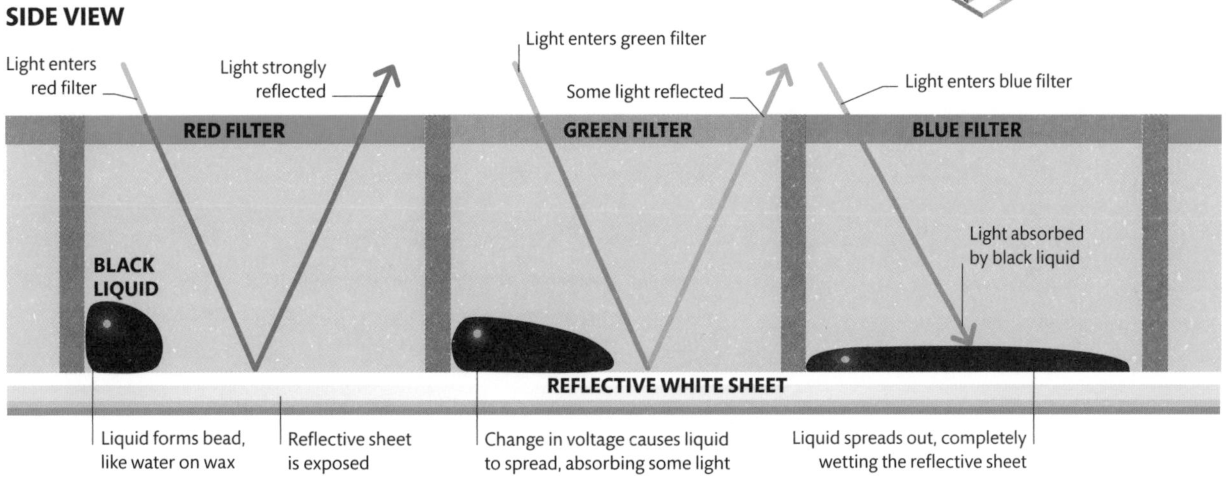
SIDE VIEW
Light enters red filter
Light strongly reflected
Light enters green filter
Some light reflected
Light enters blue filter
RED FILTER
GREEN FILTER
BLUE FILTER
Light absorbed by black liquid
BLACK LIQUID
REFLECTIVE WHITE SHEET
Liquid forms bead, like water on wax
Reflective sheet is exposed
Change in voltage causes liquid to spread, absorbing some light
Liquid spreads out, completely wetting the reflective sheet

FARMING AND FOOD TECHNOLOGY

Row planters

Traditionally, planting a field involves several processes: readying the soil, making furrows, scattering seed, and then covering it up, all of which risks compaction and erosion of the soil, as well as being time-consuming. Modern planters can do much of this work in just one pass, depositing seed in perfectly aligned rows and applying fertilizer and insecticide. The most advanced machines can plant up to 72 rows at a time, each precisely oriented in the field using GPS.

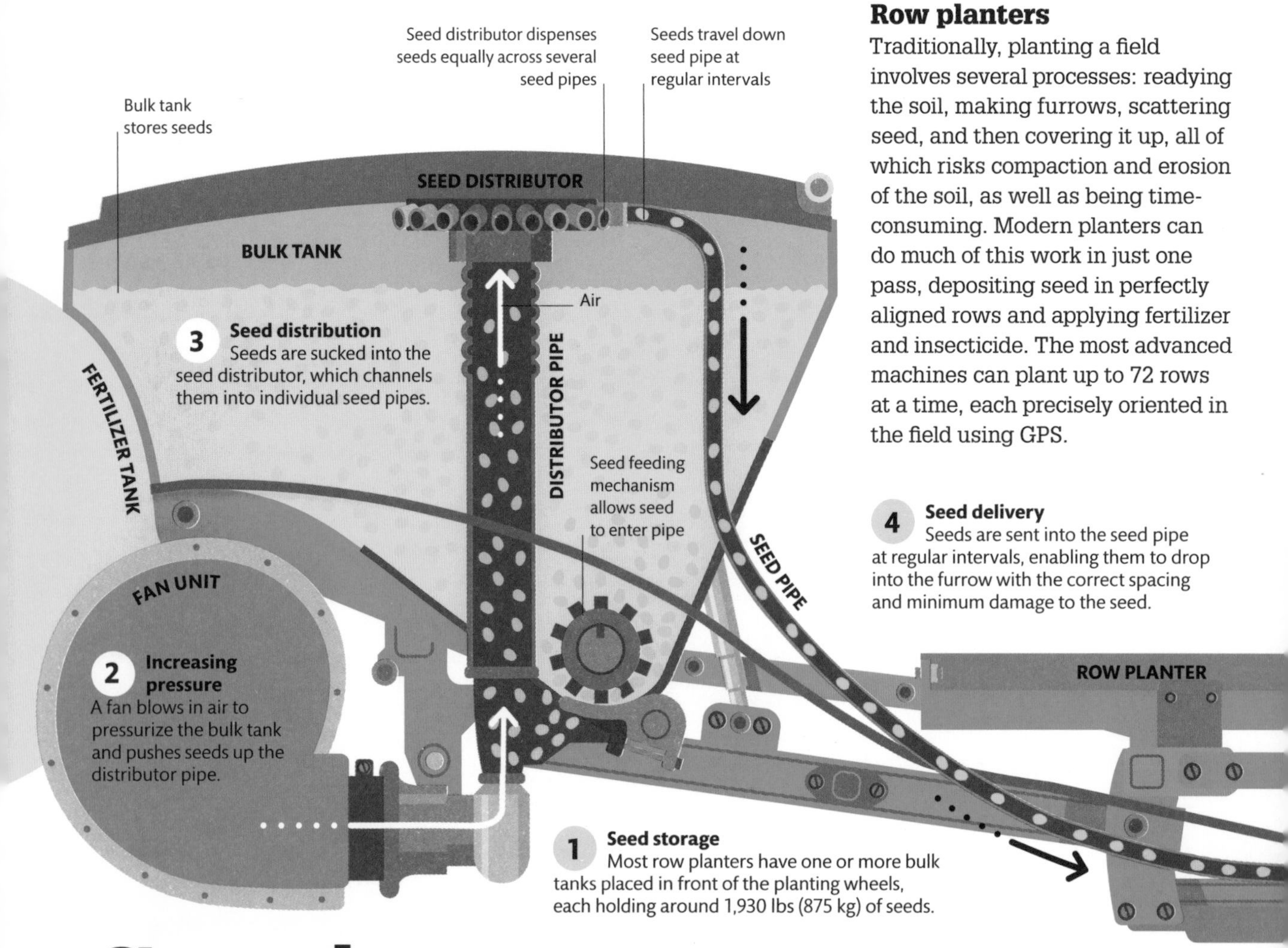
Seed distributor dispenses seeds equally across several seed pipes
Seeds travel down seed pipe at regular intervals
Bulk tank stores seeds
SEED DISTRIBUTOR
BULK TANK
Air
3 Seed distribution
Seeds are sucked into the seed distributor, which channels them into individual seed pipes.
FERTILIZER TANK
DISTRIBUTOR PIPE
Seed feeding mechanism allows seed to enter pipe
SEED PIPE
4 Seed delivery
Seeds are sent into the seed pipe at regular intervals, enabling them to drop into the furrow with the correct spacing and minimum damage to the seed.
FAN UNIT
2 Increasing pressure
A fan blows in air to pressurize the bulk tank and pushes seeds up the distributor pipe.
ROW PLANTER
1 Seed storage
Most row planters have one or more bulk tanks placed in front of the planting wheels, each holding around 1,930 lbs (875 kg) of seeds.

Growing crops

Machines for planting seeds have existed for hundreds of years. However, modern planters have increased in size and capacity so that wide stretches of field can be covered in one sweep, vastly speeding up the time it takes to sow crops.

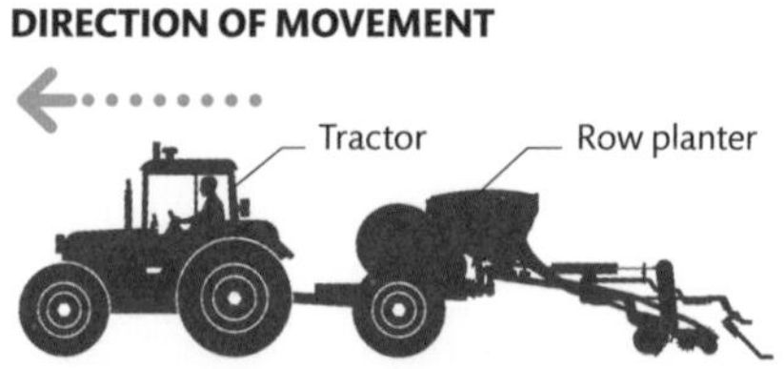
DIRECTION OF MOVEMENT
Tractor
Row planter

2.9 BILLION TONS

(2.6 BILLION METRIC TONS)—THE TOTAL AMOUNT OF GRAIN PRODUCED IN THE WORLD EACH YEAR

Irrigation

Some farmers rely on natural rainfall to water their crops, but in some climates, crops may also require irrigation systems. These vary from simple gravity-fed methods to those that apply water directly to the roots of each plant. Irrigation can be problematic: water can be wasted, crops can be contaminated if untreated waste water is used, and salinity can build up in the soil. Smart technology can be used to deliver water where it is needed most, rather than providing blanket coverage.

Surface
Water floods the entire surface or runs down furrows through gravity or pumping. This is highly labor-intensive, much water is lost to evaporation and runoff, and there is a risk of waterlogging.

Drip
A drip system uses pipes, either made of a porous material or perforated, which are placed on or under the ground to apply water directly to the roots of crops.

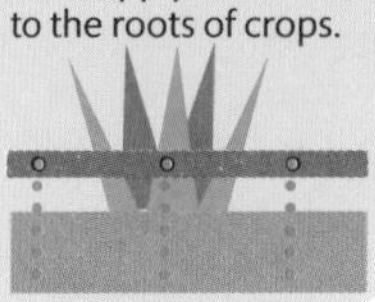

Center pivot
Sprinklers move on wheeled towers in a circular motion. This method can water a large area in a relatively short time.

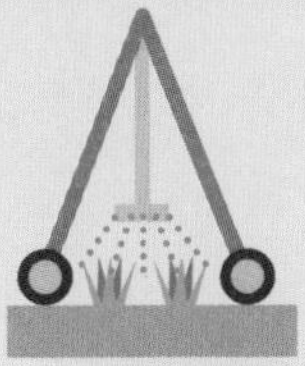

Sprinkler
Water is distributed by high-pressure overhead sprinklers or guns on moving platforms. However, water can be lost when sprayed up in the air.

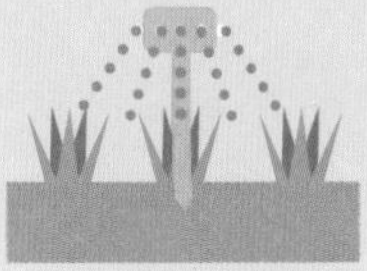

Subirrigation
An underground porous-pipe system is used to raise the water table or to discharge water directly into the root zone.

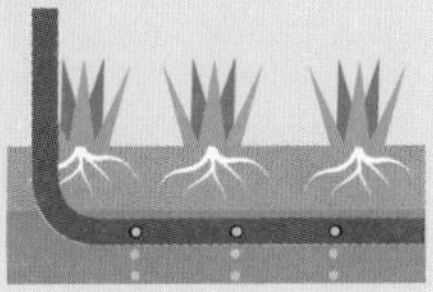

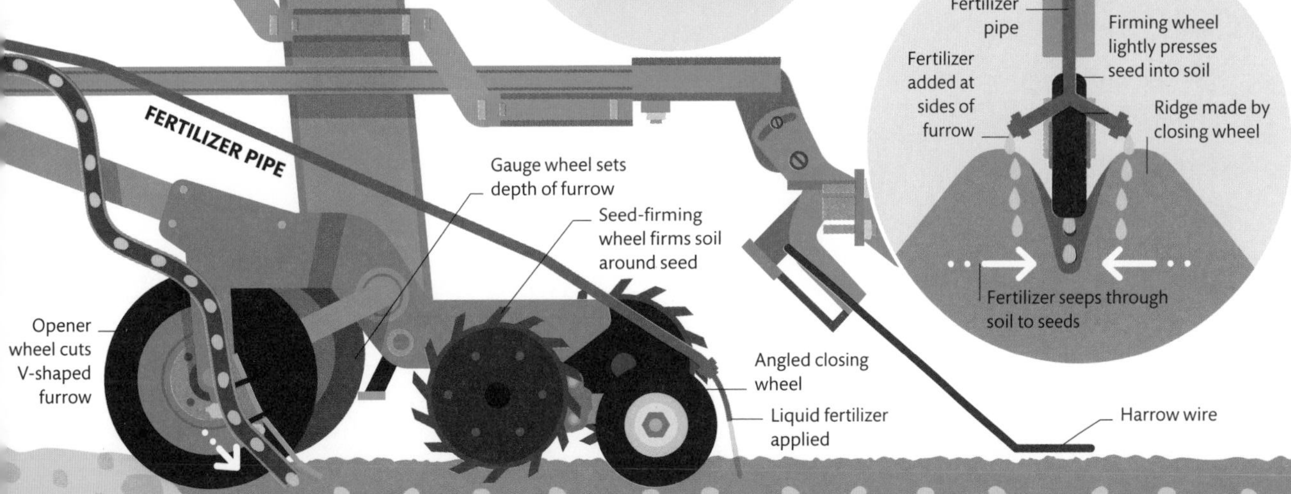
FERTILIZER PIPE
Gauge wheel sets depth of furrow
Seed-firming wheel firms soil around seed
Opener wheel cuts V-shaped furrow
Angled closing wheel
Liquid fertilizer applied
Harrow wire
SEED FURROW
Fertilizer pipe
Firming wheel lightly presses seed into soil
Fertilizer added at sides of furrow
Ridge made by closing wheel
Fertilizer seeps through soil to seeds

5 Cutting the furrow
Furrows are cut using wheels or blades to open the soil to the right depth and shape. Seeds are dropped at regular intervals behind the opener wheel. Fertilizer and insecticide are sometimes added.

6 Firming the seeds
A seed-firming wheel presses the seed into the furrow using a rolling or sliding action to improve its contact with the soil and moisture at the bottom of the furrow. It also stops the seed from bouncing out.

7 Closing and fertilizer delivery
Angled closing wheels press soil firmly around the seed. If fertilizer is not applied with the seed, it is now added to one or both sides of the furrow. The surface is then leveled by a roller or harrow wire.

Animal farming

Livestock farmers use sophisticated systems to manage their animals and enhance productivity. In dairy farming, technology has greatly increased efficiency by enabling several cows to be milked simultaneously.

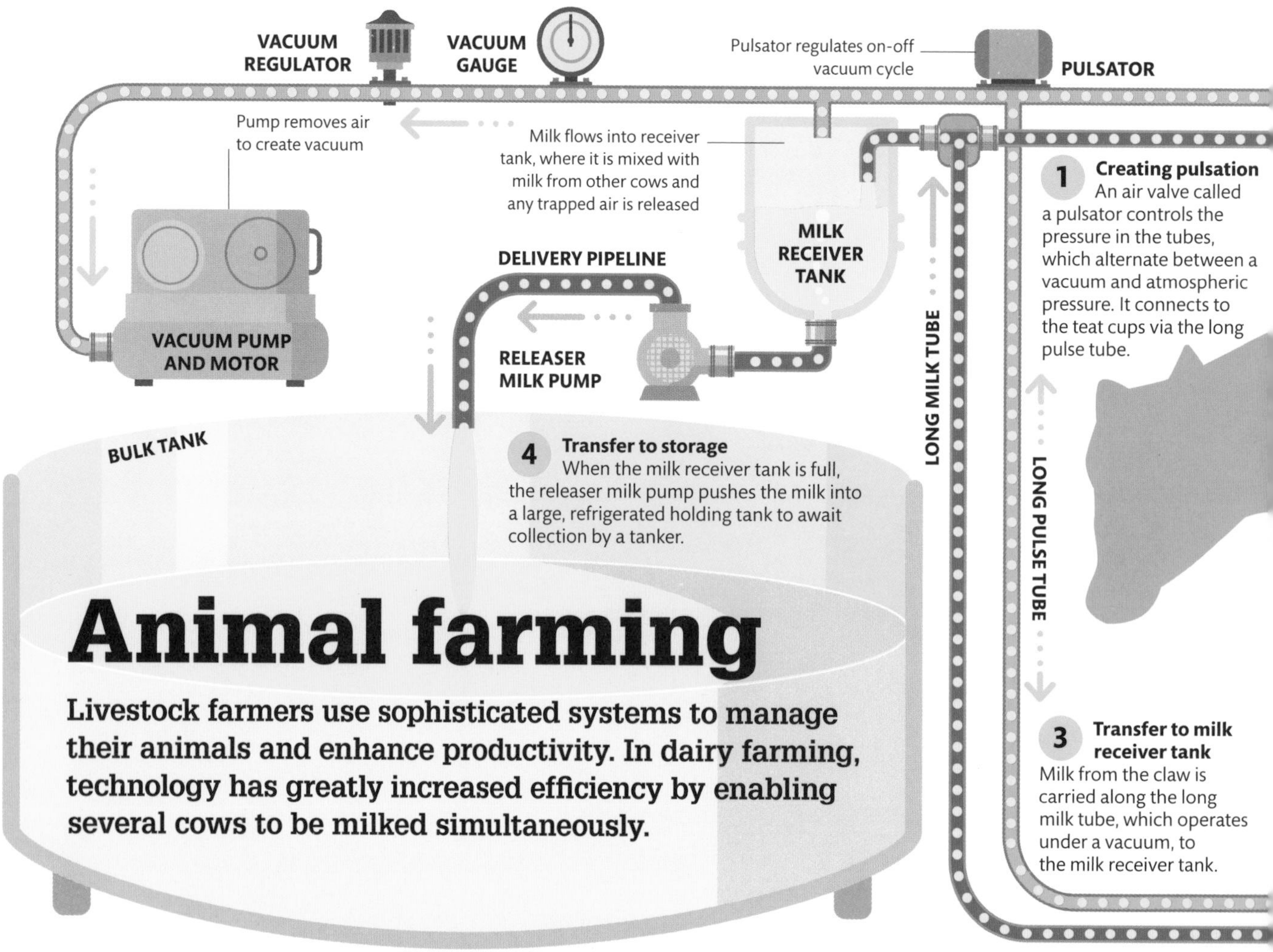
VACUUM REGULATOR
VACUUM GAUGE
Pulsator regulates on-off vacuum cycle
PULSATOR
Pump removes air to create vacuum
Milk flows into receiver tank, where it is mixed with milk from other cows and any trapped air is released
MILK RECEIVER TANK
DELIVERY PIPELINE
VACUUM PUMP AND MOTOR
RELEASER MILK PUMP
LONG MILK TUBE
LONG PULSE TUBE
BULK TANK
1 Creating pulsation
An air valve called a pulsator controls the pressure in the tubes, which alternate between a vacuum and atmospheric pressure. It connects to the teat cups via the long pulse tube.
4 Transfer to storage
When the milk receiver tank is full, the releaser milk pump pushes the milk into a large, refrigerated holding tank to await collection by a tanker.
3 Transfer to milk receiver tank
Milk from the claw is carried along the long milk tube, which operates under a vacuum, to the milk receiver tank.

Methane from manure

Dairy farming produces a lot of waste, including manure, soiled bedding, and milking waste water. Like agricultural waste matter from fruit and vegetable crops, it requires disposal. Many large farms use anaerobic digesters to turn their waste into sterile sludge, which can be used as fertilizer, or methane gas, which can be used for fuel, heating, and generating electricity. Some farmers grow additional crops, such as corn, that they add to the digester to boost its gas production and increase its energy output.

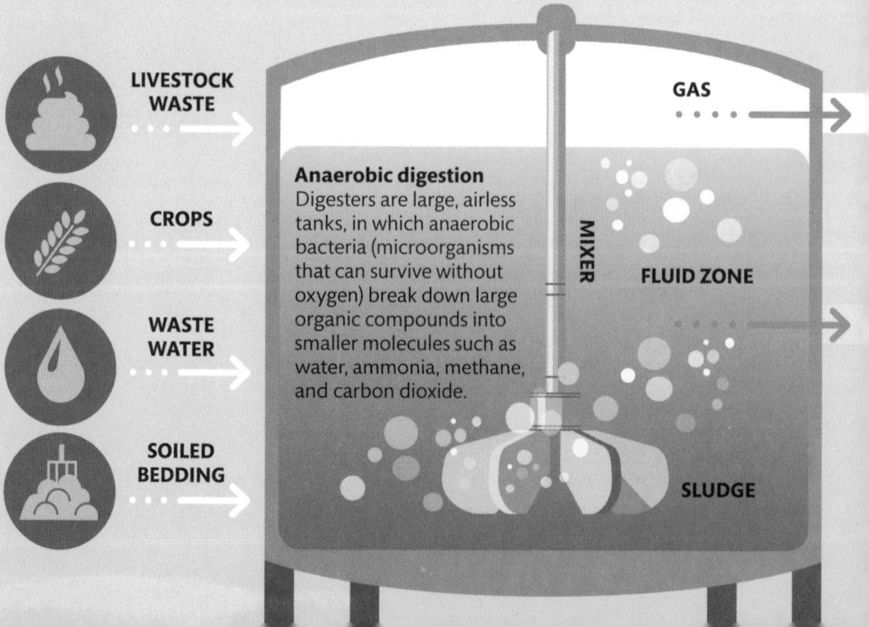
LIVESTOCK WASTE
CROPS
WASTE WATER
SOILED BEDDING
GAS
MIXER
FLUID ZONE
SLUDGE
Anaerobic digestion
Digesters are large, airless tanks, in which anaerobic bacteria (microorganisms that can survive without oxygen) break down large organic compounds into smaller molecules such as water, ammonia, methane, and carbon dioxide.

Milking machines

A milking machine uses a vacuum pump to gently draw milk from a cow's teats. The milk is drawn into four teat cups lined with silicone or rubber. The liners form a seal between the teat and the short milk tube, which transports the milk to a claw. From here, the milk is transported through the long milk tube to a milk receiver tank and then into a bulk tank.

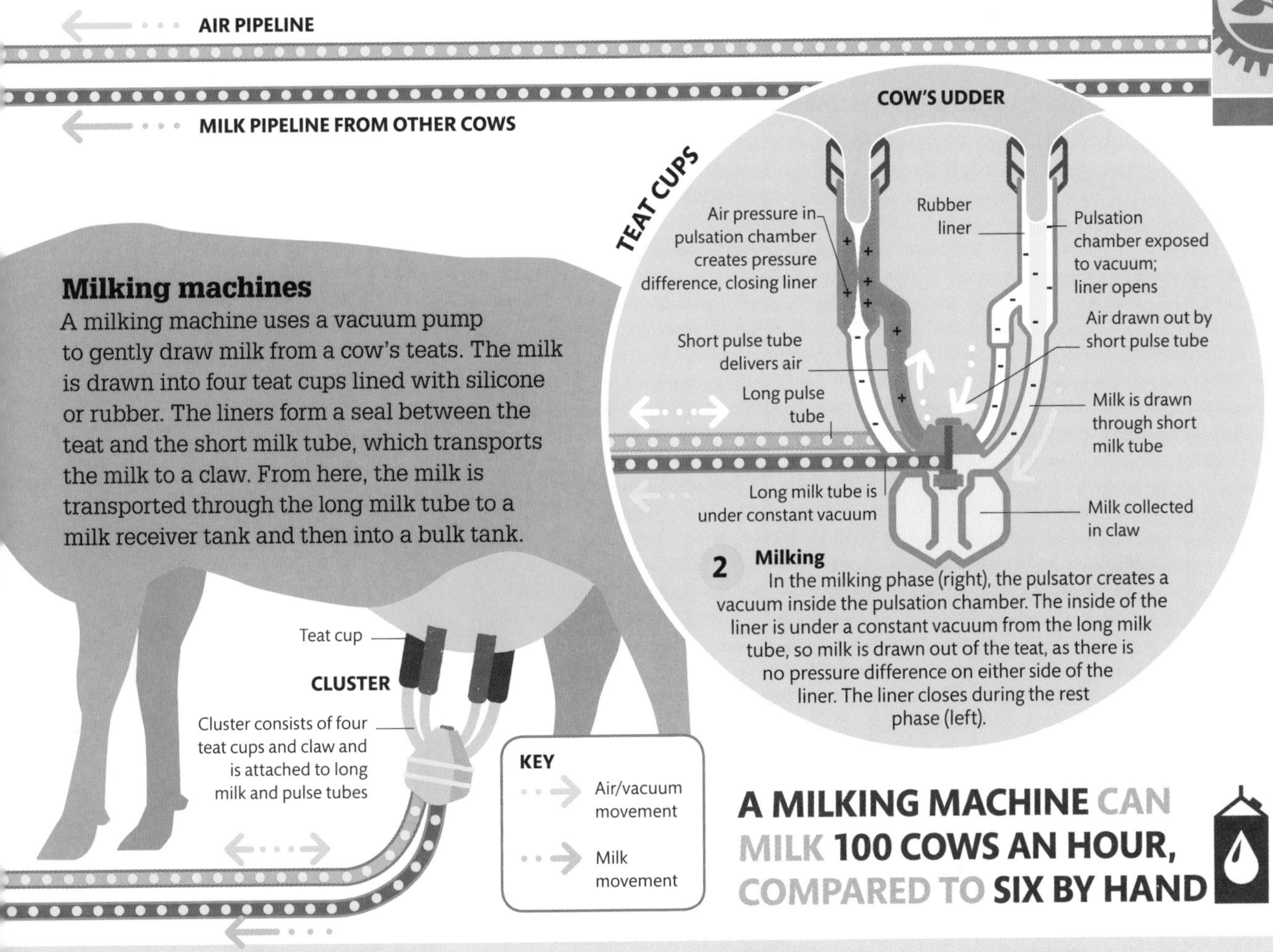
AIR PIPELINE
MILK PIPELINE FROM OTHER COWS
COW'S UDDER
TEAT CUPS
Air pressure in pulsation chamber creates pressure difference, closing liner
Rubber liner
Pulsation chamber exposed to vacuum; liner opens
Air drawn out by short pulse tube
Short pulse tube delivers air
Long pulse tube
Milk is drawn through short milk tube
Long milk tube is under constant vacuum
Milk collected in claw
Teat cup
CLUSTER
Cluster consists of four teat cups and claw and is attached to long milk and pulse tubes
KEY
Air/vacuum movement
Milk movement

2 Milking
In the milking phase (right), the pulsator creates a vacuum inside the pulsation chamber. The inside of the liner is under a constant vacuum from the long milk tube, so milk is drawn out of the teat, as there is no pressure difference on either side of the liner. The liner closes during the rest phase (left).

A MILKING MACHINE CAN MILK 100 COWS AN HOUR, COMPARED TO SIX BY HAND

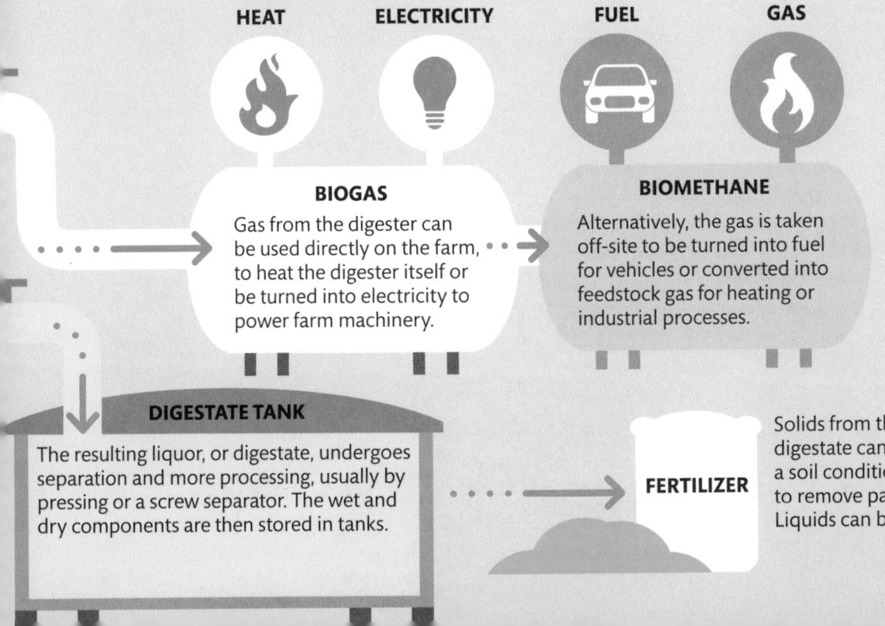
HEAT
ELECTRICITY
FUEL
GAS

BIOGAS
Gas from the digester can be used directly on the farm, to heat the digester itself or be turned into electricity to power farm machinery.

BIOMETHANE
Alternatively, the gas is taken off-site to be turned into fuel for vehicles or converted into feedstock gas for heating or industrial processes.

DIGESTATE TANK
The resulting liquor, or digestate, undergoes separation and more processing, usually by pressing or a screw separator. The wet and dry components are then stored in tanks.

FERTILIZER
Solids from the digestate can be used as a soil conditioner or, oncetreated to remove pathogens, as animal bedding. Liquids can be sprayed onto fields.

HOW ARE ROBOTS USED IN DAIRY FARMING?

Sensors are used to scan a cow's ID tag to detect whether she has been milked recently, and robotic arms can apply and remove the teat cups.

Harvesters

Using a machine to harvest a large crop avoids the need for manual labor. The newest machines use robotic technology to harvest crops such as fruits and vegetables that, until recently, were picked by hand.

Combine harvesters

One of the biggest pieces of farm machinery is the combine harvester, capable of collecting around 77 tons (70 metric tons) of grain an hour. Combine harvesters, or simply combines, take their name from the combining of three separate harvesting actions in a single machine: reaping (cutting), threshing (spinning a crop to separate the grain), and winnowing (blowing air to remove plant husks, or chaff). Finally, the combine deposits the straw back onto the field.

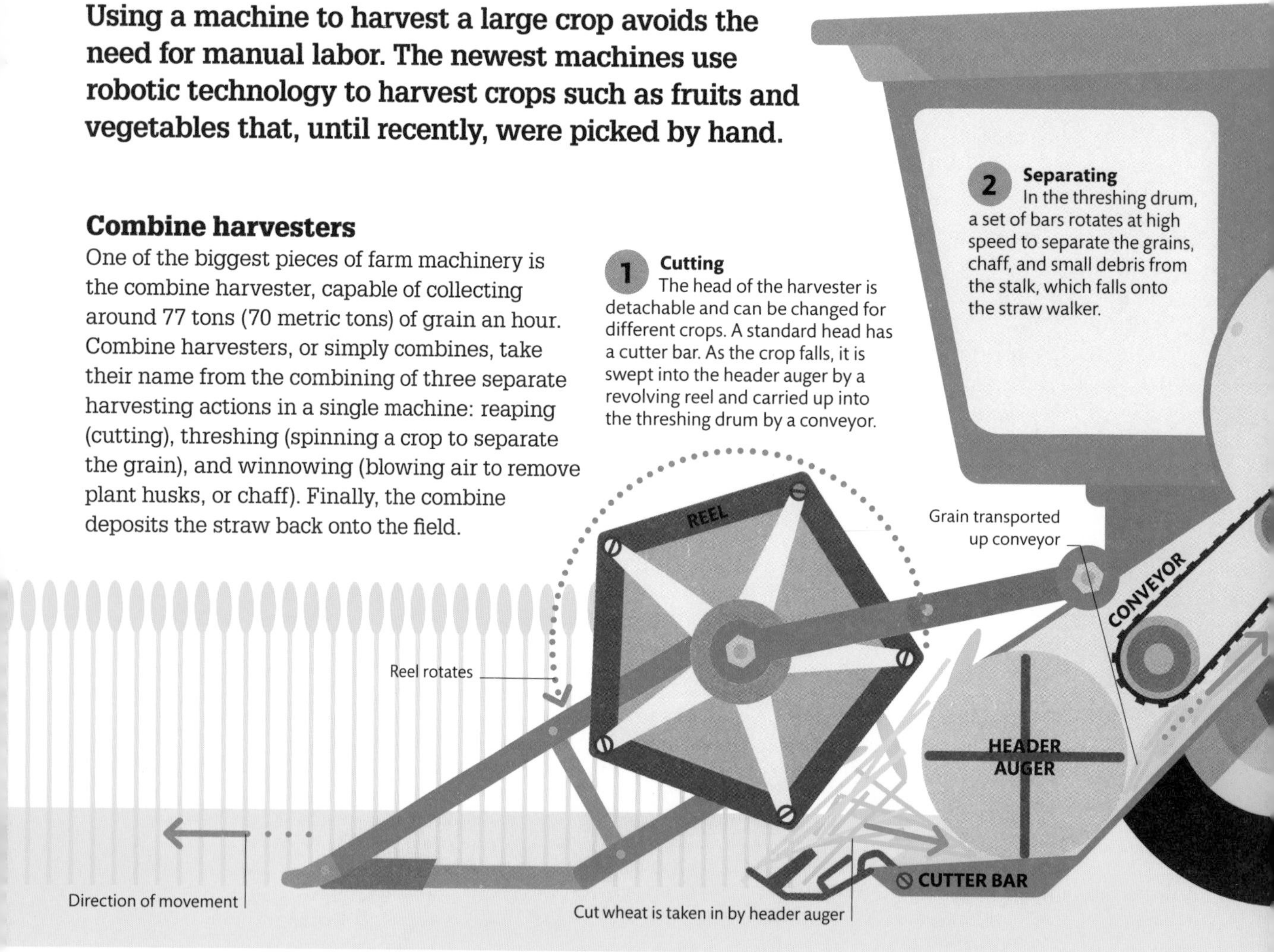

1 Cutting The head of the harvester is detachable and can be changed for different crops. A standard head has a cutter bar. As the crop falls, it is swept into the header auger by a revolving reel and carried up into the threshing drum by a conveyor.

2 Separating In the threshing drum, a set of bars rotates at high speed to separate the grains, chaff, and small debris from the stalk, which falls onto the straw walker.

The future of harvesting

Robots could be the future of fruit and vegetable harvesting. Some prototype robot pickers use sensors to assess whether a crop is ready to harvest. Others combine this with a camera that detects the color of the crop. Picking produce requires delicate handling; for fruits such as apples, robot pickers use a vacuum arm to suction the fruit, while other robots use implements to carefully snip the fruit or vegetable off its stem.

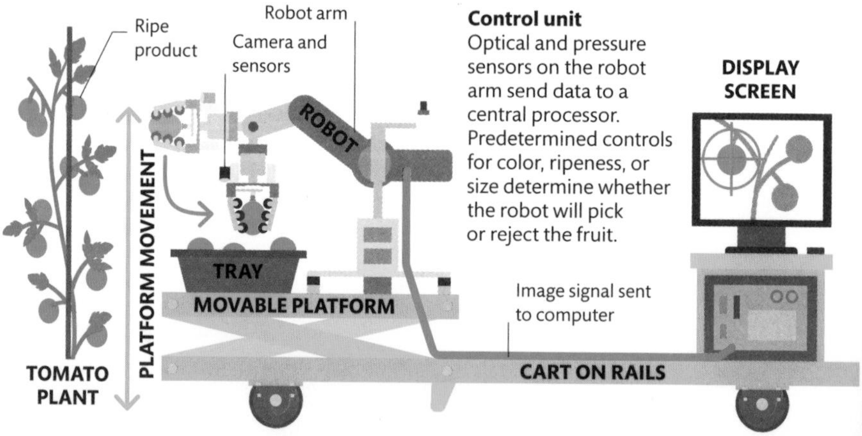

Control unit
Optical and pressure sensors on the robot arm send data to a central processor. Predetermined controls for color, ripeness, or size determine whether the robot will pick or reject the fruit.

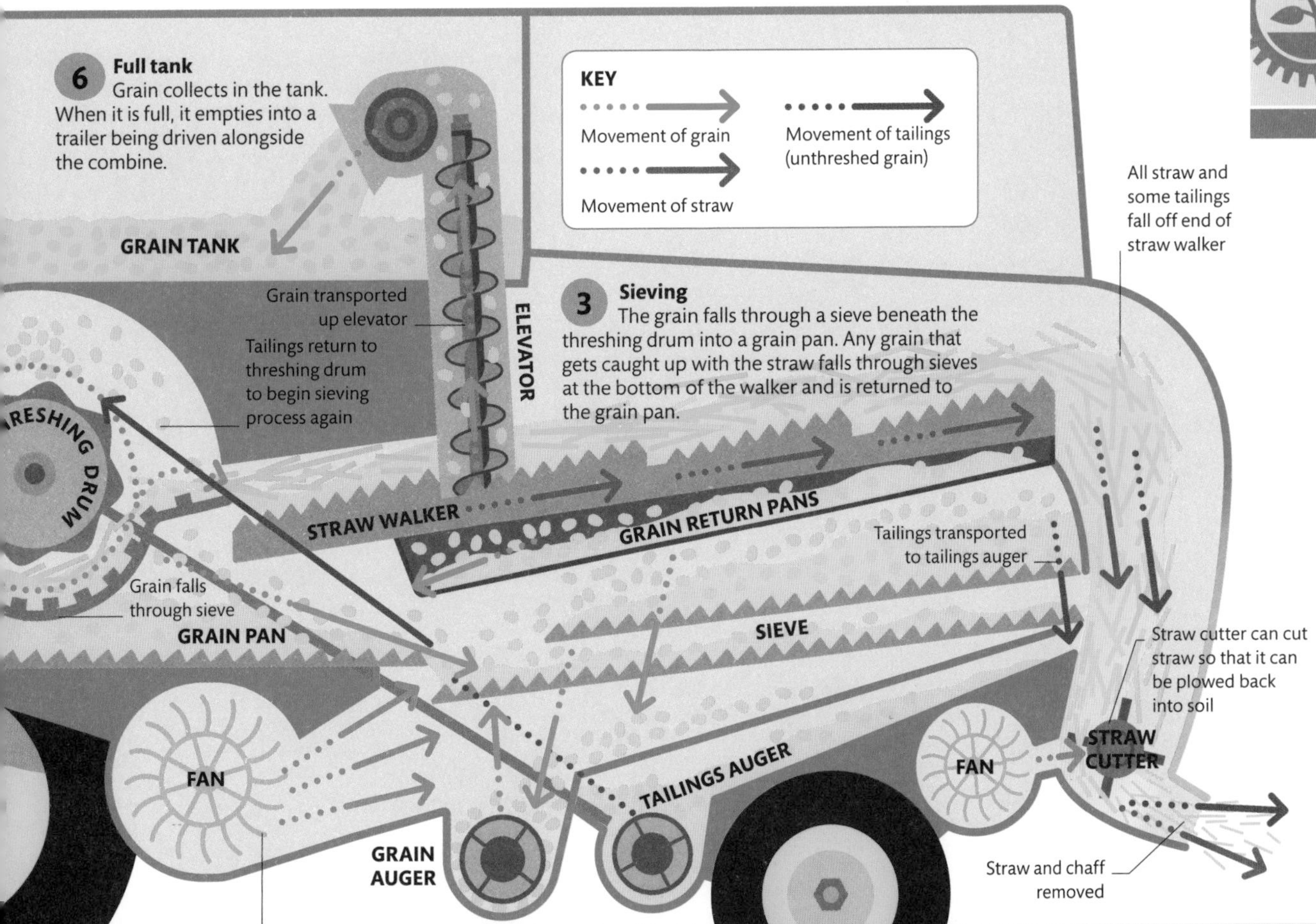
KEY
Movement of grain
Movement of tailings (unthreshed grain)
Movement of straw
GRAIN TANK
Grain transported up elevator
ELEVATOR
Tailings return to threshing drum to begin sieving process again
RESHING DRUM
STRAW WALKER
GRAIN RETURN PANS
Tailings transported to tailings auger
All straw and some tailings fall off end of straw walker
Grain falls through sieve
GRAIN PAN
SIEVE
Straw cutter can cut straw so that it can be plowed back into soil
FAN
TAILINGS AUGER
FAN
STRAW CUTTER
GRAIN AUGER
Straw and chaff removed
Fans blow chaff off wheat and out back of combine

6 Full tank
Grain collects in the tank. When it is full, it empties into a trailer being driven alongside the combine.

3 Sieving
The grain falls through a sieve beneath the threshing drum into a grain pan. Any grain that gets caught up with the straw falls through sieves at the bottom of the walker and is returned to the grain pan.

5 Grain fed to tank
From the grain pan, the grain passes over a number of other sieves before being pulled up into the grain tank by a grain auger and elevator.

4 Removing straw and chaff
Straw and chaff are blown out of the machine by fans. Straw may be chopped by a straw cutter first.

COMMON MECHANICAL PICKERS

Cotton harvester
There are two types of cotton-harvesting machines. A cotton picker picks the cotton heads from the plant using revolving spindles or prongs. A cotton stripper pulls up the entire cotton plant, and another machine then removes unwanted material.

Sugar beet harvester
Blades remove the leaves, and then wheels lift the beets onto the harvester. The crop passes through cleaning rollers to brush away soil before being lifted into a holding tank.

Mechanical tree shaker
To harvest olives, nuts, and other less-bruisable fruit, mechanical tree shakers are often used. These machines use a hydraulic cylinder to grip the trunk and shake the tree. The crop is then collected.

42
THE NUMBER OF **LOAVES** OF BREAD THAT CAN BE MADE FROM ONE **BUSHEL** OF **WHEAT**

Farming without soil

As demand for food increases, farmers are devising more efficient methods for growing crops. Farming without soil allows farmers to grow plants almost anywhere, carefully controlling growing conditions and minimizing environmental impact.

A HYDROPONIC FARM USES JUST 10 PERCENT OF THE WATER USED ON A CONVENTIONAL FARM

Hydroponics

In a hydroponic system, crops are grown without soil and are fed by nutrients dissolved in water that is usually delivered by a pump. Nutrient levels can be tailored to the plant type, and lighting, ventilation, humidity, and temperature are easy to control. There are several different kinds of hydroponic systems.

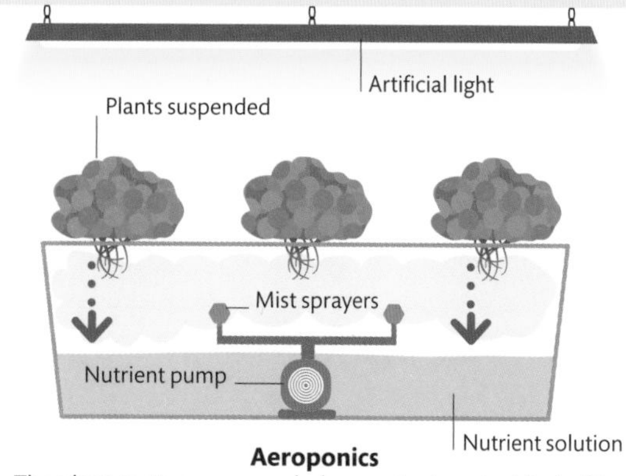

Aeroponics
The plant roots are suspended over a tank and misted with nutrient solution delivered by a nutrient pump. They are misted every few minutes to prevent the roots from drying out.

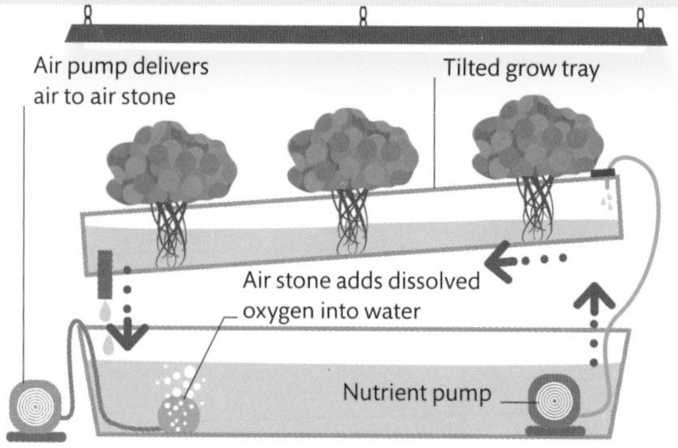

Nutrient film technique
The nutrient solution is pumped into a grow tray and continually flows over the tips of the roots. The tray is set at an angle so that the water flows back into the tank under gravity.

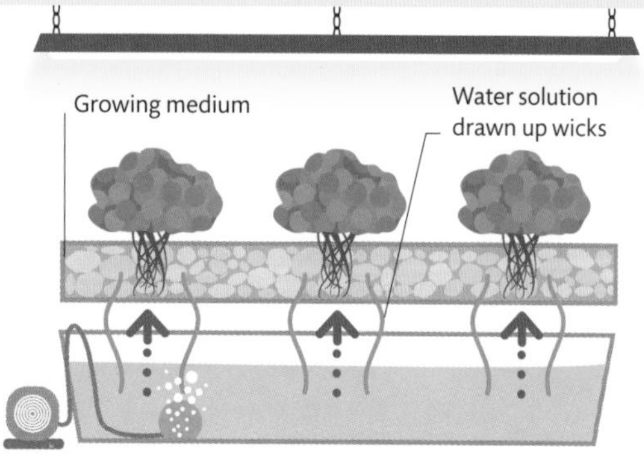

Wick system
Plants are grown in a medium of perlite, coir, or vermiculite. The nutrients are drawn up from the tank to the growing medium by the capillary action of absorbent wicks.

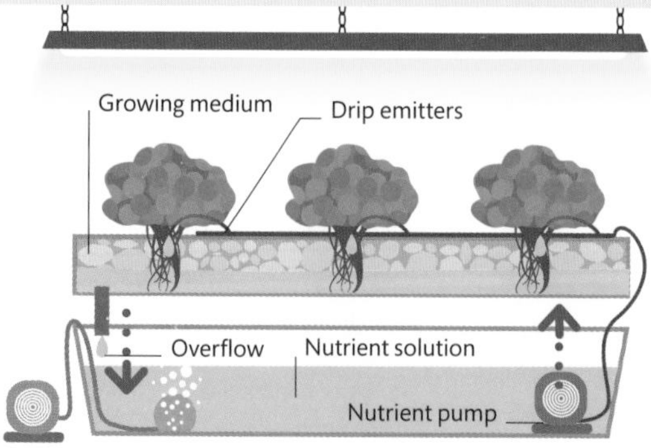

Drip system
Nutrients are dripped onto the growing medium around each plant at regular intervals. Excess solution runs off and can be recycled back into the system.

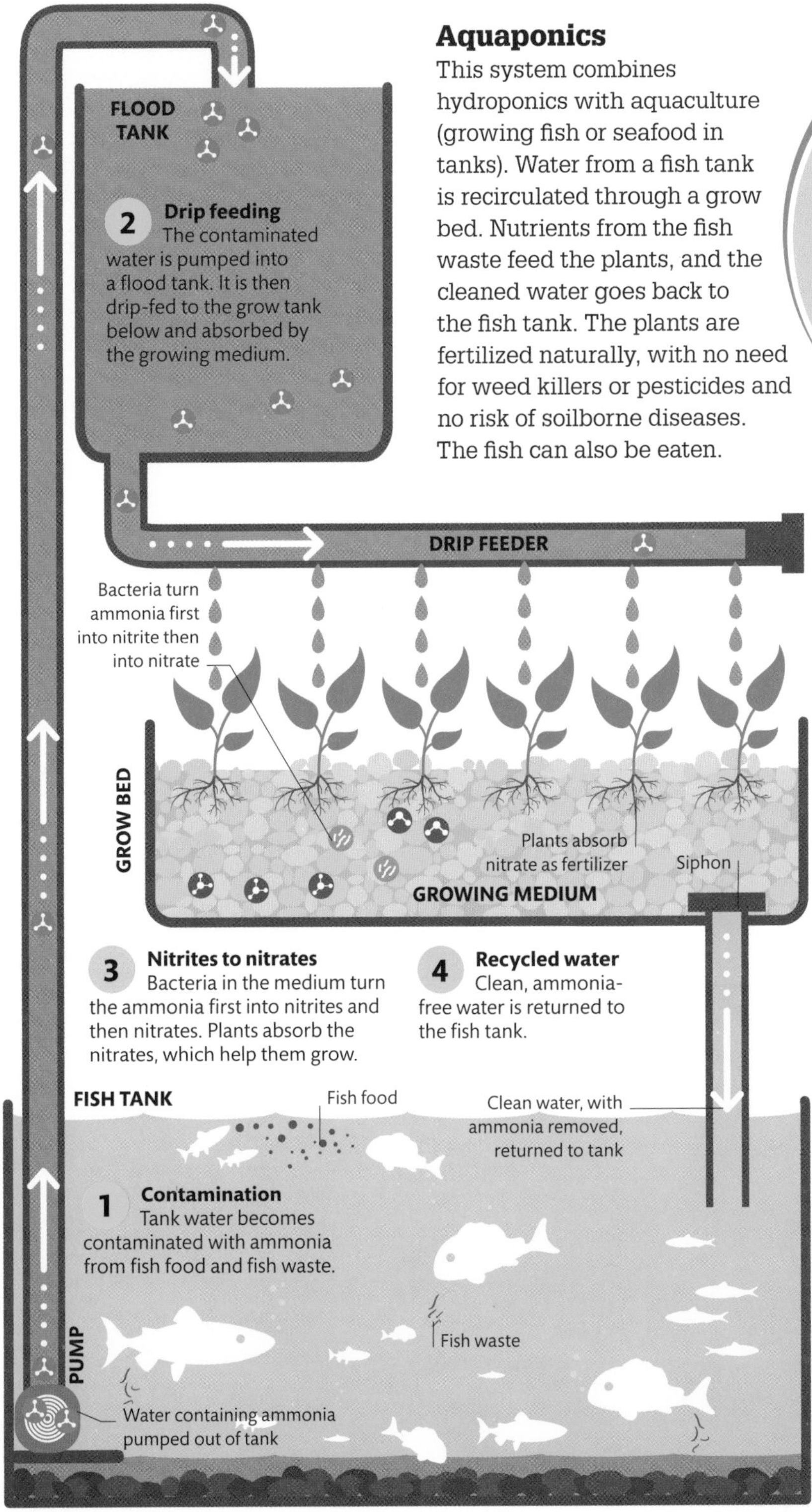
FLOOD TANK
2 Drip feeding
The contaminated water is pumped into a flood tank. It is then drip-fed to the grow tank below and absorbed by the growing medium.
DRIP FEEDER
Bacteria turn ammonia first into nitrite then into nitrate
GROW BED
Plants absorb nitrate as fertilizer
Siphon
GROWING MEDIUM
3 Nitrites to nitrates
Bacteria in the medium turn the ammonia first into nitrites and then nitrates. Plants absorb the nitrates, which help them grow.
4 Recycled water
Clean, ammonia-free water is returned to the fish tank.
FISH TANK
Fish food
Clean water, with ammonia removed, returned to tank
1 Contamination
Tank water becomes contaminated with ammonia from fish food and fish waste.
Fish waste
PUMP
Water containing ammonia pumped out of tank

Aquaponics

This system combines hydroponics with aquaculture (growing fish or seafood in tanks). Water from a fish tank is recirculated through a grow bed. Nutrients from the fish waste feed the plants, and the cleaned water goes back to the fish tank. The plants are fertilized naturally, with no need for weed killers or pesticides and no risk of soilborne diseases. The fish can also be eaten.

HOW MUCH SPACE DOES HYDROPONIC FARMING SAVE?

Farmers can plant between 4 and 10 times the number of plants hydroponically in the same space as a traditional farm.

VERTICAL FARMING

Urban farms may one day feature soilless systems in skyscrapers. Crops could be grown in vertical shelf systems or on lightweight decks. Robots would tend and harvest the crops, while sensors would monitor crop growth.

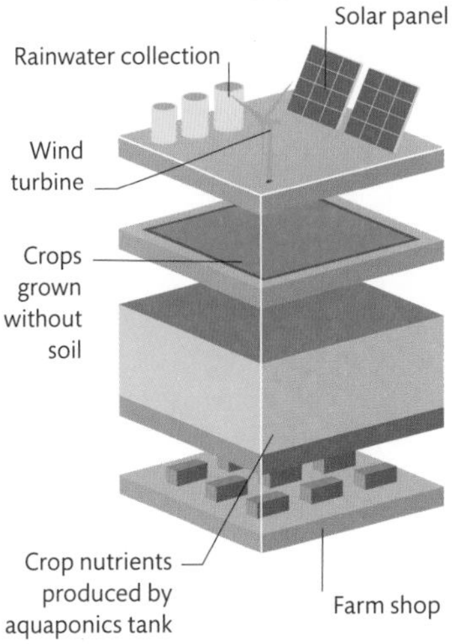
Solar panel
Rainwater collection
Wind turbine
Crops grown without soil
Crop nutrients produced by aquaponics tank
Farm shop

KEY

Ammonia
Bacteria
Nitrite
Nitrate

Precision agriculture

Farming is increasingly going digital. Farmers can now use communications and computer technology to collect data from their crops and animals, which they can then use to manage their farms more efficiently and remotely control their machinery.

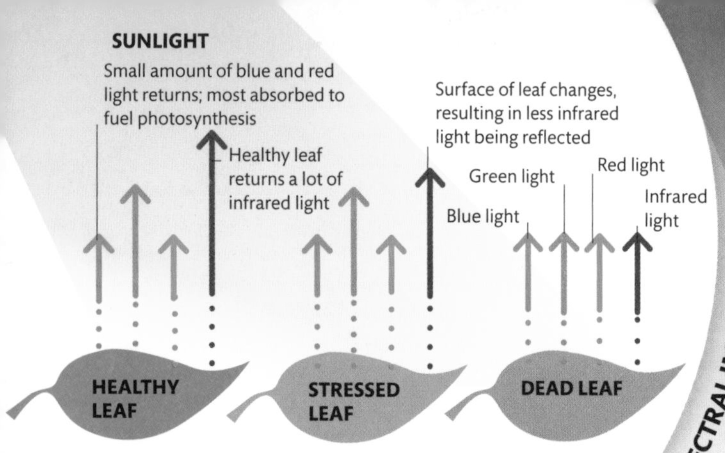

SUNLIGHT
Small amount of blue and red light returns; most absorbed to fuel photosynthesis
Healthy leaf returns a lot of infrared light
Surface of leaf changes, resulting in less infrared light being reflected
Green light
Red light
Blue light
Infrared light
HEALTHY LEAF
STRESSED LEAF
DEAD LEAF

A leaf's surface reflects light differently depending on its physical state. A healthy leaf absorbs most of the blue and red light to use as energy for photosynthesis but reflects back a lot of green and infrared light. However, when a plant becomes stressed (because of disease or dehydration), its physiology changes, and it reflects less green and infrared light.

1 Remote imaging
Drones are used for aerial surveys of the land. Many drones use multispectral cameras, which have multiple lenses. These allow them to survey in infrared and visible wavelengths to detect water levels in soil and chlorophyll content of crops.

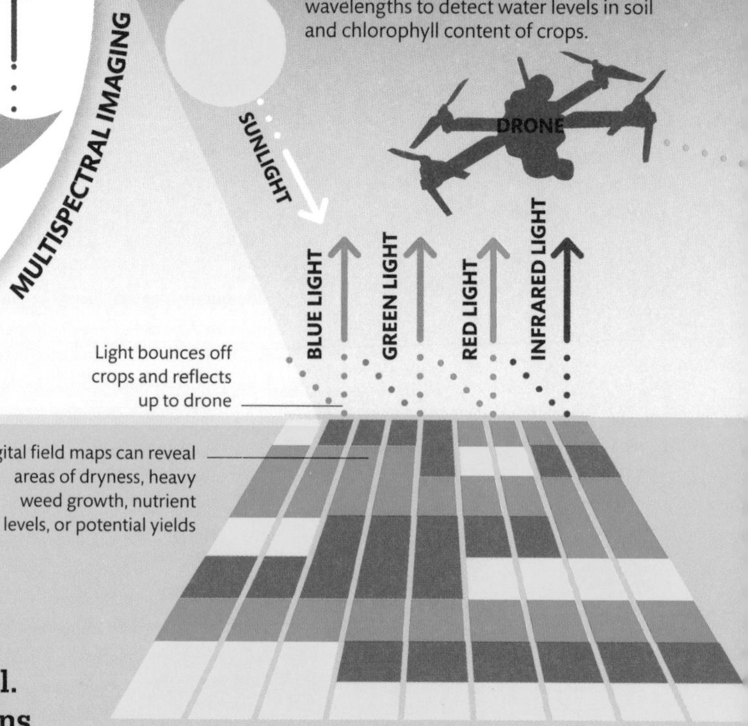

MULTISPECTRAL IMAGING
SUNLIGHT
DRONE
BLUE LIGHT
GREEN LIGHT
RED LIGHT
INFRARED LIGHT
Light bounces off crops and reflects up to drone
Digital field maps can reveal areas of dryness, heavy weed growth, nutrient levels, or potential yields

Monitoring crops

Precision agriculture enables farmers to use data from various sources, from sensors in a field to drones and satellites, to improve crop yields and reduce waste. GPS data allows exact locations to be calculated so that each part of the field can be managed precisely. Farmers can download information about a specific site on a field, such as weed distribution or pH levels in the soil, and treat each location individually. Agricultural equipment connected to the Internet allows a farmer to manage his or her farm remotely.

MONITORING LIVESTOCK

Livestock can have various sensors attached to them that provide useful information for farmers. Chips and tags enable tracking, useful for farmers searching for lost animals, or allow precise identification of animals through regulatory and retail systems. Sensors can also alert the farmer to any medical problems or indicate whether an animal is ready to mate or give birth.

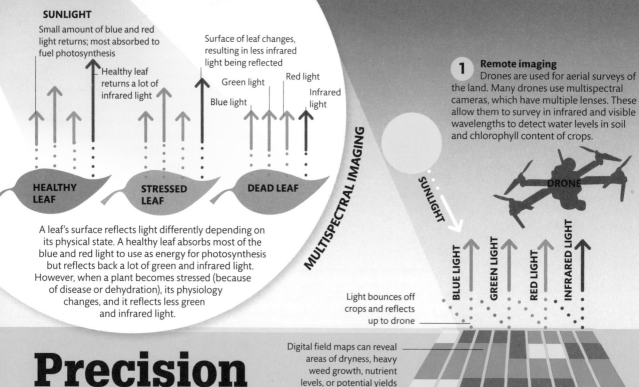

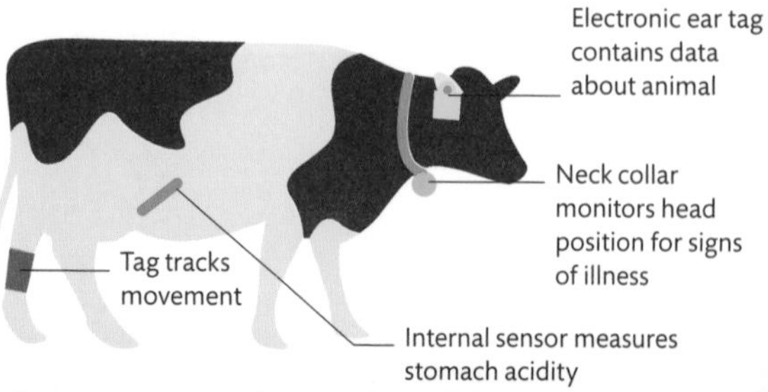

Electronic ear tag contains data about animal
Neck collar monitors head position for signs of illness
Tag tracks movement
Internal sensor measures stomach acidity

3 Gathering all the data
Data from the farmer's drones and various sensors, such as ground soil sensors, is sent to a central data collection hub.

4 Satellite information
Data from GPS (see pp.194–195) and weather satellites is also sent to the cloud (see below). This information can help the farmer plan the optimal time for planting, watering, and harvesting crops or predict when industry may have increased demands for produce.

5 Data analysis and storage
Input from the sources is analyzed and stored in the cloud—a remote server accessed via the Internet. Records can be automatically updated, alerts given, and data used to provide the farmer, regulatory agencies, and other collaborators with information that would take hours to compile by hand.

6 Farmer receives data
Encoded instructions from the farm office or directly from the cloud are uploaded to the machinery. These can then deliver precise amounts of water, fertilizer, or weed killer to the exact part of the field that needs it.

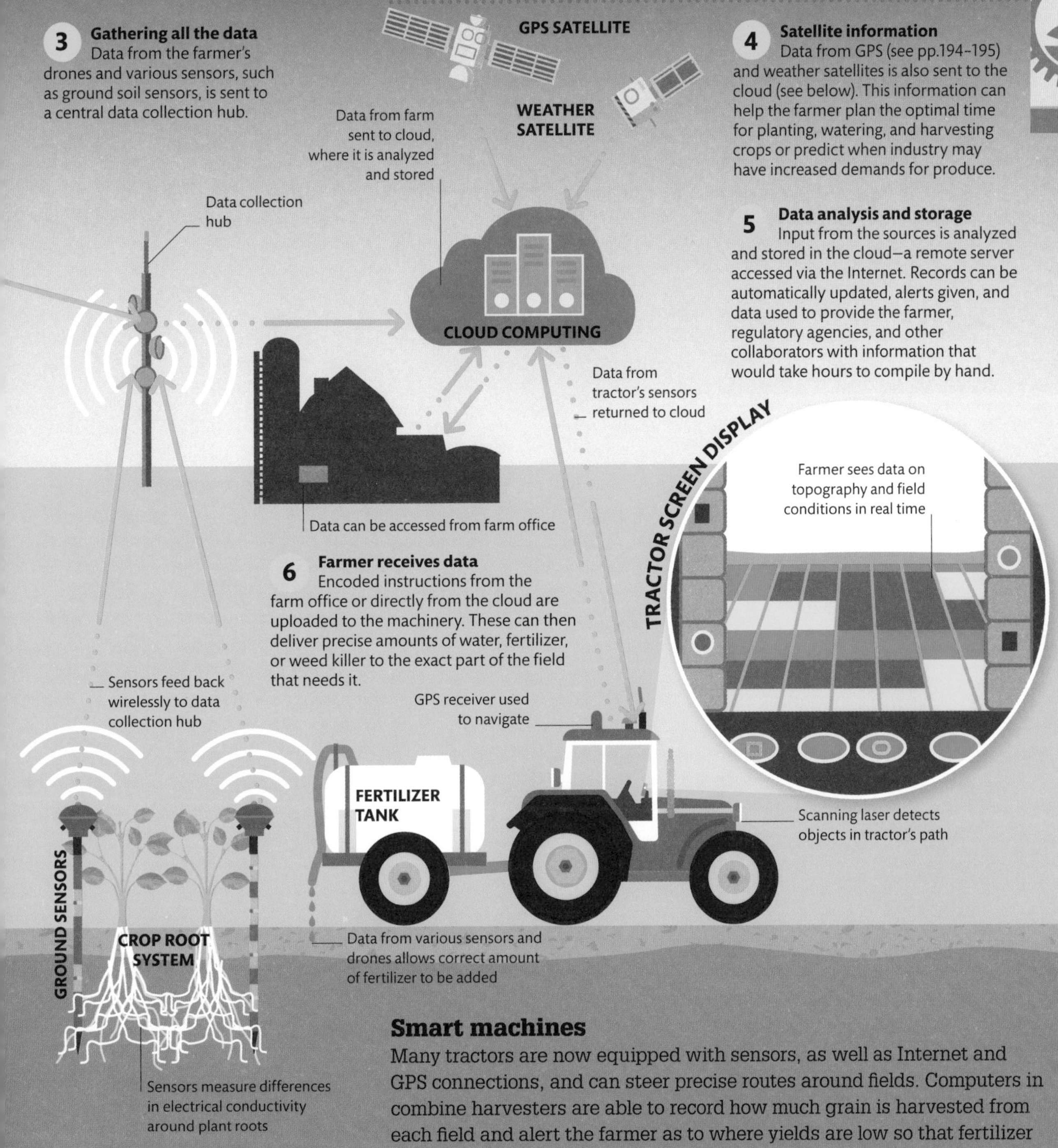
GPS SATELLITE
WEATHER SATELLITE
Data from farm sent to cloud, where it is analyzed and stored
Data collection hub
CLOUD COMPUTING
Data from tractor's sensors returned to cloud
Data can be accessed from farm office
TRACTOR SCREEN DISPLAY
Farmer sees data on topography and field conditions in real time
Sensors feed back wirelessly to data collection hub
GPS receiver used to navigate
FERTILIZER TANK
Scanning laser detects objects in tractor's path
GROUND SENSORS
CROP ROOT SYSTEM
Data from various sensors and drones allows correct amount of fertilizer to be added
Sensors measure differences in electrical conductivity around plant roots

2 Collecting data from the ground
Ground sensors can be used to monitor water, nutrient, and fertilizer levels in the soil. They work by detecting ions that indicate changes in chemical composition. Others detect soil compaction and aeration.

Smart machines

Many tractors are now equipped with sensors, as well as Internet and GPS connections, and can steer precise routes around fields. Computers in combine harvesters are able to record how much grain is harvested from each field and alert the farmer as to where yields are low so that fertilizer can be applied. In the future, fleets of agribots—robot farming machines—might be used, which would have the potential to work day and night. Cultivation could be tailored to each plant. Water and fertilizer could be applied according to need, weeds lasered instead of using herbicides, and only selected parts of the crop harvested rather than the whole plant.

Sorting and packing

Once crops have been harvested, they have to be prepared for their destination. To meet modern quality-control standards, crops have to be sorted, washed, graded, and packed to arrive in peak condition.

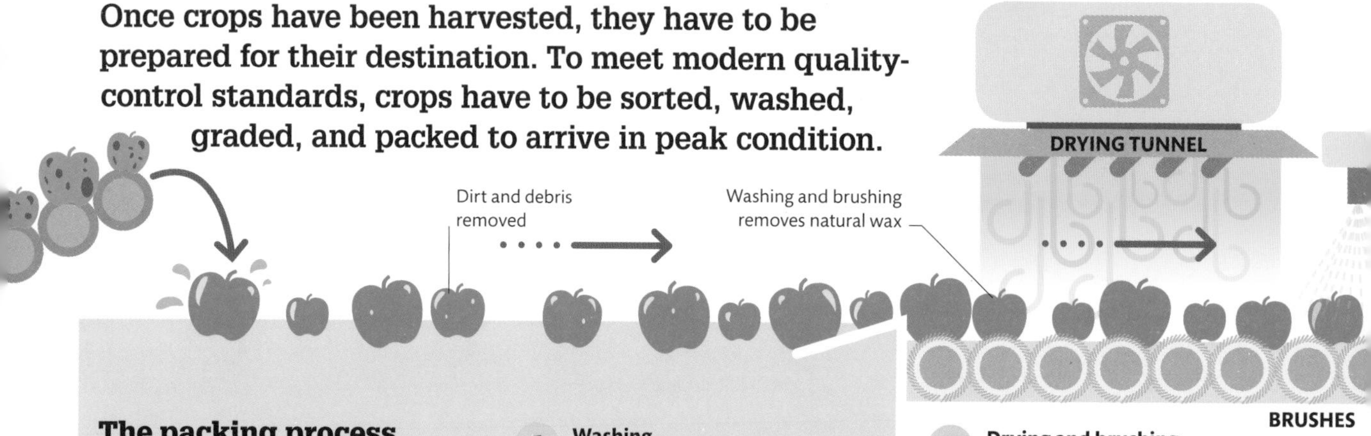
DRYING TUNNEL
Dirt and debris removed
Washing and brushing removes natural wax
BRUSHES

The packing process

Getting a fresh product to customers involves various cleaning, grading, and packing processes. Automation is replacing this once labor-intensive process with optical recognition systems and sorting devices. The process can be adapted to all kinds of fruits and vegetables, from heavy, muddy potatoes to delicate grapes.

1 Washing
Washing is done by submersion in tanks or by overhead sprayers. Mild detergent is added to help remove farm chemicals, pathogens, and dirt.

2 Drying and brushing
The crops are dried as they pass over rotating brushes, which dislodge any surface deposits that were not removed during the washing process.

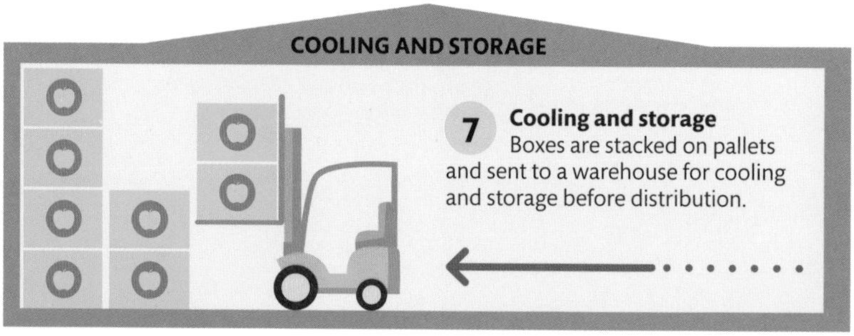
COOLING AND STORAGE

7 Cooling and storage
Boxes are stacked on pallets and sent to a warehouse for cooling and storage before distribution.

Optical sorting

Packing houses often use optical sorters to process crops. The items pass over or under sensors, either on a conveyor or, in freefall sorters (right), while falling. The sensors are connected to an image-processing system. The passing objects are compared with predefined criteria for selection. Rejected items trigger the separation system: a blast of compressed air for small items or mechanical retrieval for larger produce. The rejected material is dumped, while the rest carries on for further processing.

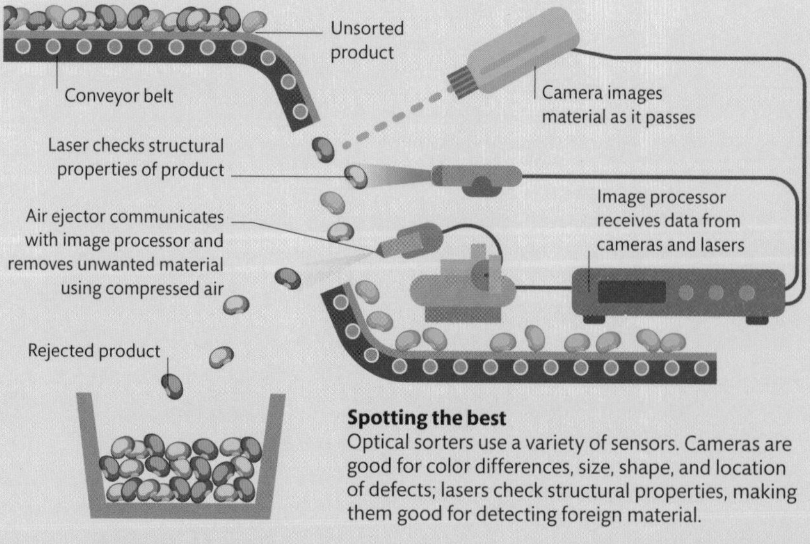
Unsorted product
Conveyor belt
Camera images material as it passes
Laser checks structural properties of product
Image processor receives data from cameras and lasers
Air ejector communicates with image processor and removes unwanted material using compressed air
Rejected product

Spotting the best
Optical sorters use a variety of sensors. Cameras are good for color differences, size, shape, and location of defects; lasers check structural properties, making them good for detecting foreign material.

3 Waxing
Waxing replaces natural waxes on fruits that are lost in washing. Produce may also be dipped in a fungicide or irradiated to reduce organism growth.

4 Hand sorting
Damaged or diseased fruits are picked out by experienced workers, who also remove underripe and misshapen items.

OPTICAL SORTERS CAN PROCESS UP TO 38 TONS (35 TONNES) PER HOUR

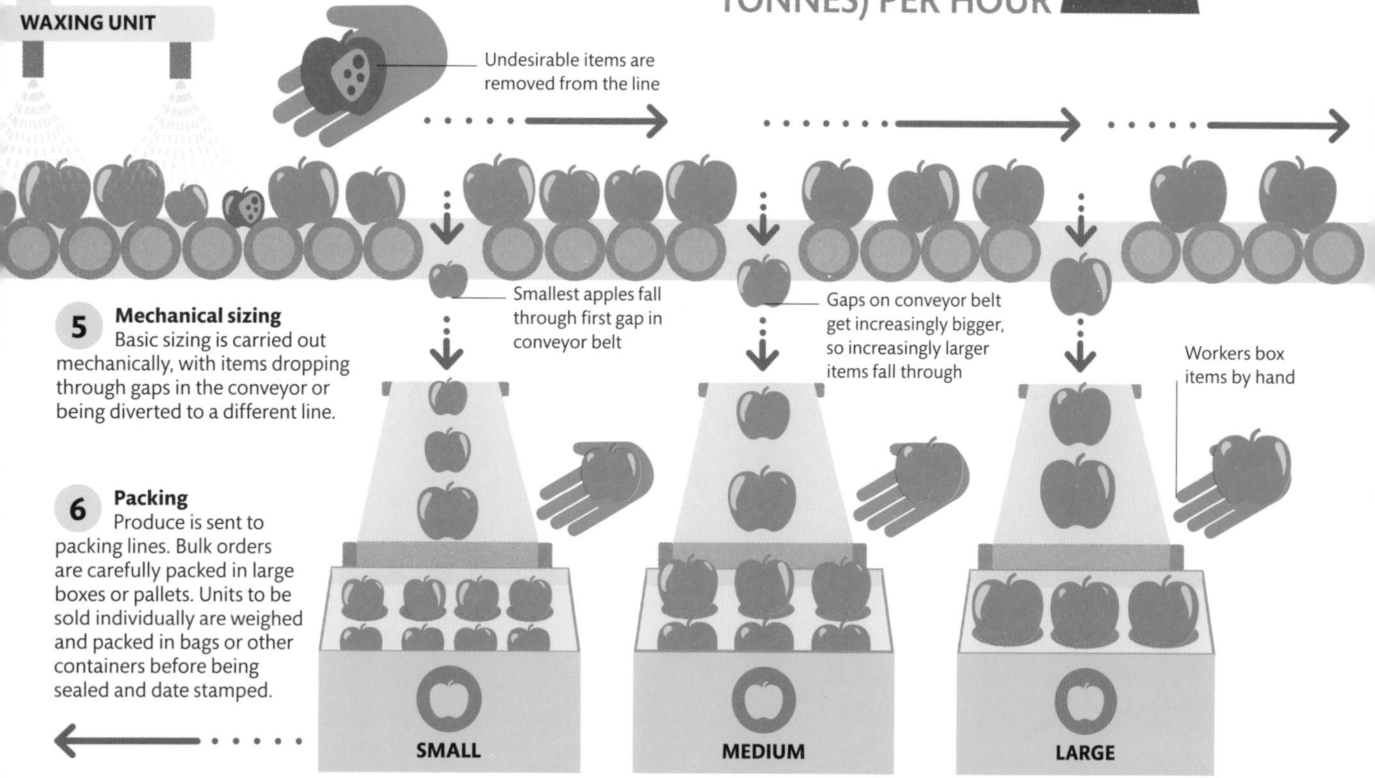
WAXING UNIT
Undesirable items are removed from the line
Smallest apples fall through first gap in conveyor belt
Gaps on conveyor belt get increasingly bigger, so increasingly larger items fall through
Workers box items by hand
SMALL
MEDIUM
LARGE

5 Mechanical sizing
Basic sizing is carried out mechanically, with items dropping through gaps in the conveyor or being diverted to a different line.

6 Packing
Produce is sent to packing lines. Bulk orders are carefully packed in large boxes or pallets. Units to be sold individually are weighed and packed in bags or other containers before being sealed and date stamped.

WHEN WERE CARDBOARD BOXES FIRST USED FOR PACKAGING?

Cardboard was invented in 1856, but it was not until 1903 that it was first formed into boxes and used for packing.

MODIFIED ATMOSPHERE PACKAGING

Some fruits and vegetables have high respiration rates or give off ripening gases that decrease their shelf life. Changing the atmosphere inside the package can slow this down. Vacuum packing removes air, which helps reduce enzyme reactions and bacterial growth. Gas flushing replaces air with a modified gas mixture that prevents spoilage. Permeable bagging materials can be used to allow gases created by produce to diffuse out and equalize with ambient levels.

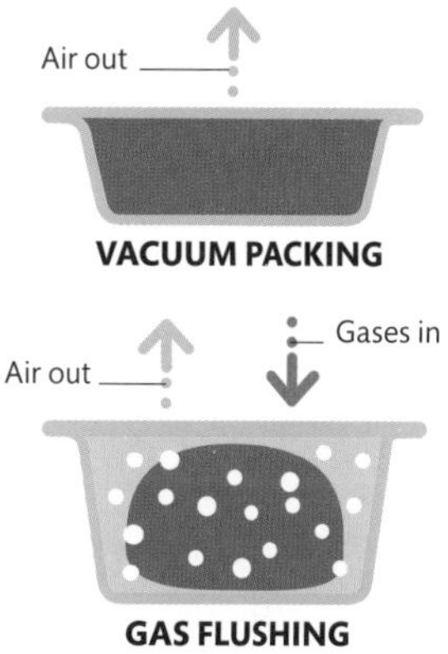
Air out
VACUUM PACKING
Air out
Gases in
GAS FLUSHING

Food preservation

As soon as food is harvested, it is under attack from microscopic organisms, such as bacteria, and enzymes. These degrade the food until it becomes inedible. Over thousands of years, various methods have been developed to hold these processes at bay for as long as possible.

Pasteurization

Pasteurization is a preservation process used for liquids such as milk, sauces, and fruit juices. The liquid is heated at a high temperature for a short length of time before being cooled. The higher the temperature, the shorter the period of time that the liquid must be heated. The heat is sufficient to kill pathogens, yeasts, and molds and deactivate enzymes that would otherwise start to break down the liquid. Products such as milk change consistency if heated for too long so must be kept refrigerated after pasteurization.

1 Raw milk stored
Raw milk is stored in a balance tank. The milk is kept at around 39–41°F (4–5°C) before pasteurization.

2 Initial heating
A pump draws the milk into a heat exchanger called a regenerator. The incoming cold, raw milk is preheated by pipes above that contain heated milk that is further along in the process.

3 Secondary heating
The raw milk passes through a heating section, where pipes filled with hot water, supplied by a hot water pump, heat it further. The long, looped pipe ensures that the milk is kept at the correct temperature for long enough.

4 Heated milk checked
The milk flows into a holding tube, where it is kept for a period of time. A flow diversion pump at the top of the tube makes sure only pasteurized milk leaves. If the milk is hot enough, it begins the cooling process.

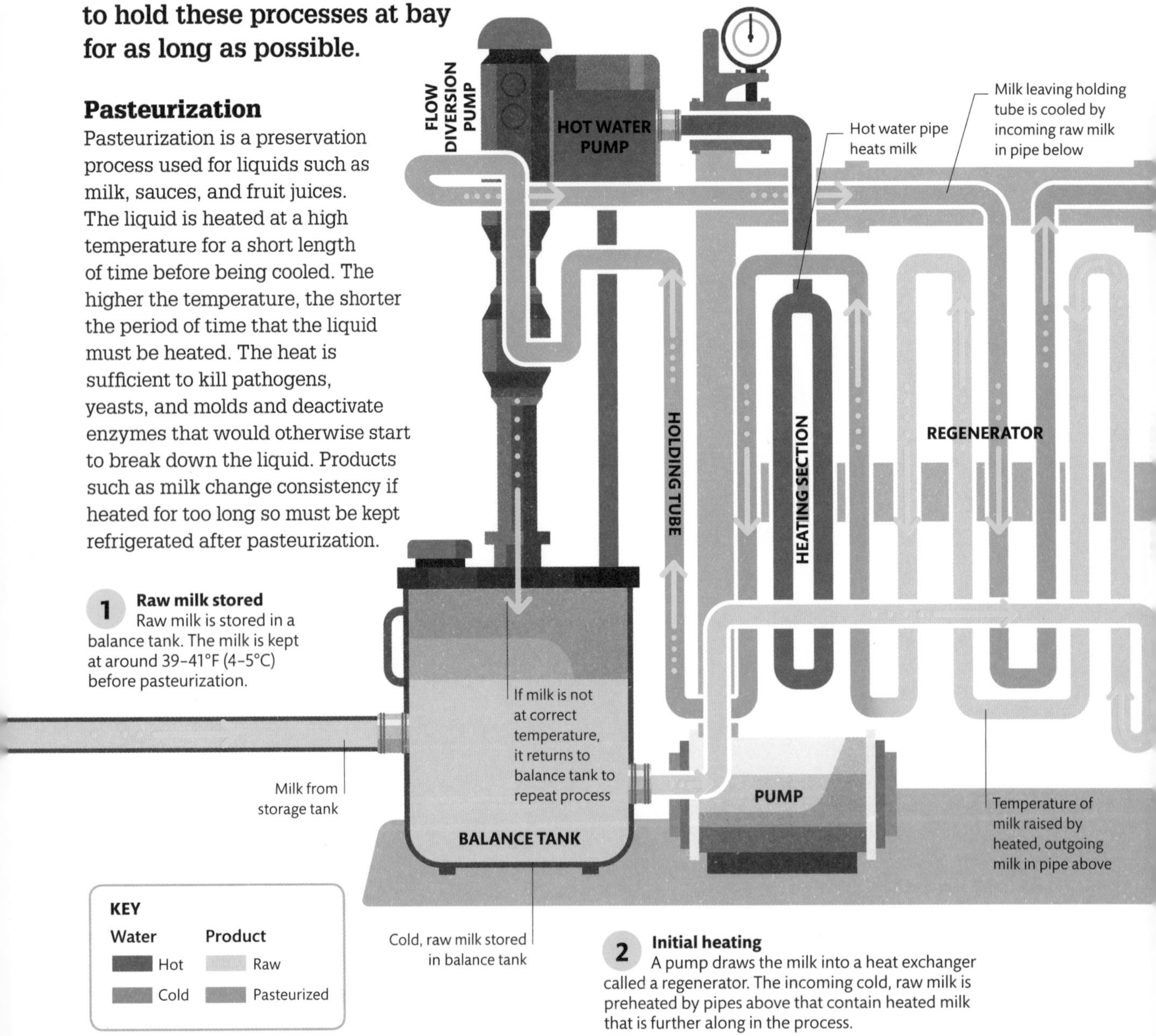

FLOW DIVERSION PUMP
HOT WATER PUMP
Hot water pipe heats milk
Milk leaving holding tube is cooled by incoming raw milk in pipe below
HOLDING TUBE
HEATING SECTION
REGENERATOR
If milk is not at correct temperature, it returns to balance tank to repeat process
Milk from storage tank
PUMP
BALANCE TANK
Temperature of milk raised by heated, outgoing milk in pipe above
Cold, raw milk stored in balance tank
KEY
Water
Hot
Cold
Product
Raw
Pasteurized

6 Flash cooling The treated milk is then rapidly cooled in the cooling section by pipes filled with cold water provided by a cold water pump.

WHAT IS BOTULISM?

Botulism is caused by a toxin released by bacterial spores found in food that has been improperly preserved. It can be fatal.

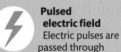

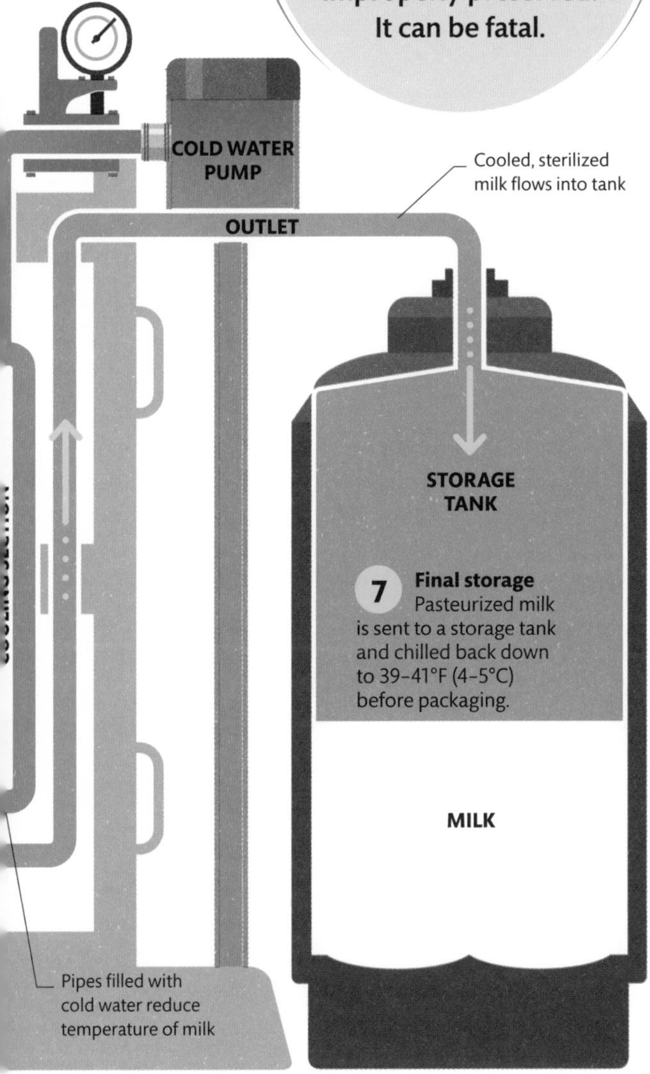
COLD WATER PUMP
Cooled, sterilized milk flows into tank
OUTLET
STORAGE TANK
MILK
Pipes filled with cold water reduce temperature of milk

7 Final storage Pasteurized milk is sent to a storage tank and chilled back down to 39–41°F (4–5°C) before packaging.

5 Initial cooling The heated milk travels from the holding tube to the next section of the regenerator. It is cooled by the incoming cold milk in the pipe below.

Preservation methods

Some methods of preserving food have been used since ancient times and are still in use today. Pickling, sugaring, fermentation, smoking, curing, cold storage, salting, freezing, canning, and even burial all create conditions that are hostile to spoilage organisms. In recent years, though, commercial processing has led to the development of new preservation technology.

Irradiation Ionizing radiation kills molds, bacteria, and insects, sterilizes food at high doses, and slows the ripening of fruits.

Vacuum packing Food is sealed in plastic pouches in a vacuum. This protects foods against oxidation and suffocates any bacteria.

Pressurization Sealed food is placed in a container that is then filled with a liquid, creating high pressure. This deactivates microbes.

Food additives Substances such as antimicrobials and antioxidants are added to the product to inhibit growth of organisms and prevent spoilage.

Modified atmospheres Replacing air with CO_2 (carbon dioxide) or nitrogen prevents the growth of microbes and suffocates insects.

Biopreservation Naturally present microbes or antimicrobials are used to preserve the foodstuff. The method is often used in meat and seafood processing.

Hurdle technology A series of challenges are set for organisms to overcome, such as high acidity, additives, and oxygen removal.

Pulsed electric field Electric pulses are passed through the foodstuff. This enlarges the pores of bacterial cells and kills them.

GRAIN STORED IN A **CO_2 ATMOSPHERE** REMAINS EDIBLE FOR **FIVE YEARS**

WHAT IS THE OLDEST FOOD ADDITIVE?

Salt has been used for around 10,000 years as a preservative for meat and vegetables, as well as to bring out the flavor in food.

Food processing

Most food for sale has undergone some sort of processing, usually to prolong its shelf life or to turn it into something more useful to the customer. Even fresh produce goes through basic processing.

Making a lasagne ready meal

The epitome of processed food is the ready meal, where the main course and any accompanying side dishes are ready to heat and eat from a tray. Ready meals involve heavily automated production lines where the ingredients are prepared, cooked, and packed in a continuous process. Several lines are needed to create an elaborate dish, such as a lasagne.

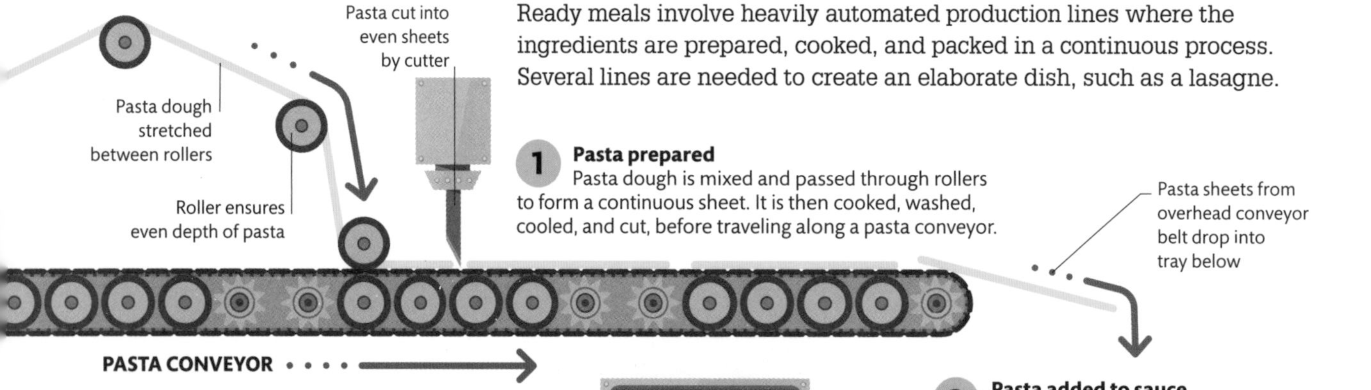

Pasta cut into even sheets by cutter
Pasta dough stretched between rollers
Roller ensures even depth of pasta
Pasta sheets from overhead conveyor belt drop into tray below
PASTA CONVEYOR

1 Pasta prepared
Pasta dough is mixed and passed through rollers to form a continuous sheet. It is then cooked, washed, cooled, and cut, before traveling along a pasta conveyor.

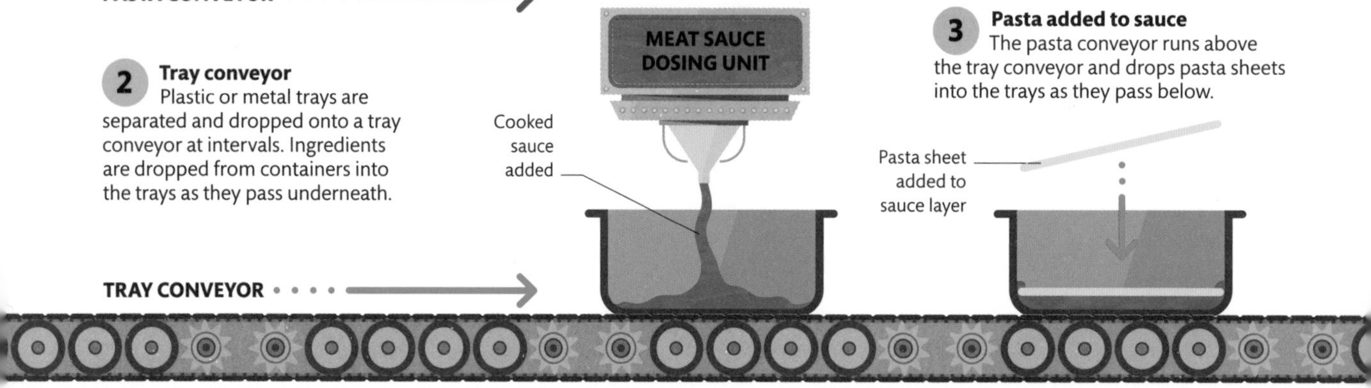

MEAT SAUCE DOSING UNIT
Cooked sauce added
Pasta sheet added to sauce layer
TRAY CONVEYOR

2 Tray conveyor
Plastic or metal trays are separated and dropped onto a tray conveyor at intervals. Ingredients are dropped from containers into the trays as they pass underneath.

3 Pasta added to sauce
The pasta conveyor runs above the tray conveyor and drops pasta sheets into the trays as they pass below.

6 Packaging
The tray passes under a roller that applies a thin sheet of film that is then heat-sealed onto the tray and trimmed. The tray is then wrapped in a cardboard sleeve or box that includes the date produced and ingredients it contains.

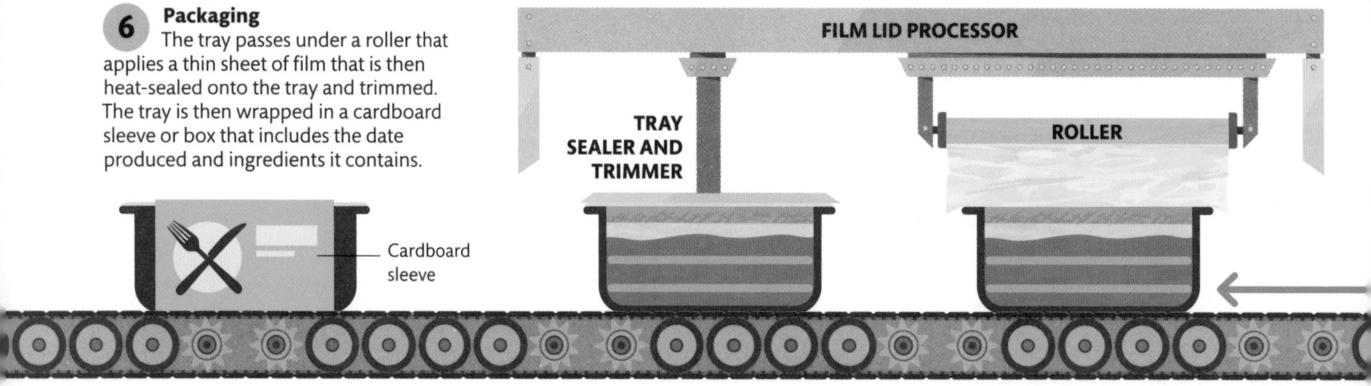

FILM LID PROCESSOR
TRAY SEALER AND TRIMMER
ROLLER
Cardboard sleeve

Additives

Food additives are often viewed as a bad thing, but many are necessary to preserve the appearance, taste, and shelf life of processed foods. Processing can also destroy nutrients and natural colors and flavors, so these have to be added back in. Common additives include bulking agents, preservatives, thickeners, acidulants (which increase acidity), sweeteners, and colorants. Many additives are natural products, and all additives have to meet regulatory standards.

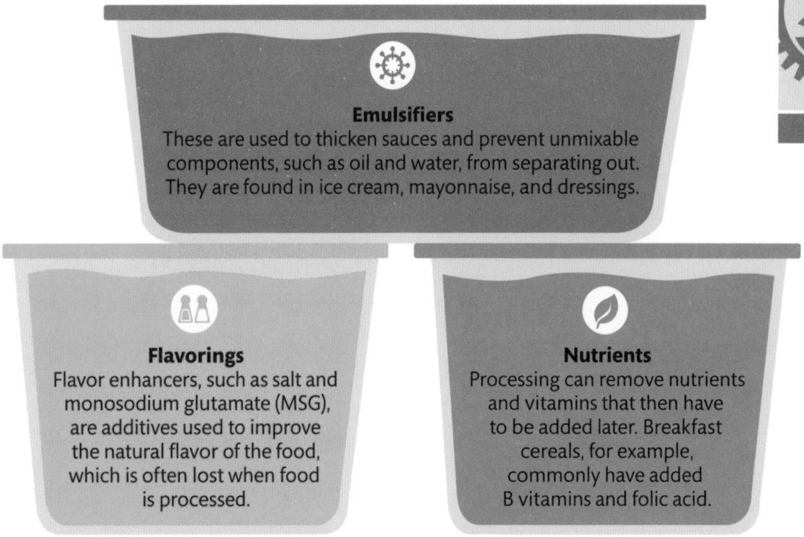
Emulsifiers
These are used to thicken sauces and prevent unmixable components, such as oil and water, from separating out. They are found in ice cream, mayonnaise, and dressings.
Flavorings
Flavor enhancers, such as salt and monosodium glutamate (MSG), are additives used to improve the natural flavor of the food, which is often lost when food is processed.
Nutrients
Processing can remove nutrients and vitamins that then have to be added later. Breakfast cereals, for example, commonly have added B vitamins and folic acid.

4 Dosing units add sauce
The trays continue along the conveyor, passing underneath dosing units, which provide layers of sauce, and the pasta conveyor, which adds more pasta.

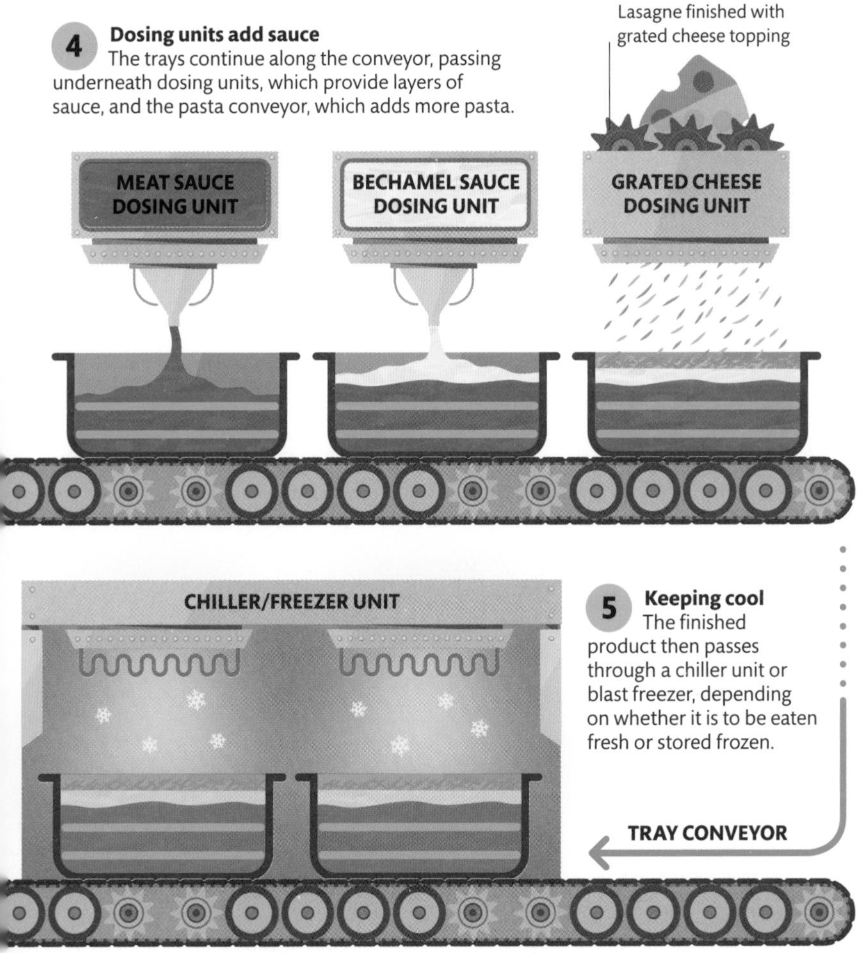
Lasagne finished with grated cheese topping
MEAT SAUCE DOSING UNIT
BECHAMEL SAUCE DOSING UNIT
GRATED CHEESE DOSING UNIT
CHILLER/FREEZER UNIT
TRAY CONVEYOR

5 Keeping cool
The finished product then passes through a chiller unit or blast freezer, depending on whether it is to be eaten fresh or stored frozen.

THE FIRST READY MEAL WAS CREATED IN 1953 TO USE UP LEFTOVER THANKSGIVING TURKEY

AIRPLANE FOOD

Extra additives have to be added to in-flight ready meals, since our ability to smell and taste food at high altitude diminishes. Salt and sugar become particularly difficult to taste in the low pressure and humidity of aircraft cabins. Spices are often added to increase flavor.

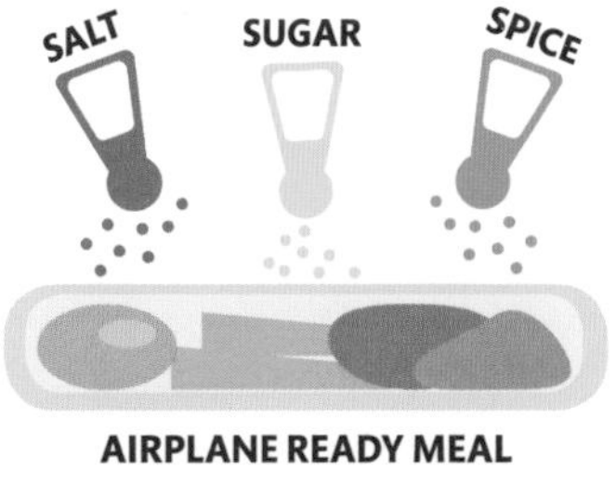
SALT
SUGAR
SPICE
AIRPLANE READY MEAL

Genetic modification

Genetic modification of crops and animals has made significant inroads into farming. Its use is controversial in many parts of the world, although it is often argued that it is the only way to feed an ever-growing population.

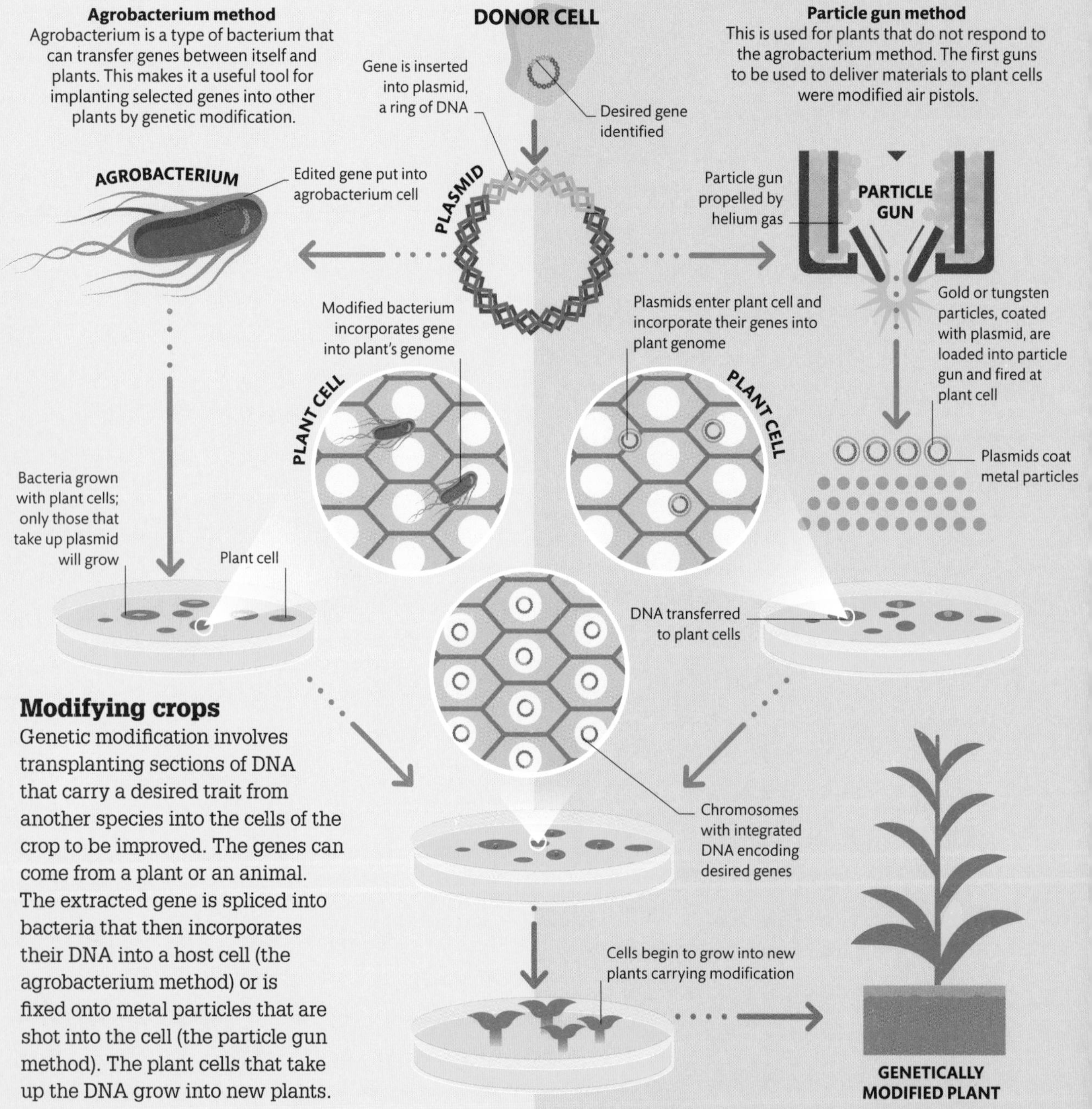

Agrobacterium method
Agrobacterium is a type of bacterium that can transfer genes between itself and plants. This makes it a useful tool for implanting selected genes into other plants by genetic modification.
DONOR CELL
Gene is inserted into plasmid, a ring of DNA
Desired gene identified
Particle gun method
This is used for plants that do not respond to the agrobacterium method. The first guns to be used to deliver materials to plant cells were modified air pistols.
AGROBACTERIUM
Edited gene put into agrobacterium cell
PLASMID
Particle gun propelled by helium gas
PARTICLE GUN
Gold or tungsten particles, coated with plasmid, are loaded into particle gun and fired at plant cell
Modified bacterium incorporates gene into plant's genome
Plasmids enter plant cell and incorporate their genes into plant genome
PLANT CELL
PLANT CELL
Plasmids coat metal particles
Bacteria grown with plant cells; only those that take up plasmid will grow
Plant cell
DNA transferred to plant cells
Chromosomes with integrated DNA encoding desired genes
Cells begin to grow into new plants carrying modification
GENETICALLY MODIFIED PLANT

Modifying crops

Genetic modification involves transplanting sections of DNA that carry a desired trait from another species into the cells of the crop to be improved. The genes can come from a plant or an animal. The extracted gene is spliced into bacteria that then incorporates their DNA into a host cell (the agrobacterium method) or is fixed onto metal particles that are shot into the cell (the particle gun method). The plant cells that take up the DNA grow into new plants.

GENETICALLY MODIFIED ANIMALS

While genetically engineered crops are already being grown commercially in some parts of the world, most modified animals are still at the research stage. Genetically modified (GM) livestock are being bred to improve commercially important traits such as better growth rate, disease resistance, meat quality, or offspring survival rate. GM salmon, for example, have been bred to grow twice as fast as conventional salmon.

THE **FIRST GENETICALLY MODIFIED CROP** TO BE SOLD WAS A **TOMATO**

Transgenic animals

A few products are already produced by transgenic livestock, with others under development. Transgenic animals are animals that have had a gene from another species inserted into their DNA. One use of transgenic animals is to produce pharmaceutical goods. Raising animals is cheaper than setting up a pharmaceutical production line to make drugs, but developers are currently restricted to products that can be extracted from milk, eggs, or other products that do not harm the animal. The use of urine also has potential for investigation as it is not dependent on the sex or age of the animal.

ANIMAL	USES
Cow 	Transgenic cows can be used to create several products, such as milk containing human lactoferrin, a protein that can be used to treat infections. Scientists have also created genetically modified cows' milk that has a lower lactose content, making it suitable for people who are lactose intolerant.
Pig 	Scientists are researching how the genes of pigs could be edited so that the animals' organs could be made suitable for use in human organ transplants. Pigs have been genetically modified so that they produce phytase, an enzyme that reduces the pig's excretion of phosphorus, making its waste less polluting.
Goat 	Goats have been genetically modified to produce human antithrombin, a protein that prevents blood clotting (see below). Scientists have also created goats that are capable of producing silk in their milk by inserting the silk protein gene found in spiders into the goat's DNA.
Sheep 	Scientists have produced sheep that have high levels of omega-3 fatty acids in their meat by inserting a roundworm gene, linked to the production of the fatty acids, into the sheep genome. Sheep have also been genetically modified to carry the gene for Huntington's disease, allowing scientists to study the illness.

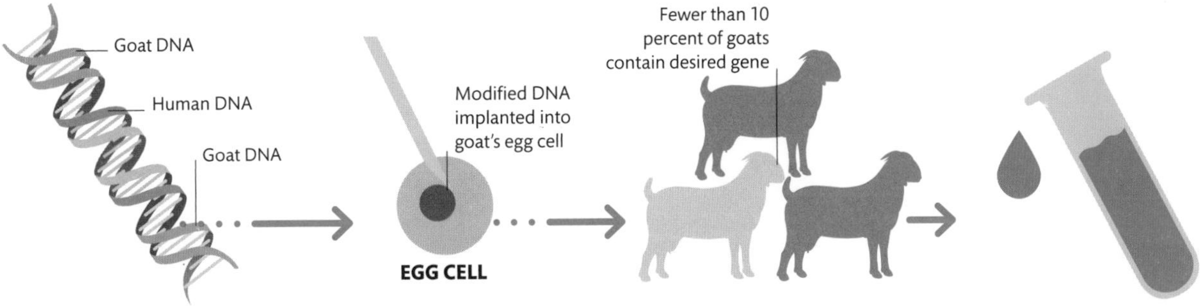

1 Modifying the DNA A section of human DNA containing the code for the blood hormone antithrombin (which reduces clotting) is inserted into goat DNA.

2 Implanting the DNA The modified strand of DNA is injected into the nucleus of a fertilized goat egg. It is then implanted into a female goat, who carries the embryo to term.

3 Testing the offspring The offspring are tested to see whether they carry the antithrombin gene. Those that do are bred to form a herd of modified goats.

4 Extracting the protein Milk from the modified goats is filtered and purified. In a year, a single goat can produce as much antithrombin as 90,000 blood donations.

MEDICAL TECHNOLOGY

Pacemakers

A pacemaker is a battery-powered device implanted in the chest that corrects heartbeat abnormalities by sending electrical impulses to the heart.

Normal heart activity

Heartbeats occur when nerve signals make the heart muscle contract. The signals come from patches of nerve tissue in the heart called nodes. Each heartbeat begins with a signal from the sinoatrial node, the "natural pacemaker," to make the upper chambers (atria) contract. The signal then passes to the atrioventricular node and down to the lower chambers (ventricles), making them contract.

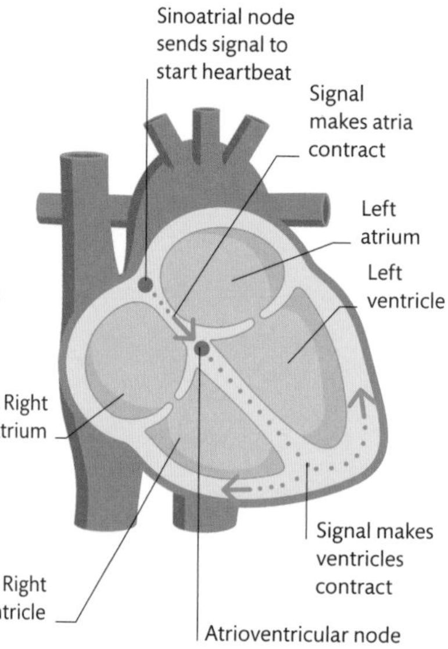

Sinoatrial node sends signal to start heartbeat
Signal makes atria contract
Left atrium
Left ventricle
Right atrium
Signal makes ventricles contract
Right ventricle
Atrioventricular node

CAN I USE A CELL PHONE IF I HAVE A PACEMAKER?

Yes, but the phone should be kept at least 6 in (15 cm) away from the pacemaker. There is no evidence that Wi-Fi or other wireless Internet devices interfere with pacemakers.

How pacemakers work

In some heart disorders, the heart's nodes do not work properly, so the heart beats too slowly, too fast, or with an abnormal rhythm. A pacemaker may be implanted into the patient's chest, to take over the role of the nodes and regulate the heartbeat. Some pacemakers act on one chamber of the heart, while others act on two or three chambers to ensure that the chambers work in a normal rhythm.

Leadless pacemakers

Some pacemakers no longer need wires in order to work. These tiny devices are implanted directly into the right ventricle of the heart using a catheter. They contain a battery and a microchip that senses and, if necessary, corrects the heart rhythm. The microchip also transmits data to electrodes on the skin, enabling heart activity to be monitored externally.

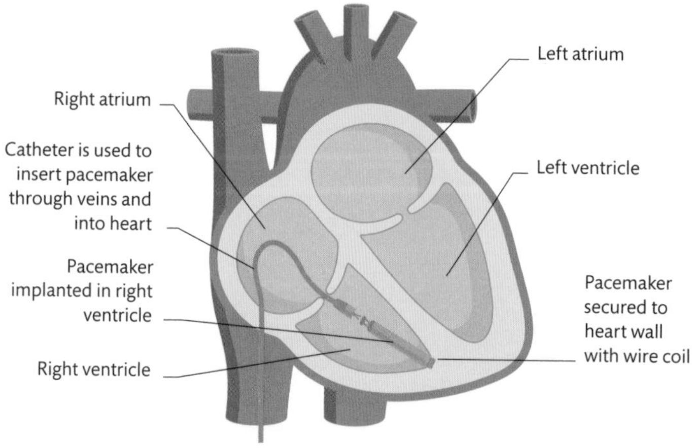

Left atrium
Right atrium
Catheter is used to insert pacemaker through veins and into heart
Left ventricle
Pacemaker implanted in right ventricle
Pacemaker secured to heart wall with wire coil
Right ventricle

Biventricular pacemaker

This device is used for people with disorders, such as heart failure, in which the ventricles fail to contract at the same time. The pacemaker has three leads and sends signals to the right atrium and to both ventricles at once to synchronize contractions of the chambers. Treatment with a biventricular pacemaker is also sometimes called cardiac resynchronization therapy (CRT).

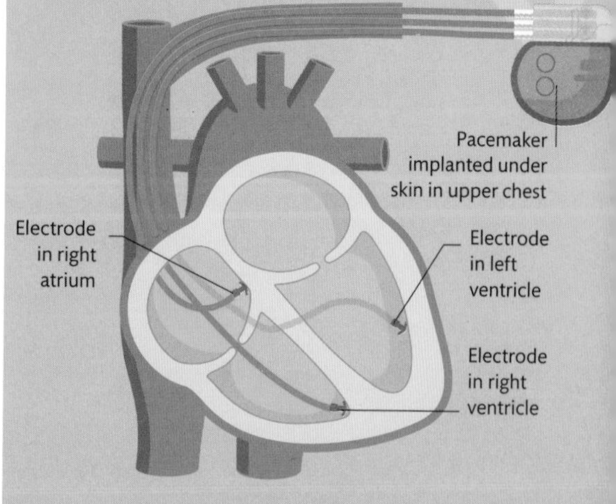

Pacemaker implanted under skin in upper chest
Electrode in right atrium
Electrode in left ventricle
Electrode in right ventricle

WORLDWIDE, MORE THAN 1 MILLION PACEMAKERS ARE IMPLANTED EVERY YEAR

Dual-chamber pacemaker
This device has two leads: one for the right atrium and one for the right ventricle. It is used to correct faulty signals from the heart's nodes that cause an abnormal heartbeat rhythm. By sending out corrective signals, the pacemaker makes the heart's chambers contract in a normal rhythm.

ICD

An implantable cardioverter defibrillator (ICD) is fitted to people at risk of life-threatening abnormal heart rhythms. Like a pacemaker, an ICD can detect very fast or chaotic heartbeats; in such cases, the ICD gives the heart a small electric shock (cardioversion) or a larger shock (defibrillation) to reestablish a normal heart rhythm. Sometimes, an ICD may be combined with a pacemaker.

1 Pacemaker monitors heart
Electrodes inside the heart's chambers constantly monitor electrical signals in the heart and send data about this activity to a microprocessor inside the pacemaker. The microprocessor is programmed to recognize when signals are abnormal or missing.

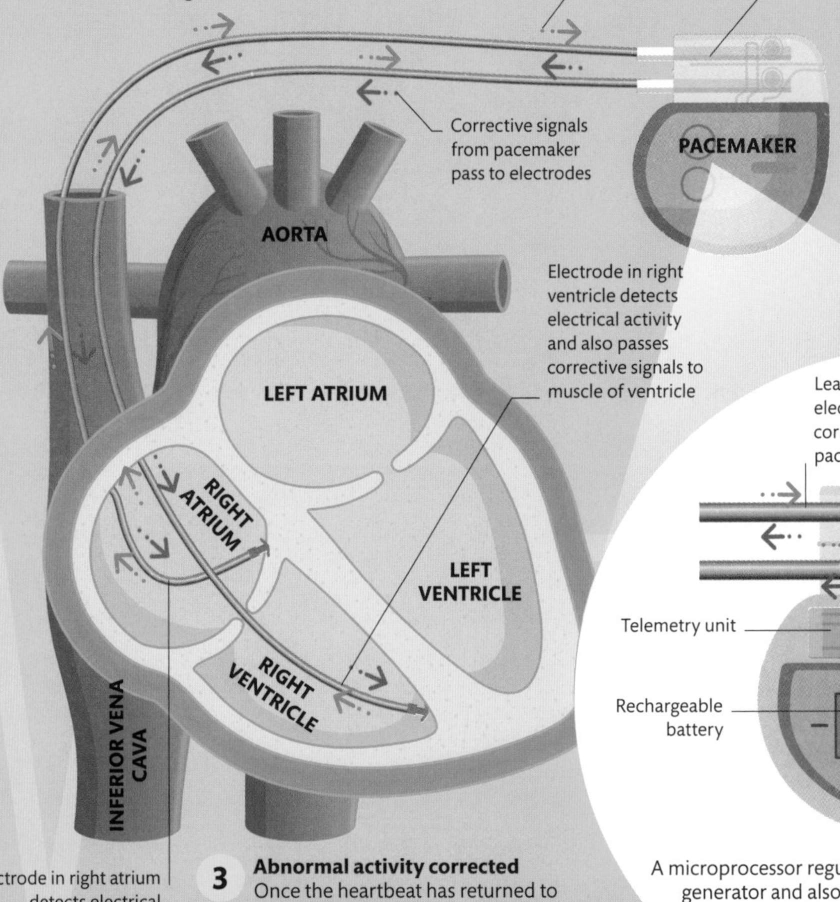
Information from electrodes passes to pacemaker
Pacemaker implanted under skin in upper chest
Corrective signals from pacemaker pass to electrodes
PACEMAKER
AORTA
Electrode in right ventricle detects electrical activity and also passes corrective signals to muscle of ventricle
LEFT ATRIUM
RIGHT ATRIUM
LEFT VENTRICLE
RIGHT VENTRICLE
INFERIOR VENA CAVA
Electrode in right atrium detects electrical activity and also passes corrective signals to muscle of atrium

2 Pacemaker detects abnormal activity
When the microprocessor identifies abnormal signals, it instructs the pulse generator in the pacemaker to transmit low-voltage electrical pulses to the electrodes in the heart. The pulses stimulate the muscle in the heart's chambers to contract.

3 Abnormal activity corrected
Once the heartbeat has returned to normal, the pacemaker stops sending electrical pulses. However, it continues to monitor the heart and collect data. This information can be relayed to an external computer, enabling doctors to assess how well the pacemaker is working.

INSIDE A PACEMAKER

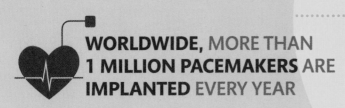

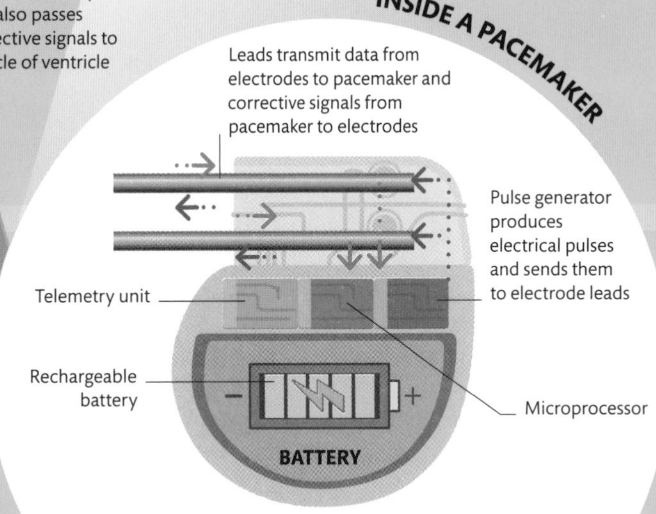
Leads transmit data from electrodes to pacemaker and corrective signals from pacemaker to electrodes
Pulse generator produces electrical pulses and sends them to electrode leads
Telemetry unit
Rechargeable battery
Microprocessor
BATTERY

A microprocessor regulates electrical pulses sent out by a pulse generator and also contains a memory and a monitor to collect data about the heart's activity. Connected to the microprocessor is a telemetry unit, which exchanges data with an external computer. Power is provided by a rechargeable battery.

X-ray imaging

The most familiar type of medical imaging, X-rays are used to view internal body tissues and detect disorders, such as fractures or tumors. X-ray imaging is generally simple and painless, although it does involve exposure to radiation.

DO X-RAYS INCREASE THE RISK OF CANCER?

Yes, although the risk depends on the type of X-ray. On average, a single plain X-ray of the chest, limbs, or teeth gives you an additional risk of getting cancer of less than one in a million.

Digital X-ray imaging
The patient is positioned between an X-ray generator and a detector. X-rays from the generator pass through the patient's body to the detector, which converts the X-ray pattern it captures into digital signals. These signals are then processed by computer into an image that is displayed on a monitor.

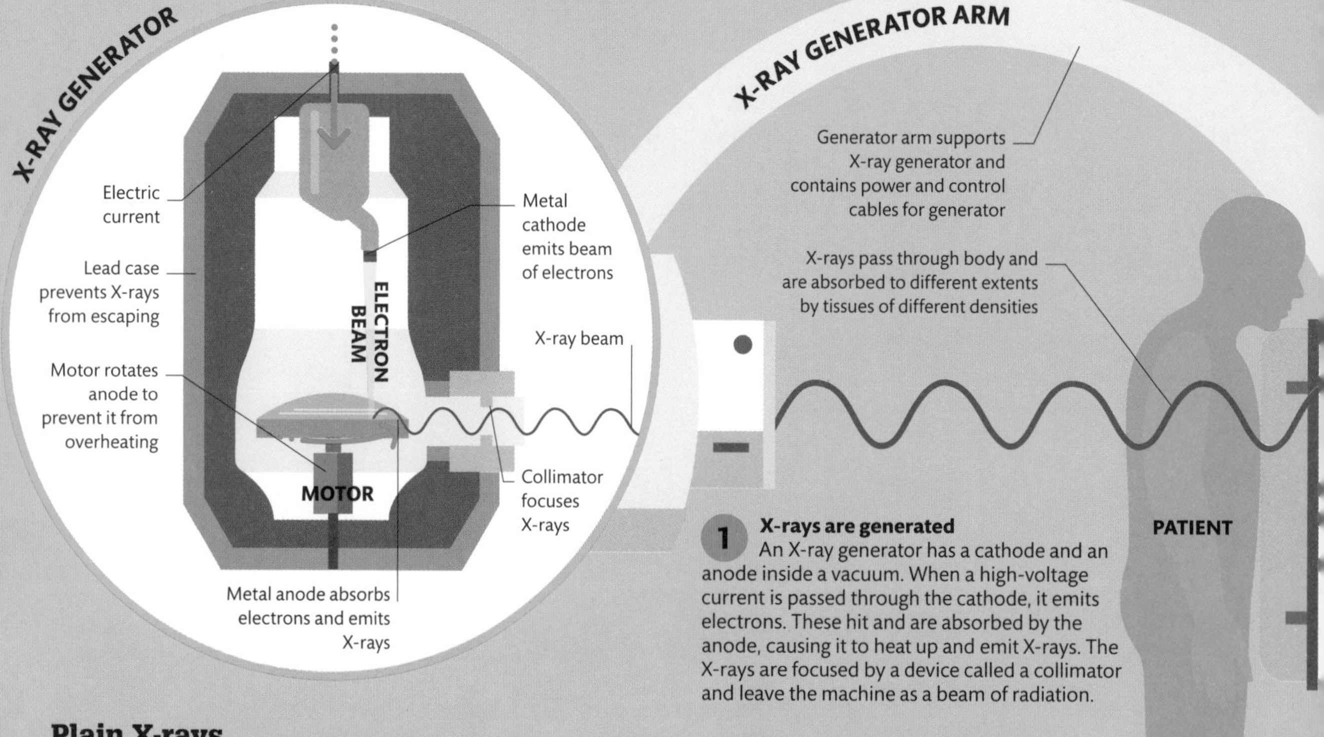

X-RAY GENERATOR
Electric current
Lead case prevents X-rays from escaping
Motor rotates anode to prevent it from overheating
MOTOR
ELECTRON BEAM
Metal cathode emits beam of electrons
X-ray beam
Collimator focuses X-rays
Metal anode absorbs electrons and emits X-rays
X-RAY GENERATOR ARM
Generator arm supports X-ray generator and contains power and control cables for generator
X-rays pass through body and are absorbed to different extents by tissues of different densities
PATIENT

1 X-rays are generated
An X-ray generator has a cathode and an anode inside a vacuum. When a high-voltage current is passed through the cathode, it emits electrons. These hit and are absorbed by the anode, causing it to heat up and emit X-rays. The X-rays are focused by a device called a collimator and leave the machine as a beam of radiation.

Plain X-rays

X-rays are a type of electromagnetic radiation, like light, but they are invisible (see p.137). They are also much higher in energy than light and so can pass through body tissues. When X-rays are directed at the body, they pass easily through softer, less dense tissue, such as muscle and lung tissue, but much less readily through dense tissue, such as bone. In digital X-ray imaging, the X-rays that pass through the body are picked up by a special detector, and the image data is then processed by computer into an image. Traditional X-ray imaging uses photographic film, but this method is now rarely used.

LEAD IS VERY **DENSE** AND SO IS PARTICULARLY EFFECTIVE AT **SHIELDING** AGAINST **X-RAYS**

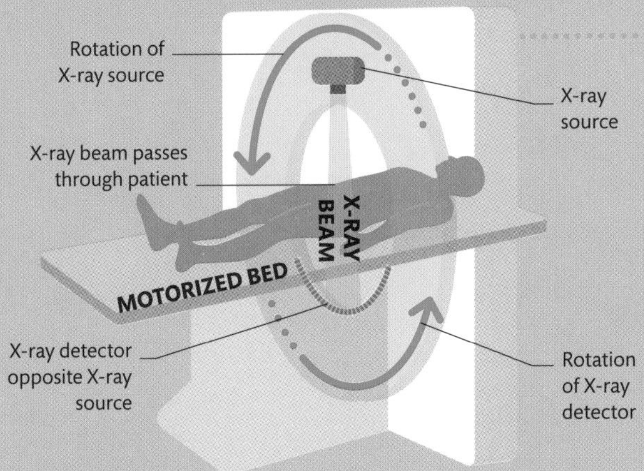

CT scans

Computer tomography (CT) scanning is a type of X-ray imaging that produces cross-sectional images ("slices") through the body. During a CT scan, the X-ray source and detector rotate around the patient, who lies on a motorized bed that moves forward with each scan. The detector is extremely sensitive, and the image data from it can be processed by computer to create highly detailed or even 3-D images of body tissues.

OTHER TYPES OF MEDICAL X-RAYS

In addition to plain X-rays and CT scans, there are various specialized X-rays, some of which use a contrast medium (a substance opaque to X-rays) to highlight specific tissues.

Dental X-rays
Low-dose X-rays of the teeth and jaws to reveal dental problems such as decay, abscesses, or disorders of the gums or jawbone.

Bone-density scanning
Low-dose X-ray scanning to reveal any areas of low bone density; usually carried out on the spine or pelvis to check for osteoporosis.

Mammography
Low-dose X-ray imaging of the breasts to detect any abnormalities such as tumors; frequently carried out to screen for breast cancer in women.

Angiography
X-ray imaging of the heart and blood vessels using an injected liquid contrast medium to show the interior of these structures clearly.

Fluoroscopy
X-rays directed onto a fluorescent screen to give real-time views of moving body parts or track the movement of medical devices through the body.

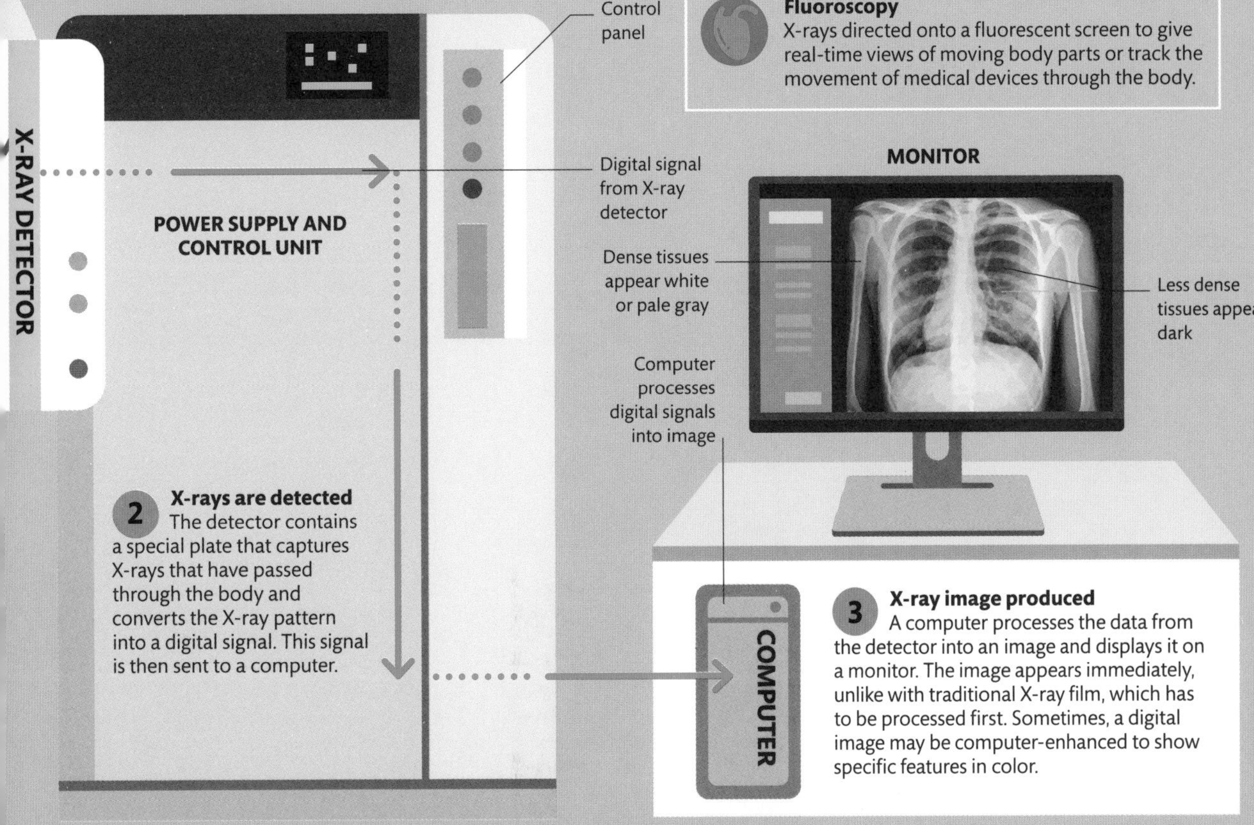

2 X-rays are detected
The detector contains a special plate that captures X-rays that have passed through the body and converts the X-ray pattern into a digital signal. This signal is then sent to a computer.

3 X-ray image produced
A computer processes the data from the detector into an image and displays it on a monitor. The image appears immediately, unlike with traditional X-ray film, which has to be processed first. Sometimes, a digital image may be computer-enhanced to show specific features in color.

MRI scanner

Magnetic resonance imaging (MRI) is a technique in which a powerful magnetic field and radio waves are used to produce detailed images of the body's internal structures.

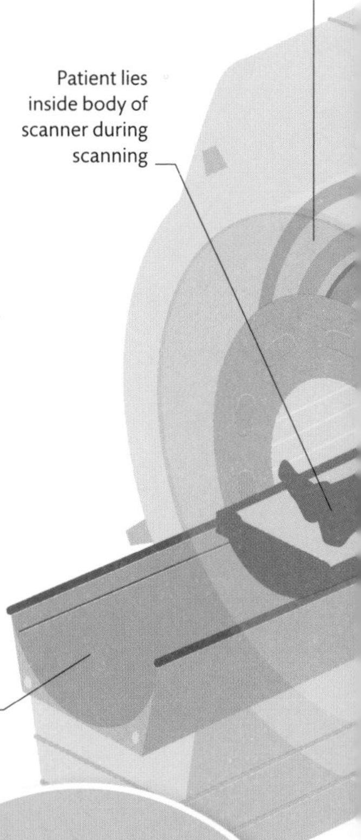

ELECTROMAGNETS

Passing an electric current through a wire creates a magnetic field, turning the wire into an electromagnet. The stronger the current, the stronger the magnetic field. The superconducting electromagnet in an MRI scanner is supercooled with liquid helium to give almost no electrical resistance, allowing very high currents to flow through the electromagnet and produce an extremely strong magnetic field.

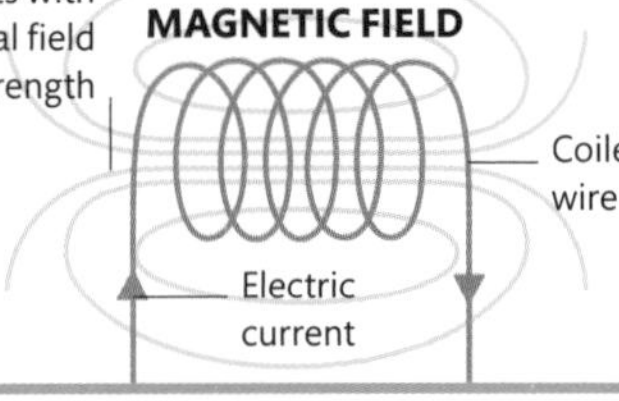

How an MRI scanner works

An MRI scanner contains magnets and a radiofrequency coil. A motorized bed moves the patient inside the machine. The main electromagnet generates a very strong magnetic field that makes protons (positively charged particles in atoms) inside body cells align. Gradient magnets alter the field in order to select the specific area of the body to be imaged. The radiofrequency coil emits pulses of radio waves to excite the protons. Radio signals from the protons are then detected by the radiofrequency coil and sent to a computer, which processes the radio-signal data into an image. The MRI image is similar to an X-ray or CT scan (see pp.234–235) but shows more detail, especially in soft tissues.

The scanning process
MRI acts on protons that make up the nuclei of atoms of hydrogen, one of the most abundant elements in the body. It works by making the protons align with a strong magnetic field then exciting them with radio waves and detecting the energy they give off when they return to their previous positions.

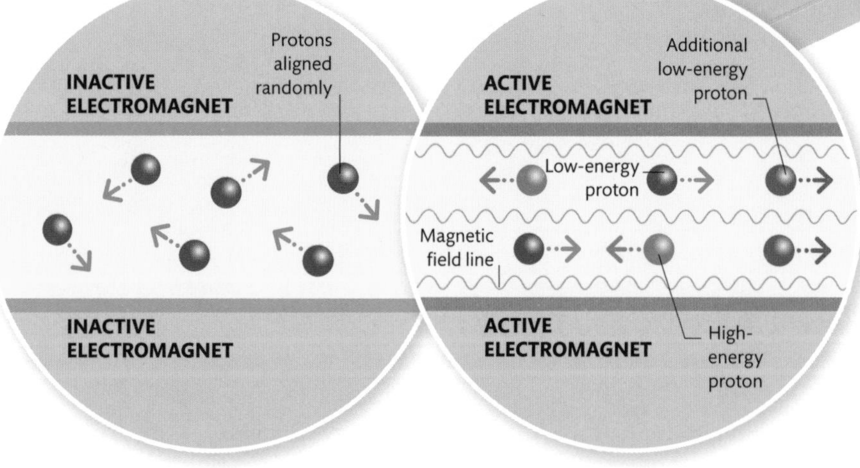

1 Normal state of protons
The nucleus of each hydrogen atom consists of a proton. Each proton has a tiny magnetic field, and it spins around the axis of the field. Normally, the protons spin in random directions.

2 Electromagnet turned on
When the electromagnet is on, the protons align along the magnetic field. They may lie in the same direction as the field (a low-energy state) or be counteraligned (a high-energy state). There are slightly more aligned protons than counteraligned ones.

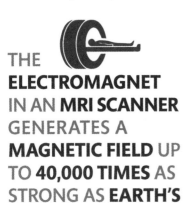

THE **ELECTROMAGNET** IN AN **MRI SCANNER** GENERATES A **MAGNETIC FIELD** UP TO **40,000 TIMES** AS STRONG AS **EARTH'S**

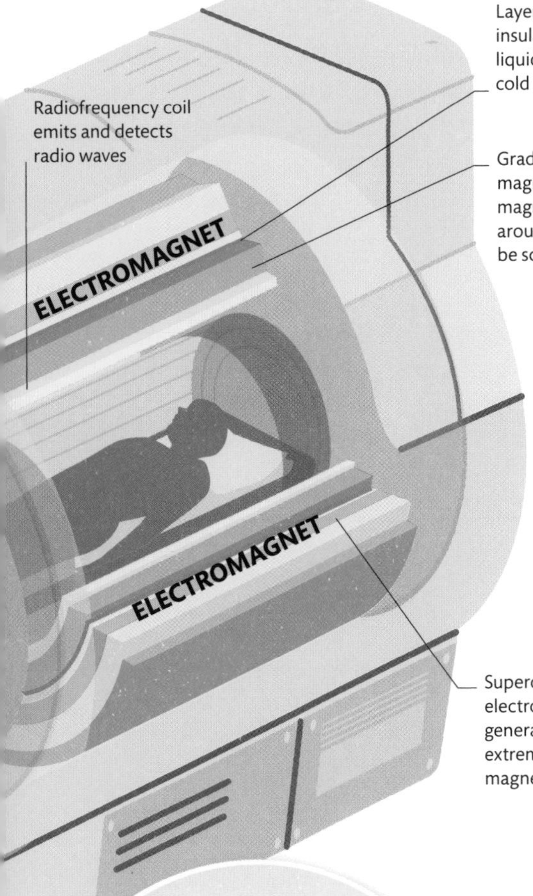
Layer of thermal insulation keeps liquid helium cold
Radiofrequency coil emits and detects radio waves
Gradient magnets focus magnetic field around area to be scanned
ELECTROMAGNET
ELECTROMAGNET
Superconducting electromagnet generates extremely strong magnetic field

Specialized uses of MRI

Specific types of MRI can be used to give extra information about body tissues. For example, a contrast material (a substance that appears white on scans) may be used to highlight specific tissues. Other types of MRI may be used to show the function of certain tissues or physical processes in real time.

Type	Uses
Magnetic resonance angiography	A contrast material is injected into the blood to highlight the interior of blood vessels and reveal any areas that are blocked, narrowed, or damaged.
Functional MRI	Also known as fMRI, this technique detects the flow of blood in the brain; areas with high blood flow indicate high brain activity, and vice versa.
Real-time MRI	Multiple MRI images are taken to give a continuous record of body processes as they happen, such as the heart beating or movements of the joints.
MRI and PET (positron emission tomography)	PET scanning uses injected radioactive substances to show tissue activity. A combined MRI and PET scan shows both the structure and activity of tissues.

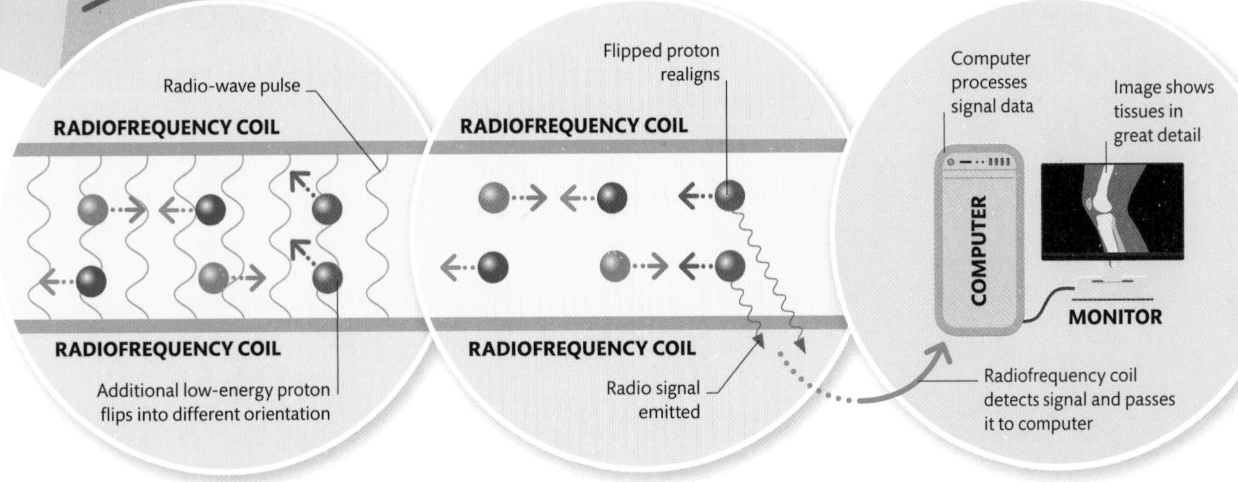
Radio-wave pulse
RADIOFREQUENCY COIL
RADIOFREQUENCY COIL
Additional low-energy proton flips into different orientation
Flipped proton realigns
RADIOFREQUENCY COIL
RADIOFREQUENCY COIL
Radio signal emitted
Computer processes signal data
Image shows tissues in great detail
COMPUTER
MONITOR
Radiofrequency coil detects signal and passes it to computer

3 Radio-wave pulse emitted The radiofrequency coil emits a radio-wave pulse that makes the protons flip their alignment. All the protons flip but the additional low-energy protons take on a different orientation to the other protons.

4 Protons emit radio signals After the stimulating radio pulse has stopped, the flipped protons return to their low-energy state and realign. In doing so, they release their absorbed energy as radio signals, which are picked up by the radiofrequency coil.

5 Signals processed into image The signals are passed to a computer, which processes them into an image. The protons in different body tissues produce different signals, so the image can show the tissues distinctly and in great detail.

Keyhole surgery

Keyhole surgery involves performing operations through tiny incisions rather than large, open cuts. Surgery can also be done via a flexible endoscope—a thin tube inserted through a natural opening, such as the mouth.

WHEN WAS THE FIRST KEYHOLE SURGERY PERFORMED?

The first keyhole operation was carried out in 1901, on dogs. The first keyhole surgery on humans was performed in 1910.

How keyhole surgery works

Small incisions are made in the skin, and hollow instruments called trocars are inserted into the incisions to keep them open for the endoscope and other instruments. A rigid endoscope transmits light to the operation site. It also enables the surgeon to view the operation site, either directly through an eyepiece or, if a video camera is attached to the eyepiece, on a monitor. Surgical instruments are inserted through separate incisions for tasks such as cutting or stitching tissue or clamping blood vessels.

Rigid endoscope

A rigid endoscope contains fiber-optic cables that transmit light to the operation site and lenses that relay images from the site to an eyepiece. Often, a video camera is attached to the eyepiece, and the image is transmitted to a monitor to provide a clear view for the surgeon.

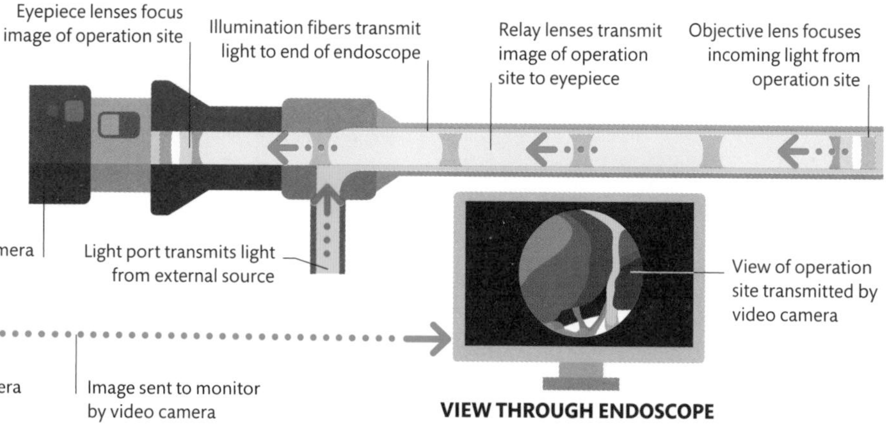
Eyepiece lenses focus image of operation site
Illumination fibers transmit light to end of endoscope
Relay lenses transmit image of operation site to eyepiece
Objective lens focuses incoming light from operation site
Video camera
Light port transmits light from external source
View of operation site transmitted by video camera
Image sent to monitor by video camera
VIEW THROUGH ENDOSCOPE

Keyhole surgery on abdomen

Called laparoscopy, keyhole surgery on the abdomen is performed by using a rigid endoscope. Carbon dioxide gas is pumped into the abdomen to make space around the organs, and the surgeon inserts the laparoscope to provide a view of the operation site. Instruments to carry out the surgery are inserted through other small incisions in the abdomen.

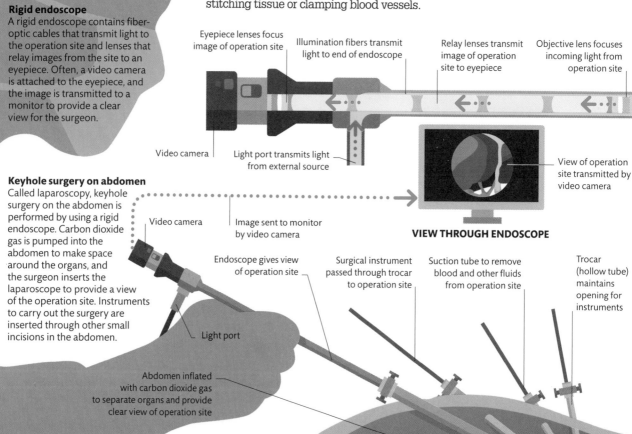

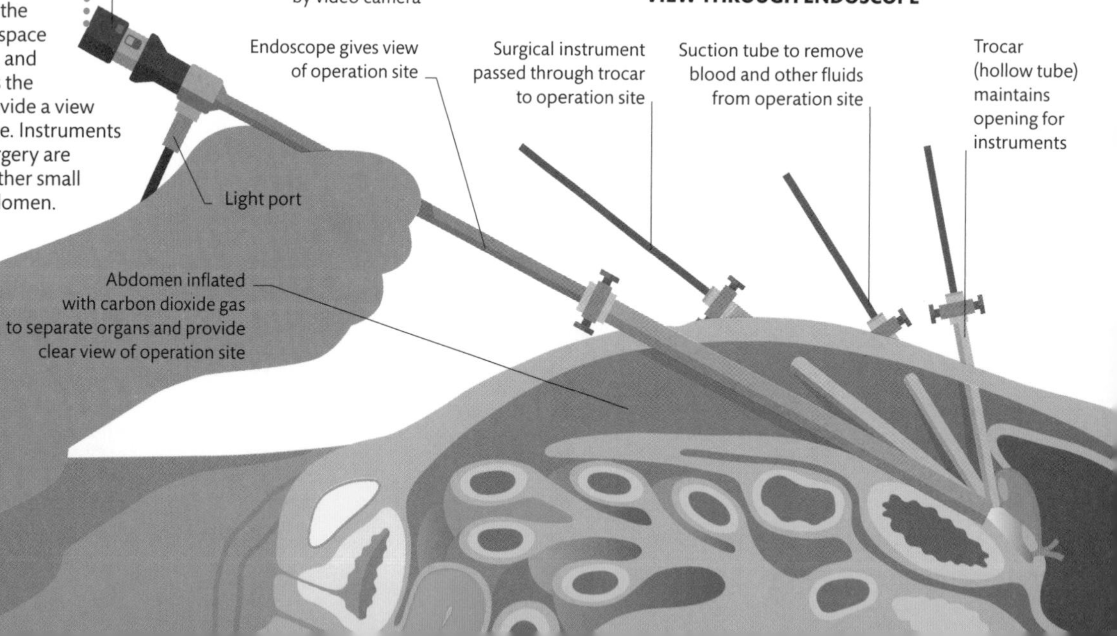
Video camera
Endoscope gives view of operation site
Surgical instrument passed through trocar to operation site
Suction tube to remove blood and other fluids from operation site
Trocar (hollow tube) maintains opening for instruments
Light port
Abdomen inflated with carbon dioxide gas to separate organs and provide clear view of operation site

Flexible endoscopy

In this form of surgery, a flexible endoscope is introduced into a body cavity, such as the windpipe or intestine, through the mouth or other natural opening. The endoscope contains optical fibers to transmit light to the operation site and a video camera at the tip to send images from the site back to a monitor. It also has channels to pass air, water, and surgical instruments to the operation site.

10,000
THE NUMBER OF **OPTICAL FIBERS** IN SOME FLEXIBLE **ENDOSCOPES**

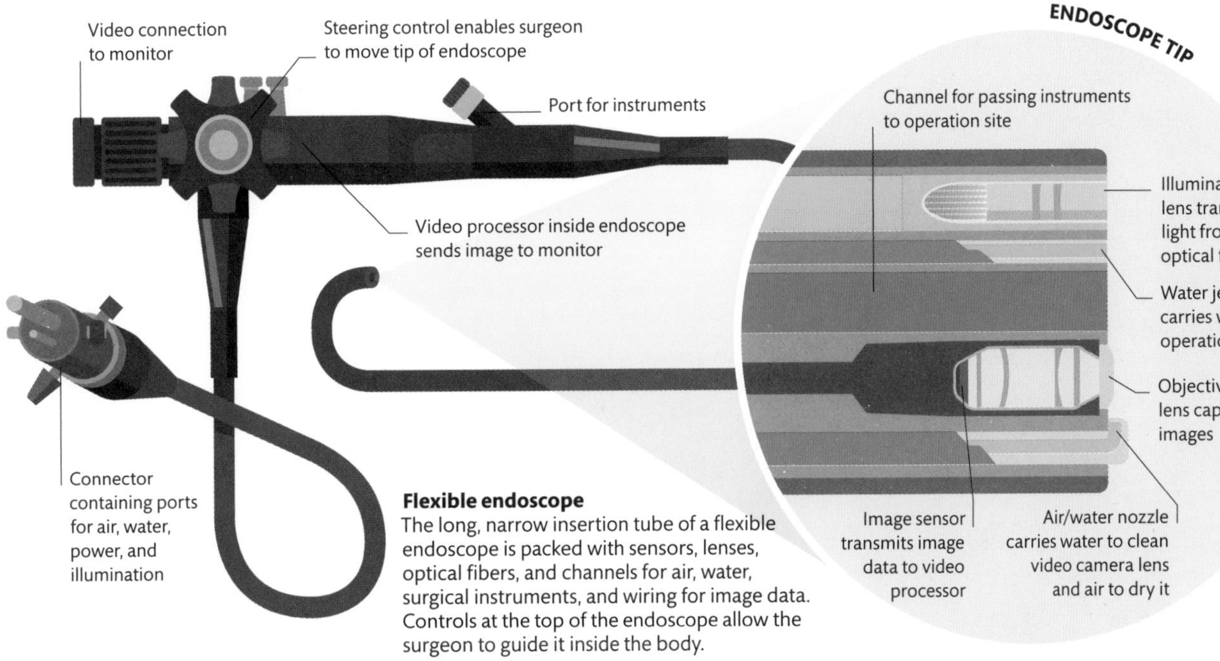
Video connection to monitor
Steering control enables surgeon to move tip of endoscope
Port for instruments
ENDOSCOPE TIP
Channel for passing instruments to operation site
Illumination lens transmits light from optical fibers
Video processor inside endoscope sends image to monitor
Water jet outlet carries water to operation site
Objective lens captures images
Connector containing ports for air, water, power, and illumination
Image sensor transmits image data to video processor
Air/water nozzle carries water to clean video camera lens and air to dry it

Flexible endoscope
The long, narrow insertion tube of a flexible endoscope is packed with sensors, lenses, optical fibers, and channels for air, water, surgical instruments, and wiring for image data. Controls at the top of the endoscope allow the surgeon to guide it inside the body.

Robot-assisted surgery

Some forms of keyhole surgery can now be performed with the help of a robotic system. Robotic arms are mounted on a cart beside the patient. An endoscope on one arm transmits views from inside the body to the surgeon's console and to a video monitor. The other arms hold surgical instruments. The surgeon uses hand controls in the console to move the instruments inside the patient. One of the advantages of robotic surgery is that the robotic system can scale down the surgeon's movements, enabling more precise control of the instruments.

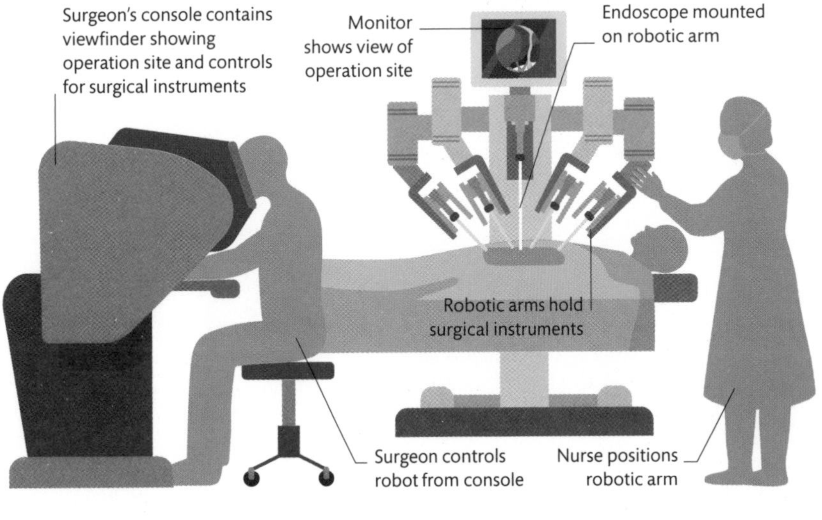
Surgeon's console contains viewfinder showing operation site and controls for surgical instruments
Monitor shows view of operation site
Endoscope mounted on robotic arm
Robotic arms hold surgical instruments
Surgeon controls robot from console
Nurse positions robotic arm

Prosthetic limbs

A prosthetic limb is a device designed to replace a missing limb and help the user perform normal activities. Prostheses range from relatively simple mechanical devices to sophisticated electronic or robotic limbs that interact with the user's own nervous system.

How a myoelectric lower-arm prosthesis works
Electrodes detect electrical signals from muscles in the residual arm. The signals are transmitted to a microprocessor, which converts them into data to instruct motors in the wrist and hand to move.

Prosthetic arms

The simplest prosthetic arms are mechanical, operated by cables running to the opposite shoulder and with a metal hook for gripping objects. More sophisticated myoelectric prostheses use electrodes to pick up muscle impulses from the remaining limb and convert these to electrical signals, which drive a motor to move the prosthetic arm and hand. For people who are missing most or all of their arm, targeted muscle reinnervation may be used. The nerve supply to a lost arm muscle is rerouted into a different muscle, such as a chest muscle; when the user thinks about moving the arm, the chest muscle contracts, and sensors placed over this muscle transmit signals to the prosthesis.

TOUCH SENSORS

Various prosthetic hands are being developed to restore a sense of touch to the user. These systems relay signals not just from the user's muscles to the prosthesis but from the prosthesis back to the brain. Sensors in the fingertips detect pressure or vibrations and relay this data to a computer chip. This converts the data into signals that are relayed to implants attached to nerves in the user's arm, which send impulses to the brain.

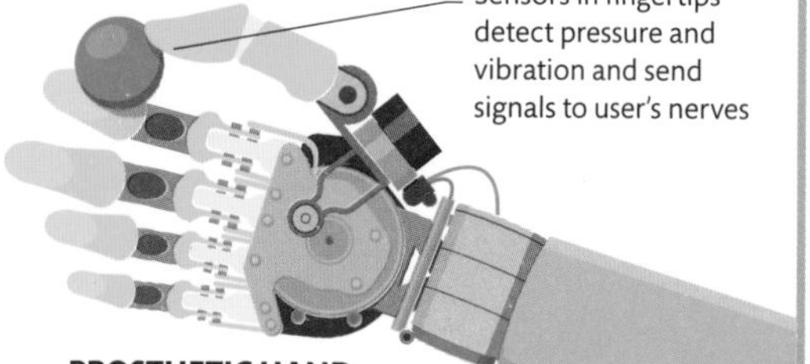
Sensors in fingertips detect pressure and vibration and send signals to user's nerves
PROSTHETIC HAND

1 Sensors detect electrical signals
Sensors on the inner surface of the prosthesis socket, or implanted into the muscles of the residual arm, detect electrical signals from the arm muscles. These signals are given off by the muscles when they contract after being stimulated by nerve signals from the brain.

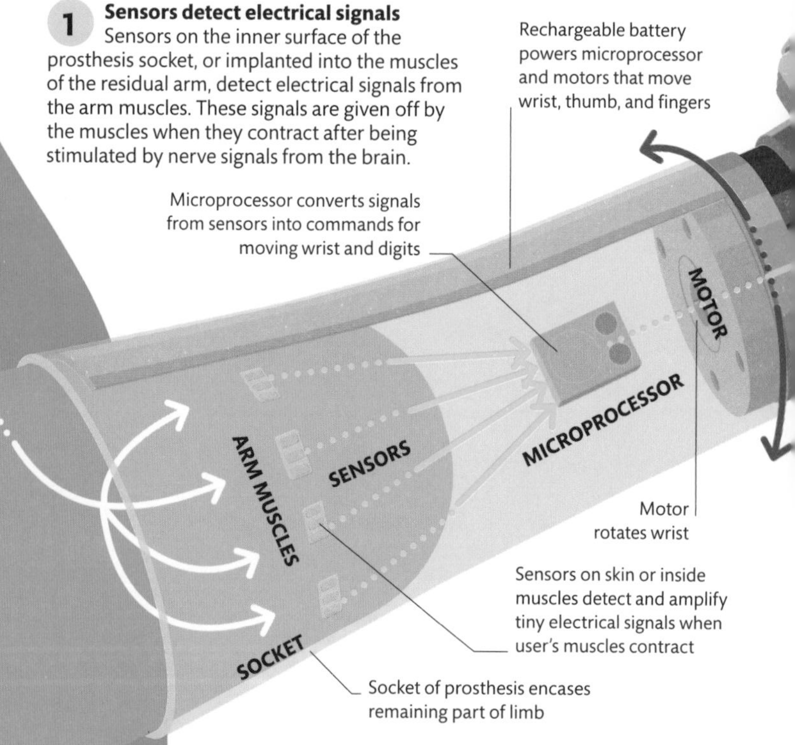
NERVE SIGNALS FROM BRAIN TO ARM MUSCLES
Rechargeable battery powers microprocessor and motors that move wrist, thumb, and fingers
Microprocessor converts signals from sensors into commands for moving wrist and digits
MOTOR
MICROPROCESSOR
SENSORS
ARM MUSCLES
SOCKET
Motor rotates wrist
Sensors on skin or inside muscles detect and amplify tiny electrical signals when user's muscles contract
Socket of prosthesis encases remaining part of limb

ATHLETES USING **RUNNING BLADES** HAVE TO **MOVE CONSTANTLY** TO STAY **BALANCED**

WHEN WERE PROSTHESES FIRST USED?

Artificial body parts were used at least 3,000 years ago. The oldest surviving prosthetic body part is a toe made of wood and leather found on an ancient Egyptian mummy.

3 Hand movements
The wrist, fingers, and thumb are moved by motors. Some types of prostheses allow the digits to move in unison for powered grips or in a coordinated way for precision tasks.

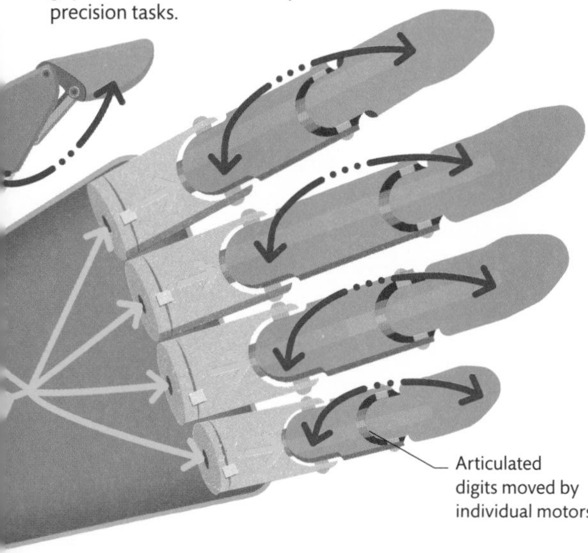
Articulated digits moved by individual motors

2 Data sent to microprocessor
The signals from the muscles are sent to the microprocessor, which translates this data into commands to activate motors in the hand and wrist. Different muscle signals can enable different types of grips.

RUNNING BLADES

Used by athletes, running blades are made of layers of carbon fiber bonded together, making them light but strong and flexible. The soles have treads or spikes for traction. The blade bends as the runner lands on it, and then as the "foot" rolls, the blade rebounds, releasing energy to power the athlete forward.

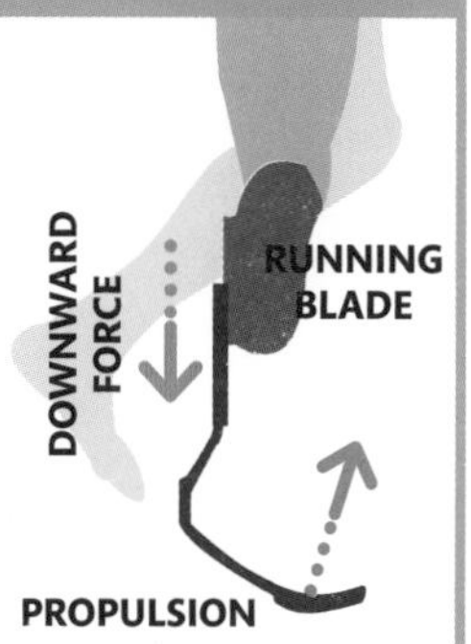
DOWNWARD FORCE
RUNNING BLADE
PROPULSION

Prosthetic legs

Prosthetic legs not only support the user but also emulate some of the functions of a natural leg. They are made of lightweight material such as carbon fiber. In some types, the user's weight is taken on a titanium pylon, while in other types, a hard outer shell bears the weight. Extra features may include an energy-storing foot for propulsion and a computer-controlled knee to regulate movement and stability.

Above-knee prostheses
Most prostheses have a flexible knee and ankle. The simplest joints are mechanical. Others have sensors and a microprocessor that operates a hydraulic or pneumatic system to control the prosthesis.

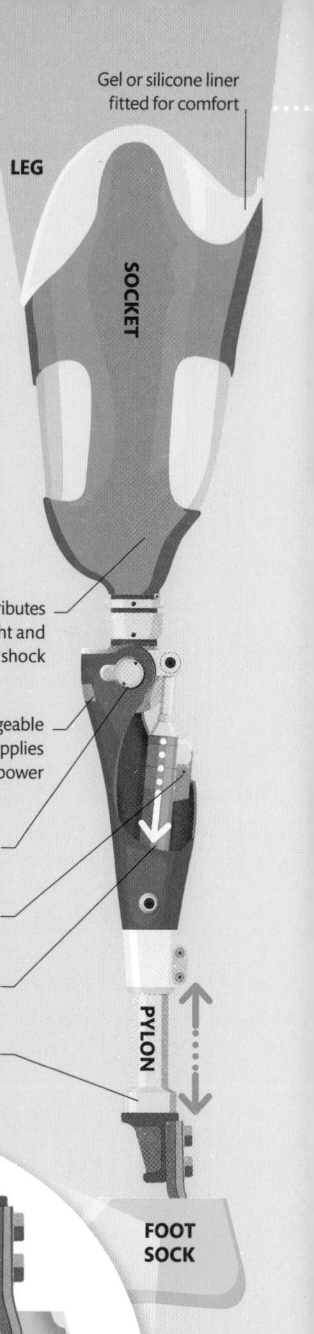
Gel or silicone liner fitted for comfort
LEG
SOCKET
Socket distributes user's weight and absorbs shock
Rechargeable battery supplies power
Sensors detect angle of knee and its speed of movement
Microprocessor controls release of fluid or air into piston
Piston absorbs shock and gives support
PYLON
Pylon can be adjusted for user's height
FOOT SOCK

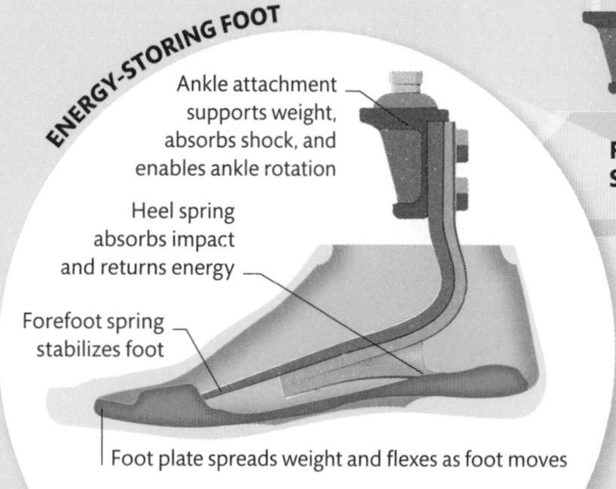
ENERGY-STORING FOOT
Ankle attachment supports weight, absorbs shock, and enables ankle rotation
Heel spring absorbs impact and returns energy
Forefoot spring stabilizes foot
Foot plate spreads weight and flexes as foot moves

An energy-storing foot has a springlike structure in the heel. As the user puts weight on it, the spring compresses; when the heel lifts, the spring releases the energy to push the user forward.

Brain implants

A brain implant is an artificial device embedded in the brain that works with one or more other devices to improve or restore brain function in people disabled by injury or illness. A sensory implant interfaces with the brain via the nervous system and may help to restore hearing or vision. The technology of implants is still in its early stages.

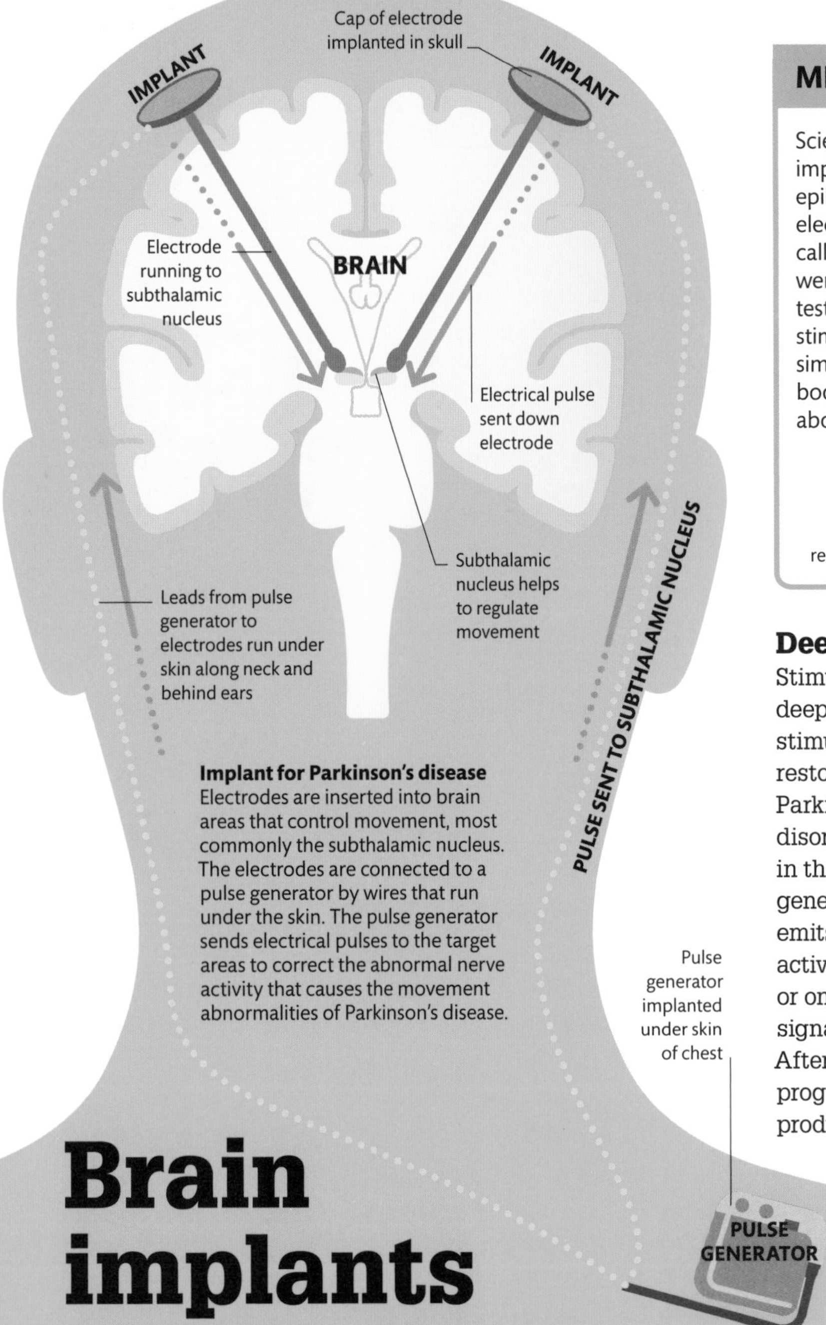

Cap of electrode implanted in skull
IMPLANT
IMPLANT
BRAIN
Electrode running to subthalamic nucleus
Electrical pulse sent down electrode
Subthalamic nucleus helps to regulate movement
Leads from pulse generator to electrodes run under skin along neck and behind ears
PULSE SENT TO SUBTHALAMIC NUCLEUS
Pulse generator implanted under skin of chest
PULSE GENERATOR

Implant for Parkinson's disease
Electrodes are inserted into brain areas that control movement, most commonly the subthalamic nucleus. The electrodes are connected to a pulse generator by wires that run under the skin. The pulse generator sends electrical pulses to the target areas to correct the abnormal nerve activity that causes the movement abnormalities of Parkinson's disease.

MEMORY IMPLANTS

Scientists are developing brain implants to improve memory. In one study, people with epilepsy who already had brain implants had electrodes inserted into an area of the brain called the hippocampus. Their brain signals were recorded while they completed memory tests. Later, the same brain signals were used to stimulate their brains while they carried out similar tests. The stimulation boosted memory by about a third.

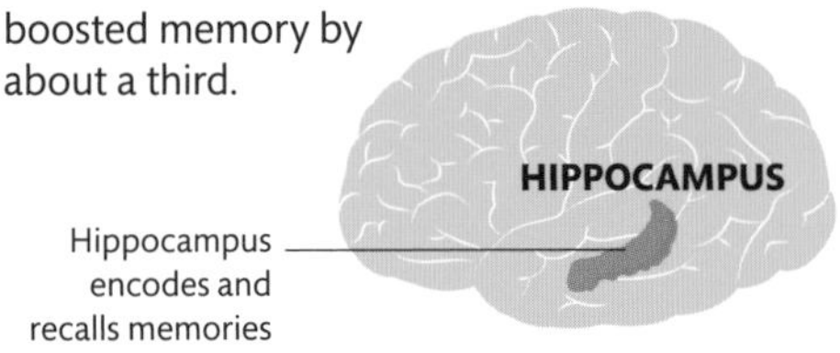

HIPPOCAMPUS
Hippocampus encodes and recalls memories

Deep brain stimulation

Stimulation of specific groups of nerve cells deep in the brain—known as deep brain stimulation, or DBS—may be used to restore normal brain activity in people with Parkinson's disease, certain other movement disorders, or epilepsy. Electrodes are implanted in the brain, and a device called a pulse generator, implanted in the chest or stomach, emits electrical pulses to regulate the brain's activity. The device may operate continuously or only when electrodes detect abnormal nerve signals (such as if an epileptic seizure starts). After the system has been fitted, a specialist programs the pulse generator so that it produces pulses only when necessary.

WHAT ARE BRAIN ELECTRODES MADE OF?

Electrodes implanted in the brain are made of substances such as gold or platinum-iridium, which conduct electrical impulses well and do not harm brain tissue.

1 Video camera captures scene
The user wears glasses with a miniature video camera fitted to the bridge of the glasses. The camera captures images and transmits them via a wire to a portable video processing unit (VPU) worn by the user.

3 Data transmitted to retinal implant
The transmitter relays signals to a receiver inside the eye socket, on the side of the eyeball. The receiver comprises an antenna, which detects the signals, and an electronic unit, which sends impulses to stimulate the retinal implant.

4 Implant sends data to brain
The implant consists of an electrode array attached to the retina. The electrodes stimulate the retina's remaining cells to send signals along the optic nerve to the brain, where visual perception occurs.

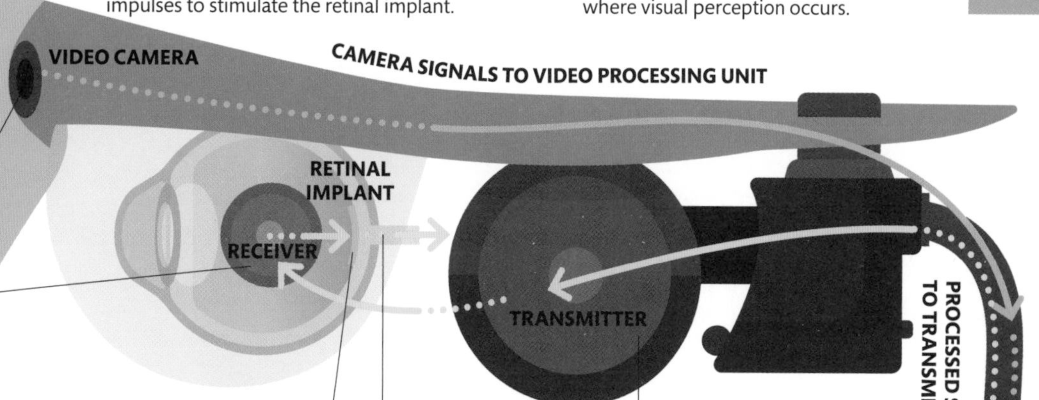
VIDEO CAMERA
CAMERA SIGNALS TO VIDEO PROCESSING UNIT
RETINAL IMPLANT
RECEIVER
TRANSMITTER
PROCESSED SIGNALS TO TRANSMITTER
Camera sends signal to processor
Receiver relays signals from transmitter to retinal implant
Retinal implant produces electrical impulses to stimulate retina
Nerve impulses from stimulated retinal cells travel along optic nerve to brain
Transmitter sends signals wirelessly to receiver on side of eyeball

Bionic eye
Damage to cells in the retina (the light-sensitive layer at the back of the eye) can result in loss of vision. Retinal implants, such as the "bionic eye" system, can convert light patterns into data and bypass the damaged retinal cells to send the data to the brain.

Sensory implants

Some brain implants are used to restore vision or hearing in people whose nerves are not sending information efficiently to the brain. Retinal implants can help to restore vision by stimulating the optic nerve to send nerve impulses to the brain. Implants inside the cochlea, in the inner ear, stimulate the auditory nerve to transmit nerve impulses from the inner ear to the brain. If the auditory nerve does not work, an auditory brainstem implant may be fitted directly to the brainstem, to stimulate cells to send signals to the brain.

2 Video data from camera processed
The VPU converts the video signals into a pixellated "brightness map," which it then encodes as digital signals. It sends these signals to a transmitter mounted on the side of the user's glasses.

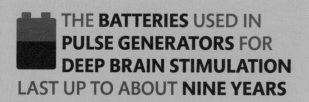

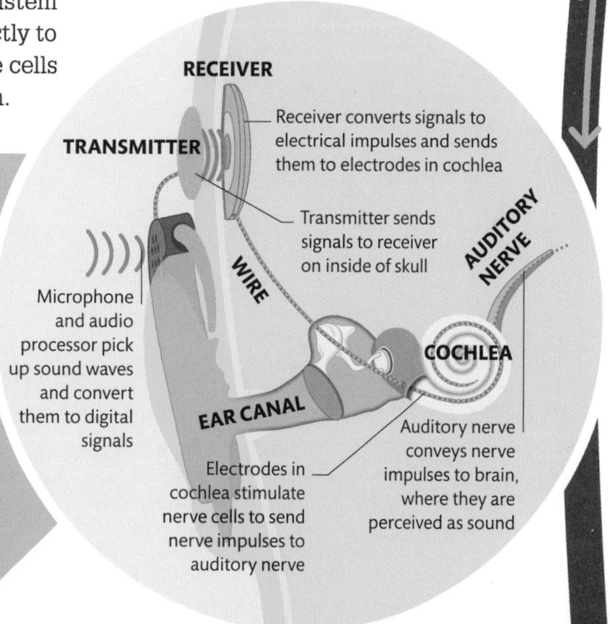
RECEIVER
TRANSMITTER
Receiver converts signals to electrical impulses and sends them to electrodes in cochlea
Transmitter sends signals to receiver on inside of skull
AUDITORY NERVE
WIRE
Microphone and audio processor pick up sound waves and convert them to digital signals
COCHLEA
EAR CANAL
Electrodes in cochlea stimulate nerve cells to send nerve impulses to auditory nerve
Auditory nerve conveys nerve impulses to brain, where they are perceived as sound

Cochlear implants
In normal hearing, sound vibrations are transmitted via the eardrums and middle ear bones to the inner ear. Hair cells inside a structure called the cochlea turn these vibrations into electrical signals, which pass along the auditory nerve to the brain. If the internal ear structures are not working properly, an implant can be fitted inside the cochlea to carry signals directly to the auditory nerve.

THE **BATTERIES** USED IN **PULSE GENERATORS** FOR **DEEP BRAIN STIMULATION** LAST UP TO ABOUT **NINE YEARS**

Genetic testing

Genes are segments of DNA—the molecule in our cells that provides the code telling the body how to develop and function. Genetic tests are done to identify any problems that may cause genes to give faulty instructions, including any disorders that may be passed on from parents to children.

HUMAN CELLS ARE THOUGHT TO CONTAIN ABOUT **20,000 GENES**

CHROMOSOMES AND GENES

The nucleus of each body cell contains 23 pairs of chromosomes, subdivided into genes. Each gene is made from units called nucleotides. These units have a sugar-phosphate backbone and one of four bases: adenine (A), cytosine (C), guanine (G), or thymine (T). Adenine always pairs with thymine, and cytosine with guanine. The sequence of bases makes up the code of DNA.

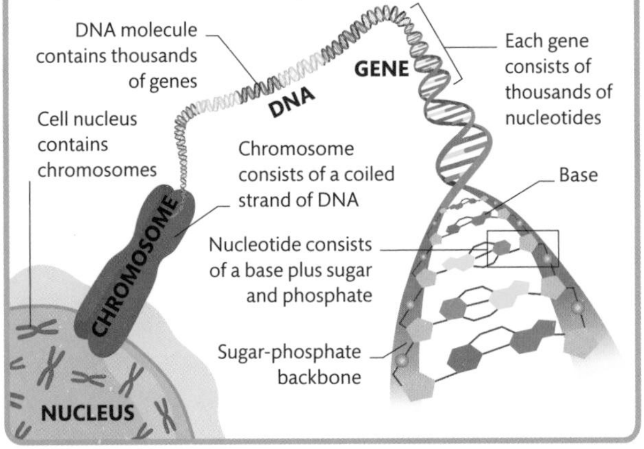
DNA molecule contains thousands of genes
DNA
GENE
Each gene consists of thousands of nucleotides
Cell nucleus contains chromosomes
Chromosome consists of a coiled strand of DNA
Base
CHROMOSOME
Nucleotide consists of a base plus sugar and phosphate
Sugar-phosphate backbone
NUCLEUS

Chromosome testing

Each human body cell has 46 chromosomes—half inherited from the mother and half from the father. Scientists may study the full set of a person's chromosomes, called a karyotype, to see if there are any extra, missing, or abnormal chromosomes.

Preparing a karyotype

In karyotyping, chromosomes are studied as cells are dividing to form new cells, when the chromosomes are coiled up into distinctive "X" shapes. The chromosomes are stained, paired, and arranged in order of size to produce the karyotype.

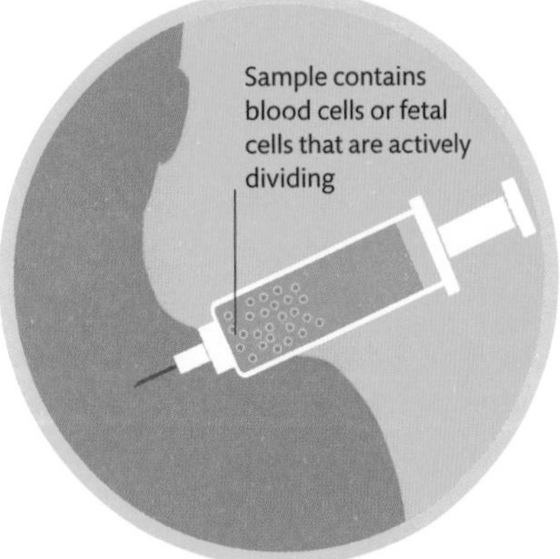
Sample contains blood cells or fetal cells that are actively dividing

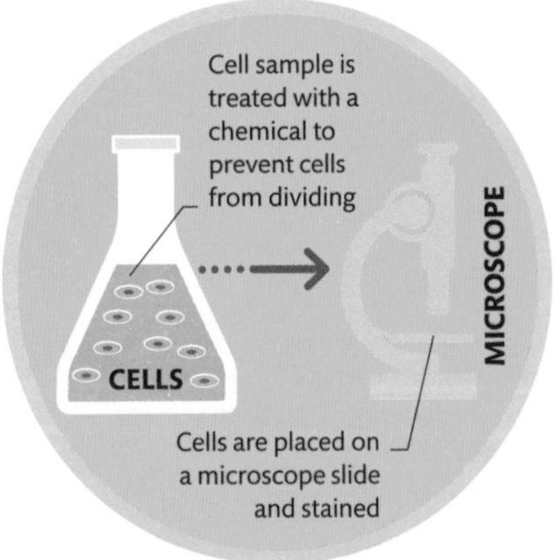
Cell sample is treated with a chemical to prevent cells from dividing
CELLS
MICROSCOPE
Cells are placed on a microscope slide and stained

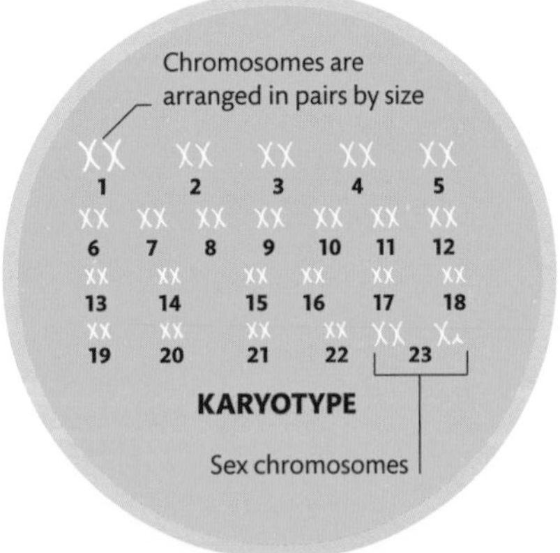
Chromosomes are arranged in pairs by size
1
2
3
4
5
6
7
8
9
10
11
12
13
14
15
16
17
18
19
20
21
22
23
KARYOTYPE
Sex chromosomes

1 Cell sample is collected
A sample of cells is taken from a person's blood or bone marrow. For genetic testing of a fetus, cells are taken from the amniotic fluid or placenta of a pregnant woman.

2 Chromosomes are extracted
The dividing cells are treated with a chemical that prevents them from dividing at a point when the chromosomes are coiled up. The cells are placed on a microscope slide and stained to highlight the chromosomes.

3 Chromosomes are sorted
The chromosomes are sorted and matched into the 22 pairs of autosomes (nonsex chromosomes) and the single pair of sex chromosomes (XX for a female, or XY for a male) to produce the karyotype.

Gene testing

Some tests allow scientists to detect abnormalities in individual genes, such as extra or missing material, or bases in the wrong place. The samples are examined by a method such as DNA sequencing, which reveals the order of nucleotides in a section of DNA. The presence of an abnormality does not always indicate a problem; it may be a variation that has no ill effects. However, some abnormalities can affect health, so expert interpretation of test results is important.

DNA sequencing
In one widely used method of DNA sequencing, modified fluorescent nucleotide bases (see opposite) are added to the ends of DNA strands to highlight each base in a DNA strand. There are four types of fluorescent tags—one for each type of nucleotide base (A, T, C, or G).

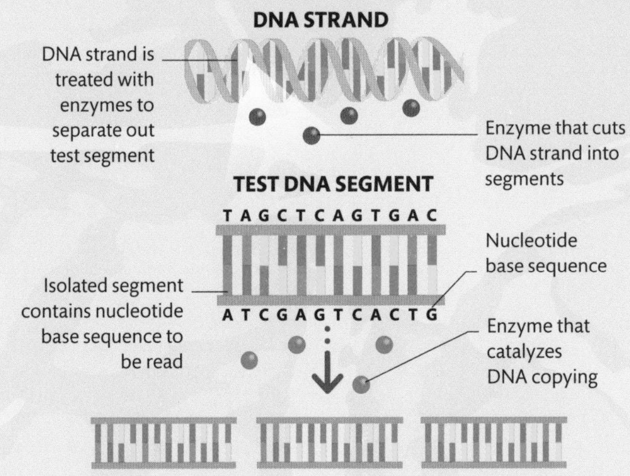

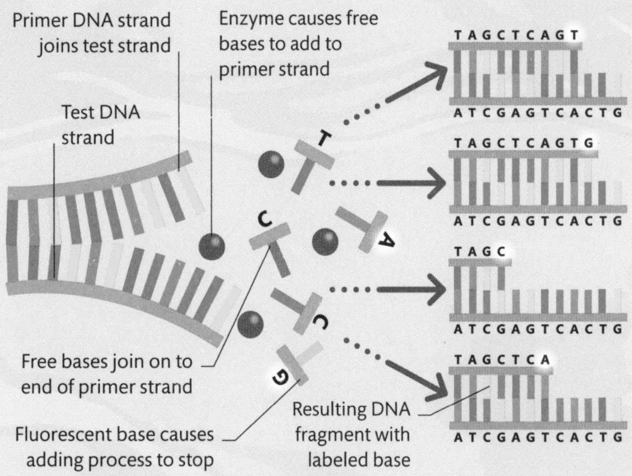

1 Isolating test DNA segment
A DNA sample may be obtained from various sources, such as cheek cells, saliva, hair, or blood. The sample is treated with an enzyme that cuts the DNA into segments in order to isolate the segment of DNA to be analyzed. Using another enzyme, this test DNA segment is then copied hundreds of times to produce a sample that is large enough for analysis.

2 Labeling bases in the test DNA
The test DNA sample is mixed with "primer" DNA, an enzyme, free nucleotide bases, and nucleotide bases labeled with a fluorescent marker. The primer joins onto the test strand, and the free bases join onto the ends of the primer. This process stops when a fluorescent base is added. Each resulting DNA fragment ends up with a labeled base corresponding to one base on the test DNA.

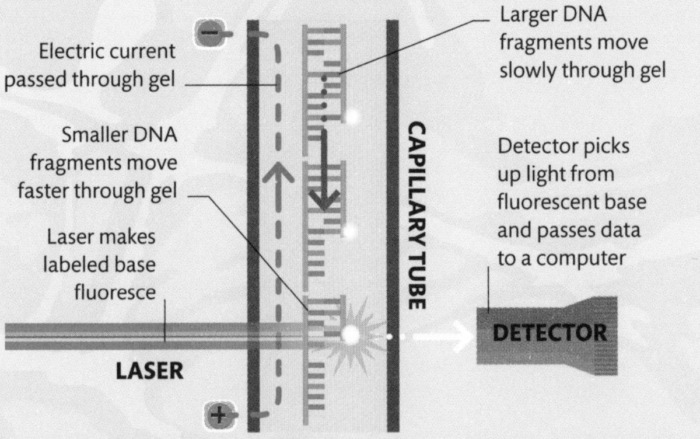

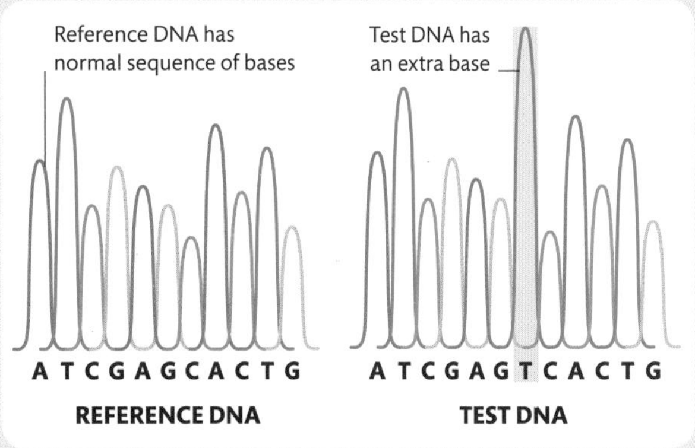

3 Detecting labeled bases in the test DNA
The DNA fragments are passed through gel in a thin tube (capillary). An electric current makes the fragments move, and they end up sorted by length; the order of the labeled bases reflects the order of bases on the test strand. As each fragment passes the laser, its labeled base fluoresces, and the detector reads each of these in order.

4 Computer analysis
The detector passes the sequence of bases in the test sample to a computer. The computer uses the data to produce an image called a chromatogram, in which the nucleotide sequence is shown as an image and as letters. The test DNA chromatogram is compared to one of a normal reference DNA sample to identify any differences.

In vitro fertilization

Often referred to as IVF, in vitro fertilization is any technique in which a woman's egg is fertilized outside the body. It may be performed to treat either male or female infertility. The woman is given drugs to make her ovaries produce more eggs than usual. The eggs are collected and mixed with sperm in a laboratory. If any eggs are fertilized, they are left to develop for a few days then placed in the woman's uterus. Any extra fertilized eggs may be frozen for use later.

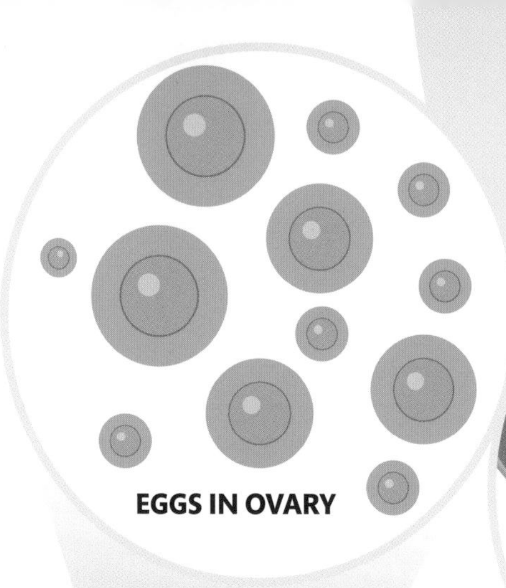

1 Hormonal stimulation
Drugs are given to stimulate the follicles in the ovaries to ripen and develop eggs. When enough eggs are ready, another injection makes the follicles release them.

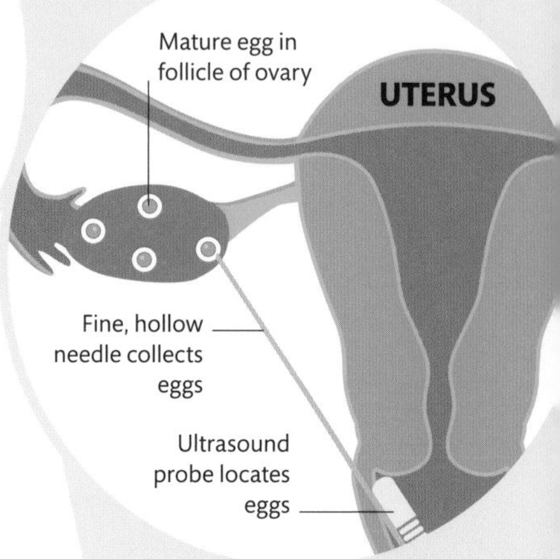

2 Eggs are collected
An ultrasound probe is inserted into the vagina to identify the mature eggs, and between 8 and 15 eggs are collected with a very fine needle.

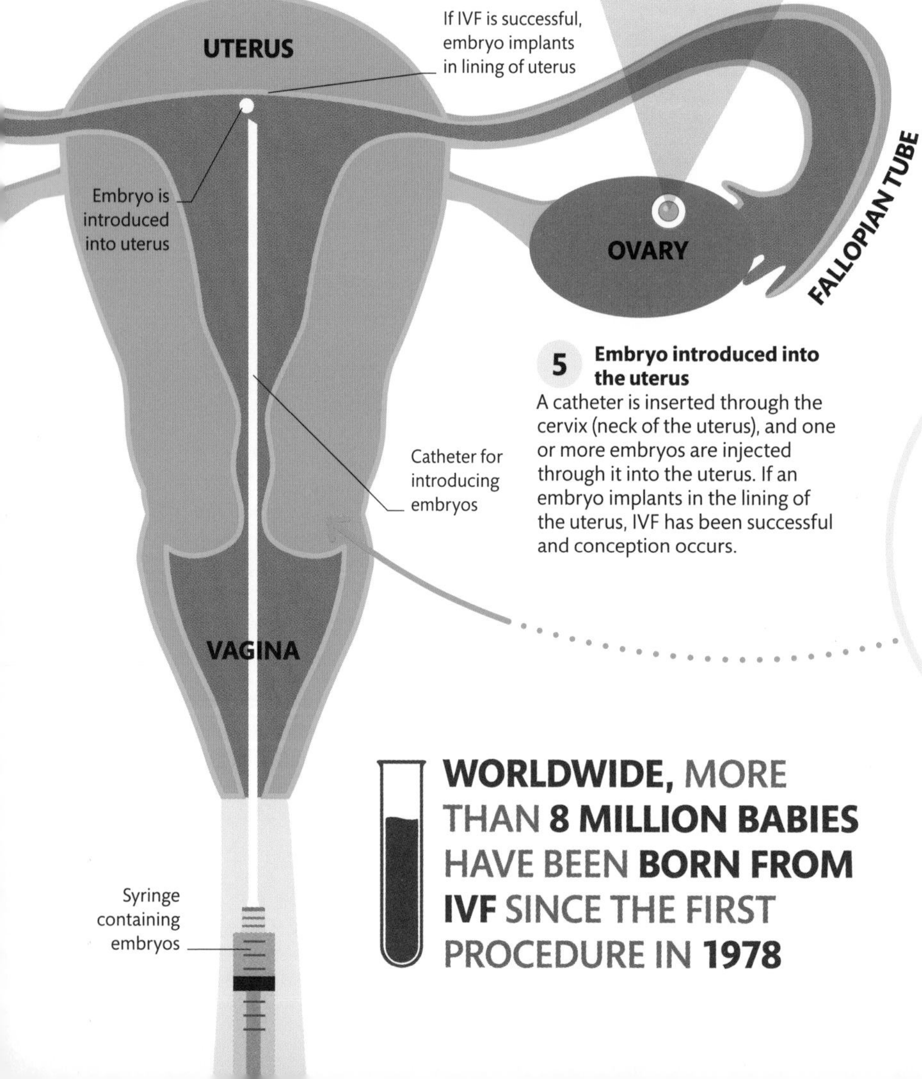

5 Embryo introduced into the uterus
A catheter is inserted through the cervix (neck of the uterus), and one or more embryos are injected through it into the uterus. If an embryo implants in the lining of the uterus, IVF has been successful and conception occurs.

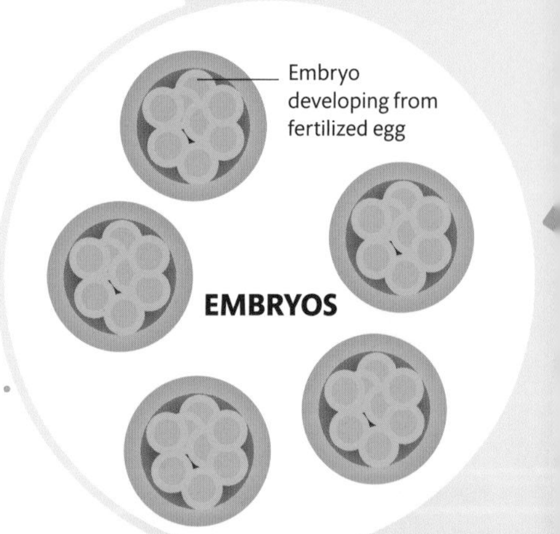

WORLDWIDE, MORE THAN **8 MILLION BABIES** HAVE BEEN **BORN FROM IVF** SINCE THE FIRST PROCEDURE IN **1978**

4 Fertilized eggs grow
The fertilized eggs are left for three days to grow into cell clusters. To maximize the chance of successful implantation in the uterus, the clusters need to grow to around eight cells (called embryos) before they can be transferred into the woman.

Procedure for IVF
Eggs are collected from the woman and sperm from the man. The sperm and eggs are brought together in a laboratory. Alternatively, a sperm is injected into an egg to ensure fertilization, a techique known as intracytoplasmic sperm injection, or ICSI (see panel, below right). The fertilized egg (called an embryo) is then introduced into the uterus, where it may implant in the lining of the uterus.

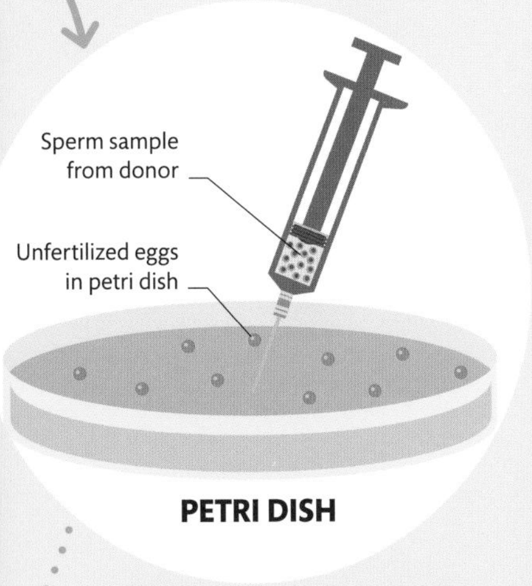

3 Sperm combined with eggs
The eggs are checked for quality then mixed with sperm and incubated at body temperature (98.6°F/37°C) in a petri dish. The mixture is checked the next day to see if any eggs have been fertilized.

HOW DOES AGE AFFECT FERTILITY?

After about the mid-20s, a woman's fertility declines with age, with the biggest decrease occurring from the mid-30s. Male fertility also declines from about the 20s but less sharply.

Assisted fertility

Assisted fertility techniques are used to help people conceive a healthy baby. The most common methods are intrauterine insemination (IUI) and in vitro fertilization (IVF), or "test-tube" fertilization.

Intrauterine insemination

Normally, fertilization occurs when a sperm fuses with an egg in the fallopian tube after sexual intercourse. The resulting fertilized cell then implants in the lining of the uterus to become an embryo. In intrauterine inseminination, sperm are introduced into the uterus through a catheter (a thin, hollow tube). IUI may be recommended if the woman cannot become pregnant naturally, if the man has insufficient healthy sperm, or if donated sperm are used.

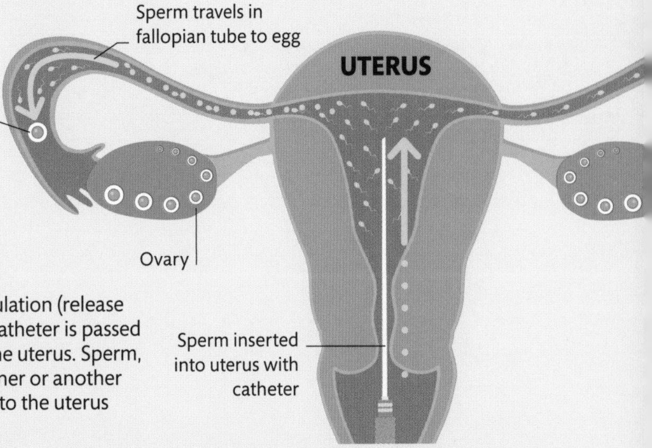

Procedure for intrauterine insemination
IUI is carried out just after ovulation (release of an egg from the ovary). A catheter is passed through the cervix and into the uterus. Sperm, either from the woman's partner or another donor, are then introduced into the uterus through the catheter.

ICSI

In intracytoplasmic sperm injection (ICSI), a man gives a sperm sample, and a single, healthy sperm cell is selected. This sperm is then injected directly into an egg removed from the woman. ICSI is usually done if a man has too few sperm or very few healthy sperm.

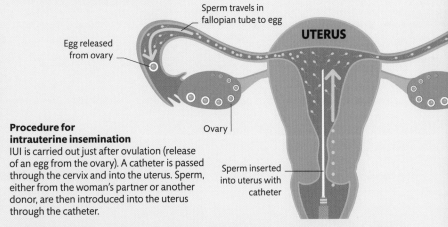

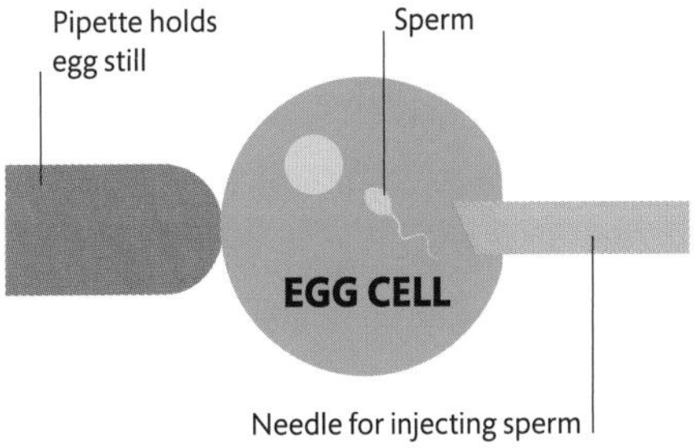

Index

Page numbers in **bold** refer to main entries.

D

E

F

Q

R

S

Acknowledgments

DK would like to thank the following people for help in preparing this book: Joe Scott for help with illustrations; Page Jones, Shahid Mahmood, and Duncan Turner for design help; Alison Sturgeon for editorial help; Helen Peters for indexing; Katie John and Joy Evatt for proofreading; Steve Connolly, Zahid Durrani, and Sunday Popo-Ola for their comments on the Materials and Construction Technology chapter; and Tom Raettig for his comments on engines and cars.